The eruption of Kilauea crater, the island of Hawaii, by Thomas Moran, 1871. ***From*** **SCRIBNER'S MAGAZINE,** ***vol. II, no. 6, October 1871 (New York: Charles Scribner's Sons, 1871), reprinted with permission.***

"BRANCHING STREAMS OF LIQUID, GORGEOUSLY BRILLIANT FIRE"

Mark Twain describes an eruption of Kilauea crater on the island of Hawaii, 1866

"The greater part of the vast floor of the desert under us was as black as ink, and apparently smooth and level; but over a mile square of it was ringed and streaked and striped with a thousand branching streams of liquid and gorgeously brilliant fire!

"Here and there were gleaming holes twenty feet in diameter, broken in the dark crust, and in them the melted lava—the color a dazzling white just tinged with yellow—was boiling and surging furiously; and from these holes branched numberless bright torrents in many directions, like the 'spokes' of a lady's fan, and kept a tolerably straight course for a while and then swept round in huge rainbow curves, or made a long succession of sharp worm-fence angles, which looked precisely like the fiercest jagged lightning. These streams met other streams, and they mingled with and crossed and recrossed each other in every conceivable direction, like skate tracks on a popular skating ground. Sometimes streams twenty or thirty feet wide flowed from the holes to some distance without dividing—and through the opera glasses we could see that they ran down small, steep hills and were genuine cataracts of fire, white at their source, but soon cooling and turning to the richest red, grained with alternate lines of black and gold. Every now and then masses of the dark crust broke away and floated slowly down these streams like rats down a river. Occasionally the molten lava flowing under the superincumbent crust broke through—split a dazzling streak, from five hundred to a thousand feet long, like a sudden flash of lightning, and then acre after acre of the cold lava parted into fragments, turned up edgewise like cakes of ice when a great river breaks up, plunged downward and were swallowed in the crimson caldron. Then the wide expanse of the 'thaw' maintained a ruddy glow for a while, but shortly cooled and became black and level again. During a 'thaw,' every dismembered cake was marked by a glittering white border which was superbly shaded inwards by aurora-borealis rays, which were a flaming yellow where they joined the white border, and from thence toward their points tapered into glowing crimson, then into a rich, pale carmine, and finally into a faint blush that held its own a moment and then dimmed and turned black. Some of the streams preferred to mingle together in a tangle of fantastic circles, and then they looked something like the confusion of ropes one sees on a ship's deck when she had just taken in sail and dropped anchor—provided one can imagine those ropes on fire.

"Through the glasses, the little fountains scattered about looked very beautiful. They boiled, and coughed, and spluttered, and discharged sprays of stringy red fire—of about the consistency of mush, for instance—from ten to fifteen feet into the air, along with a shower of brilliant white sparks—a quaint and unnatural mingling of gouts of blood and snowflakes!

"We had circles and serpents and streaks of lightning all twined and wreathed and tied together, without a break throughout an area more than a mile square." Letters from Hawaii, *by Mark Twain, edited and with an introduction by A. Grove Day; © 1966 by A. Grove Day; reprinted with permission of the University of Hawaii Press*

These were days when my heart was volcanic
    As the scoriac rivers that roll—
    As the lavas that restlessly roll
Their sulphurous currents down Yaanek
    In the ultimate climes of the pole—
That groan as they roll down Mount Yaanek
    In the realms of the boreal pole.

—From *Ulalume,* by Edgar Allan Poe, 1849
    Courtesy of Dover Publications, Inc.

Stories from a Heated Earth

Our Geothermal Heritage

Raffaele Cataldi
Susan F. Hodgson
John W. Lund
Editors

Geothermal Resources Council
International Geothermal Association

Sacramento, California
1999

Library of Congress Catalog Card Number: 99-71835
ISBN 0-934412-19-7

Published 1999

Published by the Geothermal Resources Council
and the International Geothermal Association

Printed in the United States of America

Distributed by:
Geothermal Resources Council
P.O. Box 1350
Davis, CA 95617-1350
U.S.A.

Telephone: 530-758-2360
Fax: 530-758-2839
E-mail: grc@geothermal.org

In Appreciation

Generous help in preparing this volume was extended to us by many people and organizations.

We would like to thank our authors, who explored mankind's relationship with geothermal energy and resources and shared their discoveries.

Dr. Marilyn Dyrud, Professor of Communications, Oregon Institute of Technology, skillfully and cheerfully edited manuscripts, ensuring coherent presentation and graceful expression. Her work was essential.

Tonya "Toni" Boyd, Research Assistant, Geo-Heat Center, Oregon Institute of Technology, used her artist's eye to arrange the pages attractively. She made endless changes adeptly and with enthusiasm.

Jim Spriggs, Research Analyst II for the Department of Conservation, Division of Oil, Gas, and Geothermal Resources, designed the cover and offered artistic advice, as did Al Zucca, Research Analyst I, who helped with illustrations.

Mary Smothers, Photographer for the Oregon Institute of Technology, carefully processed many illustrations.

Marilyn Nemzer, Executive Director of the Geothermal Education Office, proofread the final draft, to our great benefit.

Claudia Stone, of Stone Research, helped with revisions.

Dr. Valentina Svalova, Scientific Secretary of the Geothermal Council of Russia, Head of the International Projects Department, Institute of Environmental Geoscience, Russian Academy of Sciences, provided original art for three chapters besides authoring one.

Donna Gibson, Office Manager for the Geo-Heat Center, undertook varied tasks, efficiently typing and helping us all to keep in touch.

Professor Silvio Cataldi, Professor of Greek Ancient History at the University of Turin, helped find Greek and Latin sources mentioning ancient phenomena, uses, myths, customs, and other elements related to geothermal energy in the Mediterranean area.

Professor Dina Micalella, Professor of Greek Ancient Literature at the University of Turin, provided some texts and references to Greek and Latin works dealing with the subject of this volume for Chapters 2, 6, 9, 10, and 11.

Members of the California Department of Conservation, Division of Oil, Gas, and Geothermal Resources, gave fundamental support, including William F. Guerard, Jr., State Oil and Gas Supervisor; Kenneth P. Henderson, Chief Deputy; James T. Campion, Jr., Technical Services Manager; Richard P. Thomas, Geothermal Officer; and Elizabeth A. Johnson, Geothermal District Engineer.

The Geo-Heat Center of the Oregon Institute of Technology offered significant assistance, as did the U.S. Department of Energy, Office of Geothermal Technologies.

ENEL SpA/Geothermal Department provided figures for several chapters and kindly endorsed the undertaking.

Our publishers, the Geothermal Resources Council and the International Geothermal Association, and their members sustained us with aid and enthusiasm.

Thank you all. We couldn't have done it without you.

Hot spring at Cerro Prieto Geothermal Field, Mexico. *S. Hodgson*

Cupid lay by his brand, and fell asleep:
A maid of Dian's this advantage found,
And his love-kindling fire did quickly steep
In a cold valley-fountain of that ground;
Which borrow'd from this holy fire of love
A dateless lively heat, still to endure,
And grew a seething bath, which yet men prove
Against strange maladies a sovereign cure.
But at my mistress' eye Love's brand new-fir'd,
The boy for trial needs would touch my breast;
I, sick withal, the help of bath desir'd,
And thither hied, a sad distemper'd guest,
  But found no cure: the bath for my help lies
  Where Cupid got new fire,—my mistress' eyes.

—*Sonnet 153*, William Shakespeare, c. 1595
Courtesy of Dover Publications, Inc.

"I left my rheumatism in Baden-Baden. Baden-Baden is welcome to it. It was little, but it was all I had to give."

—From *A Tramp Abroad*, Mark Twain, 1880

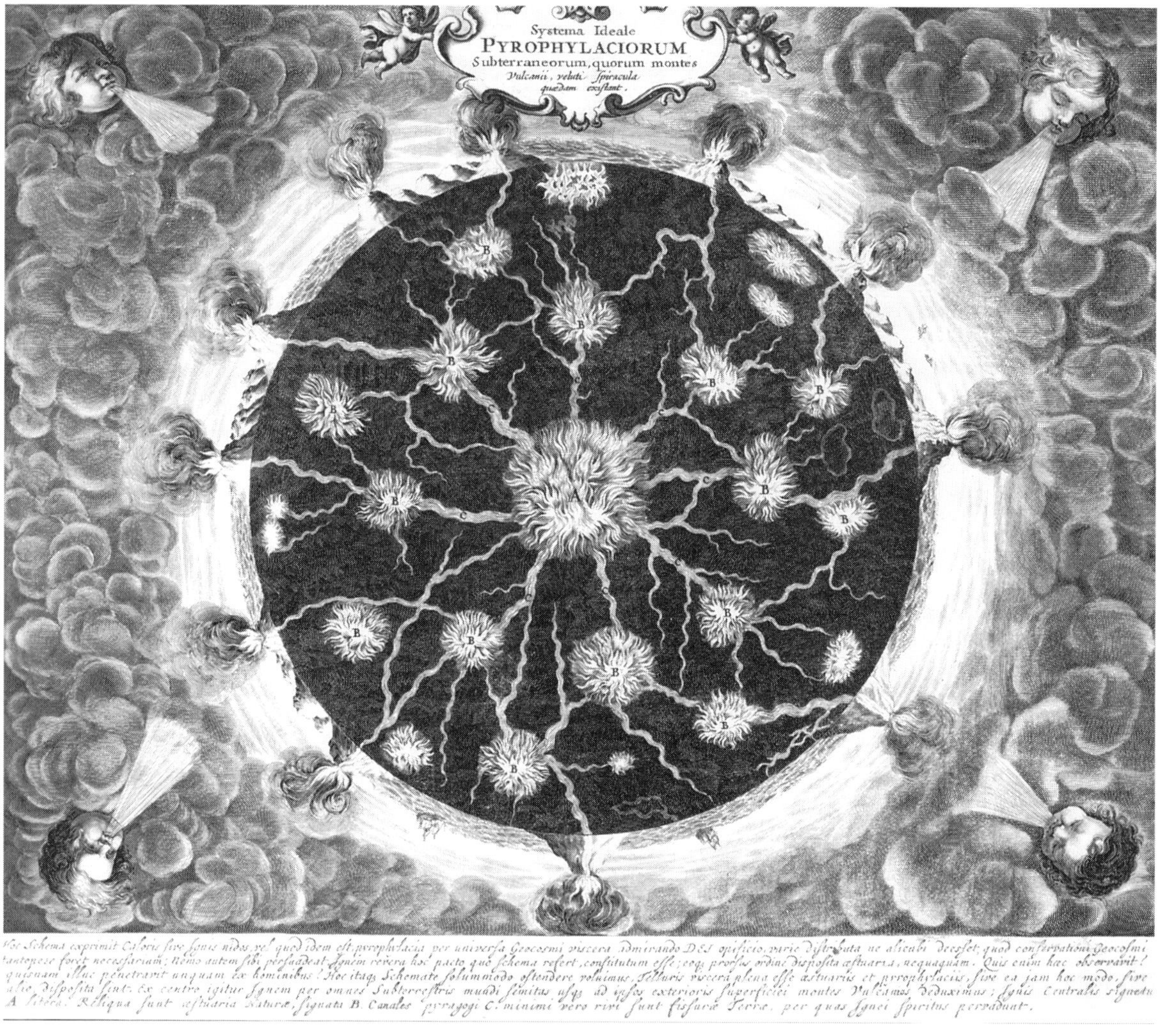

The Latin text above and beneath the engraving reads, "*Subterranean World*, an idealized model of the magma chambers and lava chimneys that feed active volcanoes continuously. The model shows the spots where igneous processes are triggered, spots that also correspond to the places where the magma chambers form. Wonderful works of God, these chambers are widely scattered inside the whole terrestrial globe to supply continuously the material indispensable for the conservation of the Earth. A: [the central ball of fire] Remnant of the primordial inner core of the internal fire of the Earth; B: [the smaller balls of fire] Magmatic chambers; C: [the linear fire trails] Fissures (or chimneys) through which magma rises up to the surface are very thin indeed, thinner than a very small trickle of water." *Athanasius Kircher*, Mundus Subterraneus, *1665; copperplate engraving by Theodor Matham; trans. by R. Cataldi; The Library of Congress*

"Athanasius Kircher's *Mundus Subterraneus,*" writes The Library of Congress, "is an early illustrated scientific treatise that describes the natural processes of the Earth from a physical standpoint within the context of his religious beliefs. A Jesuit and an accomplished linguist and natural philosopher, Kircher traveled widely, visiting the sites of earthquakes and volcanoes in Sicily and Italy. Diagrams and maps in his book convey Kircher's eclectic ideas and beliefs through a combination of natural and allegorical images and emblems. The book was published by the Amsterdam firm of Johannes Janssonius van Waesbergen and his wife Elizabeth, daughter of the famous map and atlas printer Johannes Janssonius, and is an excellent example of the fine, detailed copper engraving associated with the great Dutch printing houses of the 17th century."

CONTENTS

Islandia, 1585, by Abraham Ortelius. Hand-colored engraving, printed in *Theatro d'el Orbe de la Terra*, Antwerp, 1602, originally titled *Theatrum Orbis Terrarum* ("*The Terrestrial Globe*"). The map features an eruption of Mt. Hekla, Iceland's most famous volcano, which, writes I. B. Fridleifsson in Chapter 19, was "a site well known to the Church in Europe during the Middle Ages as one of the gateways to Hell." Polar bears still drift to Iceland on ice floes, as depicted in the upper right-hand corner. *Courtesy of the Osher Map Library, University of Southern Maine*

About the Stories, An Introduction

"Give me a condor's quill!
Give me Vesuvius' crater for an inkstand!...
To produce a mighty book, you must choose a mighty theme."

—From *Moby-Dick,* Herman Melville, 1851

STORIES ARE TRUE OR IMAGINARY ACCOUNTS: FACT AND FICTION. THE GEOTHERMAL STORIES IN this volume fit the mold and tell something never before recorded to such extent: mankind's abiding interaction with geothermal phenomena at the Earth's surface.

This is not an Earth science text or a technical treatise, and geothermal phenomena themselves are not geologically or industrially described. Instead, the focus is on people throughout time, on how geothermal features at the surface of the Earth have affected mankind all over the world, sometimes with important consequences. Each chapter relates this story, as for the first time a project so comprehensive views the history of people around the globe through a geothermal lens.

Volcanoes, hot springs, fumaroles, geysers, hot and steaming ground, and hydrothermal by-products such as sulfate, borate, silica, travertine, kaolin—along with earthquakes, volcanic explosions, crater collapses, lava flows, and ash clouds—all from the start offered joy and fear, aid and destruction, history and tales, fact and legend. They molded old customs and helped forge new experiences that through millennia became tangled in time.

From the beginning, humans were both enhanced and harmed by these phenomena, and the dichotomy continues today. Prehistoric peoples, often holding them in awe, thought of the Earth as sacred and considered terrestrial heat, sometimes called in Antiquity the "internal fire of the Earth," as one of the Earth's natural resources bestowed by the gods for the benefit of humankind. Hot springs and fumaroles were described as *Wakan Tanka* ("Great Mystery") by the American Dakota Sioux and as *tapu* ("sacred") by the Maori. These beliefs, expressed in popular customs, traditions, religious feelings, and rituals, were handed down through hundreds of generations and still influence people today.

Where did our authors find this information? Proof takes many forms, and works by archaeologists, historians, geographers, anthropologists, scientists, and engineers all played a role, as did those by artists, photographers, poets, philosophers, and literary figures. Some stories came directly from the oral tradition, never before recorded. Many were extracted from manuscripts, some over 2,000 years old, where the oral tradition is also a part of the chain of information, for ancient writers interviewed the people around them and recorded practices in the same manner as today. Lines of antique text, pottery, drawings and maps, votive statues, shrines, stelae, myths, and legends: all tell geothermal stories.

From this melange came the histories in this book: actual and imagined—fact and fiction, different and yet the same—humans reacting to geothermal features. The sheer repetition of the experiences through millennia adds veracity to the stories, as Shahrzad explains to the King:

> "'Last night,' he said, 'in my wandering
> I lit upon a story that was like one of yours, Shahrzad.'
>
> Despite her hidden sorrow, she said smiling,
> 'The fact that stories repeat themselves
> is an indication of their truth, Your Majesty.'" *

"Contemporary society accentuates the break with the past," writes Amadou-Mahtar M'Bow about the world's natural, cultural, and historical sites. "A consequence is to reduce the work of the past to mere curiosity of little relevance to everyday reality." Unfortunately, a similar break affects our historical knowledge of geothermal resources, of the past uses found for natural heat and its by-products.

And as these uses expand today, often with sophisticated electrical generation projects, an erroneous belief grows that natural heat is a new form of energy, unknown and unused by ancient peoples. Our authors clearly show this is not so, that the knowledge of geothermal phenomena, the use of natural heat with the related cultural traditions, and many scientific concepts on the origin of the Earth's energy began in ancient times.

Not only does this link with Antiquity exist, but readers will note a common heritage among peoples living near surficial geothermal resources, how civilizations flourished in such areas in ways clearly different but yet strangely similar. The volume tours the geothermal world, beginning with the small boy on the cover gazing at Mt. Fuji. On the inside cover we start at Easter Island and move around the globe, through the final chapter on the Andean region in western South America. Finally, at the inside back cover we end where we began—back at Easter Island, 2,400 miles west of South America, on the most isolated, inhabited island on Earth and one with a rich geothermal heritage.

Chapters 1 and 2 hold a special place. The first describes an imaginary creation of the universe, including our planet with internal heat. The second reenacts mankind's first encounters with geothermal heat from volcanoes, earthquakes, and hot springs. Which thermal feature was viewed first—a volcanic eruption, a hot spring, a fumarole? We will never know, but the chapters assure us they all were important.

Does soaking in thermal waters cure illnesses? From time immemorial, people worldwide have thought so. With nothing more to guide them than their own curiosity and experience, they have bathed in hot springs to alleviate the same diseases, including arthritis, rheumatism, psoriasis, and leprosy. Development by these hot springs usually flourished, although occasionally some inhabitants tried to temper it, unhappy with the arrival of so many ill people at their villages. In another approach, many North American Indian tribes kept hot springs with their health-giving properties as sites of neutral ground, open to the use of all.

Through time, geothermal resources have provided the raw materials for items both practical and artistic. Obsidian tools and arrow points once chipped for hunting are still made today in a few remote areas for the same reasons. Thermo-mineral muds collected for skin care through millennia from the edges of hot springs and other geothermal features now sell for high prices at beauty counters in exclusive stores for the same purposes. Pumice, always used for thorough cleansing, is still sold commercially as a key ingredient in "extra strength" soaps and by itself as a skin abrasive.

Forever commanding attention, volcanoes have inspired religious practices, provided fertile soils for farming and natural environments, and offered overwhelming beauty. They also have spewed vast amounts of gas, lava, rocks, and ash, causing widespread destruction to people and property. But time and again people returned to the looming slopes, as they do today, for the good outweighs the bad. Many chapters highlight this point.

We did not include dramatic photographs of volcanic eruptions for a reason. Such photos are widely available and the eruptions have existed through time, controlled by the same geological processes. The natures and consequences of the eruptions have never changed, and describing eruptions does not describe mankind's life around volcanoes. However, a brief idea of the nature of volcanic events is offered in Chapter 2, in the evocative painting of the eruption of Mt. St. Helens as viewed by an Indian group. A vivid description by Mark Twain of the paradoxical power and beauty of an eruption in Kilauea crater on the island of Hawaii is among the pre-chapter pages.

Thus we see how many applications of geothermal resources stand not alone but on an historical continuum. Exploring the cultural links between past and present has resulted in this volume, some of which was written for a geothermal historical session at the 1995 World Geothermal Congress in Florence, organized and co-chaired by Raffaele Caltaldi and John Lund; Susan Hodgson was among the presenters. The 13 papers from the Italian conference have grown into the 34 chapters, and we thank our distinguished publishers, the Geothermal Resources Council and the International Geothermal Association, for their encouragement and support of the work.

Although not all parts of the world are represented equally, and coverage for those that are is incomplete, one thing is clear: only the surface has been scratched. Just waiting to be found are thousands more *Stories from a Heated Earth.*

Raffaele Cataldi
Susan F. Hodgson
John W. Lund

Spring 1999

*Quotation from *Arabian Nights & Days* by Naguib Mahfouz, with permission of Bantam Doubleday Dell Publishing Group, Inc.

Volcanic stone sculpture from Teotihuacán, Mexico, A.D. 250 - 750, 19 in. high. The figure is Huehuetéotl, the old god of fire, one of the most ancient deities of Mesoamerica. He is often depicted with these features and in this posture, sitting crosslegged with hands on his knees, the right hand cupped open and the left one closed in a fist. During ceremonies, fire was kindled in the brazier on his head. *Natural History Museum of Los Angeles County*

1. When Geothermal Was Still in the Minds of the Gods of Teotihuacán

by
Raffaele Cataldi

Cuando aún era de noche,
cuando aún no había día,
cuando aún no había luz,
se reunieron,
se convocaron los dioses
allá en Teotihuacán!
— *Códego Matritense*

When the night was still black,
when there was no day,
when there was no light,
the gods came together
in a meeting
there, in Teotihuacán!
— *Madrid Codex*

So, where was Teotihuacán?

There was no sun, nor stars, nor space before time began. There was neither earth, nor moon, nor air, nor light before the gods of Teotihuacán drew the coordinates of the sky. There was no primordial chaos; there was only night and void, barely broken by impalpable specks of cosmic dust.

So, where was Teotihuacán?

The *Madrid Codex* does not say, nor do other ancient texts of cosmogony. Nevertheless, we would like to think Teotihuacán really existed before time began: up beyond the sky, in a place reached only by the roots of thought.

Translated with revisions and reprinted by permission from *Geotermia, Revista Mexicana de Geoenergía,* 1992, 8, no. 2, pp. 145-154.

Why the gods and goddesses met first in Teotihuacán, or why they met at all, never is explained in the *Codex*. Were they bored, wandering in the dark with only the light of their minds, or melancholic from the solitude of immense nothingness?

One god, certainly the wisest and most influential, called his brothers and sisters together up there in Teotihuacán. Perhaps after talking, learning how useless they felt, they agreed together to create the world.

The one who had called the meeting set to work. Directing his gaze on the horizon, beyond the portals of the heavens, he projected a thought and created the dawn, which gave birth to the sun.

A goddess gathered the cosmic dust that hindered the flight of the gods through the void. Then not knowing where to put it, she piled the dust in a corner of Teotihuacán. Now the heavens and the void stood limpid, but the pile of dust detracted from the perfection of Teotihuacán and the gods compressed the dust and carried it beyond the dawn.

Another god thought the inert dust might be useful and, with the help of the others, set off explosives, shattering the dust into small bodies. The fragments flew so far apart that they never came back together, a great engineering feat.

But the random form of each fragment did not please a goddess who had stayed in Teotihuacán, watching with the spirit of an architect. She knew that once the fragments were shaped into spheres and polished, they would reflect the light of the dawn. So she called together all the gods and goddesses and they began the task, finishing just as the sun left the womb of the dawn.

Morning came and light shone on Teotihuacán and in the skies around it; but it was a faded, uniform light with no contours or edges. A diaphanous light, weak and cold. Where were the shadows and contrasts? Where were heat and the colors of the rainbow? Much remained to be done.

The god who had called the meeting knew this, but since the others were satisfied, the work ended and each went back to wandering alone through the embryonic universe. The sun stood still on the horizon, motion and space were lacking, and life remained in the mind of the god who knew that work had only begun.

Teotihuacán remained deserted and time passed. How much time? Perhaps a second, perhaps a day, perhaps billions of years; who can say? There was still no Milky Way, the stars of the Bear shone with faded light, and the rhythm of time was not marked by the Earth circling the sun.

We have to think that the gods turned melancholic, roaming through the primordial universe still so cold. Summoned by the most farsighted among them, they gladly returned to Teotihuacán. This time there was less discussion. Each knew better how to complete the task, and they worked with enthusiasm.

The god who had created the dawn was first. He ordered the sun, the first star, to rise and start its journey along the vault of the sky. At first just the sun had light and heat of its own, formed as it was in the womb of the dawn.

The spectacle was fascinating and stately, and the gods and goddesses were entranced. They looked for the spheres of the architect goddess and held them, infusing them with heat and light, making more stars to fill the sky, arranging them into constellations, galaxies, and swarms of comets.

One of the gods, with the spirit of a topographer, looked for a reference to fix the positions of the stars to each other. He placed the beginning of the celestial axes in the center of Teotihuacán, tracing from there coordinates for the heavens. Pivoting on the main axis, he spun around with his arms outspread and imparted a slow rotation to the world, and motion and space were born.

Then a god noticed the sun's mane of tousled hair. He trimmed off the longest curls and threw them into space. Luckily, the wisest brother noticed. Worried about the fate of the curls, and rightly so, he gathered them together while the others finished the universe. He pressed and molded each curl into a ball and arranged the balls around the sun like a crown on the head of a king. He set the balls into dual motion, spinning each around itself as it spun around the sun, and so the planets were born.

For the planet he would call *Gaea*, meaning "Earth," he had a special plan, slightly risky but full of promise. He inserted a red-hot core into the center, covering this with shells of rock to hold the heat. He shaped the last shell with care, digging trenches for water and places to keep ice in reserve. He cloaked Gaea in a tunic of air whose folds would hold the clouds and the wind.

He patterned the rhythms of time by the seasons and established the cadences of day and night. At last, he offered Gaea a pearl in a setting outside the tunic, safe from the insults of the wind. He gave the pearl a name as dear to the gods of Teotihuacán as that of the sun—the moon. Truly, the work was finished. All life needed was ready: light, air, water, land, and heat. Gaea, the Earth with internal heat, no longer was just an idea in the minds of the gods.

One detail was lacking. Where was Teotihuacán? Should it stay outside of space, up beyond the portals of the heavens, or should a trace, a memory, come to Gaea?

The god who had called everyone together decided. He placed Teotihuacán on Gaea, at the projection of the celestial zenith. Then he set Gaea's continents in perpetual drift, letting the site move endlessly on the Earth, holding the memory of the first meeting of the gods. Now he flew far from Teotihuacán beyond the confines of the sky, where he found the others waiting.

Today Teotihuacán is on the North American continent, near Mexico City. Someday, it will be elsewhere on another continent; perhaps a future codex will mention where.

The Author:

Raffaele Cataldi
55, Via del Borghetto
56124 Pisa, Italy
Telephone and fax:
39.050.59.81.07
E-mail: rafcat@tin.it

Indians watch a volcanic eruption from a canoe. The painting, titled *Mount St. Helens* by Paul Kane, depicts this active Cascade Range volcano in the Northwestern United States. *Courtesy of the Royal Ontario Museum, Toronto, Canada*

2. The Year Zero of Geothermics

Abstract: The *year zero of geothermics* is defined as that period between humans' first experiences with geothermal phenomena and the time when the simplest uses of the products of the Earth's heat were known, and when the first core of a spiritual rapport with some geothermal phenomena was established. Terrestrial heat and humans began coexisting in the Paleolithic, when an ancestor first found an area with active geothermal phenomena. Gradually, people learned to use natural heat for thermal baths and cooking. They sought ways to exploit some geothermal by-products. Spiritual—or transcendental—relationships began later, perhaps 35,000 to 40,000 years ago when the practice of burying the dead began. Slowly, a religious attitude developed toward underground "powers," linking them with geothermal phenomena—some violent and destructive, some peaceful and beneficial. These relationships evolved and consolidated over many millennia, eventually forming the cultural basis in prehistoric and ancient times for the flourishing of myths, legends, cults, and popular beliefs directly or indirectly related to geothermal phenomena.
Revised, translated, and reprinted with permission from *Geotermia, Revista Mexicana de Geoenergía,* 1992, 8, no. 2, 155-175.

by
Raffaele Cataldi

Let us enter the unknown
to encounter the new.
—Baudelaire

Foreword

THE FIRST OBSERVATIONS OF VOLCANIC ERUPTIONS AND INITIAL uses of natural heat and its by-products most likely date back to remote times in the Paleolithic; however, concrete evidence dates from 8,000 to 10,000 years ago. Therefore, speculation on early human relationships with geothermal phenomena and products of the Earth's heat can be made only on the basis of indirect considerations. As Febvre explains:

> "History is made as much as possible by using written documents; unfortunately, though, they are not always available. In this case, history should be made on the base of any indirect information enabling historians 'to produce honey even in the absence of the flowers they normally use....' In such cases the historical sources should be tales, signs, vestiges, landscapes, bricks, field boundaries, seeds, lunar eclipses, chemical analyses on swords and findings of a different nature, geological information, and any other elements related to the development of humankind; ...all elements

that display traces of the presence, activity, behavior, customs, and ways of thinking of people.... Is it not perhaps true that the most stimulating part of an historical investigation is to find ways to lead silent things to speak...?" (Febvre, 1976).

Beginning of the Relationship between Humans and The Earth's Heat

The geological origin of geothermal heat is at one with that of our planet: a retreat to the dawn of geologic time when no life forms existed. Several billion years passed from that moment before humans evolved and observed geothermal phenomena. But when and where were the first experiences with such phenomena? Was it the moment an ancient ancestor first noticed a natural manifestation, or was it later? Perhaps it started somewhere in the night of prehistory, at a place that we can only imagine, while an ancient ancestor was roaming in search of food.

We cannot say if this ancestor was male or female, still *homo erectus* or already *homo sapiens.* However, he or she was already able to mumble word-like sounds, find shelter for the night, chip hard stones into objects, and hunt with skill.

One night, while sleeping peacefully in a shelter beneath an overhang, our ancestor awakened to a shaking bed of dry twigs and leaves. Outside the refuge, thuds resounded from falling stones and soil. Was a nocturnal beast prowling nearby? Our alert friend got up and crouched, listening and watching intently, but nothing happened.

A few hours later, the Earth shook again. In the light, our ancestor could realize that the continuing clatter of stones had not been from an animal. The shocks continued all that day and for several days after, some stronger, some weaker. It was annoying to be rocked permanently, as if the leg joints had given way, but our friend grew accustomed to it. Eventually, the quaking stopped for a few days.

After the moon had traversed perhaps half the sky, while foraging for food in an area distant from the usual shelter, our ancestor noted a cone-shaped mountain topped by a pale, thin, and mysterious "cloud." What was that cloud, oddly flared at the top, and how did it get there? For the summer sky was immaculately blue. Was it smoke from a lightning fire? At first our friend did not pay much attention to that cloud as the mountain was so far away. Certainly that forager had no idea that the cloud was steam from a fumarole atop an awakening volcano.

In the evening, now tired and ready to enter a shelter, our ancestor turned instinctively toward the cone-shaped mountain, where the thin, whitish cloud of the morning had become a tall, wide, and grayish column of steam, with great fringed billows, iridescent in the setting sun. Suddenly, the grayish column swelled, grew more ashen and blackened the sky; at the same time, a deep rumble shook the Earth, and the ground started to leap frenetically.

In terror, our ancestor raced away, staggering on trembling ground. The flight was illuminated by glaring reddish lights on top of the cone-shaped mountain. After many hours of flight, exhausted, our forebear stopped and turned toward it. Now the mountain tip was missing, the slopes furrowed by bleeding streams. Was this a monstrous animal wounded by a thunderbolt, its whole body quivering, shaking in the pangs of death? Where was the top of the mountain? Why did the neck and body of that monster erupt in spurts of blood?

The volcanic eruption continued for many days, sometimes more quietly and other times more violently than the first night. Searching for food and shelter, our ancestor eventually met others and tried describing to them what had happened in his area: the quaking, the cloud, the fire. The others, too, had fallen against the shaking ground, and some of their fellows had been killed by balls of fire dropping from the sky. The survivors formed a small group and decided to leave the area together, searching for a safer place far from the frightening mountain.

Thus, the first contact of humans with geothermal phenomena may not have been one of disbelief or indifference, but of surprise, fear, and terror: a shocking new experience with strong feelings! In prehistoric times, similar feelings and experiences perhaps constituted the primordial cue from which legends related to volcanic eruptions and earthquakes would form in later epochs. According to two legends, volcanoes were fed from the breasts of the Titans struck down by the lightning-bolts of Zeus (Hesiod, c. 800 B.C.), whereas earthquakes were caused by the Giants defeated by the gods of Olympus and buried alive under active volcanoes (Strabo, c. 30 B.C.; Virgil, c. 30-19 B.C.).

The volcanic eruption, however, not only offered those ancient ancestors the macabre impression of a beheaded monster but also a majestic spectacle to watch, one with lava fountains, incandescent bombs hurtled into the night sky, and glaring fires on the slopes of a conical mountain. These progenitors had even stopped to observe the phenomena of that night, committing them to memory, and trying to fathom their cause. The sense of astonishment they must have felt for phenomena never seen before, and the curiosity they must have felt to watch them may have sparked the first thoughts about manifestations of the Earth's heat. Their experience may have been the first core of human knowledge about volcanic activity and related events.

First Uses of Geothermal Resources

Hundreds of generations passed after that night, but the memory of what could happen atop a cone-shaped mountain perhaps was not lost, handed down from parent to child. Through that long time, the small initial group of ancestors who experienced the effects of a volcanic eruption grew and moved into surrounding lands.

One day, one of them came to cross an especially warm stream. Surprised, this individual instinctively looked for the sun, the most likely heat source, but the weather had turned cold and the sun was hidden behind thick clouds. The water of the stream was cloudier and whiter than usual. Curious, this individual paused. Taking a few more steps in the water and finding it warmer still, this ancestor slowly moved upstream, noting how the temperature grew warmer and how a sense of well-being entered the body. Seeing a large boulder half submerged in the stream and forgetting hunger, our friend sat on it to bask and enjoy the heat of the water. This was the first human use of terrestrial heat: thermal bathing.

News spread of the warm stream and the fertile valley around it, rich with prey and lush with vegetation. More people came and settled in natural shelters on the hilly slopes nearby. Here was the first human settlement induced by thermal waters: the first place in the world where the development of a community was favored by the presence of geothermal manifestations.

Water at the source of the thermal spring was much warmer than water downstream, farther away. Another ancestor learned this one bitterly cold day while trying to warm up in a natural pool very near the source. Even with skin browned by the sun, encrusted with mud, and tough as leather, this ancestor was scalded and had to jump immediately from the pool. By retelling the misadventure through generations and by testing the water at different parts of the warm stream, people in the community started learning the different effects of thermal waters.

Certainly cooking began later, perhaps when a hunter accidentally dropped small, newly killed prey into a natural pool of nearly boiling water. Finally, when the animal was retrieved, the hunter found that cooked meat was much easier to skin and more enjoyable to eat. Thus a hunter's fumbling attempts to recover a quarry started, by sheer chance, the second use of terrestrial heat: cooking food.

Much time must have passed, perhaps millennia, before our forebears took another step to strengthen their relationship with geothermal phenomena. This may have occurred when a different group settled in an area rich with natural manifestations. These people, too, had learned how to use warm water for bathing and how to cook meat and vegetables in boiling waters. In this area, however, besides thermal springs and a small lake with warm water, wafts of whitish vapor smelling like rotting grass could be seen in several spots. These were scattered over a barren portion of land, near the edges of a lush wood. Here the ground was parched and pocked, almost without vegetation, and was furrowed by trickles of cloudy warm water flowing into the little lake. Small ravines and tortuous gorges could be seen, their mineralized edges caked with polychrome efflorescences: white to yellow, emerald green to pink, and red to violet. In some places, the shores of the lake displayed the same hues.

All this—the lush forest, the wisps of steam, the transparent water of the little lake, and the variegated efflorescence—gave the site a delightfully rich palette. Thus the spot became an attraction for roaming hunters and gatherers and a visiting place for residents of the nearby area, who enjoyed a relaxing warm bath in winter. The place eventually became a kind of open-air resort: our ancestors' first thermal spa.

The mineralized efflorescences lining the little ravines and coating the ground must have drawn the visitors' attention, but many centuries probably passed before people realized that none were edible. Meanwhile, some ancestors found that the gray-white mud deposited on the shores of the little warm lake would dry up infected sores and tone the skin. Thus, people slowly learned that the thermo-mineral mud deposited under and around hot, natural pools could be used for what we call balneotherapeutics: the third use of natural heat.

Practical Applications Only or Something More?

All of these events probably took place in geothermal areas over several thousand centuries, from an indeterminable moment in the Paleolithic to around the beginning of the Neolithic. During that time, people learned more about the different manifestations of terrestrial heat, and they put some forms to use.

At a certain point, though, besides functional uses, a deeper relationship must have evolved with geothermal phenomena, one delving into the spiritual, with a transcendental nature. We are prone to think that such religious relationships started to take shape in the human mind toward the end of the Mousterian period (35,000 to 40,000 years ago), when people began to bury the dead. In that period, people probably began thinking that the spirits of the dead would be "preserved" somehow under the surface of the Earth, perhaps believing that an afterlife might exist in this realm. Moreover at that time, our ancestors probably sensed that some manifestations of terrestrial heat, such as hot springs, fumaroles, and volcanic eruptions, were somehow linked to, and in some way controlled by, subterranean "forces." If so, people may have imagined that a living, dynamic, underground world could exist, one inhabited by "powers"—sometimes peaceful and benevolent, other times violent and destructive—able to rule over the heat beneath and at the surface of the Earth.

These ideas must have formed in a period much earlier than the epoch when myths, legends, and popular beliefs were developed to explain the different manifestations of terrestrial heat. At any rate, the idea of an underworld as a common gathering place for dead souls grew long after the first burials; thus the concept of a subterranean realm and the myth of Pluto as its lord, the idea of hell as a place of expiation for souls amid eternal flames, and the figure of Lucifer as the devil date from periods much closer to our own.

Concerning the spiritual relationship established between the early inhabitants of the Larderello area in Central Italy and its many geothermal manifestations, Burgassi wrote in 1987:

> "It seems that...as early as the Neolithic, these manifestations were considered to be a sign of the existence of underground divinities who occasionally rose up through them to the surface, accompanied by hisses, rumbles and subterranean noises. It may have been then that the belief began to spread in prehistoric epochs that these divinities wanted to make people aware of their intrinsic power, and of the sign of respect they required from whoever ventured to tread on the roof of their abode.
>
> "Some think that...the area of the manifestations was the uncontested realm of infernal divinities; but this idea does not stand up, because the concept of hell...certainly came after the Neolithic period. The view must rather be held that Neolithic people were inclined to believe not in the devil and in hell, but in the existence of underground divinities, some of them powerful and dangerous, others gracious and benevolent. Thus, people had to learn gradually, at their own expense, to respect the former and not to arouse their wrath; at the same time, people must have begun to have confidence in the benevolent divinities, and to enjoy the healing gifts they could receive from them....
>
> "The heat on a human scale perennially supplied by the gracious manifestations as an antidote to the cold, and the therapeutic properties of certain muds and waters, must therefore have

> represented a largesse bestowed by benign deities of the underworld, and certainly not from malevolent angels, as the concept of hell would lead one to think...."

Perhaps all early communities settling in areas with active geothermal manifestations felt this way. These attitudes probably developed over long periods of time and at different rates from place to place. Certainly, a human transcendent relationship with geothermal heat would consolidate in different times in each area, depending on people's intellectual capacity to correlate natural phenomena with root causes and on their degree of spiritual maturity. Did it also begin toward the end of the Paleolithic—the worship of underworld divinities?

Conclusions

In prehistory, humans developed both a utilitarian, rational relationship with geothermal phenomena and a transcendental, spiritual one. Both relationships evolved at different times in different areas and included a series of aspects that are represented in the following figures as "likely events" that occurred in temporal sequence. To allow for differences of the periods in which these various events may have occurred in different areas of the world, two time intervals are shown. The inductive reconstruction of these events is made by logical extrapolation, moving back from 8,000 to 10,000 years ago (the period with the most ancient data on the presence of people in geothermal areas) into older times.

Despite uncertainty over when specific events may have taken place and how long they lasted, the period in which they occurred in prehistoric epochs we call the *year zero of geothermics*. Begun that far-off day when an ancient ancestor glimpsed a thin plume of steam atop a cone-shaped mountain, the *year zero* may have lasted many thousands of centuries, until the approach of historical times. Over the course of that symbolic year, human relationships with geothermal phenomena were close and intense, evolving slowly as observations on volcanic eruptions and natural manifestations accrued and ways were found to use the Earth's heat. The first uses were, of course, the most natural and immediate: taking hot baths, cooking, and practicing basic therapeutics.

Indeed, if we think of the experience gained by people over that long span of time while trying to understand the beneficial effects of some geothermal manifestations and the dangerous effects of others, if we bear in mind that the discovery of numerous thermal localities in distant parts of the world occurred independently and at different times, if we consider that many such localities probably attracted permanent settlements from the moment of discovery, and if we think of the contribution that the observation of volcanic eruptions gave to the territorial knowledge of many areas on Earth, then we must conclude that the *year zero of geothermics* formed the basis for much of the later development of terrestrial heat and its by-products.

To ignore or dismiss the difficult and sometimes distressing human experiences in that *year zero* is to disregard the logical framework, cultural roots, and conceptual continuity from which geothermal development has grown since the dawn of humankind. With such a framework, we can explain the formation of a more complex human-geothermal relationship that began in the Upper Paleolithic and involved cultural and spiritual aspects from which myths, legends, and cults formed and flourished in protohistoric and ancient times.

With the same framework, we can also understand the spirit of observation exhibited by Lower Neolithic peoples and their attention to such detail in wall paintings depicting natural landscapes and volcanic eruptions. The wall painting in the final figure was found in a prehistoric settlement of Southeastern Anatolia dating from the 7th millennium B.C. (Mellaart, 1967).

The onset of the Metal Age ended that long but fertile period of initial human contacts with the Earth's heat, the *year zero of geothermics*. From there, geothermal development would continue through the centuries, emerging from the darkness of time into the age of recorded history.

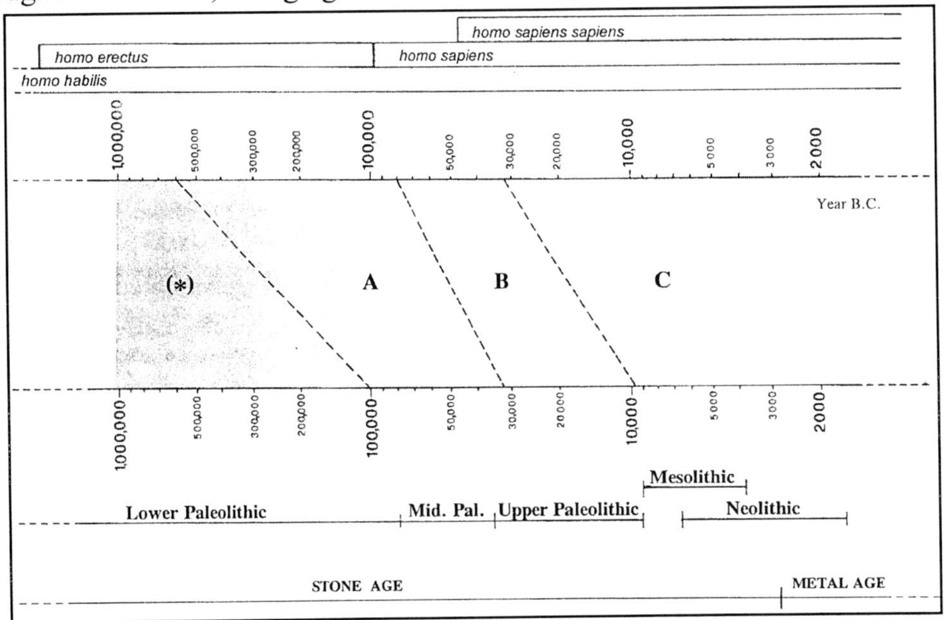

Chronological reconstruction through prehistory of the initial uses of geothermal resources and by-products: (*) people's initial contacts with geothermal phenomena; (A) thermal bathing; (B) cooking in boiling waters; (C) therapeutic application (drying sores, stopping bleeding, healing and toning skin).

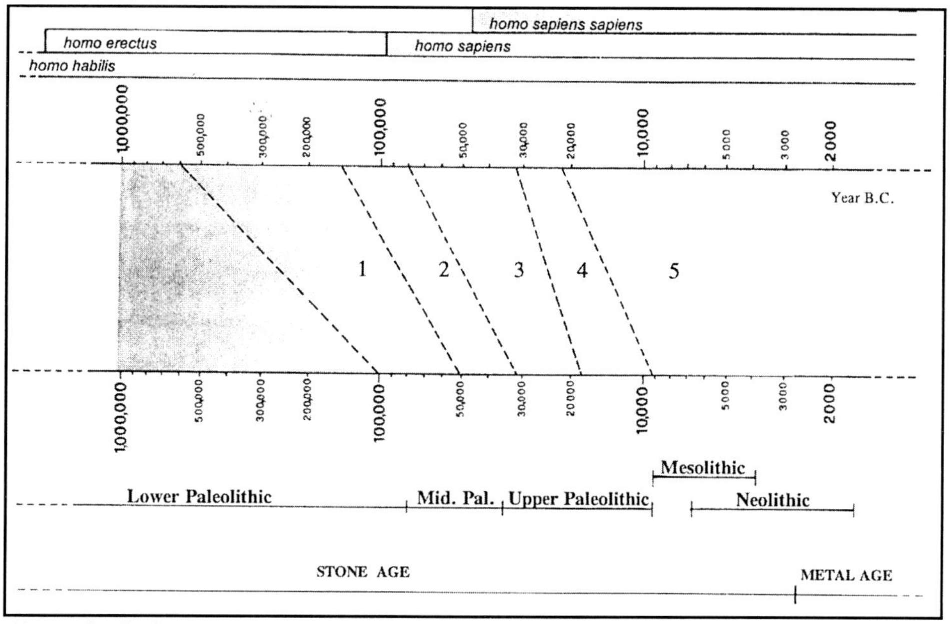

Chronological reconstruction through prehistory of human rational relationships with geothermal manifestations: (1) first formation of a collective memory and establishment of first "cores" of knowledge about geodynamic and volcanic events; (2) comparisons among different types of geothermal manifestations; (3) isolated human settlements near natural manifestations; (4) first comparisons of different effects of thermal waters; (5) visits to sites with manifestations for recreational purposes.

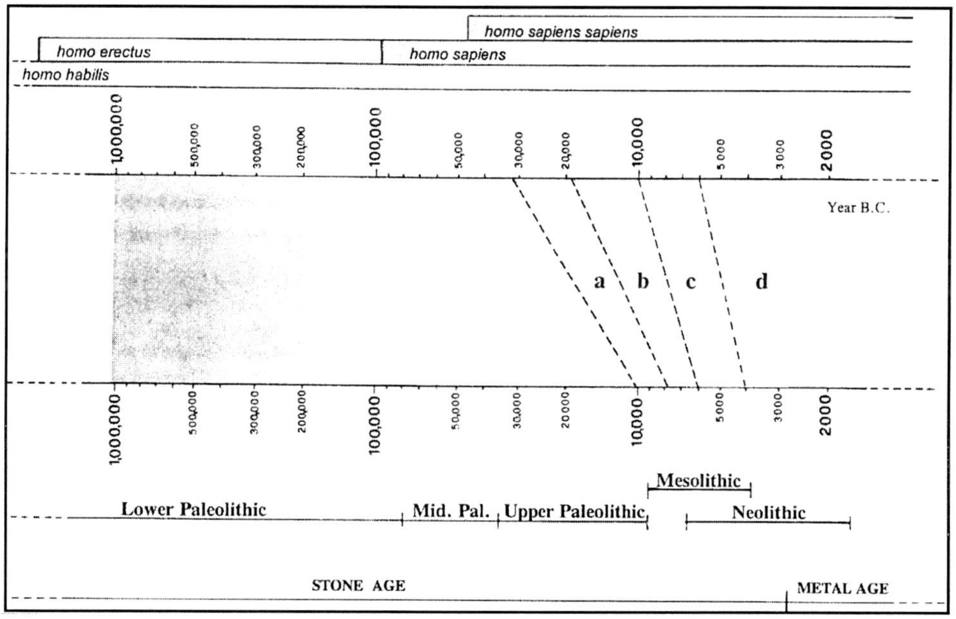

Chronological reconstruction through prehistory of initial human religious relationships with geothermal manifestations: (a) first correlations between natural manifestations, volcanic eruptions, and subterranean "forces." Initial belief in different "powers" dwelling beneath the surface and embryonic formulation of two logical equations: manifestations/subsoil/underground "powers," and death/burial/afterlife; (b) initial relationship established between underground "powers" and subterranean entities; (c) beginning of cult of subterranean divinities; (d) cult of thermal waters probably begun.

(a)

(b)

(c)

Neolithic village in Southeastern Anatolia, Turkey, with the erupting volcano Hasan Dağı (?) in the background (about 6200 B.C.): (a) wall painting found in "level vii" of the Çatal Hüyük Neolithic settlement (*Mellaart, 1967*); (b) reconstruction of the original wall painting (*Mellaart, 1967*); (c) interpretation of the explosive phenomena shown in the painting (*Cataldi, 1992*).

Cited References

Burgassi, P. D. 1987. Historical outline of geothermal technology in the Larderello region. *Geothermal Resources Council Bulletin,* 16, no. 3, 3-18. (See Chapter 13 in this volume).

Cataldi, R. 1992. El año cero de la geotermia. In *La geotermia el el período Precolombino en las áreas Mediterránea y Mesoamericana. Geotermia, Revista Mexicana de Geoenergía,* 8 no. 2, 155-175.

Febvre, L. 1976. *Problemi di metodo storico.* Einaudi Edit., Turin.

Hesiod. c. 800 B.C. *Theogony.* In *Hesiod.* Reprint 1972, M. Mazòn, trans. 8th ed. Les Belles Lettres Edit., Paris.

Mellaart, J. 1967. Çatal Hüyük: A Neolithic town in Anatolia. In *New aspects of antiquity.* M. Wheeler, ed. Thames and Hudson Ltd. and The Camelot Press, London and Southampton.

Strabo. c. 30 B.C. *Geography, vol. 2.* In *The geography of Strabo, vol. 2.* Reprint 1967, L. Jones, trans. The Loeb Classical Library, London-Cambridge.

Virgil. c. 30-19 B.C. *Aeneid.* In *Virgil.* Reprint 1916, R. Rushton, trans. The Loeb Classical Library, London and New York.

The Author:

Raffaele Cataldi
55, Via del Borghetto
56124 Pisa
Italy
Telephone/Fax: 39.050.59.81.07
E-mail:rafcat@tin.it

At Mtagata hot springs in Uganda in the 1800s. *From* With Stanley in Africa *by Captain McClure, 1891, Worthington Company, New York*

3. From the Beginning, Life in the Great East African Rift

by
John W. Lund

Abstract: The East African Rift is associated with the origins of our human ancestors, starting at least four million years ago. This Great Rift traverses most East African nations, from Mozambique to Ethiopia, with over 200 hot springs, active volcanoes, and volcanic deposits all along its margins.

Certain hot springs were long believed to have curative properties and volcanoes were the dwellings of wild animals or the devil.

Lake Nakuru and Lake Natron are inhabited by flamingos who feed in the hot acidic water and raise their young in the warm mud. By eating the blue-green algae, they obtain a reddish color, and legend has it that they are the resurrection of the phoenix.

In the highlands of Ethiopia are 1,000-year-old churches carved from cemented volcanic ash. Local legends tell that these churches were built on divine command with the aid of angels.

The top of a volcanic monolith in the same area, Amba Wahni, was a prison for royal relations of the king who might usurp the throne.

"*BASALT*, originally an African word. 'The Ægyptians also found in Æthyopia another kind of Marble which they call *Basaltes*, resembling yron as well in colour as hardnes.'
—*Pliny* XXXVI,vii, §11"
Second Edition, 1989, by permission of Oxford University Press

Early Humans

Africa is recognized as the very cradle of the human race (Weaver, 1985). Prehuman remains of *Australopithecus anamensis*, the earliest known human ancestors to walk erect and dated between 4.07 and 4.17 million years old, have been recently discovered in Northern Kenya (Wilford, 1998). Remains of *Australopithecus afarensis* have been found in the Afar badlands of Ethiopia. The most famous find is Lucy, an adult female touted as the oldest known hominid, having emerged as early as 3.6 million years ago. These people died close to rivers or lakes and probably were familiar with the hot springs of the area. Numerous human fossil remains and footprints have been preserved in the volcanic ash (tuff) of the Rift Valley, the most famous at Olduvai Gorge in Northern Tanzania. This is a 40-kilometer long, river-carved canyon about 90 m deep, dropping abruptly from the Serengeti Plain. It offers well-exposed strata of lakebed sediments and volcanic ash going back nearly two million years. Here Louis and Mary Leakey found examples of *Homo habilis* dated at about 1.8 million years old (Weaver, 1985). Tools made from volcanic rock are associated with this find, mainly as cobbles used as hammer stones and choppers. Better specimens of *Homo habilis* were found on the eastern side of Lake Turkana in Northern Kenya.

Homo erectus, dated at about 1.6 million years ago, was found on the western bank of Lake Turkana. Remains from more than 200 early *erectus* humans have been found along the slopes

The Great East African Rift System

The East African Rift System, or The Great Rift, runs parallel to Africa's eastern coast for 5600 km (Stager, 1990). This furrow, formed from a series of cracks in the Earth's crust and the resulting highlands, has two branches. The western branch starts in Mozambique to the south and runs north along the border of Zambia and Tanzania, through Burundi and Rwanda, and into Uganda. The eastern branch starts in Central Tanzania and runs north through Kenya and Ethiopia, ending on the continent at Djibouti and Eritrea at the Afar Triangle bordering on the Red Sea and Gulf of Aden. The rift can be traced further north to the Dead Sea, Turkey, and the Horn of Africa through the Gulf of Aden. Enormous troughs, up to 2 km deep and 90 km wide, are formed along these cracks. Elongated lakes, also formed by the rift system along the western branch, include Malawi, Rukwa, Tanganyika, Kivu, Edward, and Albert. Shallow saline lakes along the eastern branch include among the larger Eyasi, Natron, Turkana, Abaya, Abbe, and Assal. Massive Lake Victoria lies between the two branches. The volcanoes Kilimanjaro, Africa's highest peak at 5895 m, and Mount Kenya at 5199 m, stand adjacent to the rift near the center. Hot springs, many at boiling temperatures, lie all along the rift.

The stretching of the Earth and deposits of volcanic material form bold escarpments and dramatic valleys. The Ethiopian Highlands began some 20 million years ago. The rift system is fueled by hot magma upwelling from the asthenosphere, thinning and stretching the crust. Rising domes create the high elevations of Kenya and Ethiopia. Vast outpourings of lava coat the land and faults fracture the crust. At it highest point, the rift reaches 1900 m above sea level, and it is 156 m below sea level near the Red Sea. With time, movement of tectonic plates may split away the eastern portion of Africa from the main part of the continent along the rift, forming a separate land mass. The break is taking place at the rate of up to two centimeters a year.

There are numerous hot springs and fumaroles in the rift, and most are assumed to have been present in the past. Sixty-four hot springs are reported for Kenya, varying in temperatures from 22° to 98° C (Kamondo, 1988), and all of these are associated with the rift system. A 45-megawatt geothermal power plant is in operation at Olkaria just northwest of Nairobi, the first geothermal electrical power in Africa. In Ethiopia, Djibouti, and Eritrea, there are at least 65 hot springs and fumaroles in the rift system, including the Afar area with reported temperatures of 40° to 96° C (Endeshaw, 1988). Over 30 hot springs each exist in Uganda, Tanzania, and Northern Zambia. They are associated with the rift system.

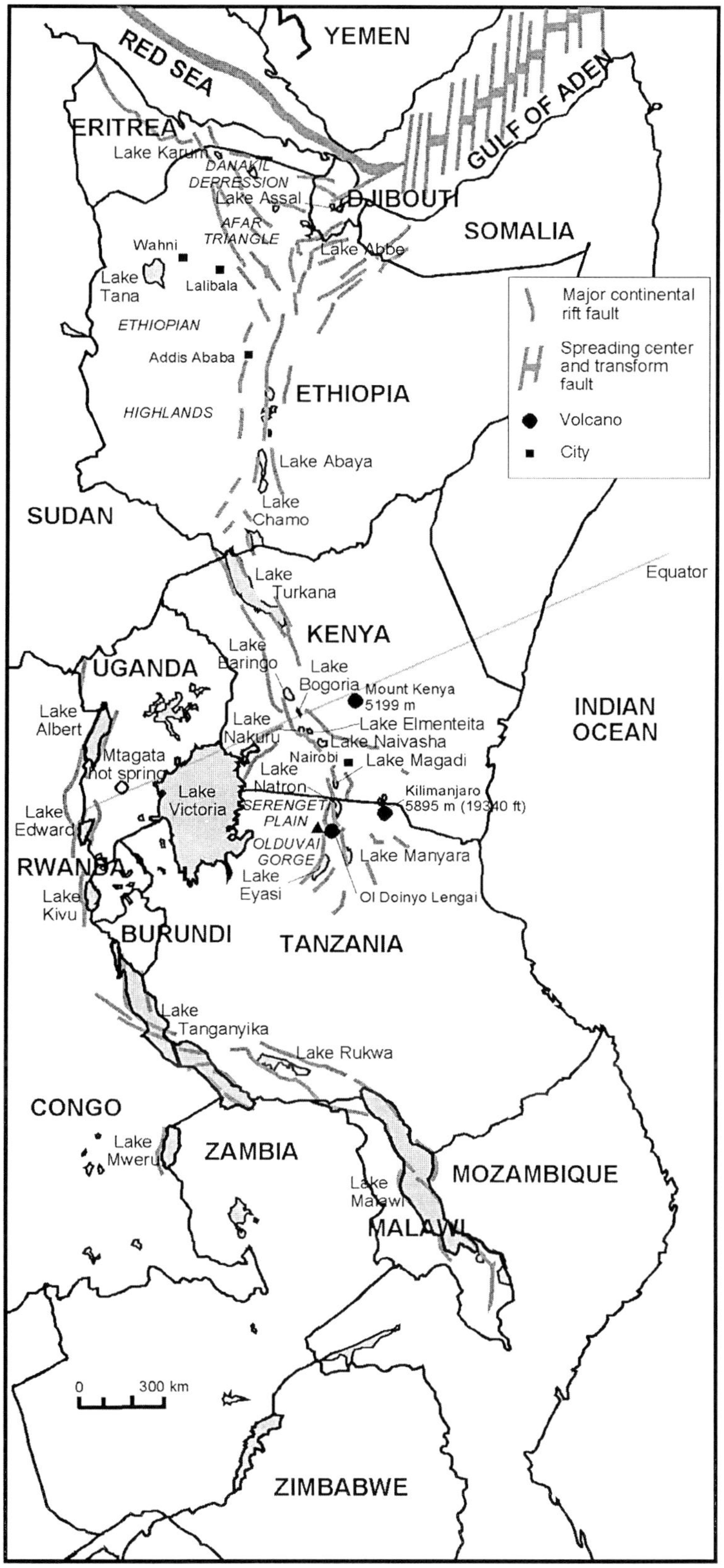

Map of the East African Rift System.

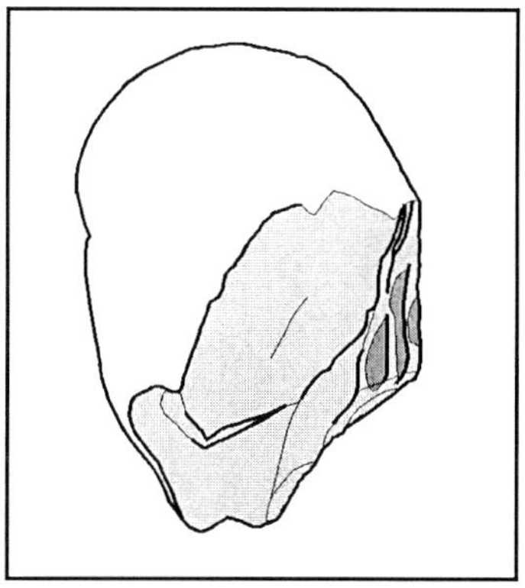

Interpretation of a hammering tool chipped from volcanic rock.

and gullies bordering Lake Turkana (Leakey and Walker, 1985). *Homo sapiens*, the successor to *erectus*, entered the scene about 600,000 to 1,000,000 years ago, based on a recent discovery in Eritrea. Both *erectus* and *sapiens* migrated out of Africa, probably starting between one and two million years ago. Remains of *erectus* have been found in India, China, and Southeastern Asia, including the Heidelberg, Peking, and Java man. *Erectus* remains have not been found in Europe, although *sapiens* finds are numerous. The latter species is the ancestor of the Neanderthal, appearing about 125,000 years ago in Europe and replaced by the Cro-Magnon about 30,000 to 40,000 years ago. Modern humans appeared about 10,000 years ago.

LEGENDS

FEW LEGENDS CONCERNING HOT SPRINGS, FUMAROLES, AND VOLCANOES HAVE BEEN RECORDED OR passed down orally. As for the ancient traditions, the only picture that is available must be derived from archival records and surviving orally transmitted accounts. Generally speaking, African myths may be understood as imaginative traditions not only about the nature, history, and destiny of the world but also about the gods, humanity, and society. Each ethnic or tribal group in Africa has unique religious beliefs. While much of African mythology is religious in nature, as the tales convey sacred truths, not all myths are necessarily religious. Some may be primarily social and historical; even the gods may play somewhat secondary or inferior roles in them, or no role at all. These myths clarify the history and rationale of a social group, institution, tradition, custom, or social development (Mbitu and Prime, 1997).

The origin of mankind is usually linked immediately to the cosmogony in a creation myth. In these myths, humans may be placed on the Earth by a creator god, or they may arrive from the heavens or the sea. Other mythic groups include trickster myths, dilemma tales, proverbs and riddles, myths of destruction, and hero myths (Mbitu and Prime, 1997). African myths are not only lighthearted stories for children or pre-scientific explanations of the cosmos; they are insights—related through narrative, poetry, and ceremonial language—to a reality representing the experiences of unique, varied modes of existence. In short, they have been woven out of the substance of human experience: struggles with the land and elements, movements and migration, wars between kingdoms, conflicts over pastures and waterholes, and wrestlings with the mysteries of existence, with life and death (Courlander, 1996).

Few of these myths relate to experiences with hot springs, fumaroles, and volcanoes. But some do. The myth that follows is of the Menasa Bet-Abrehe, the Cushitic people of Northern Ethiopia descended from ancient immigrants from Arabia. Here, certain years are remembered for the wonders and signs that occurred, and one year from the time of Emperor John and of Ras Alula (1870-1890) is called the "Year of the Earthquake" (Courlander, 1996):

> "In this year there was an earthquake in the whole Tigre country. About noontime the Earth was torn asunder and trembled much; and the mountain fires [volcanoes?] were kindled. And many large boulders that are now in the plain fell down from the mountains at that time, they say. And with the stones that fell down, some people were wounded in the country of Menasa Bet-Abrehe.

> And in the clefts of the Earth fire was also seen, they say. And there are persons who have seen a cow swallowed by the Earth. And the earthquake stopped after a short while. And also the fires were soon extinguished. Even those that were born in that year are living as young men. And this time is very well known."

Ol Doinyo Lengai, situated 200 km to the west of Kilimanjaro near Lake Natron, is one of numerous volcanoes scattered over the bottom of the East African Rift Valley. Lengai has two special characteristics: it is the only constantly active volcano east of the rift, and the composition of its lava flows is unique and geologically important. The magma moves out of the African mantle through ample beds of limestone, and it is the source of an extraordinary lava, silica-deficient but calcium- and sodium-rich. Since the lava contains no silica, it appears as a kind of molten limestone or carbonatite. Lengai was considered beneficial by local tribes, who believed that eruptions heralded the arrival of wild animals for hunting and farming. *Ol Doinyo* in the Masai language means "God's Place" (McQuire and Kilburn, 1995).

Erta Ale volcano is situated at the bottom of the Danakil Depression, 120 m below sea level, in Ethiopia. The lava is basalt that forms a lake of 4000 square meters. For the Ethiopians, Erta Ale is the mountain dwelling of the devil. A recent explorer told his local guides that he had met no one at the bottom of the crater. They replied with humor that there were only little timid devils there, but the big one, the real one, is there behind the crest of the caldera, and it would do no good to go and see him (McQuire and Kilburn, 1995)!

The Masai, a pastoral and hunting people of Kenya and Tanzania, have myths and traditions for many natural phenomena such as the sun and moon, the stars, sunrise and sunset, the rainbow, comets, lightning, night and day (Courlander, 1996). When the Masai feel a shock from an earthquake, some say that a number of warriors are going on a raid, others that a mountain is trembling. When smoke or steam issues from the Earth, as from the active volcano Donyo Engai (Ol Doinyo Lengai) and steam jets near the Gilgil River northwest of Nairobi, the Masai say that there is a large deposit of chalk lying beneath the surface, that what one sees is dust. They are referring to the carbonatite lava, which, on contact with air and while cooling, changes rapidly from black to chestnut—and within a few hours from chestnut to white. The white volcanic material, in the form of fine cinders and dust, covers the foot of the volcano and the Masai land (McQuire and Kilburn, 1995).

Geothermal Features and Use

Uganda

In many parts of Uganda, the inhabitants have long believed that certain springs have curative properties. Some of the springs are visited by people traveling from long distances to reach them. The most noted are those at Kitagata, otherwise known as Mugabe's Bath. Here there are two springs with temperatures of about 40° C, one of which is fenced off for the use of the Omugabe (or local ruler) and some chiefs, while the other is used by ordinary people. Since the waters contain only about 0.025 grams of dissolved salts per liter (25 ppm), the belief in their medicinal properties seems unlikely to have much foundation in fact (Dixon and Morton, 1970). (This same paper shows a contradictory value of 1,500 ppm for this spring.) Other springs thought to have curative properties are located at Panyamur and Palabek.

Henry M. Stanley, the famous African explorer, visited a hot spring in Uganda during his travels in 1876 (McClure, 1891). Captain McClure writes of Stanley's aborted trip to Lake Albert and the visit to the Mtagata hot springs:

> "After parting with his cowardly escort, he proceeded to explore the streams and lakes to the north and west [of Uganda]. Then, returning to his generous host, he asked for guides to take him to the hot springs of Mtagata, the healing properties of which he had heard of far and wide from the natives. These were cheerfully given, and after a march of two days he reached them....
>
> "He found a crowd of diseased persons here, trying the effect of the water. Naked men and women were lying around in the steaming water, half-asleep and half-cooked, for the water showed a temperature of one hundred and twenty-nine degrees [F, or 54° C]. The springs were, however, of different temperatures. The hottest one issued from the base of a rocky hill, while four others twenty degrees [F] cooler [about 43° C] came bubbling up out of black mud, and were the favorites of the invalids. Stanley camped here three days, and bathed in the water and drank it, but could perceive no effect whatever on his system."

Salt has been obtained at Lake Katwe since well before Stanley's visit over 100 years ago. During the 1960s the output ranged from 2,900 to 9,900 tonnes per year. Based on recent drilling, it is estimated that the lake contains over 20 million tonnes of mixed salts (Dixon and Morton, 1970).

Kenya and Tanzania

Situated in the rift valley in Southern Kenya and Northern Tanzania are a series of elongated lakes from Lake Nakuru to Lake Natron. Volcanic ash collected in these lake basins turns the water basic or alkaline. These are the soda lakes, forbidding places of extreme heat, little fresh water, and meager food supply—bodies of water so full of soda or sodium carbonate that they burn almost anything that tries to enter. Hot springs on the shores and in the lakes can reach 60° C, hot enough to scald and kill almost any living creature.

Amazingly, the lesser flamingo (*Phoenicopterus minor*) is well suited to this hostile environment. The region's four million flamingos form tribes thousands of birds strong that live by feeding in the volcanic lakes, which, although caustic, teem with microscopic spirolina (*Spirulina platensis*), a type of blue-green algae (*Nature*, 1998). The flamingos must adapt to each lake they inhabit and be able to "read" the water in order to survive, for a mistake can bring a scalding death.

Legend relates that the fire bird built a nest of frankincense and set it aflame. The bird remained in the nest and was consumed by the fire. Then something stirred in the ashes, and a brilliant bird reappeared more beautiful than before—the phoenix—the immortal bird resurrected from fire to live another 500 years. It is the algae of the lakes that contain scarlet pigment that gives the flamingos their firebird color, thus the association with the phoenix. The adult birds incubate their eggs in the 60° C mud for a four-week period. Long ago, when people saw the chicks emerging from the lake, they thought they were the resurrection of the phoenix, coming for a drink at the shore. The Phoenicians traded these scarlet-colored flamingos, believing they would bring immortality.

The most alkaline of these shallow, corrosive bodies of water is Lake Natron in Tanzania, just south of the Kenyan border. High mineral concentrations of sulfur, chlorine, phosphorus, and soda cause the water to turn caustic. The pH is as high as 10.5, which the flamingos can withstand. However, the skin underneath their feathers needs an occasional rinsing in clean water, which can be obtained from streams running into the lakes. If the soda doesn't wash off, it can solidify on the birds' feathers and prevent them from flying. Lake Nakuru in Kenya is also home to many flamingos. However, these and adjacent lakes have almost dried up in the past, threatening the birds' habitat. The lake depths are quite variable, with a maximum depth of about three meters. Agriculture and livestock raised in the lake catchment area and effluents from sewage treatment plants threaten to pollute the lake.

In addition to providing food and breeding space for millions of flamingos, the soda lakes of this portion of the Rift Valley also hold great commercial value. In the past and yet today, salt has been extracted from the lake. When a lake dries up, the lake surface becomes a crystalline mass and workers can walk directly upon it to harvest the crust for use by the local population. Today, sodium carbonate is a major ingredient in producing glass, ceramics, and paper, as well as a source of pink table salt. The Elmola tribe of less than 500 members fishes these saline lakes for tilapia.

Ethiopia

Menelik II, Emperor of Ethiopia from 1889 to 1909, transformed the country from a collection of semi-independent states into a united nation. As ruler of the kingdom of Shoa, in Central Ethiopia, he conquered the Oromo people to the south and annexed their land. He founded the city of Addis Ababa, which became the national capital in 1889. One of the reasons for selecting this site on the Ethiopia plateau was the location of the hot springs. The springs had been used by the local people for many years and legend has it that Menelik's queen, Taytu, liked this location because she, too, enjoyed the thermal waters.

In the highlands of Ethiopia are over 50 Christian churches, many carved in natural rock. Those that remained hidden were spared destruction by armies and rulers. Most are almost 1,000 years old and have been used for worship by Ethiopia's Orthodox Christians since the Middle Ages. The most famous are the 11 churches of Lalibala, carved from volcanic rock and decorated with brilliant Byzantine-flavored frescoes depicting the lives of Christ and the saints of the Ethiopian Church (Gerster, 1970). The churches of Lalibala are just as impressive today as they were when the Portuguese Chaplain Francisco Alvares visited them in the 1520s, despairing when his accounts were received with disbelief.

The complex of churches and chapels at Lalibala were carved about 800 years ago. The roof line of the churches was the top of a bank of pink tuff (cemented volcanic ash) on the shoulder of a mountain. Workers freed large rectangular blocks of tuff by trenching straight down into this naturally soft rock. Artists sculptured these blocks into houses of worship, removing material from the inside to construct vaulted chambers. Four of the shrines come together as a single monolith in a maze of tunnels, galleries, shafts, and alleyways. One of the churches, the elegant Church of St. George (*Bieta Georgis* or the "House of George"), is built as a cruciform shape in the surrounding rock.

The area is named for King Lalibala, who may have ruled toward the end of the 12th century. He was of the Zagwe dynasty and could trace his lineage to Moses. He was a usurper, a mountain war lord who seized power from the reigning king when the kingdom was weak. He gained his reputation by manufacturing a large number of iron tools for cutting and hewing stone and paid for the construction of the churches. Local legend says he built the churches on divine command and with the aid of angels. The church also had a military side, as excavations have yielded remnants of fortifications. At least two of the rock-hewn structures used as churches are believed to have first served as a royal palace and reception hall (Gerster, 1970).

The elegant Church of St. George, *Bieta Georgis*, Lalibala, Ethiopia. The entrance to the structure, which is several stories high, is at the bottom right. *Postcard reprinted with permission of the Embassy of Ethiopia*

Speculation is that the workers might have been foreigners, particularly Egyptian Copts. These structures are similar to Aksumite rock shrines and Egypt's Abu Simbel, even resembling ancient constructions in Central Anatolia and India. Legend, on the other hand, gives a different story, one that says King Lalibala was taken to heaven and shown the design of the churches God wanted him to construct. Upon returning to Earth, the angels joined in the labor of the construction of 10 earthly replicas of the churches, which is reported in the *Royal Chronicles* (Courlander, 1996):

> "When he began to construct these churches angels came to help him in each of the operations; there were thus a company of angels at work as well as a company of men, for angels joined the workers, the quarrymen, the stonecutters, and the laborers. The angels worked with them by day and by themselves at night. The men would do a cubit's work [about 18 inches] during the day, but would find a further three cubits completed on the morrow for the angels had worked through the night. Seeing this the workers exclaimed, 'How wonderful! We did a cubit yesterday and today we have four!' They doubted whether angels were doing this work because they could not see them,

> but Lalibala knew, because the angels, who understood his virtue, did not hide from him; the angels were his companions and for that reason did not hide from his sight."

Today on major religious holidays, pilgrims visit these churches, tripling Lalibala's population of about 9,000 people. Residents and visitors trade goods, exchange gossip, and crowd themselves into the sunken courtyards of the churches, which reverberate with their chants and the beating of drums.

Amba Wahni, the 550 m Ethiopian mountain where Princes of the Blood were imprisoned.
By Lyle Mabbott, from In Search of Sheba *by Barbara Toy, with permission*

Northwest of Lalibala at Wahni is a volcanic monolith with an interesting history. Amba Wahni was the last of the Princes' Prison Mountains, a high peak where princes of royal blood, all eligible to ascend the throne, were imprisoned to remove any threat they might pose to the existing king. Only when the oldest was to be crowned king was a Prince of the Blood brought down from the lofty heights; all the rest remained imprisoned (Toy, 1961).

The front of the church on Amba Wahni, with a maskal carved in tuff inserted in the wall. *From* In Search of Sheba *by Barbara Toy, with permission*

Amba Wahni is a pinnacle of volcanic tuff, rising with sheer faces from the valley floor to a height of about 550 m. Its top is rounded and against its sides are placed the buildings, walls, and passageways of the prison. The sides are of the pinnacle are too steep to climb, and a stairway with a guard tower at the top was used to gain access to the prison. Today the steps have eroded away and the only access to the ruins is by helicopter.

In the 17th century, King Fasilidas chose Amba Wahni as a prison for his relations, and it was still in existence when James Bruce came to the country in the 18th century. Bruce heard many tales about the mountain, although he never set eyes on it. Johnson based his *Rasselas, Prince of Abyssinia* on the concept of Amba Wahni, although he later severely criticized Bruce's account, implying that even Johnson himself thought this was just another Ethiopian legend or traveler's tale.

Many of the princes spent their whole lives on the mountain, leaving only if they were chosen to be king. It is reported that as many as 400 princes lived on its summit at one time. Heavily guarded, they had little contact with the outside world. Soon after Bruce's departure from the country, the custom was abandoned due to civil wars. All the princes were released from the

Remains of the church roof on top of Amba Wahni. Between the buttresses, square tiles of tuff are placed, similar to those in the palaces of Gondar. *From* In Search of Sheba *by Barbara Toy, with permission*

mountain, some as young boys, the others as old men. Amba Wahni was deserted and forgotten, only recently rediscovered and visited by Barbara Toy (1961).

Cameroon

THE GREATEST CONCENTRATIONS OF GEOTHERMAL PHENOMENA ARE FOUND IN THE EAST AFRICAN Rift Valley, but known geothermal springs are also found in Northwestern Africa: Algeria, Tunisia, and Morocco. A phenomenon that received widespread attention was the eruption of a lethal jet of carbon dioxide from Lake Nyos in Cameroon, West Africa. Carbon dioxide escapes from hot rock into groundwater and enters lakes where it is concentrated and, under pressure, can escape in an explosive eruption. The 1986 release killed 1,700 villagers and 3,000 cattle (Stager, 1987).

Several legends exist about such explosions. One story, reported by an anthropologist, concerns a king who hanged himself after being tricked by the ruler of a rival people. The deceased leader's body formed a lake that later exploded, killing many of the enemy. Other legends tell of the sudden death of cattle, of fish raining down from exploding lakes. There are more than 30 crater lakes in Cameroon that have the potential for similar eruptions. Lake Kivu in Rwanda, perhaps the world's most gaseous lake, holds huge volumes of carbon dioxide as well as highly flammable methane gas.

The Future Is the Past

HUMANS HAVE PROBABLY USED GEOTHERMAL HOT SPRINGS AND STOOD IN AWE OF VOLCANIC eruptions along the Great Rift for well over one million years. This area is the origin of humankind and its first encounters with geothermal phenomena. Recent volcanic eruptions, the continuous flow of hot springs, and modern geothermal wells—all indicate that the African people continue interacting with geothermal forces as natural features while developing spas and agricultural, heating and cooling, and electrical power generation projects.

> "The fool would say, 'This world is a virgin girl.' The wise man knows the world is old."
>
> —*From a Hausa poem*

REFERENCES

Courlander, H. 1996. *A treasury of African folklore.* Marlow and Company, New York.

Dixon, C. G. and W. H. Morton. 1970. Thermal and mineral springs in Uganda. *Geothermics, Special Issue 2,* 2, part 2, 1035-38. (U.N. Symposium on the Development and Utilization of Geothermal Resources, International Institute for Geothermal Research, Pisa).

Endeshaw, A. 1988. Current status (1988) of geothermal exploration in Ethiopia. *Geothermics,* 17, nos. 2/3, 477-488.

Gerster, G. 1970. Search out Medieval churches in Ethiopia's wilds. *National Geographic,* 138, no. 6, 856-884.

Kamondo, W. C. 1988. Possible uses of geothermal fluids in Kenya. *Geothermics,* 17, nos. 2/3, 489-501.

Leakey, R. and A. Walker. 1985. *Homo erectus* unearthed. *National Geographic,* 168, no. 5, 624-629.

Mbitu, N. and R. Prime. 1997. *Essential African mythology.* Harper Collins Publishers, San Francisco.

McClure, Capt. 1891. *With Stanley in Africa.* Worthington Company, New York.

McQuire, B. and C. Kilburn. 1995. *Volcanoes of the world.* Thunder Bay Press, San Diego, California.

Nature. 1998. Fire bird. Washington, D.C. (National Geographic program).

Stager, C. 1987. Silent death from Cameroon's killer lake. *National Geographic,* 172, no. 3, 404-420.

Stager, C. 1990. Africa's great rift. *National Geographic,* 177, no. 5, 2-41.

Toy, B. 1961. *In search of Sheba.* John Murray, London.

Weaver, K. F. 1985. The search for our ancestors. *National Geographic,* 168, no. 5, 560-623.

Wilford, J. N. 1998, 13 May. New data on Kenya fossils roll back human antiquity. *The Oregonian,* B12.

Note: The world map printed in this and succeeding chapters is courtesy of Corel Corp. Ltd.

The Author:

John W. Lund, Director
Geo-Heat Center
Oregon Institute of Technology
3201 Campus Drive
Klamath Falls, Oregon 97601 USA
Telephone: 541.885.1750
Fax: 541.885.1754
E-mail: lundj@oit.edu

"The Bath of the Queen of Sheba" at Hammei Tiberias, taken before her marriage to King Solomon, a legend recounted in the text. From the *Book of Astrology,* by Abu Ma'ashar al-Balkhi, in Arabic. Iran or Iraq, 14th century. *Bodleian Library, University of Oxford. MS. Bodl. Or. 133, folio 35 b, reprinted with permission*

4. Geothermal Energy Utilization in the Jordan Valley between Lake Kinneret and the Dead Sea: A View from Antiquity

by
Felice C. Jaffé
Esti Dvorjetski
Dov Levitte
Rober Massarwieh
Ali Swarieh

Abstract: In the Hellenistic, and mainly in the Roman and Byzantine periods, four major thermo-mineral spring systems were utilized intensively for bathing and therapeutic purposes in the Jordan Rift Valley, between the Dead Sea to the south and Lake Kinneret (Sea of Galilee) to the north. Two of them, Tiberias and Hammat Gader, are still (or again) in use, while a new spa has been developed recently in the Zarqa Ma'in Wadi near the abandoned spring cluster of Zara (Kallirhoe). Along with a short outline of their geological situation, this chapter discusses the utilization of these springs in Antiquity within the framework of their historical, cultural, and religious backgrounds.

Introduction

An original Middle Eastern explanation of the nature of geothermal energy is suggested in one of the numerous, charming and almost always meaningful, timeless, and archetypal Solomonic legends, some of which are still told by mothers and grandmothers to their children in this region.

One of the legends relates that, once upon a time, while preparing for his wedding with the Queen of Sheba, King Solomon wanted her to take a hot bath. So he sent one of his demons to a large cold spring near the shore of the Lake (or "Sea") of Tiberias, also called Lake Genesareth or Lake Kinneret, and instructed him to heat the water (photo at left). The Queen traveled to Tiberias, took her warm nuptial bath, married the wise king, and, so the legend goes, they lived happily ever after. The demon enjoyed his work so much that he remained in Tiberias and has been heating the water of the same spring ever since.

Hammei Tiberias was, and still is, only one of the thermo-mineral springs used for bathing in the Jordan Valley (see map). Along with a short review of their geological background, the purpose of this chapter is to describe how thermal springs were used in ancient times for bathing and medicinal purposes in this history-laden part of the world.

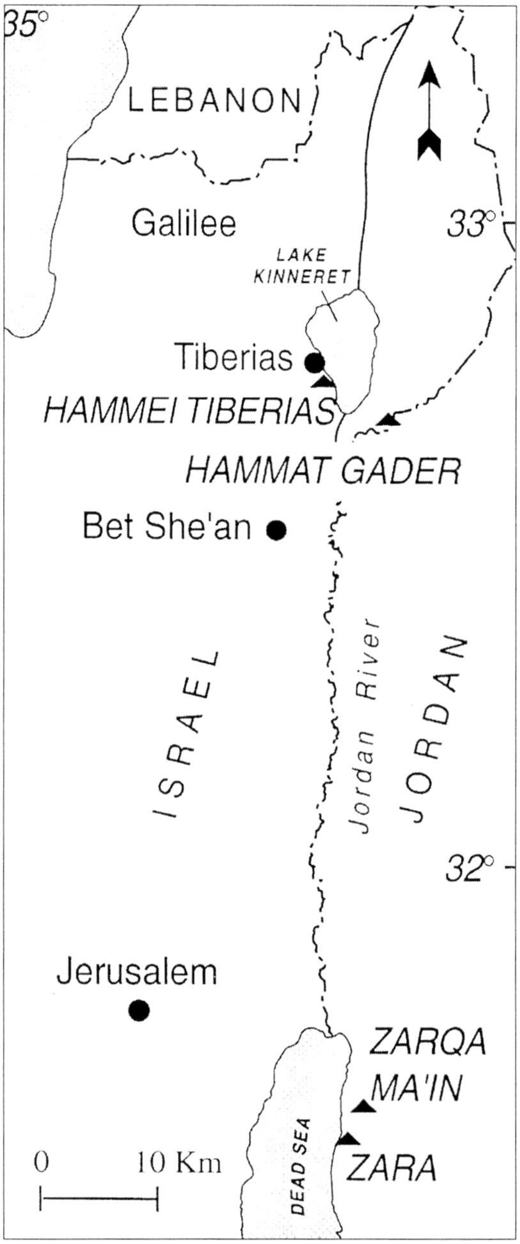

Location of the main thermo-mineral springs used in Antiquity in the Jordan Valley.

Utilization of Thermo-Mineral Springs in Antiquity: A General Outlook

The main thermo-mineral springs in the Jordan Valley (see map) were utilized essentially for bathing and medicinal purposes during the Hellenistic, Roman, Byzantine, and Islamic periods (Dvorjetski, 1992a, 1992b, 1996a, 1996b; Weber, 1997). At present, Hammei Tiberias, Hammat Gader, and Zarqa Ma'in are still actively utilized for the same purposes, while Zara (Kallirhoe) has been abandoned since the Byzantine period.

Even in earlier periods, people living in this region most likely understood the peculiar nature and healing properties of the hot springs and may have developed cultic and ritual practices besides bathing. But archaeological proofs of this are not available for prehistoric times. However, based on findings from the early excavations made in Hammat Gader in the Yarmuk Valley, it can be said that the Tell Bani ("the mound of the bath"), the top of which has remains of a 5th century A.D. synagogue, was already inhabited during the Bronze Age I-III (Sukenik, 1935). Also considering the fact that the valley of the Yarmuk River is a particularly lush and fertile region, it is likely that the hot springs were known already to early settlers and possibly used by them (see End Note on the Yarmukian Neolithic culture).

Regional Geological Setting and Thermo-Mineral Springs

The Jordan Valley forms a part of the Dead Sea Transform fault system (DST) that affects the northernmost part of the Afro-Arabian Rift (AAR). The DST is a major transform fault that extends from the northern tip of the Gulf of Eilat-Aqaba (Red Sea) to the Beqa'a Valley in Lebanon (Barjous and Mikbel, 1990; Bender, 1974; Freund and Garfunkel, 1981; Girdler, 1990; Hatcher et al., 1981; Picard, 1970; Quennell, 1982). Taken as a whole, the Jordan Valley represents the deepest depression on Earth, reaching -210 m (below sea level) at Lake Kinneret and -400 m at the Dead Sea. Hence, all the hot springs considered in this chapter (see map) are well below sea level.

In the Biblical *Old Testament*, the Jordan Rift Valley was referred to as the "Great Deep" (*Genesis* 7:11). The DST marks an interplate boundary along which the Arabian Plate slips to the north relative to the Israel-Sinai part of the African Plate. An 107 km sinistral strike-slip displacement has accumulated there since the Late Cretaceous, and its motion is widely believed to have continued through the Holocene and Recent. Long-term mean rates of total slippage range from 6 to 12 mm/y. While details of segmentation are not clear, geological and seismological evidence indicates a complex pattern of strain accumulation and release along the DST; therefore, rather than a single dislocation, the DST appears to consist of co-linear or offset segments with bends and stopovers. Transgressive raised blocks and transtensional rhomb-shaped grabens have been identified at the restraining and releasing bends, respectively. Two rhomb-shaped grabens (the Dead Sea and the Sea of Galilee) form the most prominent features of the DST.

The last stage of the rift development occurred in the past 4-5 million years (Freund et al., 1970; Freund and Garfunkel, 1981), during which the change in the poles of plate motion led to the development of oblique stresses and to the formation of the DST trough-morphology.

The rift-fill, consisting of clastic, fluviatile, limnic, and evaporitic sediments, is more than 10 km thick (Schulman and Bartov, 1978). A remarkable amount of clastic sediments has been discharged into the recent structure of the rift by numerous streams, as evidenced by the large alluvial fans seen along the margins of the depression; however, the rate of subsidence in the rift is greater than the rate of filling (Begin and Zilberman, 1994).

Recent fault traces run across the alluvial fans. In the northern part of the rift, this tectonic activity was accompanied by large-scale basaltic volcanism, which occurred mainly at the uplifted shoulders. The volcanism began during the Middle to Upper Miocene in the west and continued in the Pliocene on both the eastern and western shoulders of the rift. Volcanism, however, took place in younger times in the northeast, away from the rift. The geological features on the two flanks of the DST display an interesting asymmetry, which cannot be explained entirely by the significant north-south displacement. In the Dead Sea area, for instance, the eastern flank is raised 1-2 km.

On the western shoulder, the sediments are essentially Cretaceous in age. East of the Dead Sea, however, the sedimentary formations are considerably older than the Cretaceous: they range from Jurassic-Triassic to Paleozoic-Mesozoic in the Nubian sandstones, and even to Lower Paleozoic (Cambrian) in the Zarqa Ma'in region. Basalt flows on the western flank almost reach Bet She'an in the south, while on the eastern flank similar flows can only be seen near Lake Kinneret. East of the lake, however, in the Golan Plateau, these flows reach their maximum development. In the Zarqa Ma'in region, eight post-Eocene to Recent olivine basalt plugs and several flows of similar rocks are to be found, while across the Dead Sea, volcanic formations are unknown.

cont......

Finally, major east-west faults (of which the Zarqa fault is the most prominent with its 50 km length) affect the eastern shore of the Dead Sea; these faults, however, do not exist at all on the western flank of the DST.

Numerous small and a few major hot springs are the main manifestations of geothermal energy in the Jordan Valley. Since only four major thermo-mineral springs are known to have been utilized intensively in Antiquity, this review will be restricted to their description: Tiberias on the western flank and Hammat Gader, Zara (Kallirhoe), and Zarqa Ma'in (Barou) on the eastern flank (see map).

Much information about these springs has been provided by the regional studies in the Lake Kinneret and the Dead Sea areas. The economic importance of these conspicuous fresh and saltwater springs spurred vigorous hydrological, hydrogeological, and hydrogeochemical investigations, as well as studies in related fields, such as isotope geology and heat flow (Eckstein 1979a, 1979b; Gat et al., 1979; Galanis et al., 1986; Gavrieli and Bein, 1995; Bein, 1976; Mazor et al., 1980; Serruya, 1978; Vengosh et al., 1994).

The existence of well-developed rift fracture-zones accounts for an easy ascent of the thermal waters from deep layers and for their outflow at the surface. The variability that has been documented in temperature and chemical concentrations among the various spring systems, and also among the individual springs in a given spring cluster, leads one to think that different confined aquifers with thermal water exist at depth and that a variable dilution of this water occurs with colder water of meteoric origin circulating in shallow aquifers (see map).

Preliminary studies on the low-enthalpy geothermal potential of the thermo-mineral springs in question, and of the confined aquifers existing in their surroundings, have been carried out and submitted with specific recommendations for further work, mainly on the eastern flank of the Jordan Valley. However, funds for drilling in specific target areas have not been available as yet (Abu-Ajamieh, 1980; Abu-Ajamieh et al., 1989; Allen, 1988; Galanis et al., 1986; Myslil, 1988; Tournaye, 1990, 1992).

Geological and physical-chemical characteristics of the thermo-mineral springs in the Jordan Valley

	Hammei Tiberias (a)	**Hammat Gader** (b)	**Zara (Kallirhoe)** (c)	**Zarqa Ma'in** (d)
Number of springs	17	5	45	54
Type of aquifer	Kurnub sandstone	Ajlun/Kurnub sandstone	Nubian/Kurnub sandstone	Kurnub sandstone
Age of aquifer	L. Cretaceous	U.-L. Cretaceous	Cambrian-L. Cretaceous	L. Cretaceous
Total yield (m^3/y)	1	22	17	19
Temperature (° C)	27-62	25-50	42-62	34-63
TDS (mg/l)	30,941	1,444	1,001	1,651
Chemical type	NaCa-Cl	$NaCa\text{-}HCO_3$	NaCa-Cl	NaCa-Cl

(a) Eggel, 1839; Buchmann, 1957; Bein, 1976; Mazor et al., 1980
(b) Noetling, 1886; Parker, 1970; Levitte et al., 1978; Starinski et al., 1979; Mazor et al., 1980; Mazor et al., 1981; Arad and Bein, 1986
(c) Allen, 1988; Swarieh and Massarwieh, 1996
(d) Truesdell et al., 1983; Rimawi and Salameh, 1988; Abu-Ajamieh et al., 1989; Salameh and Bannayan, 1993

Undoubtedly, however, the peak utilization of thermo-mineral waters took place during the Roman Empire. To understand the love for bathing during this period, it may be appropriate to quote a graffito, scratched on the pavement of the forum in the Roman veterans' colony of Timgad, in the heartland of Numidia, Algeria, "*venari, lavari, ludere, ridere: hoc est vivere*," which means "hunting, bathing, joking, and laughing: that is living" (Dunbabin, 1989; Yegül, 1992).

Several, often complementary, kinds of sources about the history of the Jordan Valley springs are available. References to these springs are to be found in numerous classical works by Roman and Byzantine historians, chroniclers, and geographers from the 1st century B.C. to the 6th century A.D., such as Strabo, Pliny the Elder, Josephus Flavius, Origen, Ptolemaeus, Jerome, Epiphanius, Eusebius, Eunapius, and Antoninus of Placentia. These authors described at length the healing properties of these thermo-mineral springs and their popularity for the treatment of many diseases, particularly leprosy.

After the destruction of the Second Temple and the Bar Kochba revolt, Galilee became the main region where Jews settled, prospered, and established their religious institutions. Thus, references to localities and events in this region are frequent in the *Talmud* (often referred to as "rabbinical literature"), including the description of thermo-mineral springs and their healing properties, together with recommendations for sufferers from skin diseases, leprosy, ailments of the urinary and digestive tracts, rheumatism, arthritis, and nervous diseases (Dvorjetski, 1996a, 1996b). The therapeutic properties of the *piloma*—a kind of thermo-mineral mud or "fango"—were also recognized and utilized in Hammei Tiberias (Buchmann, 1960).

After the end of the Byzantine epoch, at the beginning of the Early Islamic period in the 7th century A.D., the area under review lost its geopolitical importance and the whole region underwent impoverishment. For many centuries, however, the spas continued to be visited—but only by local people. This is attested to by many well known Arab and Jewish travelers and geographers, such as Al Ishtahi and Al Muqaddasi (10 century), Benjamin de Tudela and Ali Mehirat (12th century), Yaqut (13th century), Bassola (16th century), and Hasselquist and Yerushalmi (18th century).

From the beginning of the 19th century, historical exploration in the Middle East, particularly the Holy Land, became fashionable. Many European voyagers traveled in the region, accurately observing and recording ancient remains (Buckingham, 1821; Seetzen, 1859; Guérin, 1868;

Conder and Kitchener, 1881; Tristram, 1873). Particular mention should be made of Burckhardt (1822), well-known for the discovery of Petra, a very important town in pre-Christian times and the unsuspected capital of the Nabataean Kingdom in the 4th century B.C.

Finally, recent archaeological excavations, coupled with the study and the interpretation of epigraphic material, have shed new light on the structure of bathing establishments in the region and also on its religious, social, and cultural background and significance. Some highlights follow for each of the main thermo-mineral springs of this area.

Hammei Tiberias

In the *Talmud*, Tiberias is identified with the Biblical Hammath, a fortified city in the territory of the tribe of Naphtali, which was located in the northern part of ancient Palestine (*Book of Joshua* 19:35). However, this identification is not certain, as no remains older than the Hellenistic period were discovered during archaeological excavations conducted in the area. What is certain, though, is that the town proper of Tiberias was founded only around the year 20 A.D. as the capital of Galilee by Herod Antipas, who named it after the Emperor Tiberius. The thermo-mineral springs are situated just at the southern edge of the town. Their appearance toward the end of the 19th century is seen in the illustration.

Tiberias was a Jewish town that gradually assumed great spiritual and religious importance. The Academy, which succeeded the *Sanhedrin* (or *Synhedrion*) as the supreme religious institution, was founded in Tiberias; it continued to function well after the Arab invasion in the Early Islamic period, probably until the 10th century. The bulk of the Palestinian (Jerusalem) *Talmud* was written in this institution.

It is thus not surprising that, while the baths essentially had a curative purpose, religious factors were also of great importance. The strong influence exerted by religion on balneological practices can be easily drawn from many documents of the period, which contain in-depth discussions on the religious opportunity to bathe, especially on the Sabbath (Saturday), and on the moral consequences of taking baths in thermal establishments where the indiscriminate and contemporaneous presence of men and women was allowed. In particular, the rabbis strongly opposed the practice of thermal bathing, not only for religious reasons but also because it was customary to erect idolatrous statues inside the spas, such as the statue of the Emperor Hadrian. The *jeunesse dorée* of Tiberias reacted to such limiting factors by preferring to visit the more festive Hammat Gader thermal spa, where men and women were allowed to bathe together in the same pools.

Excavations also attest to the strong impact of the religious way of life prevailing in Tiberias: several superimposed synagogues were found on at least three levels in the spring area. Beneath the synagogues was a public building whose function is not clear (Dothan, 1993).
The synagogues date from the 4th to the middle of the 7th centuries, at the beginning of the rule of the Umayyad dynasty.

The thermal establishment of Hammei Tiberias toward the end of the Ottoman period (Wilson, 1880, vol. 2, p. 65): (a) general view; (b) detail of the interior of the public bath. Note the pierced dome.

The floors of these synagogues are decorated with mosaics inscribed in Aramaic and Greek. Two beautiful mosaics of high artistry were discovered on the synagogue floor on the level labeled IIA; one depicts the Torah ark with various cultic objects. However, the most beautiful and relatively well preserved mosaic floor on this level portrays Helios, the Greek representation of the sun deity, riding on his chariot surrounded by a zodiac. The free artistic expression displayed in the nude representation of the signs of the zodiac, as well as the frequent use of Greek in the accompanying inscriptions, clearly indicate a strong Hellenistic-Roman influence in the life of the Jewish capital of Galilee.

A modern, fully equipped spa now is operating at Hammei Tiberias.

Hammat Gader

The Hammat Gader thermo-mineral establishment (see map), recently excavated by the Hebrew University of Jerusalem, is one of the largest baths in the Roman world, second only in importance to Baiae, in the Phlegraean Fields near Naples (Hirschfeld, 1984, 1993 and 1997; Yegül, 1992). Patients flocked here from distant regions, as faraway as Athens. It was an imposing rectangular architectural complex, covering 3500 (50 x 70) m^2. The overall plan of the establishment indicates a functional juxtaposition of five major and one minor hall, all interconnected and each containing a large pool. One of the most intriguing areas of the establishment is

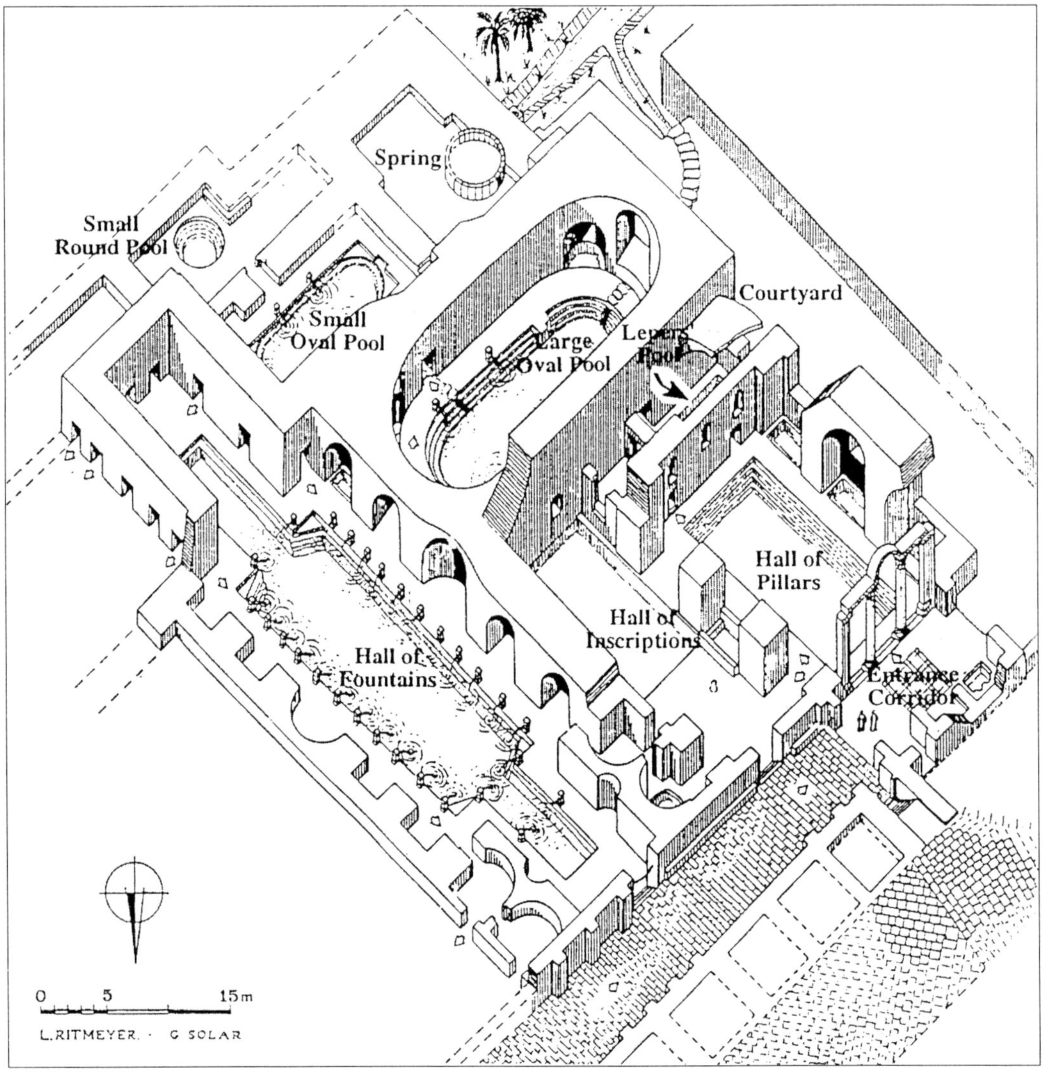

The Hammat Gader bathing establishment, axonometric restoration. *By Ritmeyer and Solar in Yegül 1992; slightly modified and reprinted with permission*

the narrow pool between the Oval Hall and the Hall of the Pillars. It may have been used as a special unit for the treatment of leprosy. The spa was well known as a place of licentiousness and frivolity, a unique and rather successful combination of fair, amusement center, and social meeting place where prostitution was a common phenomenon (Dvorjetski, 1994a, 1997a, and 1997b).

At the end of the Byzantine period in the 7th century, the baths were severely damaged by a major earthquake. An inscription in Greek commemorates the renovations undertaken by the Caliph Mu'awiya, the founder of the Umayyad dynasty, shortly after he ascended the throne in Damascus and took over the region from Byzantine rule in the 7th century. The text of this inscription has been interpreted as a clear sign of the tolerance prevailing at the beginning of the Early Islamic period. However, the importance of the baths gradually diminished, although local inhabitants continued to frequent them for cures during the following centuries.

A modern balneological establishment, opened in recent years, is functioning successfully today at Hammat Gader.

Zara (Kallirhoe) and Zarqa Ma'in (Barou)

Zara is better known under its historical Greek name of Kallirhoe, "the beautiful spring," and both names will be used in this section. Dunbabin (1989) suggests that the meaning of Kallirhoe is also well reflected by the German term "Schönbrun." Due to its extreme summer climate and its relative inaccessibility on the eastern shore of the Dead Sea, Zara was never a well developed spa. Its claim to fame stems from the description given it by the renowned Jewish historian and military commander Josephus Flavius (1st century A.D.) in relation to the death of Herod the Great. The importance of Kallirhoe is also attested to by its depiction on the famous mosaic of Madaba (near Mount Nebo in the Amman area). The mosaic, which covers the floor of a Byzantine church built toward the second half of the 6th century (Avi Yonah, 1954; Piccirillo, 1993; Dvorjetski, 1994b), represents a geographical map of the Holy Land. It is somewhat paradoxical to note that at the time of the execution of the mosaic, the use of Kallirhoe had practically halted and that it has not been in use since.

In his usual accurate and elegant style, Josephus Flavius (*De bello judaico, I,* 656-657) narrates the death of Herod the Great, an extremely cruel but fascinating tyrant. This king was already a 70-year-old, despondent, and moribund ruler when he decided to visit the Kallirhoe springs, whose waters were "potable and sweet," in the hope of some relief from his various illnesses, of which a realistic but somewhat unpleasant list is provided by Josephus Flavius. Learned commentaries have been written on this subject, and modern medical opinions are divided on the possible diagnosis of the illnesses affecting Herod the Great. These illnesses could have been any one or a combination of cancer of the bowels, cirrhosis, and severe diabetes, as well as arteriosclerosis and thrombosis in the abdominal region, complicated by terminal cirrhosis. Under these circumstances, the stay of the king at Kallirhoe apparently was rather unsuccessful. So, he sailed again across the Dead Sea and returned to his winter palace in Jericho, where he arrived in a "melancholy state." However, he went on plotting as usual, as he had done most of his life; he still managed to have his third son, Antipater, killed after a new and terrible fit of anger overtook him in his poor state of health, and he finally died five days later, in 4 B.C.

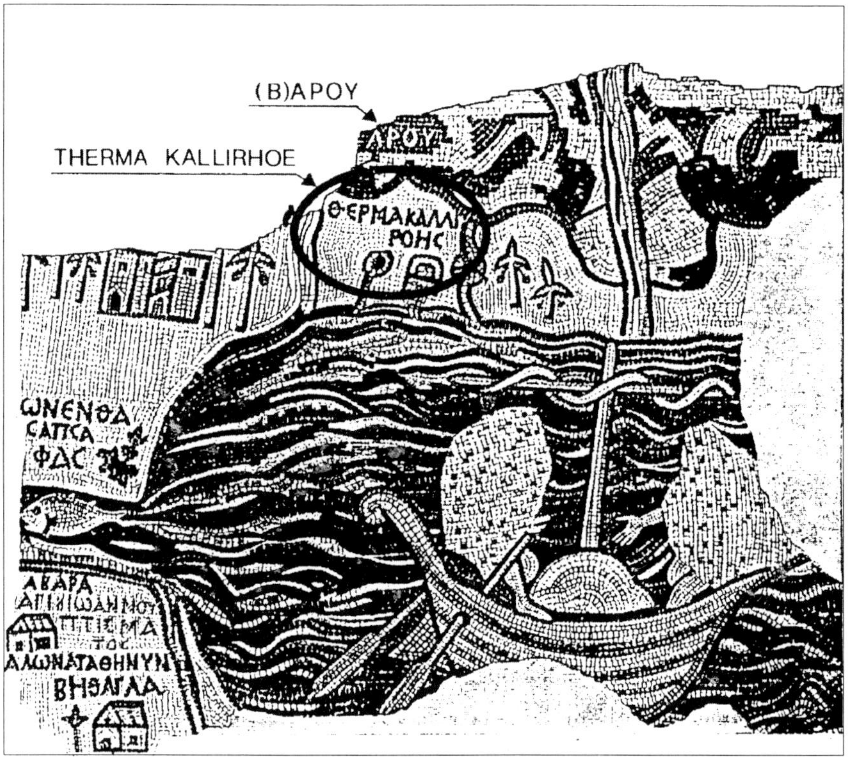

Kallirhoe on the Madaba mosaic. Above the name "Therma Kallirhoe," the locality of (B)apoy (pronounced "Barou") can be seen. It may well correspond to the Zarqa Ma'in (Barou) springs. Note that the faces of the persons sailing on the Dead Sea were expunged during an iconoclastic episode. *Piccirillo, 1993*

The Zara region was studied extensively and excavated by German archaeologists (Clamer, 1989 and 1997; Donner, 1963; Strobel, 1966 and 1989; Strobel and Clamer, 1986; Kraus, 1995, map of Kallirhoe; Weber, 1997). It appears that in Antiquity Kallirhoe was never a popular spa but rather a luxurious and exclusive holiday residence with private bathing facilities for the king and the aristocracy during the reigns of Herod the Great and his son Herod Antipas, the Tetrarch of Galilee. An ancient Kallirhoe pier, still visible at Zara, supports the evidence that Kallirhoe was connected by boat to the western shore of the Dead Sea and from there to the Jericho region.

A road connected Zara with the Zarqa Ma'in thermo-mineral springs; it probably continued from there to the fortress of Machaerus, where at present the village of Mekawir is located, about 8 km south-southeast of Kallirhoe (Khouri, 1988; Dvorjetski, 1996b). Machaerus was Herod's hilltop palace, built to protect his kingdom against the incursions of the Nabataeans. It was the palace where Salome danced in front of Herod Antipas so successfully that she obtained the head of St. John the Baptist.

Archaeological evidence indicates that Kallirhoe was occupied during Roman times but abandoned in the 2nd century after the Bar Kochba revolt, during which the Machaerus castle was destroyed. It was reoccupied in the 4th century by the Byzantines and was used in trade on the Dead Sea, as the Madaba mosaic map indicates (see illustration). Kallirhoe was probably reabandoned in the 5th century, or shortly thereafter, thus falling into oblivion.

The Zarqa Ma'in spring area is situated in a deep gorge, some 3.5 km northeast of the Zara (Kallirhoe) springs (see map). The area is easily accessible from Amman by a good road, which, however, does not continue to the Dead Sea shore; therefore, the two spring systems of Zarqa Ma'in and Zara are practically without communication between them.

The Zarqa Ma'in springs may have been visited for balneological purposes during Antiquity, but archaeological and historical evidence of their use is lacking. For a long time, these springs have been confused with those of Zara (Kallirhoe), but now it is believed that they rather correspond to the ancient springs of Baris (or Barou; [B]apoy on the mosaic), an important group of springs about which very little is known.

Recently, the main Zarqa Ma'in springs have been harnessed to serve a well frequented hotel with bathing and curative facilities. They supply at present the largest thermo-mineral establishment in the whole Middle East.

Concluding Remarks

Obviously, no quantitative data are available on the utilization in Antiquity of thermo-mineral springs in the Jordan Rift Valley. There is, however, sufficient information to surmise that conditions and applications have not changed significantly over the last 2,000 years. Quite to the contrary: the spas just described were far more important in Antiquity than in the last few centuries and at present.

It can be reasonably assumed, though, that a major change may take place in the next century. Then the geothermal potential of the springs area and of the whole surrounding region is expected to be harnessed, not only for classical balneological purposes but for new and challenging low-enthalpy applications.

[*End Note:* The Neolithic village of Sha'ar Hagolan, on the northern bank of the Yarmuk River, is located less than 5 kilometers downstream from the Hammat Gader springs. It is one of the most important Neolithic settlements in the entire Middle East and the type locality of the Yarmukian culture that flourished some 8,000 years ago (6400 to 5800 B.C.). This culture is known from at least 20 other localities in Israel, Jordan, and Lebanon (Garfinkel, 1993 and 1999).]

Acknowledgments

The authors want to thank Dr. Rachel Milstein, Hebrew University of Jerusalem, for providing a copy of the miniature, "The Bath of the Queen of Sheba," and Ms. Doris Nicholson, Bodleian Library, University of Oxford, for granting permission to publish it. They are also indebted to Father Michele Piccirillo, OFM, Convent of the Flagellation, Jerusalem, for the interesting information he provided and his comments on the Madaba mosaic. Dr. Youseph Hamarneh, National Resources Authority, Amman, most kindly established the necessary contacts and followed the preparation of this chapter with competence and interest. Finally, thanks are due to Dr. Ronnie Bogoch, Geological Survey of Israel, Jerusalem, for his critical but constructive review of the manuscript.

References

Abu-Ajamieh, M. 1980. *The geothermal resources of Zerka Ma'in and Zara.* Natural Resources Authority, Amman.

Abu-Ajamieh, M., F. Bender, and R. Eicher. 1989. *Mineral resources of Jordan.* Natural Resources Authority, Amman.

Allen, D. 1988. *Preliminary evaluation of the geothermal potential of Jordan and recommendations for further studies.* British Geological Survey. (Unpublished report).

Arad, A. and A. Bein. 1986. Saline versus freshwater: Contribution to the thermal waters of the Northern Jordan Rift Valley. *Israel. Journ. of Hydrol.*, 83, nos. 1-2, 49-66.

Avi Yonah, M. 1954. *The Madaba mosaic map.* Israel Exploration Society.

Barjous, M. and S. Mikbel. 1990. Tectonic evolution of the Gulf of Ababa-Dead Sea Transform fault system. *Tectonophysics*, no. 180, 49-59.

Begin, B. and E. Zilbermann. 1994. The formation of relief in Israel. In *Encyclopaedia Hebraica, vol. 6.* Eretz Edit., Jerusalem.

Bein, A. 1976. *A critical review of salinity studies of Lake Kinneret.* Israel Geological Survey. (Unpublished report in Hebrew).

Bender, F. 1974. *Geology of Jordan.* Bornträger Edit., Berlin.

Buchmann, M. 1957. The hot springs of Tiberias. *Journ. Medical Assoc. of Israel,* no. 75, 1-12.

Buchmann, M. 1960. *Tiberias und seine Heilquellen.* Tiberias Hot Springs Co., Tiberias, Israel.

Buckingham, J. S. 1821. *Travels in the countries of Basham and Gilead.* Longmann Edit., London.

Burckhardt, J. L. 1822. *Travels in Syria and the Holy Land.* John Murray Edit., London.

Clamer, C. 1989. 'Ain Ez-Zara excavations 1986. *Annual of the Department of Antiquities of Jordan,* 33, 217-225.

Clamer, C. 1997. *Fouilles archéologiques de Ain ez-Zara-Calliroé, villégiature hérodienne.* Institut Français d'Archéologie du Proche Orient (IFAPO), Paris.

Conder, C. and R. Kitchener. 1881. *The survey of Western Palestine.* Palestine Exploration Fund, London.

Donner, H. 1963. Kallirrhoë. Das Sanatorium Herods des Grossen, Zeitsch. *Deutschen Palaestina Verein,* no. 79, 59-89.

Dothan, M. 1993. Hammath-Tiberias. In *The new encyclopaedia of archaeology excavations in the Holy Land, vol. 2.* E. Stern Edit.

Dunbabin, K. 1989. Baiarum grata voluptas (Pleasures and dangers of the baths). *Papers of the British School at Rome,* no. 57, 6-46.

Dvorjetski, E. 1992a. Medicinal hot springs in Eretz-Israel and in the Decapolis during the Hellenistic, Roman and Byzantine periods. *Aram,* 4, no.1/2, 425-449.

Dvorjetski, E. 1992b. *Medicinal hot springs in Eretz-Israel during the period of the Second Temple, the Mishna and the Talmud, I-II.* Hebrew University of Jerusalem. (In Hebrew).

Dvorjetski, E. 1994a. Medicinal hot springs in Eretz-Israel in Antiquity: Sacred places or popular sites of healing? *Jerusalem Studies in Jewish Folklore,* no. 16, 7-27. (In Hebrew).

Dvorjetski, E. 1994b. *The thermo-mineral springs and the ships' load in the Dead Sea according to the Madaba map.* Barkai and Shiler Edit., Ariel Press. (In Hebrew).

Dvorjetski, E. 1996a. Properties of the therapeutic baths in Eretz-Israel in Antiquity. In *Illness and healing in ancient times.* University of Haifa, The Reuven and Edith Museum, catalogue 13.

Dvorjetski, E. 1996b. The healing qualities of the thermo-mineral baths in Eretz-Israel and Jordan in classical Antiquity. *Proceedings of the 35th International Congress on the History of Medicine,* Kos, Greece.

Dvorjetski, E. 1997a. The inscriptions from the synagogues of Hammat Tiberias and Hammat Gader and their contribution to the thermo-mineral springs research. In *Eretz-Israel in the Byzantine period. Galilee Studies 1996.* Safrai Edit., Bar Ilan University. (In Hebrew).

Dvorjetski, E. 1997b. The relationship between Jews and Gentiles. In *The thermo-mineral springs in Eretz-Israel during Roman and Byzantine periods.* Oppenheimer and Mor, eds. University of Haifa and Tel Aviv, Yad Ben Zvi Publ., Jerusalem.

Eckstein, Y. 1979a. Review of the heat flow data from the Eastern Mediterranean region. *Pure Applied Geophysics,* 117, no. 1/2, 150-159.

Eckstein, Y. 1979b. Heat flow and the hydrological cycle: Examples from Israel. In *Terrestrial heat flow in Europe.* Verman and Rybach, eds. Springer, Berlin.

Eggel, W. 1839. Chemische Untersuchung der heissen Quelle von Ammaus am Galileer Meere. *Inaugural Dissert. I,* Ludwig Friedrich Fues, Tübingen.

Freund, R. and Z. Garfunkel. 1981. The Dead Sea Rift. *Tectonophysics,* no. 80.

Freund, R., Z. Garfunkel, I. Zak, M. Goldberg, T. Weissbrod, and B. Derin. 1970. The shear zone along the Dead Sea Rift. *Phil. Trans. Roy. Soc. of London,* 267, 107-130.

Galanis, S. P., J. H. Saas, R. J. Munroe, and M. Abu-Ajamieh. 1986. *Heat flow data at Zerka Ma'in and Zara, and a geothermal reconnaissance of Jordan.* USGS Open File Report no. 86-631.

Garfinkel, Y. 1993. The Yarmukian Culture in Israel. *Paléorient,* no. 19/1, 115-134.

Garfinkel, Y., ed. 1999. *The Yarmukians—Neolithic art from the Sha'ar Hagolan.* Bible Lands Museum, Jerusalem.

Gat, J., E. Mazor, and Y. Tsur. 1979. The stable isotopes composition of mineral waters in the Jordan Rift Valley. *Journ. of Hydrol.,* no. 7, 334-352.

Gavrieli, I. and A. Bein. 1995. Sulphur isotopes in the Dead Sea and in thermal-saline springs along the shores. *Proceedings of the IAEA Meeting on Isotopes in Water Resources Management,* Vienna, 174-176.

Girdler, R. 1990. The Dead Sea transform fault system in geological and tectonic processes of the Dead Sea Rift zone. *Tectonophysics,* no. 180, 1-13.

Guérin, M. V. 1868. *Description géographique, historique et archéologique de la Palestine.* Imprimérie Impériale, Paris.

Hatcher, R., I. Zietz, R. Regan, and M. Abu-Ajameh. 1981. Sinistral strike-slip motion on the Dead Sea Rift: Confirmation from new magnetic data. *Geology,* no. 9, 458-462.

Hirschfeld, Y. 1984. Sumptuous Roman baths uncovered near Sea of Galilee. *Bibl. Arch. Review,* no. 10, 22-40.

Hirschfeld, Y. 1993. Hammat Gader. In *The new encyclopaedia of archaeology excavations in the Holy Land, vol. 2.* E. Stern Edit., Jerusalem.

Hirschfeld, Y. 1997. *The Roman baths of Hammat Gader. Final report.* Israel Exploration Society.

Josephus Flavius. *De bello judaico (The Jewish War).* Reprint 1982, G. Cornfeld, trans. Zondervan Edit., Grand Rapids, Michigan.

Khouri, R. 1988. *The antiquities of the Jordan Rift Valley.* Al Kutba Edit., Amman.

Kraus, W. 1995. *Zurück zu den Anfängen, Beiträge und Schriftenverzeichniss von Professor Dr. A. Strobel.* Flacius Verlag Edit., Bayern. (With map of Kallirhoë).

Levitte, D., A. Olshina, and D. Wachs. 1978. Geological and geophysical investigations in the Hammat Gader springs area. *Geological Survey of Israel,* rep. hydr. no. 4/78.

Mazor, E., A. Kaufman, and I. Carmi. 1981. Hammat Gader (Israel): Geochemistry of a mixed thermal spring complex. *Journ. of Hydrol.,* no. 18, 289-303.

Mazor, E., D. Levitte, A. Truesdell, J. Healy, and A. Wissenbaum. 1980. Mixing models applied to warm (up to 60° C) springs: Jordan Rift Valley. *Israel. Journ. of Hydrol.,* 45, nos. 1/2, 1-19.

Myslil, V. 1988. *Evaluation of geothermal potential of Jordan.* National Resources Authority, Amman.

Noetling, F. 1886. Geologische Skizze der Umgebung des el Hammi, *Zeitschr. Deutschen Palaestina Verein,* no. 9, 58-88.

Parker, D., UN and FAO. 1970. *Investigation of the sandstone aquifer of East Jordan: The hydrogeology of the Mesozoic-Cenozoic of the western highland and the plateau of East Jordan.* AGL: SF/Jor 9. (Technical report, based on the work of D. H. Parker.)

Picard, L. 1970. On Afro-Arabian Graben tectonics. *Geol. Rundschau,* no. 59, 337-381.

Piccirillo. M. 1993. *The mosaics of Jordan.* American Center of Oriental Research, Amman.

Quennell, A., ed. 1982. *Rift Valleys: Afro-Arabian benchmark papers in geology.* Hutchison Ross, Stroudsburg, Pennsylvania.

Rimawi, O. and E. Salameh. 1988. Hydrochemistry and ground water system of Zerka Ma'in-Zara thermal field. *Jordan. Journ. of Hydrol.,* no. 98, 147-163.

Salameh E. and H. Bannayan. 1993. *Water resources of Jordan: Present status and future potential.* Friedrich Ebert Stiftung Edit., Amman.

Schulman, N. and Y. Bartov. 1978. Tectonics and sedimentation along the Rift Valley. In *Sedimentology in Israel, Cyprus and Turkey. Proceedings of the 10th International Congress on Sedimentology, Guidebook, Part II,* Jerusalem, 37-94.

Seetzen, U. I. 1859. *Reisen durch Syrien, Paläestina, Phönikien, die Transjordanische Länder, Arabia Petraea und Unter-Aegypten.* Reimer Edit., Berlin.

Serruya, C. 1978. *Lake Kinneret.* Junk Edit., The Hague.

Starinski, A., A. Katz, and D. Levitte. 1979. Temperature-composition-depth relationship in Rift Valley hot springs: Hammat Gader, Northern Israel. *Chemical Geology,* no. 27, 233-244.

Strobel, A. 1966. Zur Ortslage von Kallirrhoë, Zeitschr. *Deutschen Palaestina Verein,* no. 82, 149-162.

Strobel, A. 1989. Die Grabungskampagne in ez-Zara (Kallirrhoë) am Toten Meer (Jordanien). *Jahrbuch Deutschen Instituts für Altertumswissenschaft des Heiligen Landes,* no. 1, 173-176.

Strobel, A. and Cr. Clamer. 1986. Excavations at Ez-Zara. *Annual Department of Antiquities of Jordan,* no. 30, 381-384.

Sukenik, E. 1935. El Hammeh: The site, its history and its remains. In *The ancient synagogue of El-Hammeh.* R. Mass Edit., Jerusalem.

Swarieh, A. and R. Massarwieh. 1996. *Geothermal water in Zara and Zarqa Ma'in area.* Natural Resources Authority, Amman.

Tournaye, D. 1990. *Etude du potentiel et des utilisations des ressources géothermales de Jordanie.* Companie Française de Géothermie, Orléans, France. (Unpublished report).

Tournaye, D. 1992. *Etude de l'utilisation des ressources géothermiques pour le chauffage des serres, zone sud d'Amman.* Companie Française de Géothermie, Orléans, France. (Unpublished report).

Tristram, H. 1873. *The land of Moab. Travels and discoveries on the east side of the Dead Sea and the Jordan.* J. Murray Edit., London.

Truesdell, A., W. Duffield, and M. Abu-Ajamieh. 1983. Origin and health sources of thermal waters at Zerqa Ma'in and Zara, Jordan. *Proceedings of the 4th International Symposium on Water-Rock Interactions,* Misasa, Japan. (Abstract).

Vengosh, A., A. Starinksy, Y. Kolodny, and A. Chivas. 1994. Boron isotope geochemistry of thermal springs from the Northern Rift Valley. *Israel. Journ. of Hydrol.,* no. 162, 165-169.

Weber, T. 1997. Thermal springs, medical supply and healing cults in Roman-Byzantine Jordan. *Studies in the History and Archaeology of Jordan,* 6, 331-338.

Wilson, C. W. 1880. *Picturesque Palestine, Sinai and Egypt.* 4 Volumes. London.

Yegül, F. 1992. *Baths and bathing in classical Antiquity.* The Architectural History Foundation, New York; MIT Press, Cambridge, Massachusetts, and London.

The Authors:

Felice C. Jaffé
Section of Earth Sciences
University of Geneva
13, Rue des Maraîchers
1221 Geneva 4, Switzerland
Telephone: 41.22.702.6624
Fax: 41.22.320.5732
E-mail: Felice.Jaffe@terre.unige.ch

Esti Dvorjetski
Department of General Studies
Technion, Israel Institute of Technology
32000 Haifa, Israel
Telephone: 97.24.876.9406
Fax: 97.24.877.6035
E-mail: dvorjet@tx.te.ac.il

Dov Levitte
Geological Survey of Israel
30, Malchei Israel Street
95501 Jerusalem, Israel
Telephone: 97.22.531.4211
Fax: 97.22.538.0688
E-mail: levitte@mail.gsi.gov.il

Rober Massarwieh and
Ali Swarieh
Natural Resources Authority
P.O. Box 7
Amman, Jordan
Telephone: 96.2.857.600
Fax: 96.2.681.1866

16th century Ottoman miniature of a Turkish bath. The dome on the right arches over the entrance to the bath, the cold section where customers arrived. Customers would disrobe in the narrow rectangular room on the lower right, seated on raised banks of earth that were covered with blankets and rugs. They left their clothing in piles, including any clean clothing brought for later. Seven bundles of clothing are depicted, including turbans. Customers also carried their own towels and dried themselves here after the bath.

The man in the small room above is sitting with his bag containing items needed for the bath, such as a razor, loofah, and soap. An employee hangs items out to dry on the roof.

The hot bath itself is on the left, covered by a dome encrusted with colored glass balls where light streams through. The customers are here, enjoying their bath. Interpretation by the College of Fine Arts. ***Printed with permission of the Topkapı Palace Museum, İstanbul***

5. The Geothermal History of Anatolia, Turkey

by
Mehmet Emin Özgüler
Ahmet Kasap

Abstract: Anatolia has long been a bridge between the East and the West, a place of settlement for European and Asian invaders. It has been the home of empire builders, beginning with the Hittites around 1450 B.C. The Hittites were followed by a long procession of colonizers, including the Phrygians, the Persians, the Greeks, the Romans, the Byzantines, and the Turks. Many civilizations flourished under the influence of European and Asian cultures and a little part of each became the whole that created Anatolia.

Anatolia, an Asian part of Turkey, is located in a zone of intense volcanic and tectonic activity rich with geothermal manifestations such as hot springs, fumaroles, mud pools, and mineral resources. Since prehistoric times, the people of Anatolia have used these geothermal and volcanic resources. But archaeological findings indicate the use of geothermal resources started with the Romans, followed by the Turks, for bathing in geothermal waters was such an essential part of life in these civilizations. Today, the Turkish bath is very well known all over the world.

Introduction

ANATOLIA, AN ASIAN PART OF TURKEY, IS AN AREA OF ancient settlements that has seen many changes throughout history. Many civilizations flourished in Anatolia under the influence of eastern and western cultures. Hot springs and volcanic products were used by Anatolian inhabitants from prehistoric times to the present. Radiocarbon dating indicates the earliest evidence of agriculture in Anatolia goes back to 7040 B.C. The following are dates for important civilizations in Anatolia (Akurgal, 1978):

- Prehistoric period, 7000-2000 B.C.
- Hattian culture, 2500-2000
- Old Hittite Kingdom, 1750-1450
- Hittite Empire, 1450-1200
- Anatolia's Dark Age, 1180-750
- Neo-Hittite civilization, 1200-700
- Urartian civilization, 900-600
- Phrygian civilization, 750-300

- Lydian, Lycian, and Carian civilizations, 600-200
- Hellenistic age, 300-30
- Roman age, 300 B.C.-395 A.D.
- Byzantium period, 390-1453
- Seljukian civilization, 1071-1308
- Ottoman Empire, 1299-1922

Due to its geological circumstances, Anatolia has been greatly influenced by geothermal resources, which it has used throughout time. Although Anatolia has many hot spring and volcanic areas with an historic past, only a few are mentioned in this chapter. Generally the hot springs, often considered a luxury, have been used for recreation and therapeutic treatments since the Hittite civilization. Over 700 hot spring areas and spas contain mineral components useful for curing rheumatic, sciatic, gynecological, kidney, and psychiatric problems (İzbırak, 1983). Archaeological findings reveal that the urbanization of an apparent sanctuary site tied to hot mineral springs occurred in the Roman period.

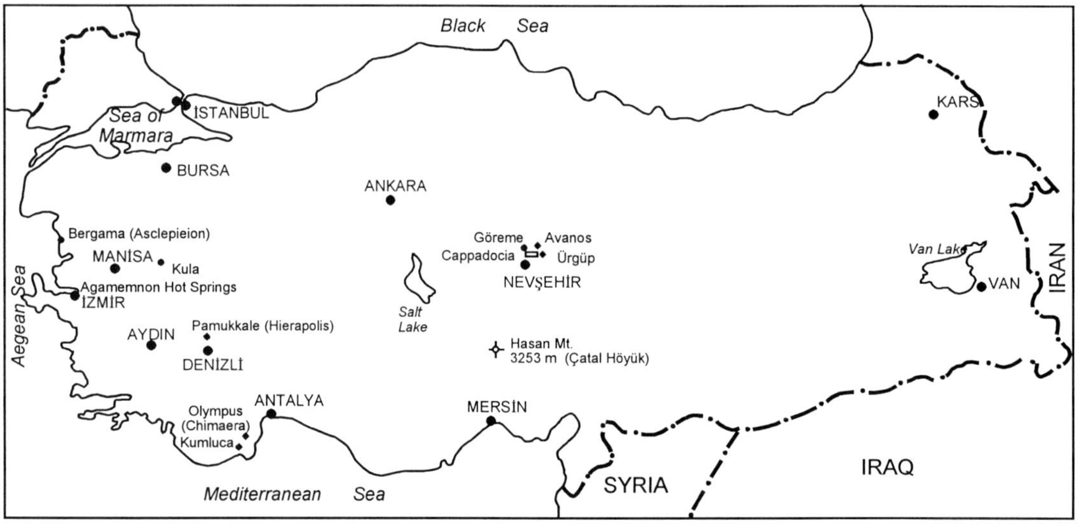

Map of Turkey.

Use of Volcanics in Anatolia

The First Contacts

The most ancient sign of contact between Anatolian peoples and geothermal resources dates back 12,000 years, to the footprints found in volcanic rocks at Kula. These emphasize the close relationship between people and volcanic phenomena (Tekkaya, 1976).

About 150 kilometers east of Kula is the city of Konya. Archaeological excavations conducted at Çatal Höyük (Çatal Mound), about 60 km south of Konya, revealed that an ancient town was built at the foot of a volcano called Hasan Dağı. The Çatal Höyük settlement that evolved here was one of the earliest civilizations. Twelve different periods of habitation, dating from about 6500 to 5650 B.C., have been distinguished in this unique prehistoric center of culture, where humans created some of the first great works of art. The most important cultural works displaying the relationship between the Anatolian people and geothermal phenomena are mural paintings and painted relief sculptures that adorned the walls of ancient houses, caves, and shrines. Some of these are on exhibit in the museum at Ankara.

One mural found at Çatal Höyük by Mellaart in 1967 represents the oldest archaeological artifact showing the interest of Anatolian people in geothermal phenomena. The painting depicts an erupting volcano in the vicinity of the Neolithic town and clearly illustrates the volcano Hasan Dağı (3253 m) because it shows the two distinctive peaks formed by the volcanic eruption. The painting also shows the particulars of the eruption: explosive plumes, flames, volcanic ashes and bombs shot from the vent, lava flows at the foot of the mountain, and volcanic fragments on the slopes. The mural shows that the painter, an eyewitness to the eruption, had a keen spirit of observation and a capacity for describing natural phenomena (Göncüoğlu, 1981). Radiocarbon testing indicates that this painting and Hasan Dağı volcanic activity date back to 6200 B.C.

In addition, the magnificent obsidian mirrors, monochrome pottery, and stone tools of Neolithic age discovered at Çatal Höyük prove that the inhabitants of the area knew and valued the importance of volcanic products, which added to the prosperity of the town of Çatal Höyük.

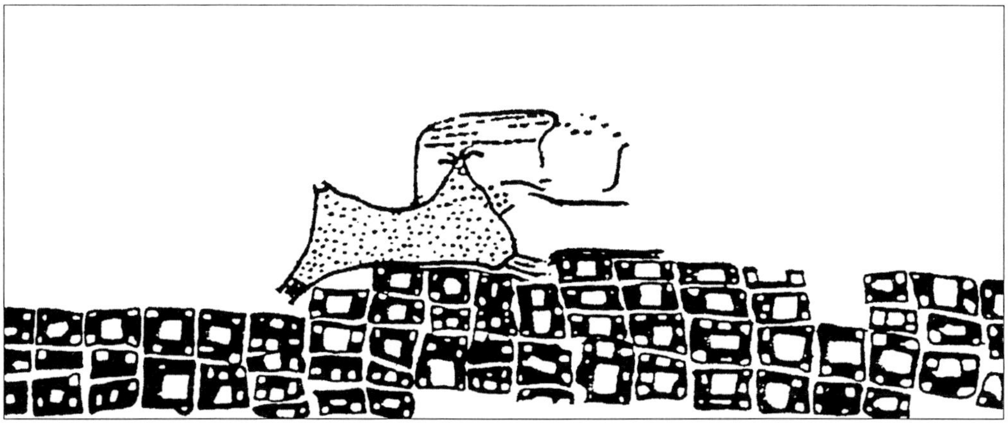

Çatal Höyük wall painting, 6200 B.C. The volcano rises behind rows of houses. This settlement, evolving in a volcanic region, was created by one of the earliest civilizations. Twelve habitation levels have been distinguished from about 6500 to 5650 B.C. This painting is one of the earliest works of art known. *From Göncüoğlu, 1981, with permission*

Land made fertile by the decomposition and disintegration of volcanic rocks was another feature that attracted Anatolian settlers to the volcanic area. Settlers also appreciated the many natural hot springs, freshwater springs, and mud pools.

Cappadocia, The Land of Fairy Chimneys

The province of Cappadocia, in Central Turkey, is famous both for its natural beauty and as a good example of the use of volcanic products by ancient Anatolian inhabitants. The province includes the towns of Ürgüp, Avanos, and the Göreme Valley. According to a widespread myth in Anatolia, the fairy chimneys of Cappadocia are the corpses of invaders who, after attacking the unarmed and humble inhabitants of the area, were turned to stone by a mighty god to whom the inhabitants had offered prayers.

More scientifically, these stones are volcanic tuffs that assumed their chimney-like shapes after softer outer materials were eroded from all sides, particularly along the joints, leaving the harder core of the rocks to stand in an amazing variety of mysterious angular remnants. From dozens to 200 secret underground cities may have been excavated here, for the tuff is soft until exposed to air. Perhaps the first city was constructed by the Hittites when they came under attack by the Phrygians in the 14th century B.C.

Cappadocia became the refuge of persecuted Christians during the early days of the Roman Empire. Underground cities were cut with amazingly intricate systems of tunnels into the rocky mountains during this period (Toksöz, 1977). The Christian Anchorites and monks who settled in the Göreme Valley realized the possibilities offered by the soft volcanic tuff. Trying to make themselves as inconspicuous as possible to hide their religious zeal, they chose to live in houses and worship in churches dug from these soft tuffs. The ancient name for Göreme is *Korama*. Here Saint Hieronymus, a Christian vine grower, hid in a cave dug from the rock while trying to elude capture. His pursuers, however, found the unfortunate fellow, dragged him out, and he was martyred.

Cappadocia is also famous for about 3,000 rock churches dating from this same period, although many churches in the Göreme Valley were built in the 7th to 13th centuries. The most important churches carved out of the volcanic tuff and decorated with vivid frescoes are the Apple Church and the Dark, Çarıklı, Barbara, Yılanlı, and Tokalı Churches. Frescoes in the churches usually depict scenes from the life of Christ corresponding to the great religious feasts such as the Nativity, the Crucifixion, and the Resurrection.

Underground dwellings in Cappadocia were excavated from volcanic tuffs ejected from ancient Mount Argaeus. From these tuffs, which are soft until exposed to air, as many as 200 secret underground cities were carved, some for up to 50,000 residents. A 10th century history says the dwellers were called "troglodytes," for they lived underground. The area has about 3,000 rock churches. *Turkish Tourist Office*

Other interesting sites in the region include the picturesque troglodyte village with family dwellings built right into the rock formations and the spectacular subterranean city of Kaymaklı, which was cut out of the volcanic tuff between the 6th and 10th centuries A.D. Dwellings such as these are known to have existed in the region as early as the 5th century B.C. Another underground city, situated not far from Kaymaklı, is found at

Derinkuyu. This one goes down to eight different living levels, with tunnels extending for several kilometers. This city could accommodate more dwellers than Kaymaklı, perhaps as many as 50,000 people. The shelters were last used in 1839 by locals hiding from an invading Egyptian army.

Olympus-Chimaera, The Eternal Flame

Chimaera was thought to live in Lykia, a monster who was part lion, goat, and serpent, who issued flames from the mouth and nose. The site of Chimaera is located a few kilometers northwest of the picturesque ruins of the ancient city of Olympus. Here flames issue eternally, burning gas escaping from fractures and fissures in the volcanic ophiolitic rocks of harzburgite, dunite, and pyroclastics. Chemical analyses indicate that the gas is 82.96 percent methane, 14.5 percent nitrogen, and 1.00 percent carbon dioxide. Originally, the flames emerged from seven places, but three have been extinguished by torrents, soil cover, and similar processes, although gas still issues from the vents. The sight of this miraculous natural phenomenon is very impressive at night.

The famous ancient city of Olympus was built near this eternal flame. The most attractive structure at Olympus is the cella-door of a temple that has recently been discovered by Ümit Serdaroğlu. The inside dimensions of the door are 2.90 by 7.85 m. From the inscription of a statue base at the door's foot, we learn that a statue of Marcus Aurelius stood somewhere in the temple. Thus the sanctuary may have been erected during the reign of this Roman emperor (161-180).

Use of Hot Springs in Anatolia

Roman and Turkish Baths

The antique ruins found around most of the hot springs in Anatolia indicate that they were constructed during the Roman, Seljukian, and Ottoman Empires. The widespread use of thermal springs in Turkey started with the Romans because public baths were an essential part of their life. Men and women bathed at established times, women in the morning and men in the mid- to late afternoon. Both could participate in light exercise or games in the *palaestra*

"I am often asked whether the traditional Turkish baths are a continuation of the Roman baths of antiquity: the answer is yes. Turkish baths are the direct and only descendant of this long line of baths and bathing culture which started with Classical prototypes (Japanese baths and the Finnish sauna come from different roots). Of course, there are also important differences.

"The avenues through which the baths and the bathing culture of Classical antiquity passed on to the historical Turkish-Islamic communities were numerous. Between the 11th and 15th centuries, the Turkish conquerors of Asia Minor came in direct contact with the vestigial representatives of small public baths in Byzantine cities. The contact through Constantinople during the centuries preceding and the decades following the conquest of that city in 1453 must have been particularly important. Even more important as an effective and vital source, however, were the Arabic-Islamic baths throughout Syria and Egypt with which the Turks became closely acquainted. In either case, large cultural centers, such as Damascus and Cairo, acted as sources of diffusion. In both regions and in both cities, the Arabic bath had already taken root through contact with the Classical, Byzantine, and late Ptolemaic models and had achieved centuries of internal development. As Turkish society patterned much of its urban life after the models provided by high Arabic culture, the assimilation of baths was only a natural process.

"Like the Byzantine and early Islamic baths, Turkish baths do not have palaestrae. The notion of exercise connected with bathing is entirely absent (in fact, the suggestion to a conservative Turk of exercise in connection with bathing would be nothing short of absurd). Also absent is the conventional frigidarium. With the exception of thermomineral baths and spas, communal pools of hot or cold water are outside the traditions of the Turkish bath. Following the development already started in late antiquity, bathing is done in individual basins, using plenty of water, but without total immersion.

"The functions of the frigidarium were assumed by a lounge-apodyterium combination in which the patrons could spend considerable time after their hot baths resting, sitting up and enjoying a cold drink, chatting with friends, or lying all wrapped up in towels. Often, in the center of the tall, well-illuminated lounge (really, the "bath hall" of the ancient and late antique baths) is a small, ornamental pool with a fiskiye (that is, a jet of water),...perhaps a vestige of the great pools in the Roman frigerdaria.

"The heating technology of Turkish baths closely follows their Byzantine and Roman prototypes, employing a fully developed hypocaust system with wood-burning furnaces serviced from exterior courts. Bathing attendants and masseurs available in Turkish baths can also be compared with those of Roman baths, although the oil massage and perfume treatment of the latter has given place to soapsud massage and a thorough rubdown with goat-hair mittens. The massage usually takes place in the middle of a large, domed space, the main hot bathing hall (the equivalent of the ancient caldarium), over a pleasantly heated marble platform. Small sweat chambers that open into the main bathing hall are comparable to the Classical laconicum. Rambunctious behavior is hardly a problem in Turkish baths although the patrons may occasionally be treated to an unsolicited concert under the well-echoing dome.... Finally, eating and nonalcoholic drinking in the baths are perfectly in order. This is particularly true for women, who bathe in a separate section. The preparation of a rich variety of delectables for a women's traditional bathing party can start several days in advance.

"Unlike the Roman tradition, bathing never became a daily routine in Turkish culture. Although it varies from person to person and from season to season, visiting the baths anywhere between once a week to once a fortnight is considered quite normal. Once there, however, cleansing is thorough. Although the advance of modern plumbing and the convenience of taking a quick, hot shower at home have rendered the use of public baths unnecessary for a large section of the population in today's Turkey, and although the ancient vie aux basins can hardly be considered the 'focus of civilized existence' anymore, the *hamam*, as an urban amenity and as a social and civic institution, is essentially but precariously alive in Turkey, and still offers a pleasant alternative to mechanized and modernized bathing."

From *Baths and Bathing in Classical Antiquity*, by F. K. Yegül, published by The MIT Press (exercise yard) before bathing. The bathers would pass from the unheated areas into rooms with more heat, proceeding gradually toward the *caldarium*, the main hot water room. After the *caldarium* they would move into the *frigidarium*, where the main cold pools awaited them. While this was the normal progression, the bathers also enjoyed diversions. At any time during the circuit, a bather might take a sweat bath in the *laconium* (a hot dry-steam sweat chamber), enjoy a massage, or stop to chat with friends. People mingled while washing, wading, exercising vigorously, swimming in the unheated *natitio* in the *frigidarium*, enjoying a professional massage or a simple back rub, and eating and drinking in small groups.

Around many hot spring areas, the ruins of thermal baths remain, with hemispherical domes characteristic of the Roman period. This structure was developed further by architects in the Seljukian and Ottoman Empires, who enlarged the spheres of the domes to accommodate more people (Aslanapa, 1986). This occurred after the Turks entered the cities and towns of Anatolia and found baths ruined by wars, earthquakes, and negligence through the centuries. The first thing they did was to restore and modernize them, not modifying the essentials of the healing springs and spas, just reconstructing and developing them with typical Turkish architecture.

A Turkish bath is not limited to an *apodyterium* ("changing room"), *frigidarium*, and *caldarium*; unlike the Roman baths, it has higher ceilings with larger hemispherical domes and running fresh-water fountains beside the pools. An impressive example of Ottoman baths can be seen in Northwestern Anatolia at the Çekirge hot spring area in Bursa.

Bathing is an old and important Turkish tradition. Even today, every city or town in Turkey has several public baths with all sorts of comforts, warmed either by hot springs or by central heating. In fact, the Turkish bath is common all over the world (Özbey, 1979; Başar, 1973).

Pamukkale-Hierapolis, The Cottoncastle

THE HIERAPOLIS HOT SPRING AREA IS CALLED PAMUKKALE TODAY BECAUSE THE SITE IS COVERED by cottonlike white travertine, formed by the precipitation of calcium carbonate from the lime-charged hot springs and streams in the region. *Pamukkale* means "cottoncastle," whereas the ancient name *Hierapolis* means "Holy City," chosen for the many temples in the city. The name Hierapolis appeared officially on ancient coins (*Hierapolis Di Frigia*, 1987).

16th century Ottoman miniature of women walking to a public bath, following an escort. Their escort glances back at them, making sure all is well. Women with faces carefully covered would go the baths in groups, staying from early morning until the midday meal. Those who went later stayed until nightfall. Women of distinguished families with baths at home did not visit public baths. The woman on the far right may carry a towel, clothing, or food in the bundle on her head. A small boy holding a toy sits on the shoulder of the fourth woman from the right.

The bath itself is in the background. The entrance hall is on the right and the hot-water area is on the left, under the dome elaborately studded with colored glass balls. Perhaps the rather intricate shading behind the dome is rising steam. *Printed with permission of the Österreichische Nationalbibliothek, Vienna*

View of the hot spring terraces at Pamukkale, near the ruins of Hierapolis, founded in 190 B.C. by Eumenes II, King of Pergamon, and bequeathed by Attalus II to Rome. The city was leveled by an earthquake in 17 A.D. but was soon rebuilt, reaching its height as a Roman thermal bathing center in the 2nd and 3rd centuries. *Turkish Tourist Office*

The site where Pamukkale stands is composed of a solid calcareous mass deposited over the ages. The deposition of travertine continues today and has buried the lower parts of many ancient buildings to a depth of several meters.

The spring that feeds the sacred pool of the hot spring area has never been located, but it lies, presumably, somewhere upslope, toward the ancient theater. Water issuing from the ground has a temperature a little below blood heat, making the pool comfortable for bathing. By the time the water reaches the plain, it has lost nearly all its heat and is used freely for drinking and irrigation.

Water passes from the sacred pool in numerous rills, mostly a foot or so wide. At this point, precipitation of lime forms a rim on either side of the rill, allowing the water to run in self-built limestone channels. Some of these rills are still flowing, but many others have been abandoned and are now dry. Where the water falls over the edge of the plateau, lime deposits form clusters of stalactites, making the whole cliff appear like a vast, slowly-changing petrified cascade. Some researchers claim that the rate of travertine precipitation has been fairly consistent and that the whole mass of the plateau may have taken about 14,000 years to accumulate.

At the beginning of the 2nd century, a monumental Roman bath was erected in the south-western part of Hierapolis at the edge of the travertine terrace during a great reconstruction project after an earthquake that may have occurred in the year 60. The plan of the bath was typical of *thermae* in Anatolia. There was a large courtyard, a second closed-rectangular area with large halls on both sides, and the bath proper, which was made up of pools enclosed in rooms located side by side. These rooms had large window openings facing the heliothermic axis. The exact outline of the courtyard is unknown. The closed rectangular area, which faces south with a large rectangular niche, most likely was the hall of the Imperial Cult. Two side corridors led into the *palaestra.*

Heat was supplied by furnaces. The central hall was heated by two furnaces located in a little barrel-vaulted room that still exists. The walls are covered by large rectangular pipes fastened vertically with metal clamps. Hot air was conducted up the pipes, eventually reaching the open air through appropriate chimney pots. The rooms and pools also were heated by the sun, which streamed in through very large windows. The windows once held glass panes mounted in marble grates or wooden frames.

It was the god Pluto who gave Hierapolis its chief claim to fame in antiquity, namely the Plutonium that was described by several ancient authors, including the Greek historian, Strabo. Writing as an eyewitness about the turn of the first millennium, Strabo said,

> "Plutonium is an orifice under a slight ridge of the hill which rises above, large enough to admit a man, and very deep. In front of it is a fenced enclosure some 50 feet square, filled with a thick mist so that the floor is barely visible. Outside of the enclosure, the air is free of the mist so long as no wind is blowing, and a man may approach safely. But for any living creature that enters inside death is instantaneous. Bulls, for example, that are taken in collapse and are brought out dead; we ourselves sent in small birds which at once fell lifeless. The eunuchs of Cyble, who practiced stopping their breath outside for awhile, were able to approach the orifice and look in and even penetrate for some distance, through not normally without holding their breath. They used this event as a propaganda for their religion. But ancient sources claim that the priests of the great mother alone are immune and could go down into the cave without being overpowered by the poisonous underground steams [of carbon dioxide]."

The Roman historian Dio Cassius visited the spot in the 2nd century and also observed how the vapor destroyed all living creatures except the eunuchs. He tested this phenomenon by sending in birds. He bent over and saw the vapor enclosed in a sort of cistern and noted that an auditorium had been built over it.

Later in the 5th century, a visit by the physician Asclepidotus is reported; he took the precaution of wrapping his cloak two or three times round his face so as to breathe in the pure air taken in with him. He was able to follow the course of the hot water for most of the way in but could not reach the end because the ground had been cut away by a stream of water too deep for a man to pass. The orifice was described as lying under the Temple of Apollo.

At the present time some people, especially tourists, try to approach the orifice and even penetrate inside. So far, anyone who has penetrated inside for quite a long distance has fallen from the poisonous steam.

Hierapolis was not merely a resort for tourists and theater-goers, and the hot springs were not used solely for bathing. Most of the area's prosperity came from industries, which were numerous and varied. Commercial companies recorded by inscriptions included guilds associated with the woolen industry and guild members, such as dyers, wool washers, and carpet weavers, who used the hot spring waters to make colors permanent.

The Asclepieion (Asclepieum)

The Asclepieion at Pergamon is one of the important therapeutic centers in Anatolia. According to Pausanias, a 2nd century Greek geographer and travel writer, the first temple of Asclepius was set up during the first half of the 4th century B.C. Recent excavations have confirmed that the sacred precinct, developed during the Hellenistic period, has existed since the 4th century. The center included statues of the gods of health and associated deities. The Temple of Asclepius, god of medicine, was Roman in form and style, but Pergamene in expression and spirit. The Asclepieion at Pergamon attained its most glorious heights in the 2nd century.

In Roman times, the Asclepieion was approached by a road called "The Sacred Way" that was 820 m long. It began in the Roman city as a narrow road lined with columns, then passed by the Roman theater and, continuing on as a wide and magnificent road, finally reached the Asclepieion. The latter sector, flanked by colonnades, is 140 m long and 8.34 m wide.

The buildings in the Asclepieion were used for religious worship and medical treatments. The methods of treatment at the Pergamon health center were noted in various inscriptions

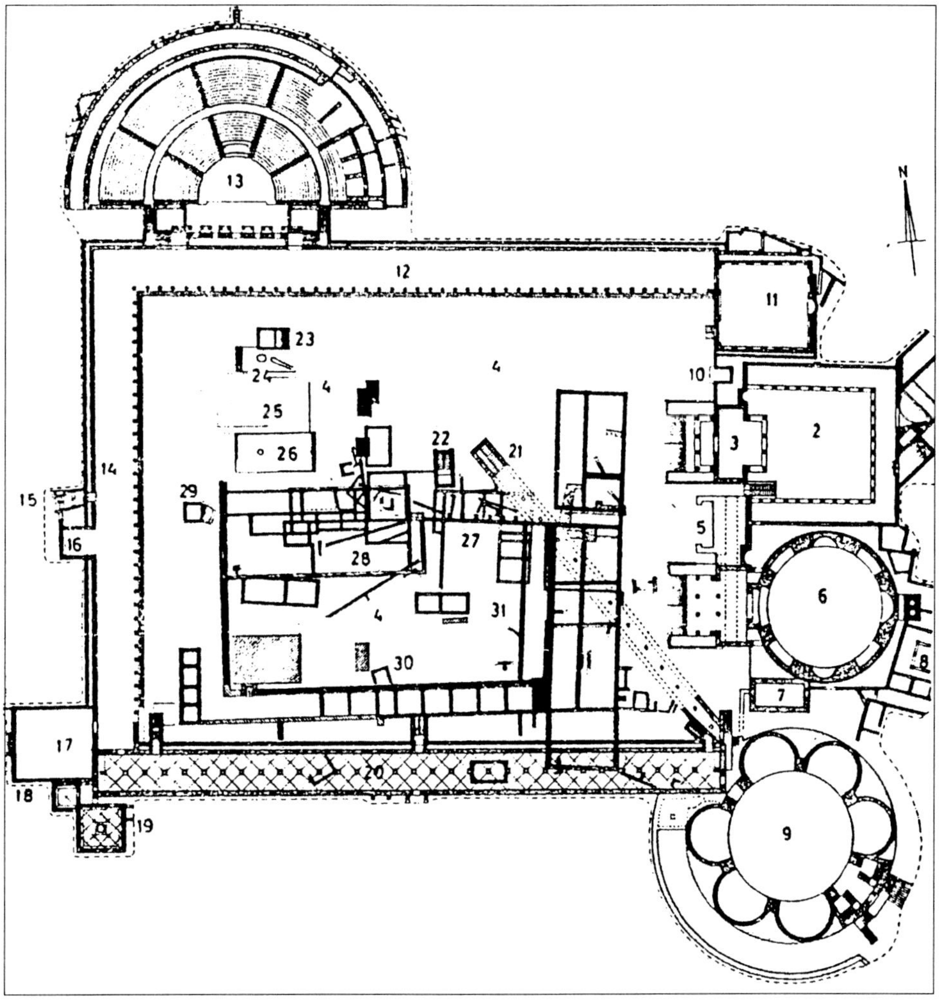

The Asclepieion of Pergamon: (1) end of colonnaded street; (2) courtyard of propylon; (3) propylon (138-161); (4) place for religious festivities; (5) niche for worship; (6) Temple of Asclepius; (7) cistern; (8) pristyle building; (9) building probably reserved for medical treatments; (10) niche for worship; (11) emperor's room; (12) northern colonnade; (13) Roman theater; (14) eastern stoa; (15) central door of the stoa; (16, 17) rooms, probably; (18, 19) lavatories for ladies and gentlemen; (20) southern stoa; (21) Roman tunnel leading to the treatment building; (22) pool; (23) fountain built in Roman times for bathing and drinking (the patients sat on the lower steps inside the pool and washed themselves with radioactive water coming from the sacred spring); (24) the crack seen in the rock today is very probably the site of the sacred spring; (25, 26) rectangular outlines in the rock probably show traces of the temples of Apollo Kalliteknos, Asclepius Soter, and the goddess Hygieia; (27, 28) sleeping rooms for incubation and auto-suggestion, which were two important methods of psychiatric treatment in Pergamon; (29) pool, probably used for mud baths; (30, 31) remains of colonnades. *Drawing and caption reprinted by permission of Professor Dr. Ekrem Akurgal*

and most especially in the writings of the orator Aelius Aristeides, who stayed there for 13 years around the middle of the 2nd century. Famous physicians such as Satyros and Galen (130?-200?) lived there and gave lessons. Methods used for physiotherapy at the Asclepieion are still applicable today. The most important of these were water and mud baths, massage, the use of medicinal herbs, and the application of ointments. Also prescribed were drinking sacred cool spring water, courses of abstention from food and drink, colonic irrigation, and running barefoot in cold weather. Autosuggestion and incubation played important roles in treatment. According to the orator Aelius Aristeides, the type of treatment was determined by a patient's dreams, and for this purpose, specially constructed sleeping rooms were provided. Rites were held in the theater within the facilities, and the patients underwent treatment involving therapy accompanied by music.

The patients believed that the god Asclepius would restore them to good health and regarded everything in the area as sacred. There were three pools or fountains at the Asclepieion designed for bathing and drinking. The patients sat on the lower steps inside the marble bath and washed themselves in water from the sacred spring. Very likely the crack in the rock was the site of the sacred spring. In ancient times, the water issuing from this spring was considered to have healing properties, like all the other geothermal springs within the vicinity, such as Beauty Baths, Kaynarca, and Dikili Kocaoba hot springs, which are still very popular. Chemical analysis has shown water at the Asclepieion to have radioactive properties and a geothermal origin.

Balçova (Agamemnon) Hot Springs

The Balçova hot springs area, in İzmir, is one of the most spectacular balneological centers in Western Anatolia. Its ancient name of *Agamemnon* is that of the Greek king of Mycenae (or Argos), leader of the Greek forces in the Trojan War. According to legend, King Agamemnon expanded the Trojan War and tried to invade the country of Mysia (Pergamon and vicinity), where he faced

Hot and Cold Springs in *The Iliad* as Literary Devices

In The Iliad *by Homer, the City of Troy in Western Anatolia is under siege by the forces of the Grecian king, Agamemnon. While the Greek warrior, Achilles, is chasing the Trojan warrior, Hector, around the walls of Troy, the men run past a hot and a cold spring. Homer may describe the springs in the following passage as a way to foreshadow the fates of the two warriors: soon Achilles, in the heat of victory, will bring Hector to icy death.*

"They raced along by the watching point and the windy fig tree
always away from under the wall and along the wagon-way
and came to the two sweet-running well springs. There there are double
springs of water that jet up, the springs of whirling Skamandros.*
One of these runs hot water and the steam on all sides
of it rises as if from a fire that was burning inside it.
But the other in the summer-time runs water that is like hail
or chill snow or ice that forms from water."

* Chief river of the Trojan plain.

The *Iliad of Homer*, xxii, 700-600 B.C.
Richard Lattimore, translator
Published by the University of Chicago Press.

severe resistance and eventually lost. He was advised by a fortune teller to send his wounded soldiers for relaxation and treatment to curing houses at today's Agamemnon hot springs (in Turkish, *Balçova Kaplıcaları*). Thus this health resort area was named for an historic event.

Agamemnon hot springs was also mentioned by the writer Philostratos, who lived in the 3rd century B.C. Philostratos mentioned in his work *Heroika* (*Work on Valiant People*) that Ionia hot springs were called Agamemnon hot springs by the native peoples of İzmir.

Aelius Aristeides, writer and famous master of oratory who lived in the İzmir-Balıkesir region, mentioned Agamemnon in his "sacred speeches," which included information about Western Anatolia during his time. He wrote that he came to these health resorts on his return from Rome to rid himself of a disease.

According to another legend, the name of Agamemnon hot springs comes from a different Agamemnon. It is believed that this Agamemnon, an offspring of the great Agamemnon, was the king of the Kyme (Aliağa) region. While waging war against his northeastern neighbor, Mysia, he sent his wounded soldiers to these hot springs for cures.

Some Ancient Legends about the Origin of Hot Springs

Legends and myths about the origin of hot springs have been transmitted orally from generation to generation. In ancient times, it was believed that hot springs formed from burning alum, sulfur, and asphaltite, which in turn heated the overlying soil and water circulating in it. This water would boil and rise to the surface in the form of a hot spring. It was also believed that when cold ground water came in contact with fire under the Earth's surface, a large explosion occurred, causing the water to take in many violent currents of air. Water loaded with such air would rise up with bubbles and noise. If such water were trapped within the cracks of a rock, it could be forced up high hills by the energy of the compressed air.

It was believed that just as boiling hot springs are formed under different conditions in different types of soil, so each hot spring has special curing properties. For instance, a sulfur-bearing hot spring heats harmful secretions on the human body, burns them, and stops muscle pains. On the other hand, an alum hot spring protects a human being from cold, supplying heat to the pores of the body.

The water in each hot spring has a different taste because each spring forms under unique conditions. Some hot spring waters with a good taste are able to dissolve stones that have developed in the human body. According to belief, this is natural because the shell of an egg put in vinegar becomes soft. Thus hot spring waters with a good taste can be drunk for good health. Hot spring waters with a bad taste can decay teeth and even cause paralysis, and people must avoid drinking them.

Waters from hot springs like Tarsus in Magnesia can make a human voice sound good. Waters from the hot springs of Hierapolis (Pamukkale) and Cappadocia deposit travertine on the surfaces of objects placed in them. This occurs because in Hierapolis there is a type of water like liquid cheese, which hardens on exposure to sun and air.

Conclusion

There is much we haven't mentioned about the ancient uses of geothermal resources in Turkey because every acre of Turkish land is full of ancient history. Many civilizations have flourished here since (and before) the beginning of recorded history. Although relatively few archaeological excavations have been made, such studies continue, uncovering more of our past every day.

Selected References

Akurgal, E. 1978. *Ancient civilizations and ruins of Turkey.* Türk Tarih Kurumu Yayını, Ankara.

Aslanapa, O. 1986. *Osmanlı devri mimarisi.* İnkılap Kitapevi, İstanbul.

Başar, Z. 1973. *Erzurum İlinde şifalı sular, Atatürk Üniversitesi yayınları,* no. 13, Turkey.

Göncüoğlu, C. 1981. Dünyanın en eski jeoloji Belgeseli. *Jnl. Yeryuvarı ve İnsan,* 6, 3-4.

Güven, S. 1990. *Vitruvius, The ten books on architecture.* M. H. Morgan, trans. Şevki Vanlı Mimarlık Vakfi Yayınları, Turkey.

Hierapolis Di Frigia. 1957-1987. Politecnico di Torino, Università di Lecce, Gruppo Editoriale, Fabbri, Bompiani, Sonzogno, Etas s.p.a. Italy.

Homer. *The Iliad.* Reprint 1951, Richmond Lattimore, trans. University of Chicago Press, Chicago.

İzbırak, R. 1983. *Geography of Turkey.* Directorate General of Press and Information, Turkey.

Özbey, S. 1979. *Şifalı Sularımız.* Yurt Haberler Ajansı, Ankara.

Tekkaya I. 1976. Insanlara ait fosil ayak izleri. *Jnl. Yeryuvarı ve İnsan,* 1, 2-12.

Toksöz, C. 1977. *A travel guide to the historic treasures of Turkey.* Mobil Oil Turk A. Ş., Turkey.

Yegül, F. K. 1986. *The bath-gymnasium complex at Sardis.* Harvard University Press, Cambridge, Massachusetts.

Yegül, F. K. 1992. *Baths and bathing in classical antiquity.* The Architectural History Foundation and The MIT Press, Cambridge, Massachusetts.

The Authors:

Mehmet Emin Özgüler and
Ahmet Kasap
MTA Ege Bölge Müdürlügü, P.K. 1
Bornova - İzmir, Turkey
Telephone: 90.232.388.8181
Fax: 90.232.388.4901

Painting depicting the nymphs of the thermal waters at the entrance to the Loutraki Spa, a modern establishment near Corinth, Greece. ***Photo courtesy of Dr. Z. Angelidis, printed with permission***

6. Geothermal Energy in Ancient Greece: From Mythology to Late Antiquity (3rd Century A.D.)

by
Michael Fytikas
Georgia Margomenou Leonidopoulou
Raffaele Cataldi

Abstract: Due to its geological and geodynamic conditions and recent and historical volcanic activity, Greece is rich in thermal springs, fumaroles, and hydrothermal minerals. Active volcanoes and natural manifestations inspired the formation of many myths and deeply influenced the beliefs of ancient Greeks. Impressive volcanic eruptions occurring in prehistoric and ancient times were recounted in a number of legends and myths or described scientifically by many Classical authors, who also attempted to interpret their origin.

Thermal springs greatly impacted the lifestyle of ancient Greeks: countless people frequented them, mainly for the waters' curative properties. To express gratitude to the healing divinities of thermal waters for successful treatments, many users left gifts or built monuments in the spring areas. In particular, temples to Asclepius were erected near many thermal springs all over Greece.

Volcanic activity on some Greek islands led to the formation of obsidian and valuable hydrothermal by-products, the exploitation of which contributed greatly to local economic development.

Introduction

Geothermal energy has been acknowledged in Greece for thousands of years. The etymology of the word "geothermal," is, in fact, Greek. Many natural manifestations were formed in recent geological and historical times, including thermal springs, fumaroles, hot grounds, and hydrothermal mineralizations. Volcanic eruptions and phreatic explosions also took place on some Greek islands during prehistoric and historic epochs; they were mentioned in mythology or described in the works of ancient and modern authors. In areas with relatively recent volcanism, many eruptive rocks (lavas, ignimbrites, some pyroclastics, and obsidian) were used by early Greeks for tools and artifacts or for building and lining materials, millstones, and other useful items. Thermal springs afforded the ancient Greeks the opportunity to establish curative centers; at the same time, the presence of natural manifestations and the occurrence of active volcanoes and frequent earthquakes spawned myths, legends, and religious sentiments.

The strong hydrothermal activity in certain volcanic areas resulted in many valuable minerals and deposits—kaolin, iron oxides, sulfur, pozzolana, perlite, and travertine—which were exploited for a number of practical uses. The abundance of these products and their increased use in Antiquity resulted in their export to nearby Mediterranean areas. Extraction and trade of hydrothermal by-products in Greece and other geothermal localities of the Southern Mediterranean date to the 3rd or 2nd millennium B.C. (Cataldi and Chiellini, 1995).

The continuous development of land and sea trade and the colonization of many such areas by the Minoans, Phoenicians, Mycenaeans, and other ancient Mediterranean peoples during early historic times facilitated the commercialization of hydrothermal products. Extensive colonization and trade during both the Classical period of Greece and Roman times enhanced contact among Mediterranean peoples, further fostered the use of hydrothermal compounds, and contributed substantially to the development of the economy and culture in many areas. Aegina, Milos, Kimolos, Thera (present-day Santorini), Nisyros, Yialí, and Kos are the main islands of the active volcanic arc of Southern Greece. Western Thrace, Samothraki, Limnos, Lesbos, Chios, Samos, and Patmos in the Northeastern and Eastern Aegean are the areas of Tertiary volcanism that display similar characteristics in their volcanic and hydrothermal products.

Active Volcanoes in Mythology and Ancient History

The Cultural Background

Volcanic activity is one of the most impressive natural phenomena that caught the attention and excited the imaginations of prehistoric people in many areas of Greece. Depending on the degree of cultural evolution and the type of phenomenon—explosive eruptions, lava fountains, fumarolic activity, volcano-tectonic earthquakes, or thermal manifestations—the early Greeks who observed them felt surprise and astonishment or fear and terror. These experiences were recounted from father to child and thus handed down orally through many generations. This resulted in a gradual distortion of what actually happened and its impact on the area concerned. Afterwards, these stories spread over the whole of ancient Greece, intermingling with similar stories originating from different geothermal areas and eventually forming a common knowledge base regarding the "significance" of these impressive natural events. A halo of

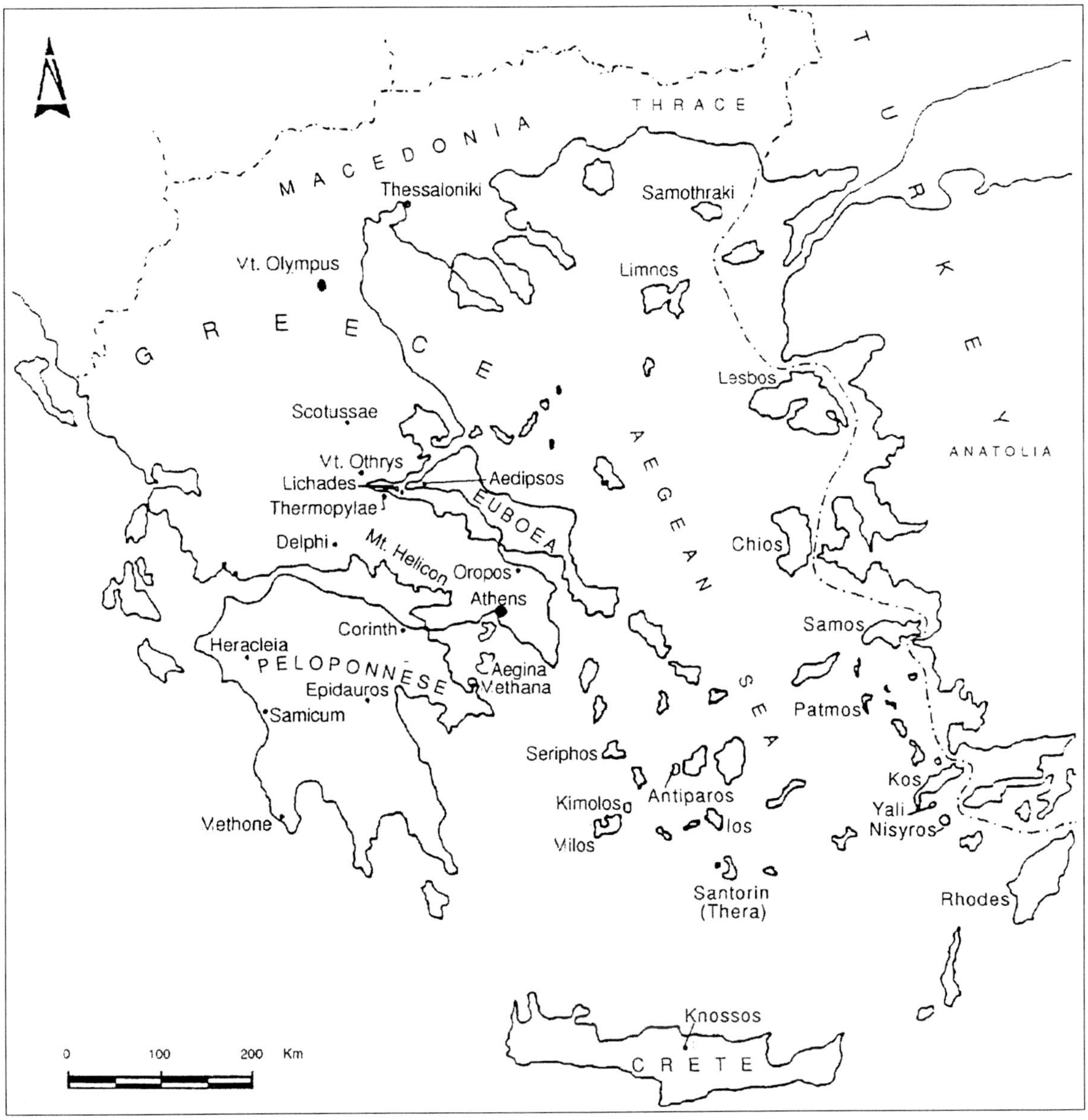

Map of the main geothermal sites in ancient Greece and other localities mentioned in the text.

grandiosity and a sense of the supernatural have thus been associated with such events since prehistoric times. Simultaneously, thick veils of mystery shrouded them in many areas of Greece.

In this way, legends and myths developed specifically related to those phenomena, substantially enriching Greek mythology. In historical Antiquity, however, besides myths and legends, scientific descriptions of the various manifestations of the Earth's heat started to appear in works of geographers and historians, while other thinkers began to speculate on the origin and nature of such phenomena.

Volcanoes in Greek Mythology

Greek mythology centers around the 12 great Olympians: Zeus (the omnipotent father), his wife Hera, Aphrodite (goddess of love and beauty), Apollo (god of the sun and music), Ares (god of war), Artemis (goddess of the hunt), Athena (goddess of wisdom and battle), Demeter (goddess of the harvest and field), Hephaestus (blacksmith to the gods), Hermes (messenger of the gods), Pluto (god of the underworld), and Poseidon (god of the sea). Except for Pluto, who lived in Hades, the gods resided on the top of Mount Olympus, about 3000 meters high, located between Thessaly and Macedonia, about 400 km north-northwest of Athens.

Hephaestus (son of Zeus and Hera) was the divinity of fire and volcanoes. Etymologically, Hephaestus means "the one who burns," "the one who shines." Perhaps less well known is the legend that Hephaestus was born rather ugly and that, after seeing him and being taken by an irrepressible wrath, Zeus violently hurled him far from Olympus. After a flight of many hundreds of kilometers, Hephaestus fell on an island which, according to one version of the legend, was Limnos in the Northern Aegean or, according to another version, the island of Sicily in Southern Italy. As a god, Hephaestus could not die, but the fall injured one of his legs and he became lame. He lived alone for many years, trained himself in the use of fire and smelting, and finally created a well equipped workshop in the deep crater that had formed as a result of his fall. At the same time, a cunning Hera eventually had her son readmitted to the Olympic court.

Beside being ugly, Hephaestus was surly. But he was not bad, was very industrious, creative, and, most importantly, had an outstanding skill for melting, forging, and casting metals. Therefore, he became the much sought metalworker and blacksmith of the gods. On special occasions, but with the necessary intercession of one of the gods, he could also manufacture arms and metallic goods for demigods and very important mortals.

Due to his metal-working ability and since he was always at the gods' disposal in his art, Hephaestus gradually gained the sympathy of all of the inhabitants of Olympus and even succeeded in winning the favor of Aphrodite, whom he married. The marriage menage, however, was all but smooth!

According to different versions of the legend, Hephaestus might have had several metal laboratories and workshops, most of which were located inside volcanoes. One version states that

Hephaestus' main workshop was on the volcanic island of Limnos; however, since only extinct volcanoes exist on this island (Fytikas et al., 1995), the most reliable version of the legend is that the main (if not the only) workshop was located inside the Etna volcano in Eastern Sicily. The popular belief in early historic and ancient times was that steam, gas, and ash escaping from the top and high slopes of certain mountains of Greece and other Mediterranean areas were nothing but the smoke formed inside Hephaestus' workshop(s), released into the atmosphere through underground chimneys. Ancient people also believed that most volcanic eruptions occurring in the Mediterranean area were caused by Hephaestus' powerful hammering on his anvil, resulting in rock fractures through which the flames and smoke, caused by the glowing embers of the god's forge, were expelled to the air.

Regardless of the site(s) where his workshop(s) may have been located, Hephaestus manufactured many excellent, specialized metal artifacts mentioned in the works of prominent Greek and Latin Classical authors, including Hesiod (8th century B.C.), Homer (7th century B.C.), Herodotus (5th century B.C.), Virgil (1st century B.C.), and Ovid (1st century B.C.-1st century A.D.). Among the most wonderful artifacts forged by Hephaestus were the thunderbolts used by Zeus to defeat the Titans, Apollo's bow and arrows, Poseidon's trident, Helios' chariot, Artemis' arrows, and Achilles' panoply of weapons used during the Trojan War.

In Greek mythology, however, besides Hephaestus, all other gods, many demigods, and some mythical heroes were also involved in volcanic, geothermal, and tectonic events. Indeed, the activity of Hephaestus was considered by the ancient Greeks the main, but not the only, triggering cause of volcanic manifestations and earthquakes.

While recognized as supernatural entities, the gods of Olympus were imagined to possess human feelings, including passions such as love, mercy, compassion, friendship, anger, jealousy, and revenge. Fierce discord could occur from time to time between the gods, and Zeus himself sometimes would forget that he had to behave as a supreme and impartial ruler. In this context, ancient Greeks interpreted volcanic events and related phenomena as evidence that the gods were fighting among themselves over some important or even frivolous matter, thus causing calamity on Earth with their fury.

In addition, the celestial hierarchy of Olympus and the order constituted by the gods on Earth were contested from time to time by groups of mythical beings endowed with exceptional physical strength and courage. This is the case with the Giants and the Titans, who warred with the gods in a vain attempt to oust them and usurp their position.

In his *Theogony* (c. 780-760 B.C.), Hesiod extensively described the battle between the gods and the Titans. The battle occurred reportedly at the foot of Mt. Orthrys in Southern Thessaly, not far from Thermopylae (which means "hot doors" or "hot pass," due to the thermal springs still found in the area), where a few hundred Greeks led by Leonidas in 480 B.C. strenuously opposed the much larger Persian army commanded by Xerxes. Since some small eruptive centers with volcanic domes and lava flows of a very recent geological age exist in this area, it is possible that memories of the volcanic activity that occurred in prehistoric epochs were handed down from generation to generation, elaborated upon, and transformed into legend during very ancient times (Papageorgiou et al., 1990). In fact, Hesiod reported that the Titans launched huge blocks of rock against the gods in heaven, while the "Earth was burning" and a "roar of earthquake" spread over a large area during the battle.

Scene of the battle between the gods of Olympus and the Titans. One of the Titans is about to throw a huge block of rock (a volcanic bomb?) at the gods. From a ceramic vase of the 5th century B.C. *Photo courtesy of Professor E. Stiros, printed with permission*

Other ancient authors also described in detail the legendary war between the gods of Olympus and the Giants. One version, reported by Apollodorus of Athens in *On the Gods* (2nd century B.C.) and recalled by Pausanias 400 years later in his *Description of Greece* (*vol. 1, Attica*), concerned Poseidon and one of the fiercest Giants, Polybotes. Poseidon decided to engage and personally punish this Giant, but in front of the god, Polybotes fled mainland Greece and sought refuge on the volcanic island of Kos in the Southeastern Aegean. Poseidon decided to use his trident to cut a huge block of rock from the southern sector of the island (near the present-day Cape Kelonas) and throw it at Polybotes. The Giant, meanwhile, fled to the south by quickly swimming across the sea. The block struck Polybotes a few tens of kilometers from the shore of Kos, seriously wounding him, but it did not kill him. The Giant remained buried under that huge rock, which disintegrated, caused tremendous earthquakes, and eventually re-emerged to form Nisyros and the volcanic islets south of Kos.

However imaginative and fantastic, this legend reflects some verifiable geological evidence:

- The existence of a semicircular feature near Cape Kelonas in the southern sector of Kos. This feature is now interpreted as the remnant of a rather young volcanic caldera.

- The young age of Nisyros and the islets nearby. They formed as the result of a repeated series of eruptions that occurred between 100,000 and 10,000 years ago.

- The intense seismic activity of this sector of the Aegean Sea (thousands of earthquakes from minor to medium intensity in only the last few years!).

- The occurrence at Nisyros of recent hydrothermal explosions, two of them in historical times.

- The petrological similarities of the rocks in Kos and Nisyros.

This legend goes on to say that, after remaining trapped under the rock and "neutralized" beneath the burden of his destiny, Polybotes endured lengthy agony punctuated by moans of pain, hisses of rage, spitting blood, and terminal quivers.

The ancient people of Nisyros used this legend to explain the numerous fumaroles, gas emissions, and hydrothermal eruptions characterizing the central sector of the island. As a matter of fact, one of the largest craters on this island (which has a diameter of some 300 meters and has been interpreted in recent years by Marini et al., 1993, as the result of a huge hydrothermal explosion) still bears the name of Polybotes (see last photo of the chapter). Another huge hydrothermal explosion occurred on Nisyros in the last century, creating a new crater with a diameter of 150 m (Gorceix, 1874).

The special rapport that the inhabitants of Nisyros established in Antiquity with the god of the sea, creator of their island, is confirmed by an imposing and sumptuous temple to Poseidon, as well as by the stamp of Poseidon's head on the ancient coins of Nisyros.

A second legend concerning the victory of the Greek gods over the rebel Giants was reported by Virgil. In the *Aeneid* (30-19 B.C.), Virgil recalled that the warrior goddess Athena, after defeating Engheladus, perhaps the strongest of the Giants, buried him alive under the Etna volcano in Eastern Sicily. In a vain attempt to free himself from that deep and secure prison, this Giant forcefully shook the mountain over him, thus causing the strong earthquakes that have frequently affected the Etna area and surrounding regions since prehistoric times. It is also worth mentioning here that, in relation to the intense seismic activity of the country, the legend of Engheladus is still alive in Greece: the name itself is a synonym for strong earthquakes.

Another legend which relates the formation of volcanic features to mythological events concerns the famous demigod Hercules (son of Zeus and Alcmene, granddaughter of Perseus), who was endowed with superhuman strength and courage. In his tragedy *Trachiniae* (c. 420-410 B.C.), Sophocles mentioned that Hercules had a servant named Lichas who, at a certain point, was persuaded by his master's wife to give him a poisoned tunic. Hercules started to suffer from tremendous pains but, after dreaming about the deception of his disloyal servant, decided to punish him with death. Hercules brought Lichas to the northwestern tip of Euboea Island and threw him into the sea. From the disintegration of Lichas' body, the volcanic islets of Lichades formed (of very recent geological age). This legend is similar to the legend of Poseidon, Polybotes, and the formation of Nisyros.

While many other legends exist about how volcanoes were considered in ancient times in Greece, those recounted are sufficient to illustrate the richness and the degree of cultural refinement of Greek mythology in relation to volcanic activity and associated phenomena.

The ancient Greeks, like other important civilizations during the early stages of formation, were not prone to using scientific thought—the observation of natural phenomena and their effects followed by an elaboration of interpretative hypotheses and theories. To the contrary, the Greeks first developed a cosmogony and a set of universal principles from a mythological or religious perspective. Then observed facts were set within the general framework of preestablished schemes. Using this type of "logic," all natural phenomena, including volcanic manifestations, always had a coherent explanation.

Volcanoes and Earthquakes in Learned Accounts

The volcanic island of Thera (known in the past also as *Stronghili*, "the round"; or *Kallisti*, "the most beautiful"; and nowadays as Santorini) was often mentioned in narratives dating back to ancient epochs, handed down orally up to historical times. Strangely, however, no acknowledgment exists of the destructive eruption occurring about 1628 B.C. that had such a tremendous impact on the entire Eastern Mediterranean, nor were memories preserved in the myths and legends of early Greece. Similarly, few descriptions or accounts of this event can be found in known historical sources. This huge volcanic eruption was named "Minoan" for a few decades by volcanologists and archaeologists, after the Minoan civilization then dominant in the Southern Aegean.

The only written reference to the eruption, although indirect, is found in Plato's *Dialogues* of *Timaeus* and *Critias*. These works give wonderful accounts of the sudden disappearance of a legendary, vast land that the author named Atlantis, the famous "lost continent." Plato claimed that the story of this disappearance was true, as it had been recounted to Solon (one of the famous seven sages of Greek Antiquity) by the priests of Saiis (ancient Alexandria) during his visit to Egypt around 590 B.C. The priests' narrative concerned a catastrophe that occurred many centuries before, resulting in the disappearance of a vast land. Upon his return to Greece, Solon mentioned this story to his relatives, and it was thus handed down orally for some generations until it reached Plato. In this way, the story might have been garbled and embellished.

Trying to find impressive reference models to support his cosmogony and philosophical conception of an orderly world, Plato further exaggerated the story of the vanished land and imagined that it could have been "larger than Libya and Asia together": an area as large as a continent. Since such a vast portion of land could not be accommodated in an area as restricted as the Mediterranean, Plato conceived that it was located in the ocean, beyond Hercules' Gates (the Strait of Gibraltar).

Besides the location, Plato described many other characteristics of his "continent," including the sophisticated level of social organization and technological development attained 8-9 millennia before his time, especially in its capital, Atlantis. Plato's detailed description of this metropolis—including multiple circular walls plated with sheets of different metals, sumptuous temples, royal palaces, and other wonderful features of an inner citadel—together with the geometric layout of the whole town and the perfect organization of its facilities, probably represents the fruit of the author's great imagination (see initial figure in Chapter 9, this volume).

However, descriptions of the continent's sub-circular form, its bays, the morphology, the cliffs, the occurrence of hot springs, some unique geologic features, and even the color of the rocks are similar to the physiographic characteristics of a volcanic island such as Thera (Vougioukalakis, 1997). In addition, Plato's attribution of a sophisticated civilization in Atlantis suggests that the author might have integrated into the *Dialogues* the high cultural level attained in early Antiquity by the Minoans on Crete and other islands to the north. The recent discovery of Minoan ruins near Akrotiri in the southwestern sector of Thera (Santorini) reveals that island inhabitants had reached a high level of civilization before the destruction by the "Minoan" eruption.

All this suggests that Plato, while fictitiously elaborating his legend of the vanished continent, might have been inspired by the story told to Solon by the Egyptian priests. They might well have referred to the catastrophic explosion on Thera during the Minoan epoch, when a huge central caldera was formed that caused a large portion of land, about 4 x 8 x 1(!) km^3, to sink (see the first two figures in Chapter 9, this volume).

Marinatos, in 1939, and other authors in the 1950s and 1960s, hypothesized that the great eruption of Thera could have caused the sudden annihilation of Minoan civilization on Crete. Marinatos' 1969 discovery of an advanced Minoan town near present-day Akrotiri on Thera seemed to prove this hypothesis; however, the age of the eruption was calculated in 1969 at about mid-15th century B.C., which coincided with the period that historians and archaeologists attributed to the end of Minoan rule. Based on the Marinatos hypothesis but adding much new data, Luce (1969) concluded that the sudden disappearance of the Minoan civilization came from a devastating *tsunami* ("seaquake"), triggered by the huge explosion of the Santorini volcano.

In the last 10 years, however, more precise scientific information has been gathered on the modalities, sequence, period, and characteristics of the volcanic phenomena and tectonic activity that occurred during the 2nd millennium B.C. at Thera and surrounding islets, as well as on the archaeological stratigraphy and findings from recent excavations. Among other studies, new absolute dates have been sought with different techniques—dendrochronological, ice-coring, and radiocarbon dating—aimed at precisely establishing the age of the "Minoan" eruption. The detailed reconstruction of the archaeological stratigraphy proves that almost everything has been preserved under the soft volcanic cover of ash and pumice. However, no human bones and very few jewels have been found at the excavation sites.

To explain this, Cioni et al. (1997) have suggested that several months or perhaps even a year before the beginning of the destructive eruption, strong earthquakes occurred, warning the people to leave the island temporarily. After some time, though, people must have returned and begun cleaning away collapsed blocks and debris from dwellings and streets, trying to continue their lives normally. Then a new series of stronger earthquakes (premonitory signs of a huge eruption, we would say today) must have struck the island, convincing all residents to collect their valuables and abandon the island forever.

In short, the results of all new studies (see bibliographical note) indicate that the huge explosion at Thera took place in 1628 B.C., plus or minus a few years. The date does not support the Marinatos hypothesis and the Luce conclusions that the collapse of the Minoan civilization, which historians and archaeologists have placed until now at around 1450 B.C., was caused by the huge explosion at Thera. Therefore, should new archaeological-historical research at Crete confirm that the Minoan civilization actually collapsed about the mid-15th century B.C., causes other than the volcanic explosion in question must account for the sudden annihilation of the Minoans.

Shortly after the catastrophic eruption on Thera, viscous magma with a limited gas content began flowing out to the sea floor from the volcanic chimney in the central part of the main caldera. Solidifying on contact with sea water, the magma slowly grew into a volcanic submarine structure, the top of which eventually formed some small islets known today as *Kamenes* ("burnt islets").

The first intracalderic submarine eruptions probably were not mentioned in any of the oral narratives handed down for tens of generations. No historical accounts exist of their occurrence, despite the fact that Thera was re-inhabited and developed by the 7th century B.C. The recent discoveries of a settlement and a cemetery about 5 kilometers north of Akrotiri date precisely to that period. The island's inhabitants were probably not aware of the volcanic activity happening at the base of the caldera almost under their feet, activity that had continued unceasingly after the destructive eruption of about 1628 B.C. and which continued to take place quietly until the 2nd century B.C.

The first record of submarine volcanic activity occurring at Thera is found in Strabo's *The Geography* (15? B.C.-7? A.D.). Recalling an event that had taken place there in 197-196 B.C., Strabo wrote that "between Thera and Therasia, the flames that leaped out of the sea for four days, making the sea bubbling and blazing, gradually, almost mechanically it could be said, pushed a little above the surface of the sea an islet of twelve stadia [2.22 kilometers] in circumference, which was composed of incandescent masses...." Therasia is an arch-shaped island in the Santorini archipelago, part of the northwestern rim of Santorini's main caldera, within which the new islet formed. Thera is, both during Strabo's time and at the present, the principal island, located east of the main caldera.

This new islet described by Strabo was named "Iera" by some people and "Automate" by others. Soon after its formation, the people of Rhodes, who then ruled the whole Southeastern Aegean, erected an altar to their protector Poseidon on the top of this islet. However, the islet was quickly flooded by seawater, and since it was composed of soft scoria and pumice, it was destroyed and washed away within a few years. Only a shallow reef, known as *bankos* ("socles"), remains in the area of this islet.

Volcanic activity on Thera reawakened in 46 A.D. Huge quantities of viscous magma erupted from the volcanic chimney below the floor of the main caldera and spouted from the sea two kilometers southwest of Iera, creating the new islet of Thia, later called *Palea Kameni* (the "old burned"). At the time of its formation, the islet had a circumference of 5550 meters, as the Roman historian Victor Sextus Aurelius wrote in his 4th century work, *Historia Romana*.

A very important author, Pliny the Elder, reported most of the volcanic phenomena that occurred in Antiquity in his *Historia naturalis XXXVII* (*Natural History in 37 Books*, 65?-79). Among other phenomena, this author so extensively and vividly described the explosion of Vesuvius in 79 that modern volcanologists named a particular type of volcanic explosion "Plinian," after him. Pliny the Elder stated that the formation of Thia within the main caldera of Thera took place on July 8, 46, whereas Seneca, who had a tendency to connect cosmic and terrestrial phenomena, in his *Naturales quaestiones* (*Natural Problems*, c. 63) wrote that the formation occurred two days earlier.

In the same work, Seneca wrote, "Who doubts that it was not the burning wind that brought to the light Thera and Therasia, and also the other islet that was born in the Aegean Sea in front of our eyes?" The author was obviously referring to Thia as the "other islet." This islet gradually acquired its present shape as the result of fragmentation at cracks and faults, followed by the collapse of its coastline in many places.

Referring to the formation of some new islands in the Aegean volcanic arch, Pliny the Elder wrote, "The land rises out from the sea.... In a different way, Nature gives back to the surface some of the Earth's pieces that it swallows, thus compensating the losses.... A number of small islands so formed in the Aegean, and among them Iera and Thia are really wonderful...." Pliny also mentioned the birth of other volcanic islands or islets of the Southern and Eastern Aegean, namely those located near the older volcanic islands of Milos and Limnos.

The volcano of Thera became active again in 726. Several explosive events took place in that year, and Theophanes the Confessor described them in his work *The Chronicles* (790-810): "In the summer of that year, steam as from a fiery furnace bubbled up from the depths of the sea for several days between the islands of Thera and Therasia.... In a short while, after it had risen and become dense due to the furious heat of the blazing fire [lava], the smoke itself began to seem like fire, and drew with it large volumes of solid matter.... Huge quantities of pumice stones were spewed out all over Asia Minor and Lesbos and Abydos, and toward those parts of Macedonia which overlook the sea...." This eruption strongly influenced the historical events of that period, because it was interpreted as a sign of the wrath of God caused by the Iconoclastic Controversy besetting the Byzantine Empire. The volcanic activity at Santorini continued intermittently in following centuries: six or seven other periods of activity occurred after 729, the most recent of which was in 1950 (Vougioukalakis et al., 1995).

Another spectacular volcanic eruption that occurred in Antiquity and was described in learned accounts was that of *Kameno Vouno* (the "burned mountain"), located on the Methana peninsula (Peloponnese), about 80 km south-southwest of Athens. The eruption and the resulting formation of this "burned mountain" very likely occurred in 250 B.C. and was reported later by Strabo in *The Geography* (15? B.C.-7 A.D.):

> "Near Methoni [the present-day Methana], at the border of the Hermionian Gulf, a mountain 7 stadia in height was formed as a result of a glowing blow. During the day it was very difficult to approach the place because of the heat and the odor of sulfur. During the night, though, it was very beautiful to stay at a certain distance and observe the shining phenomenon, which could be seen from far away.... The heat produced by the burning blocks was so strong that [when reaching the sea] the water could boil up to a distance of 5 stadia; these blocks also made the sea become turbid up to a distance of 20 stadia.... Moreover, big blocks of rock were uprooted from their positions and pulled downstream by the [lava] flow, and each block was as big as the size of the castles...."

Near Kameno Vouno some important sulfur thermal springs formed after the eruption and still exist today. One of them bears the ancient name "Thermae of Pausanias," probably recalling a visit by the outstanding geographer Pausanias in the 2nd century A.D.

Besides volcanic activity and earthquakes that occurred on islands and localities that are more or less easily identifiable today, ancient authors also gave accounts of volcanic phenomena and tectonic events which took place in marine areas, where no evidence of their occurrence can now be confirmed. These accounts, however, are probably true and refer to submarine eruptions followed by the formation of volcanic islets that were quickly eroded and dismantled by the sea, such as Iera (or Automate) at Thera.

Past accounts of volcanic or tectonic events in areas of the Aegean Sea have recently gained the attention of volcanologists, and new studies have yielded interesting results. For example, a submarine volcanic structure was recently discovered some 7 km from the northern coast of Thera, where the sea floor steeply rises to 18 m below sea level resulting in a submarine volcanic structure named the "Kolumbos cliff" (Vougioukalakis et al., 1994). This was an emerged structure in the past, and a temporary islet formed as the result of volcanic activity northeast of Thera in 1650 (Akylas, 1925).

In addition to descriptive accounts, ancient literature is rich in theoretical speculations on the origin and nature of volcanic phenomena and earthquakes (Cataldi and Chiellini, 1995). Among the skilled Greek thinkers who engaged in this type of speculation was Aristotle. In his *Meteorologica* (c. 340 B.C.), Aristotle maintained that the main cause of earthquakes was the "winds" (air streams) that penetrated underground through open cavities and, following the fast sealing of their inlets and outlets by natural causes, remained trapped in closed caverns below the surface. Thus, an increase in pressure occurred in these caverns, triggering movements of rock masses. Aristotle also maintained that the "winds so trapped have the capacity of striking sparks onto layers of sulfur and coal...thus causing their burning...." This burning, in turn, was the cause of volcanic activity for Aristotle.

In another passage of this same work, however, Aristotle specified that the strongest earthquakes were not those caused by trapped underground "winds" but those that occurred near the coasts, in localities where the marine currents could violently knock against the rock scarps.

Thermal Springs in Ancient Greece

Greece has more than 750 thermo-mineral springs, 195 of which are found on the islands of the Aegean Sea (NTO, 1966). About 25 percent of these springs are used now for balneotherapy, but in Antiquity the number of thermal localities frequented for cures and other reasons was perhaps higher than today. Many Classical authors have shed light on the influence exerted by Greek thermal springs on the formation of cults, legends, and myths, as well as the impacts on life styles and modes of thinking.

The following are among the Greek and Latin authors who mentioned thermal waters and natural manifestations in their works; the listing is chronological:

- Homer, poet (7th century B.C.)
- Pindar, choral lyrist (522?-448? B.C.)
- Herodotus, historian (484?-410? B.C.)
- Hippocrates, physician and scientist (460-377? B.C.)
- Aristophanes, playwright and philosopher (452?-385? B.C.)
- Plato, philosopher (427-347 B.C.)
- Aristotle, philosopher and scientist (384-322 B.C.)
- Demetrious of Callatis, historian and geographer (3rd century B.C.)
- Diodorus Siculus, historian (90-20 B.C.)
- Strabo, geographer and historian (66? B.C.-23? A.D.)
- Pliny the Elder, grand admiral and scientist (23-79 A.D.)
- Plutarch, biographer (46?-127? A.D.)
- Pausanias, geographer (110?-180? A.D.)
- Athenaeus, rhetorician and grammarian (170?-230? A.D.)

What these authors wrote on the origin and use of thermal manifestations is summarized as follows.

Origin and Nature of Thermal Waters

In consideration of their healing properties, ancient Greeks believed that thermal waters were special gifts bestowed upon people by the gods of Olympus and therefore sacred.

It was also believed that the gods had assigned the protection of the spring waters to the nymphs, young girls who were themselves sacred because of their divine origin, to preserve the springs' natural properties and prevent their contamination by people. Each spring, or cluster of springs, had its own group of protecting nymphs, so many of the springs were named after the nymphs ruling over them. People also believed that the "nature" (the physico-chemical characteristics) of each thermal spring was attributed to and reflected the temper of its protecting nymphs. The sulfurous springs, for instance, were protected by nymphs called Anigrids. While attributing the cause to a divine design, this way of conceiving the "nature" of the thermal waters reveals the implicit desire of the ancient Greeks to explain the different characteristics of these waters. Due to their healing properties, natural thermal waters were sacred not only to the nymphs but also to other supernatural entities, such as Asclepius (god of medicine, son of Apollo and the nymph Coronis), Hygieia (goddess of health), and the mythical hero Hercules.

Not only were the nymphs the protectors of the waters, but on special occasions they had the power to make new springs gush out from rocks. Diodorus Siculus (1st century B.C.) reported that when Hercules visited their region, the nymphs made new thermal waters spring out from dry rocks to please the goddess Athena.

Pindar was one of the first Classical writers to eulogize the thermal waters and their relation to the nymphs. In *The Odes* (470 B.C.), Pindar dedicated a hymn to Ergoteles of Himera, an outstanding athlete of *dolichos* (a long distance race), winner in three of the well-known ancient games, the most famous of which were the Olympic games. In this hymn Pindar wrote, "Now that you have gained a wreath at Olympia, two wreaths at Pytho [Delphi], and another one at the Isthmus [Corinth], you should glorify in your well-known land the thermal baths of the sacred nymphs...."

Mention of the nymphs in relation to thermal waters is found in the works of many other authors after Pindar, until the first centuries of the Christian era. One of the most recent who extensively mentioned the nymphs was Pausanias. In his *Description of Greece* (c. 161-176 A.D.), Pausanias mentioned the nymphs often, but the three passages concerning thermal waters follow:

- "In Boeotia [a region northwest of Athens], in an area near Mt. Libethrious [the present-day Mt. Helicon] images of the Muses and of the Libethrian nymphs could be seen in past times.... Two springs are to be found in that area, one named Libethrias and the other Petra.... People say that the water of these two springs is as warm and delicious as maternal milk..." (*vol. 2, Beotia*).

- "In the region of Elis [Western Peloponnese], in a locality near Samicum, not far from the river Anigrous, a cavern exists which is known as 'the cave of the Anigrid nymphs'...inside this cavern thermal healing waters spring out [the present-day Kaiafa springs].... Those who suffered from alphosis or leukae and went there for cure, after entering in the cave, customarily had to pray to the Nymphs first of all, and make offerings to them; afterwards, they could start the cure by wiping the diseased part of their body with river water; then, they had to cross the river immersed in its water up to their neck.... The disease remained blocked in the water, and the patients could come out healthy from the river..." (*vol. 3, Elis I*).

- "At a distance of some 9 km from Olympia, the Elian village of Heracleia could be found near the river Cytherus.... A spring discharging warm water into the river is said to exist there, and a sanctuary to the nymphs is to be seen near the spring.... People say that those who take a bath in that spring are cured of fatigue and all kinds of aches..." (*vol. 3, Elis II*).

Besides mythological and religious explanations, a number of skilled Greek thinkers, beginning around the 5th century B.C., started using physical concepts to describe the characteristics of thermal waters, and some of them also speculated upon the origin of the waters.

Describing the features of Thrace in his work *Melpomene* (c. 430 B.C.), Herodotus wrote that the people of that region considered the river Tearus as "the best river of the world, because its waters could heal all kind of sufferings, and especially the scab, in both men and horses...." This healing property was attributed to "thirty-eight springs, some of which cold and others hot, outflowing from the same rocks, discharged their waters into the river...," thus creating a variety of thermal regimes. "[The local people] say that, for the healing to be effective, the hot baths are to be taken during the springtime, whereas the cold baths are to be taken during the summer...."

Cave of the Anigrid nymphs at Samicum (Western Peloponnese), with the healing river Anigrous flowing from the cave. *M. Fytikas*

Hippocrates of Kos, the "father of medicine," was the first author who, in his *Airs, Waters, Lands* (c. 420 B.C.), systematically classified the waters: "It is necessary to consider the properties of each water...since the waters differ among them in heat [temperature], weight [density] and taste...; therefore, each water has its own distinctive

The Thermae of Sulla, at left, sited on Aedipsos Bay and the northwestern tip of Euboea Island (the enlarged view, upper right). Construction of the original thermal establishment on this site was promoted by the Roman General Sulla and completed c. 84-80 B.C. *M. Fytikas*

properties...." Based on these criteria and by comparing waters from different localities, Hippocrates was able to recognize empirically some elements and compounds in the natural thermal waters, such as iron, copper, silver, sulfur, asphalt, nitron, and alum.

When discussing the origin of the earthquakes in *Meteorologica*, Aristotle also tried to explain the origin of the thermal springs and maintained that the two phenomena were related to each other. To demonstrate this relationship, Aristotle (c. 340 B.C.) referred to the famous thermal springs of Aedipsos, on the northwestern tip of Euboea Island, and stated that the sea waves could penetrate through subterranean channels below the island, compress the rocks, and thus cause earthquakes. The friction of rock movements could, in turn, heat the rocks and the water contained in the underground channels, resulting in hot waters and thermal springs. We know today that the Aedipsos springs, with an outflow temperature of 78° C, clearly have a marine origin, as indicated by chemical analyses.

In another of his works, Aristotle mentioned that Scotussae, a town in Thessaly, had a small thermal spring with miraculous properties. Not only could the water of this spring "heal wounds and fractured bones of people and beasts of burden...," but it had also the power to seal breaks in pieces of wood. Aristotle noted that the local people believe, "if someone throws a piece of split wood into the water of that spring, and if the wood is not completely broken into two separated parts but only fractured, the fractures will be quickly welded, the wood will be restored and will take again its original structure and form...." We can hypothesize today that the thermal waters might have had pronounced encrusting properties.

An explanation similar to that of Aristotle for the earthquakes and the thermal springs of Aedipsos was given by Demetrious of Callatis (3rd century B.C.) for all other localities of Greece where earthquakes occur and hot manifestations exist. Writing about the geologic phenomena of the area around the Euboea Straits, Demetrious suggested that most of the volcanic islets originally constituting the small archipelago of the Lichades, as well as the promontory of Cenaeum (once forming the northwestern-most tip of Euboea), had sunk, resulting in modifications of the number and characteristics of thermal springs in the surrounding area. Supposedly, all of these phenomena were the result of intense tectonic activity that affected the area in early historic times. Demetrious' descriptions were reported by Strabo, along with many new details from his own research, in *The Geography* (c. 15 B.C.-7 A.D.): "In past times, the thermal springs of Aedipsos and Thermopylae ceased to flow for three days, but afterwards they gushed out again.... At the same time, from fractures newly formed in the rocks near Aedipsos, new thermal waters started to spring out...." The modification of hydrological patterns perhaps was caused by an earthquake, and this confirms that the high seismic activity of the area (still occurring today) dates back at least to Antiquity.

About a century later, Pausanias, in his work *Description of Greece* (161?-176 A.D.), reported on most of the thermal springs of Greece. He was especially detailed in regards to Methana and Methone and provided details about their formation and nature of the water:

- "Near a little town called Methana, in the Saronic [Hermionic] Gulf in front of Piraeus, there were important thermal baths. [The inhabitants of Methana] say that the thermal springs formed at the time when the King of Macedonia was Antigonous II the Gonatas, son of Demetrious I the Poliorcetes [between 277 and 240 B.C.].... They also say that before the appearance of the thermal springs, a volcanic eruption had occurred during which the Earth spit fire along with huge volumes of smoke and bubbles.... Only after the fire from the Earth ceased, water started to flow up.... At the beginning and for many days the water was terribly hot and salty..." (*vol. 1, Corinth*).

- "In the area of Messinia, near Methone [Southwestern Peloponnese], a temple to the goddess Athena and a sanctuary to the goddess Artemis could be seen.... [These monuments] were located near a natural hole with thermal water...and this water was darkish as if it was mixed with pitch, and it had an odor similar to that of the Cyzikos myrrh..." (*vol. 2, Messinia*).

Salus per Aquis in Greek Mythology and Antiquity

Salus (or *Sanitas*) *per aquis* ("health through waters") is a Latin expression used by ancient Romans to point out that thermal balneotherapy can help people maintain their bodies in good shape, prevent diseases, and sometimes restore function to parts of the body. Some people think that the initials from the Latin expression—*s*, *p*, and *a*—eventually became the word "spa," used today to signify a thermal spring, resort, or similar facility. To stress further the advantages of thermal bathing, ancient Romans usually completed *salus per aquis* with the motto *mens sana in corpore sano* ("healthy mind in a healthy body"). However, the Romans borrowed this motto from ancient Greeks, who used to say *noùs hygies en somati hygie*.

Indeed, thermal bathing for healing purposes is a practice that started in many Greek localities in prehistory, well before the birth of Rome, as indicated by archaeological findings and by accounts of traditions, habits, cults, and legends dating to early mythological times. Even the gods of Olympus, and especially the goddesses, practiced balneology, very likely in natural warm waters. For instance, Hera loved to take a warm bath every evening in the sacred spring of Kanathos (near Nauplia), Athena used to relax by taking a restoring bath after each of her battles on Earth, and Aphrodite cared for her beauty by bathing daily in special springs located at the southeastern foot of Mt. Olympus.

On the other hand, the nymphs of the springs were permanently "at home," and some legends describe them as taking baths or rising joyfully from pools of transparent fresh or warm waters. Many painters, mostly during the Renaissance, were inspired by these legends and created treasured works depicting them.

By far the most renowned mythological user of thermal waters was Hercules, the invincible hero who successfully completed the famous 12 labors. Although endowed with superhuman resistance, after each labor Hercules recovered his strength through a cycle of cures in one of the many Greek thermal localities or in other thermal spots in the distant lands where he found himself. Moreover, between two consecutive labors Hercules did not rest but rather traveled to discover new lands; in each he always searched for thermal localities where he could bathe to maintain his physique. This is why, according to several legends, a number of thermal localities are named after the mythical hero, such as Heracleia and several Hercules' baths in Greece, Ercolano near Naples in Italy, and the Thermae Herculanae in Western Romania.

Ancient Greek and Latin literature is rich in episodes concerning the special rapport established by Hercules with thermal waters. All stress that Hercules always bathed alone and only in natural thermal waters; he shunned man-made pools where the hot water could be contaminated by human works and fresh waters. The efficacy of balneotherapy for Hercules lay in taking advantage of the natural properties of spring waters exactly as they had been created by the gods and bestowed upon people.

In *The Clouds* (c. 423 B.C.), Aristophanes dramatized Hercules' balneotherapy, as well as his habit of always keeping himself well trained and ready to face the most arduous labors, as a dialogue between two ideal interlocutors. By using two contrary ways of reasoning (called right and wrong logic), Aristophanes let the two interlocutors discuss the reasons why none of the young people even approached Hercules' great strength and valor: "because today's youth waste their time in chatting and gossiping in comfortable thermal pools, whereas gymnasia and training grounds remain deserted" was the cutting conclusion of the dialogue!

Athenaeus, in a passage of *Deipnosophistae* (*Authorities on Banquets*), c. 193-197 A.D., also discussed this subject, concluding with the question: "Why, then, do people believe that all the hot waters rising to the surface from the Earth's depths are sacred and dedicated to Hercules?"

Apart from gods and mythical heroes, ancient Greeks frequented all known thermal localities for curative purposes. Depending, of course, on the nature of the water in each locality and of the thermo-mineral mud associated with it, one or more diseases could be treated: alphosis, psoriasis, leukae, alopecia, and eye infections among external diseases; and rheumatism, gout, urinary tract and vaginal infections, and respiratory system infirmities among internal diseases.

Hippocrates of Kos was the first physician who used thermal balneology systematically to cure his patients. In the 5th century B.C. he treated patients from all over the Mediterranean area in the famous Asclepian Center on his island. Since he also knew about the curative properties of many thermal localities, after his patients stayed on his island, Hippocrates sent them home to continue their cures at a nearby thermal locality.

Probably by the end of the 3rd millennium B.C., healed people customarily expressed their gratitude to the gods and the divinities of the hot springs by leaving a variety of offerings acknowledging the divine gift of thermal waters: coins, artifacts, pillars, and votive *tabulae* in more ancient times, and altars, statues, sanctuaries, and temples in less ancient times. This habit continued well after the beginning of the Christian era.

Of the many thermal localities frequented for cure by ancient Greeks, Aedipsos, Loutraki, Epidaurus, and Thermopylae deserve specific mention for their historical importance. The Aedipsos baths in the northwestern sector of Euboea were among the most renowned thermal stations of Antiquity in the Mediterranean area, as famous and desirable as Hierapolis in Western Anatolia and Baia (near Naples) in Southern Italy. The Aedipsos baths were used since prehistoric and early Greek times and flourished during the Hellenistic and Classical periods. They underwent a major boom during Roman times, starting in about 80 B.C.

Strabo reported that when the Roman General Sulla was in Athens in 84 B.C. (tarrying and thinking of how to take action with his 40,000-man army to put down power struggles that occurred in Rome during his military campaign against Mithradates), his feet became torpid; it was a symptom of podagra, according to Strabo. Therefore, Sulla decided to undergo a thermal treatment at Aedipsos, after which he regained his normal mobility (Plutarch, c. 105-115 A.D.). Sulla then backed the construction of a large stone and brick building at Aedipsos to function as a "modern" hydrotherapeutic thermal establishment. It was the first fully-organized establishment of this type in Greece and was named "Thermae of Sulla." It has been modernized several times since and "reinaugurated" in 1999, "fully restored and equipped with state-of-the-art technology," according to a brochure. The Thermae of Sulla was visited by many important patients in Antiquity seeking treatment, including the Roman Emperors Hadrian in 126 and Marcus Aurelius in 176.

The Thermae of Loutraki were in Northwestern Peloponnese, near Corinth. They were used in very ancient times and were revitalized around the 8th century B.C. when athletic games were organized at Corinth. The establishment was modernized in Roman times, reaching its peak in splendor and attendance. The atrium of the thermae and a beautiful painting of the nymphs of thermal waters are shown in the photograph at the beginning of this chapter.

One of the most important Asclepian baths was located in East-central Peloponnese, near Epidaurus. New thermal establishments were built by the Roman senator Sextus Iulius Maior Antoninus Pythodorus around 163 A.D., near a much older thermal center dedicated to Asclepius. Ruins of a sanctuary to the gods and temples to Apollo, Asclepius, and Hygieia have also been found. According to Pausanias, an epigraph dating to 163 indicated that many other monuments from the Classical period of Greece existed in this area (*vol. 1, Corinth*).

Thermopylae includes a group of thermal springs found in a narrow gorge near the Thessaly coast in front of the Euboea Straits. The name is known worldwide, not because of its thermal springs but because of the famous battle that took place there in 480 B.C. against Persian invaders. Apart from a memorial to Leonidas, erected in recent times, no other monument or building was placed on the site after about 150 A.D. when the famous orator Herodes Atticus (c. 101-177), a very rich man and *maecenas* ("philanthropist"), financed the construction of a beautiful thermal and swimming complex. The springs, however, had been used in Antiquity for balneotherapy well before the famous battle; this is confirmed by very ancient ruins, including an altar to Hercules. Pausanias, among others, also described the Thermopylae springs, which the local people named "women's cooking pots." Pausanias said that, from his personal experience, the hot waters supplying the Herodes Atticus resort were the most transparent and bluest he had ever seen (*vol. 2, Messinia*).

Oracular Practices in Thermal Localities

HISTORIANS THINK THAT SINCE ASCLEPIUS WAS THE MOST SKILLED PHYSICIAN OF EARLY GREECE in the campaign of the Argonauts, he lived in the 13th century B.C. In *The Iliad*, however, Homer mentioned that Asclepius' sons, Machaon and Podalerious, were the two main physicians of the Greek army during the war with Troy, which probably took place around the beginning of the 12th century B.C. The worship of Asclepius as the god of medicine was so deeply felt and diffused among the people in Antiquity that about 300 temples dedicated to him were constructed in Greece alone. The most famous and sumptuous was erected at Epidaurus, where a wonderful gold and ivory statue of Asclepius was raised inside his temple in the 4th century B.C.

Epidaurus' thermo-mineral springs with special healing properties are verified not only by ancient learned accounts and long-lasting traditions of visits but also by the first series of complete chemical analyses of their waters, carried out at the beginning of the present century by Dambergis and Komninos (1908). It is therefore no wonder that Epidaurus was considered in Antiquity the healing center *par excellence* and was the main site of Asclepian veneration. Many sick people went for healing, but the cure always was taken under the direction of the Asclepian priests.

Before beginning their cycle of baths, patients had to stay overnight in the sacred grove of Asclepius, where the mysterious recesses of the healing god were located. The Asclepian priests maintained, and people believed, that the healing properties of the thermal waters would work only after the patients had supplicated Asclepius' intercession and after they had slept and dreamed profoundly in the god's grove. The priests afterwards interpreted the supplicants' dreams in an oracular way, which enabled them to define the type of treatment and the length of cure for the patients. Those who did not dream "significantly" from the oracular viewpoint were considered by the Asclepian priests to be insufficiently pious toward the god and were not admitted for cure. The need for the patients to sleep overnight and dream "significantly" in the god's grove before starting treatment, as well as the oracular practice by the god's priests, were obligatory rituals in all Asclepian centers in the Mediterranean area, such as the famous one at Pergamon (Bergama), near the western coast of Turkey.

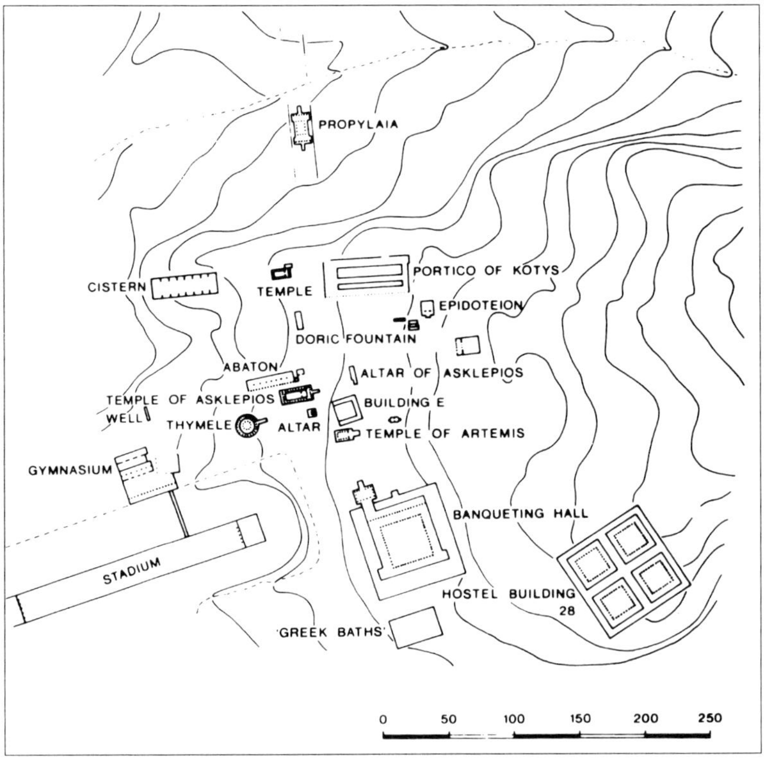

Map of the archaeological sites at Epidaurus after the excavations of 1881 to 1909. *H. Buglass, with permission*

At the end of a cure, since the beneficial effects were felt more or less immediately, the healed people expressed their gratitude to Asclepius through offerings. Many inscribed stone slabs were set by the Asclepian priests on the walls of the god's grove to show the names of important patients and often to specify the type of successful treatment that they had undergone at the Epidaurus center (Pausanias, *vol. 4, Phocis*).

The most famous oracle of ancient Greece was the one dedicated to Apollo at Delphi; it was located in the territory of Phocis at the foot of Mt. Parnassus, some 250 km northwest of Athens near a cluster of springs. The reputation of Delphi in Antiquity came from the conviction that the prophecies issued there by Apollo's women-priests were always truthful.

Artist's reconstruction of the archaeological sites at Epidaurus after the excavations of 1881 to 1909. *H. Buglass, with permission*

Strabo reported that "[The ancient Greeks] say that the oracle came out from a dark cavern full of vapor, which provoked ecstatic effects.... A high tripod was installed at the mouth of the cavern, and the Pythian priestess used to get up on the tripod in order to inhale the vapor and catch oracles and prophecies in verses or in prose.... When given in prose, the oracles or prophecies were immediately turned into verses by poets residing there at the service of the God...." This "vapor" was probably a mixture of low-pressure saturated steam and gas emitted from either a warm spring located inside a cavern or escaping through open fractures that connected the cave with a deep hot aquifer.

Pausanias offered another account of the Delphic oracle: "Inside the temple of Apollo at Delphi, an altar to Poseidon was constructed because, as [the ancient Greeks] say, the god of the sea was the first god who had issued oracles for his believers.... Near the temple, a [thermal?] spring called Cassotis had been in existence in past times; however, as [the ancient Greeks] say, at a certain moment the spring disappeared under the ground...from that moment on the women-priests of Apollo residing at the *Adyton* [the temple's inner recesses] of the god's cavern started to have the power of divination..." (*vol. 4, Phocis*).

It is clear that the natural manifestations of Delphi were not frequented for therapeutic purposes but only for divination. Usually—but not only—powerful and rich people went to Delphi for prophecies. At the end of each session, as a sign of thanks to Apollo, the visitors always offered money, jewels, or other precious objects destined for the god's treasury, which was kept by the women-priests in the *Adyton*. Very rich people, however, also financed the erection of fine monuments at Delphi, such as temples, altars, and sanctuaries, whose construction was entrusted to famous artists.

At Oropos, about 50 km northeast of Athens near the temple to Amphiaraus, a thermal spring with this divinity's name was known in Antiquity. Pausanias reported that, according to accounts of local people, Amphiaraus arose from the waters of this spring when the gods of Olympus granted him the rank of a divine being (*vol. 1, Attica*). The spring was not customarily frequented for therapeutic purposes or for ritual or purification baths; thus, no animal sacrifices were made nor were monuments constructed nearby. The spring was used as a veneration place by persons who had been healed following balneotherapeutic treatment at other thermal localities or who had experienced a prophecy or an oracle through dreams. These people visited the spring area and threw gold or silver coins into its waters as a sign of gratitude to Amphiaraus.

Obsidian and Other Geothermal By-Products

The most important geothermal by-product used by prehistoric people in areas of active or recent volcanism was probably obsidian, the darkish volcanic "glass" characterized by neat, conchoidal fractures. Because of this property, obsidian could be easily worked to produce arrowheads, knives, scrapers, and other tools.

In the Mediterranean, abundant obsidian deposits exist in several zones of Western Anatolia, the Aegean islands, and Southern Italy. All of these deposits were exploited more or less systematically during the last part of the Stone Age; however, the most ancient production zone was probably located in Western Anatolia, where exploitation of obsidian dates to the Lower Neolithic (Mellaart, 1967).

At Milos in the Western Cyclades, the exploitation of obsidian probably started in the 3rd millennium B.C.; in fact, wide use and a fine manufacturing level of obsidian objects was already in existence during the development period of the Phylakopì culture, which reached its peak between 2000 and 1600 B.C. (in the Meso-Cycladic Age). At that time, two large obsidian quarries were exploited intensively in the eastern sector of Milos, on the slopes of the Bombarda and Demeneghaki volcanic domes (Shelford, 1982). Shelford has proven that the selection and preliminary treatment of raw material was carried out at the quarry sites, where scraped blocks and chips of obsidian were left as debris deposits that can be seen today. The semi-manufactured pieces were finished by expert stonecutters in the town of Phylakopì, on the northern coast of Eastern Milos, where the remains of an obsidian laboratory have been found. A very important settlement with a high degree of civilization started to flourish at Phylakopì during the second part of the 3rd millennium B.C.

Obsidian products were used not only in Milos but were exported to almost all of the Cyclades Islands, to Crete, to the Greek mainland (mostly the Eastern Peloponnese and East-central Greece), and even to Macedonia, Thrace, the islands of the Northern Aegean, and Anatolia, where they were sold or used for barter. Their origin at Milos is undisputably confirmed by trace-element analyses, which define the exact mineralogy of Milos obsidian as compared with those from other Mediterranean volcanic areas (Shelford, 1982). The diffusion area of Milian obsidian was very large for such early times.

The wide distribution of Milian obsidian also demonstrates the intensity of sea trade in the Eastern Mediterranean during the second half of the 3rd and the 2nd millennia B.C., as well as the importance for Milos of obsidian extraction, manufacturing, and export. Indeed, most archaeologists now agree that the development of Phylakopì (its size, architectural layout, type of fortification, and population), as well as its cultural boom (potteries, textiles, house decorations, and paintings) during 2000-1600 B.C. came from the prosperity brought to the island by exporting fine obsidian goods, such as razors, knives, arrowheads, sickles, and scrapers, for which Milos had gained a monopoly in the Eastern Mediterranean. This status was due to the exceptionally good quality of Milian obsidian, particularly that excavated at Bombarda.

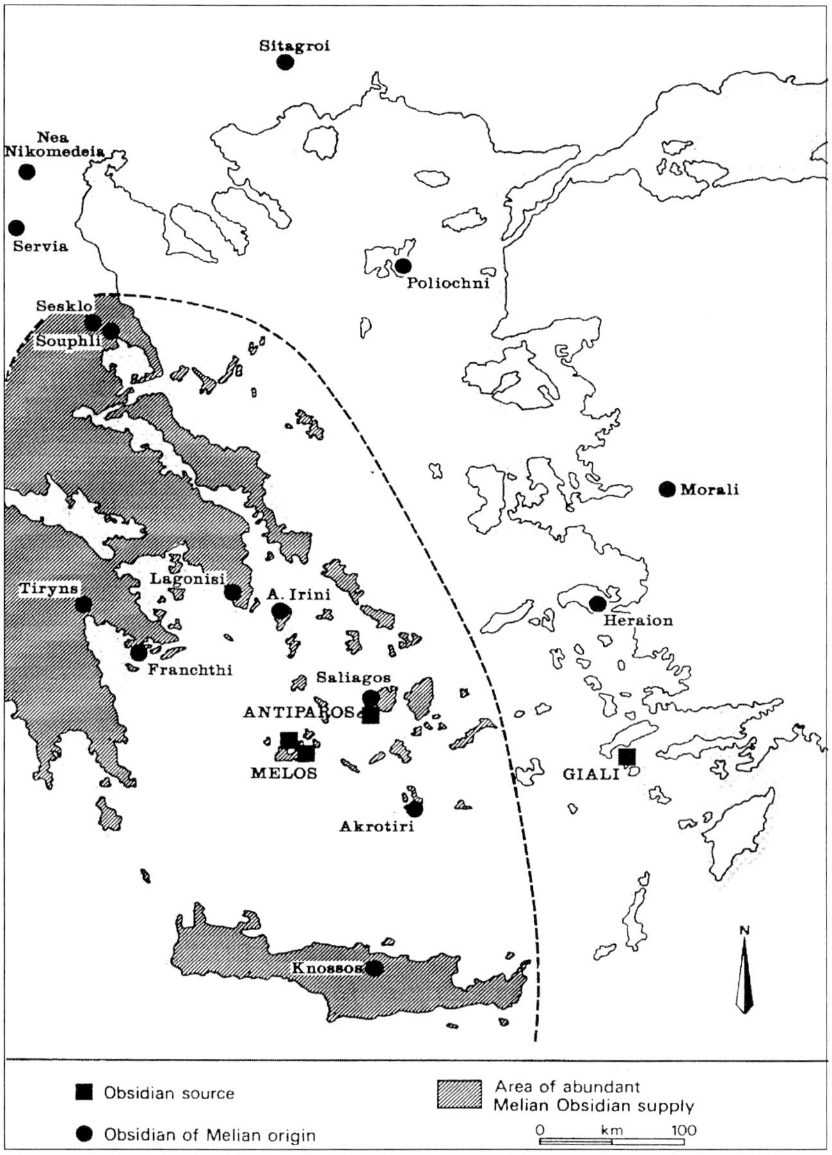

Main area of diffusion for Milian obsidian in Antiquity, from Shelford, 1982, in *An Island Polity* by C. Renfrew and J. M. Wagstaff. *Reprinted with permission of Cambridge University Press*

Beside Milos, two other volcanic islands of the Southern Aegean represented a source for obsidian in the Late Neolithic and early historic times: Antiparos, one of the Central Cyclades, and Yialí, a small and geologically very recent island of the Eastern Aegean between Kos and

Nisyros. Yialí has no resident population today, but it was certainly inhabited in ancient times as is proved by many archaeological finds, mainly tombs. Both the obsidian from Antiparos and Yialí and the manufacturing technique were of very high quality. In addition to those from Milos, several obsidian artifacts originally from Yialí have been discovered in the archaeological excavations at Knossos, extraordinary capital of the Minoan kingdom, which developed in Crete between about 2700 and 1450 B.C.

Even though most of the obsidian products were gradually replaced in early historical times by copper, iron, and bronze weapons and tools, the use of obsidian for fine artifacts did not end with the beginning of the Metal Age but continued during the whole of Antiquity. In *Melpomene* (c. 430 B.C.), Herodotus mentioned that obsidian was used to carve official seals; Teophrastus (370?-287?) reported that Milos was a very important source for obsidian in Antiquity; and Pliny the Elder, in *Natural History* (65?-79 A.D.), described the use of obsidian for fine mirrors, artifacts, and architectural decorations. During the Classical period in Greece and in Roman times, lapidaries and sculptors continued using obsidian as a semiprecious stone.

Besides obsidian, many other geothermal by-products played important roles in the economic development of some Greek volcanic areas in prehistoric and ancient times. At Milos and neighboring islets (Kimolos, Polyegos, and Antimilos), the volcanic eruptions ended 90,000 years ago; however, post-eruptive hydrothermal activity continued in prehistoric and historic epochs, mostly affecting the southeastern sector of the main island. Here, a huge phreatic explosion occurred in a year that radiocarbon dating places between 80 and 205 A.D. (Trainau and Dalambakis, 1989). The explosion resulted in the partial burial of an ancient village, probably a harbor, on the southern coast of Milos. Because of the high heat flow and the recent tectonic activity, thick hydrothermal deposits formed on the eastern sector of Milos and on Kimolos and Polyegos, including bentonite, kaolin, silica, and alunite. However, smaller deposits of other hydrothermal compounds, such as sulfur minerals (pyrite, galena, malachite, pyrolusite, and barite) are present as intercalations in the main deposits. Dispersed silver and gold granules are also common in all of these deposits (Fytikas, 1977).

Some of these hydrothermal products were known in Antiquity as "Milian (or Kimolian) earths"; they have been extracted since at least the 2nd millennium B.C., traded as raw materials all around the Mediterranean area, and used for a number of applications: pottery-making, bleaching solutions, construction, and dyeing. In *Natural History*, Pliny the Elder dealt in detail with the "Milian (or Kimolian) earths," explaining that these "earths" were used extensively in Antiquity and that some of them (smectic clays?) were particularly appreciated by the pharmaceutical industry for therapeutic properties in treating skin diseases.

Spectacular hydrothermal (phreatic) explosion craters inside the main caldera at Nisyros. One the most impressive young craters is called Polybotes, from the name of the mythical Giant. *M. Fytikas*

Due to the commercialization of these products and the exportation of finished obsidian, the prosperity of Milos continued to grow from the times of the Phylakopì culture to the Classical period. Examples of prosperity and of the sophisticated cultural level reached by the Milian people are found in countless accounts by ancient authors and in the impressive archaeological discoveries on Milos. The most important is the remains of a wonderful theater and a renowned statue of Aphrodite, known worldwide as the "Venus de Milo" (or the "Venus of Milos").

In the Southeastern Aegean, the volcanic islands of Yialí and Nisyros are to be mentioned for the importance of their hydrothermal products in early historic and ancient times. In addition to obsidian, pozzolana has been largely exploited and exported from Yialí since at least the 8th century B.C., contributing substantially to the economic development of the island. The richness of this small island in Antiquity and the flourishing there of a unique culture cannot be explained except in the light of the exportation of geothermal by-products: mainly obsidian and pozzolana.

Pozzolana was also exploited in ancient times at Nisyros. Here, moreover, the intense hydrothermal activity, which occurred in very recent geological times and still continues today, not only caused the formation of large phreatic-explosion craters and spectacular fumaroles but also resulted in the accumulation of significant deposits of native sulfur. These deposits were extensively exploited in protohistoric and ancient periods, and the product was exported as a raw material in many localities of the Mediterranean area for use in dye mixtures, skin ointments, and other pharmaceuticals. Due to the exploitation and trading of pozzolana and sulfur, Nisyros reached a high level of development in Antiquity; this is documented by numerous, beautiful archaeological findings, such as the statue to Poseidon and a great Pelasgic (pre-Greek) construction, remarkable for its dimensions and the quality of materials used.

Finally, one or more of the geothermal by-products (kaolin, silica, iron oxides, sulfur, perlite, and travertine) were also exploited in Antiquity on other volcanic islands of the Eastern Aegean (Chios, Lesbos, Samos, and Limnos), as well as at other localities of the Western Aegean and the Greek mainland [Sousaki (near Corinth), Aegina, Aedipsos, and Seriphos] where volcanic activity occurred or natural manifestations formed in relatively recent geological times.

Cited and Selected References

Bibliographical note: Readers interested in the geologic-archaeological studies conducted at Thera in the last 10 years may consult Baillie, 1990; Cioni et al., 1997; Doumas, 1990; Fytikas et al., 1995; Hammer et al., 1987; Hammer and Clausen, 1990; Johnsen et al., 1992; La March and Hirschboeck, 1994; McLelland et al., 1998; Vougioukalakis et al., 1994; Vougioukalakis et al., 1995, and Vougioukalakis, 1997.

Akylas, V. 1925. *The volcanoes and Thera Island, from older descriptions.* E. & I. Blazoudakis Edit., Athens. (In Greek).

Apollodorus of Athens. c. 150 B.C. On the gods, vol. 1. In *Apollodorus: The library.* Reprint 1921, J. G. Frazer, trans. The Loeb Series, London.

Aristophanes. c. 423 B.C. The clouds. In *Aristophanes, vol. 1.* Reprint 1950, B. B. Rogers, trans. Harvard University Press and W. Heinemann Ltd., London and Cambridge, Massachusetts.

Aristotle. c. 340 B.C. Meteorologica, book 2. In *Aristotle: Meteorologica.* Reprint 1952, H. D. P. Lee, trans. Harvard University Press and W. Heinemann Ltd., London and Cambridge, Massachusetts.

Aristotle. c. 350 B.C. De mirabilibus auscultationibus. In *The works of Aristotle, vol. 6.* Reprint 1961, L. D. Dowdall, trans. Oxford University Press, London.

Athenaeus. c. 193-197 A.D. Deipnosophistae, book 12. In *Athenaeus: The deipnosophists, vol 5.* Reprint 1955, C. B. Gulick, trans. Harvard University Press and W. Heinemann Ltd., London and Cambridge, Massachusetts.

Baillie, M. 1990. Irish tree rings and an event in 1628 B.C. In *Thera and the Aegean world III.* B.d. 3, D. A. Hardy, ed. The Thera Foundation, London.

Cataldi, R. and P. Chiellini. 1995. Geothermal energy in the Mediterranean area before the Middle Ages. *Proceedings of the World Geothermal Congress,* 373-380. (See also Chapter 11 in this volume).

Cioni, R., L. Gurioli, A. Sbrana, and G. Vougioukalakis. 1997. Precursory phenomena and destructive events related to the 1628 B.C. Minoan and 79 A.D. Plinian (Vesuvius) eruptions. Inferences from the stratigraphy in the archaeological areas. In *Volcanoes, earthquakes and archaelogy.* Geological Society of London (in press).

Dambergis, A. and T. Komninos. 1908. Curative waters. In *Pharmacography book.* P. Leoni, ed. Athens. (In Greek).

Demetrious of Callatis. 3rd century B.C. Fragments of the works. In *Fragmente der Griechischen historiker,* no. 85. F. Tacoby, ed.

Diodorus Siculus. c. 60-50 B.C. Biblioteca historica. In *The library of history, vol 1.* Reprint 1933-1967, C. H. Oldfather, trans. The Loeb Classical Library, London and New York.

Doumas, C. G. 1990. Archaeological observations at Akrotiri relating to the volcanic destruction. In *Thera and the Aegean world III.* B.d. 3, D. A. Hardy, ed. The Thera Foundation, London.

Fytikas, M. 1977. *Geology and geothermy of Milos Island.* Aristotle University of Thessaloniki, Greece. (Unpublished Ph.D. dissertation).

Fytikas, M., F. Innocenti, P. Manetti, R. Mazzuoli, A. Peccerillo, and L. Villari. 1995. Tertiary to Quaternary evolution of volcanism in the Aegean region. *Geological Society of London,* no. 17, 687-699.

Gorceix, H. 1874. Etude des fumerolles de Nisyros et de quelque-uns des eruptiòns, dont cette ile a eté le seige en 1872-73. *Ann. Chim. Phys.,* 5me Serie II, 333-354.

Hammer, C. U. and H. B. Clausen. 1990. The precision of ice-core dating. In *Thera and the Aegean world III.* B.d. 3, D. A. Hardy, ed. The Thera Foundation, London.

Hammer, C. U., H. B. Clausen, W. L. Friedrich, and H. Tauber. 1987. The Minoan eruption of Santorin in Greece, dated 1645 B.C.(?). *Nature Bd.* no. 328/6130, 517-519.

Herodotus. c. 430 B.C. Melpomene, book 4. In *Herodotus: The nine history books, vol 2.* Reprint 1971, A. D. Godley, trans. Harvard University Press and W. Heinemann Ltd., London and Cambridge, Massachusetts.

Hesiod. c. 780-760 B.C. Theogony. In *Hésiode.* Reprint 1972, P. Mazòn, trans. 8th ed. Le Belles Lettres Edit., Paris.

Hippocrates. c. 420 B.C. Airs, waters, lands, vol.1. In *Hippocrates, vol. 1.* Reprint 1957, W. H. S. Jones, trans. Harvard University Press and W. Heinemann Ltd., London and Cambridge, Massachusetts.

Homer. 7th century B.C. Rhapsody B. In *Homer: The Iliad.* Reprint 1954, A. T. Murray, trans. Harvard University Press and W. Heinemann Ltd., London and Cambridge, Massachusetts.

Johnsen, S. T., H. B. Clausen, W. Dansgaard, K. Furher, N. Gundestrup, C. V. Hammer, P. Iversen, J. Jouzel, B. Stauffer, and J. P. Steffensen. 1992. Irregular glacial interstadials in a New Greenland ice core. *Nature Bd.* no. 359, 311-313.

La March, V. C. and K. K. Hirschboeck. 1994. Frost rings in trees as records of major volcanic eruptions. *Nature Bd.* no. 307/1, 121-126.

Luce, J. V. 1969. *The end of Atlantis: New light on an old legend.* Thames and Hudson Ltd., London.

Marinatos, S. 1939. The volcanic destruction of Minoan Crete. *Antiquity,* no. 13, 425-439.

Marini, L., C. Principe, G. Chiodini, R. Cioni, M. Fytikas, and G. Marinelli. 1993. Hydrothermal eruptions of Nisyros (Dodecanese, Greece). Past events and present hazard. *Journal of Volcanology and Geothermal Research,* no. 56, 71-94.

McLelland, E., D. Kondopoulou, P. H. Sophos, and M. Westphal. 1998. Paleomagnetic estimation of emplacement temperature of Plinian airfalls. *Proceedings of the Second Workshop of European Laboratory Volcanoes; Santorin,* 125-135.

Mellaart, J. 1967. Çatal Hüyük: A Neolithic town in Anatolia. In *New aspects of antiquity.* M. Wheeler, ed. Thames and Hudson Ltd., The Camelot Press, London and Southampton.

NTO/National Touring Organisation. 1966. *Bathing sites and curative springs.* National Printing Office, Athens. (In Greek, with extended summaries in French and English).

Ovid, Publius Ovidius Naso. c. 1-16 A.D. Metamorphoses. In *The metamorphoses.* Reprint 1984, F. J. Miller, trans. 2nd ed. rev. The Loeb Classical Library, London and Cambridge, Massachusetts.

Papageorgiou, S., S. Stiros, and G. Hourmouziadis. 1990. Relations between natural environment: Geomorphological changes and history of the insettlement in ancient Thessaly. *Proceedings of the International Colloquy of Lyon, "La Thessalie: 15 ans de recherches archeol.: 1975-90,"* 21-28.

Pausanias. c. 161-176. Description of Greece. In *Pausanias: Description of Greece, vols. 1-4.* Reprint 1959-1961, W. H. S. Jones, trans. Harvard University Press and W. Heinemann Ltd., London and Cambridge, Massachusetts.

Pindar. 470 B.C. The odes. In *The odes of Pindar.* Reprint 1957, I. Sandys, trans. Harvard University Press and W. Heinemann Ltd., London and Cambridge, Massachusetts.

Plato. c. 395 B.C. Timaeus. In *Plato, vol. 7.* Reprint 1926, R. G. Bury, trans. Harvard University Press and W. Heinemann Ltd., London and New York.

Pliny the Elder. 77 A.D. Naturalis historia. In *Pliny, Natural history, vol. 10.* Reprint 1962, D. E. Eichholz, trans. The Loeb Classical Library, London and Cambridge, Massachusetts.

Plutarch. c. 105-115. Parallel lives, book 4. In *Plutarch's lives, vol. 4.* Reprint 1950, B. Perrin, trans. Harvard University Press and W. Heinemann Ltd., London and Cambridge, Massachusetts.

Seneca, Lucius Annaeus. c. 63 A.D. Naturales quaestiones. In *Seneca, vol. 10.* Reprint 1972, T. H. Corcoran, trans. The Loeb Classical Library, London and Cambridge, Massachusetts.

Shelford, P. 1982. The geology of Melos and the obsidian trade. In *An island polity: The archaeology of exploitation in Melos.* C. Renfrew and J. M. Wagstaff, eds. University Press, Cambridge.

Sophocles. c. 420-410 B.C. Trachiniae. In *Sophocles, vol. 2.* Reprint 1919, F. Storr, trans. The Loeb Classical Library, London and New York.

Strabo. c. 15 B.C.-7 A.D. The geography. In *The geography of Strabo, vols. 1 and 4.* Reprint 1949, 1954, H. L. Jones, trans. Harvard University Press and W. Heinemann Ltd., London and Cambridge, Massachusetts.

Theophanes the Confessor. c. 790-810 A.D. *The chronicles.* In *Theophanis chronographia, tom. 108.* Reprint 1975, J. P. Migne, trans. Turnhout, Belgium.

Theophrastus. c. 330-310 B.C. On the rocks. In *De lapidibus.* Reprint 1965, D. E. Eichholz, trans. Clarendon Press, Oxford.

Traineau, H. and P. Dalambakis. 1989. Mise en evidence d'une eruptiòn phreatìque sur l'ile de Milos (Greece). *Compt. Rend. Acad. Sci.,* 308, 247-252.

Victor, Sextus Aurelius. c. 360 A.D. *Historia Romana.* In F. R. Pichlmayr. Reprint 1961. Liepzig. (In Latin).

Virgil. c. 30-19 B.C. Aeneid. In *Virgil.* Reprint 1916, H. R. Fairclough, trans. The Loeb Classical Library, London and New York.

Vougioukalakis, G. 1997. The volcano of Santorin. *Korfes,* no. 124, 32-38. (In Greek).

Vougioukalakis, G., D. Mitropoulos, C. Perissoratis, A. Andricopoulos, and M. Fytikas. 1994. The submarine volcanic centre of Kolumbos, Santorin. *Proceedings of the 7th Congress of the Geological Society of Greece,* 30, no. 3, 351-360. (In Greek).

Vougioukalakis, G., M. Fytikas, P. Dalambakis, and N. Kolios. 1995. *Santorin: Guide to the volcano, Palea and Nea kameni. Volcanic activity in historic time.* ISMOSAV (Institute for the Study and Monitoring of the Santorin Volcano), Athens.

The Authors:

Michael Fytikas
Aristotle University of
Thessaloniki, Faculty of Sciences
54006 Thessaloniki, Greece
Telephone: 30.31.998.475
Fax: 30.31.998.482
E-mail: fytikas@geo.auth.gr

Georgia Margomenou Leonidopoulou
Arcadiou n. 8; Holargos
15562 Athens, Greece
Telephone: 30.1.651.4060

Raffaele Cataldi
55, Via del Borghetto
Pisa, Italy
Telephone/Fax: 39.050.59.81.07
E-mail: rafcat@tin.it

Roman thermal bath in Macedonia.

7. The Rolling Ball of History: Geothermal Resources in Macedonia

by
Kiril Popovski
Konstantin Dimitrov

Abstract: At the crossroads of eastern and western civilizations, Macedonia always has been a place where one can feel at home by discovering around oneself a great part of familiar culture. Geothermal resources are one such link. Rich in geothermal resources, Macedonia combined the old Greek and Roman use of natural hot water for public baths with the Turkish concept of "cleansing" the body and soul by using steam and hot water.

Introduction

GEOGRAPHICALLY, MACEDONIA IS SITUATED AT THE SOUTH-central part of the Balkan peninsula, which accounts for its historical and strategic importance as a crossroads for connections between Europe and the Near East and as an exit from the Central/Eastern European region to the warm waters of the Mediterranean Sea. It is the region that protected ancient Greece from northern "barbarian" invasions but also the region from which Greece finally was invaded. It was for centuries one of the most important trade and military connections for the Roman Empire (*Via Egnatia*) and, later, the Ottomans. Macedonia is also the route whereby Turkish troops tried to enter Central Europe and through which the Crusaders passed on their way to the Holy Land.

Traces of all these cultures can be found in Macedonia today, mixed and expressed in different ways. Alexander the Great invaded Greece, but at the same time he introduced Greek culture to Macedonia and paved the way for its permanent presence through religious and governmental dominance (the Eastern Orthodox Church and the Byzantine Empire). Roman culture is apparent even today, in the languages of the Albanians and Vlachos, spread along a part of the old *Via Egnatia*. Five centuries of Turkish government resulted in a great concentration of Moslems. Finally, the Macedonian nation still keeps a part of its original Slavic culture, brought 1,500 years ago from the northern part of the continent.

When speaking about geothermal resources and their rich and varied uses, one should not forget that geothermal development has been influenced by this cultural "mixture" characterizing the development of the original Macedonian culture. Rich in geothermal resources, Macedonia took the old Greek and Roman use of natural hot water for public baths and the Turkish pleasure of "cleansing" the body and soul by using steam and hot water. Many "miracle" springs can be found, offering successful medical treatment and "new youth" in the best Slavic Northern European tradition.

Unfortunately, there has been no systematic investigation of past development of geothermal resource usage. One can have only a rough orientation based on the archaeological record, rare written documents, oral tradition, and traditional local "know-how" for medical and heating purposes. This chapter presents a part of the information available, mainly collected by geologists and hydrologists at geothermal locations.

Background

Macedonian territory is characterized by a very complex geological-tectonic structure. Although the Earth's crust can be up to 40 km thick in Macedonia, some areas are very promising for the occurrence of high-enthalpy geothermal resources. This is due to the emplacement of numerous granite intrusions which, during their crystallization, supplied the surrounding rocks with enormous amounts of heat. Numerous surface manifestations can be found all around the country; nearly every village has some mineral sources.

About 200 natural mineral and thermo-mineral springs have been identified within the Republic of Macedonia (Kotevski, 1987), and the situation is similar in locations which now belong to Greece and Bulgaria. Most of the springs have been known and used for ages, mainly for health, but also for agricultural production, industry, and other heating purposes.

Four general geothermal zones can be identified: the Bosnian-Serbian-Macedonian, Pannonian, Alpine, and Dinaride. These can be subdivided into areas, each with its own particular characteristic and potential. From the geotectonic point of view, the Bosnian-Serbian-Macedonian geothermal area is situated in the internal Dinarides, the Serbian-Macedonian massif, and the Carpato-Balkanides. Plutonic and volcanic rocks, Neogene in age, are most common; thus, realistic possibilities exist for the presence of hydrogeothermal and petro-geothermal energy. Although numerous surface occurrences are known, closer definition of the geothermal resources is rather difficult and complex due to fractures, joints, and cave systems in Paleozoic and Mesozoic rocks.

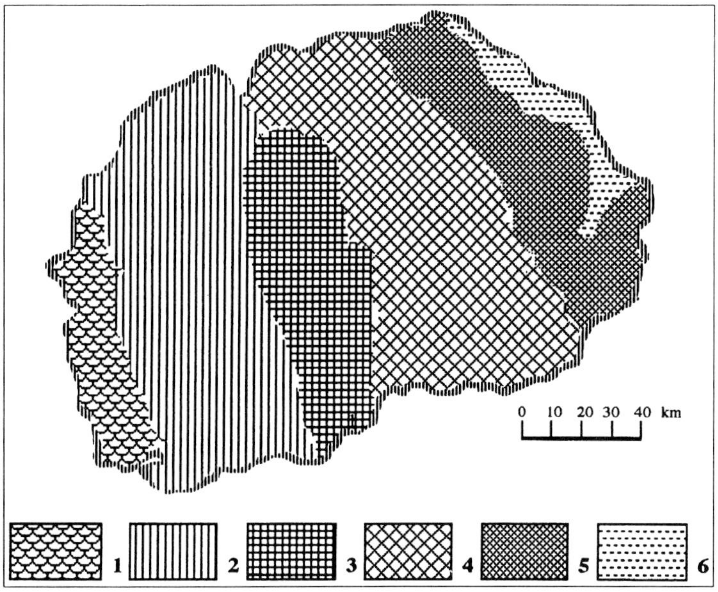

The geology of Macedonia, according to G. Kotevski, 1993: (1) Corabides; (2) West Macedonian zone; (3) Pelagonidic horst-anticlinorium; (4) Vardar zone; (5) Serbo-Macedonian massif; (6) Struma zone.

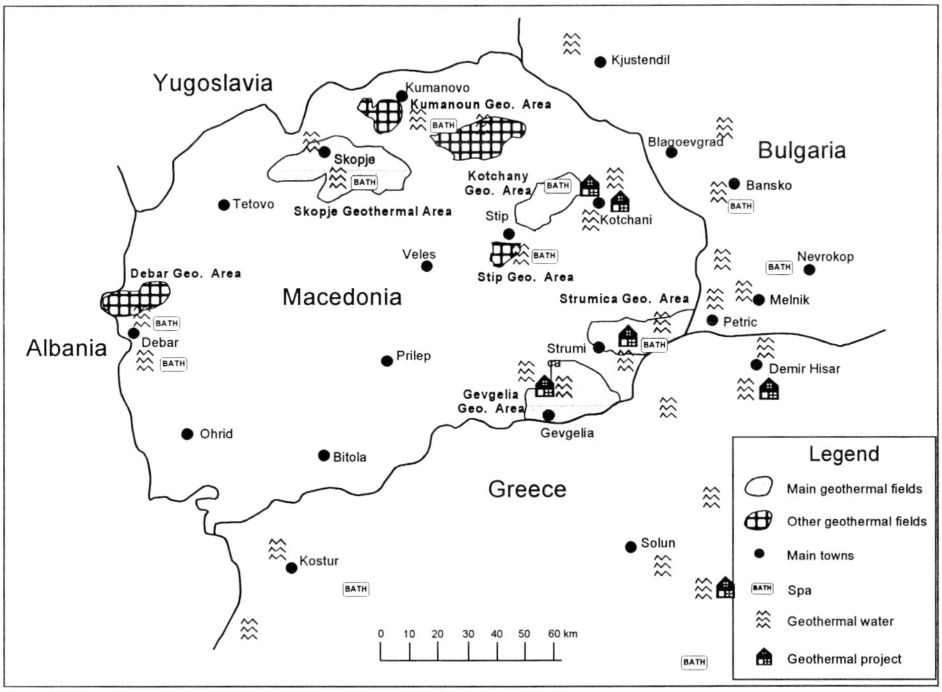

Geothermal locations in Macedonia and surrounding countries.

The seven geothermal fields found in Eastern and Northeastern Macedonia are grouped according to geotectonic divisions. The eastern and northeastern areas are part of the Macedonian-Serbian massif characterized by crystalline basement rocks. These areas are much richer than the western and southwestern regions, which are characterized by limestone.

History

There has been no systematic investigation of the history of geothermal resource use and development in Macedonia. Rough data from written documents are available, but what is known comes mostly from the oral traditions of local people and archaeological sites. The last scientist who collected data about previous uses of Macedonian thermal waters was Dr. Georgi Kotevski. His doctoral work on the hydrogeology of mineral and thermo-mineral waters in Macedonia included a list of natural sources with their characteristics and data about previous and present uses (1987). G. Velenis gave some information in his 1988 paper on the history of thermal baths in Greece, and Dr. Shterev investigated geothermal resources in the Bulgarian part of Macedonia and published information about those in the eastern part of the Republic of Macedonia (1992). Popovski published useful information, mainly on new applications (1987).

It is necessary to emphasize that all of the information about geothermal applications in older historical periods are interpretations and suppositions based on the oral traditions of local people, in combination with archaeological sites (ruins, old baths, pictures, and mosaics). Further investigation can give more precise information on specific sites and applications and can correct some of the conclusions, but the main concepts will remain.

Eight historical periods of geothermal application in the Republic of Macedonia can be identified. They are discussed in the following sections.

Ancient Macedonian and Greek Period (until 197 B.C.)

While there is no proof about geothermal uses during this period, after the introduction of Greek culture to Macedonia, a number of elite people (Aristotle, for example), as well as statesmen and traders, came to the region, bringing with them their Greek customs. Social customs included hot baths, but it's possible that Macedonians used naturally warm and hot flows before the arrival of the Greeks, since the waters are located near towns.

Roman Period (200 B.C.-395 A.D.)

Romans and their "thermae" are known as the real founders of the public use of thermal waters. Therefore, it's not strange that in Macedonia they used available hot resources, particularly during the colder months. Ruins of an old Roman bath can be found near Banja Bansko (Kotchany), old Roman coins in the ruins of the old bath in the Katlanovo Spa (Skopje), and some indications of possible applications at Bansko (Strumica).

Byzantine Empire (395-1340)

The East Roman Empire, later known as the Byzantine, followed the state organization and tradition of the Roman Empire but also was influenced by the cultures of the different nations composing it. While there is no direct evidence, during this period the use of thermal baths apparently slowly disappeared from Macedonia. The Christian religion looked unkindly on the exposure of the human body, especially to the eyes of the opposite sex or children. Also "internal" hygiene became more important than "external."

Ottoman Empire (1340-1918)

When coming to the Balkan peninsula, the Turks brought a strong military organization and a new approach to hygiene. While all Moslems had to be clean before praying in the mosque and it was a national tradition to take baths whenever possible for cleanliness, the Turks also took pleasure in being refreshed after taking a hot bath and perspiring in steam. Macedonia is littered with ruins of old Turkish thermal baths, like those in Bansko (Strumica), Gabrovo (Deltchevo), Negorci Spa (Gevgelija), Kezhovica (Shtip), Kosovrasti, and Debar Spa (Debar).

Turkish bath at Strumica, Republic of Macedonia. *J. Lund*

The Kingdom of Serbs, Croats, and Slovenians (1918-1941)

After the Balkan and First World Wars, the territory of Macedonia was divided among the four new states (Albania, Bulgaria, Greece, and SHS). The country was destroyed and major population movements resulted in a changed national structure and, therefore, a changed way of life. Many of the Turkish thermal baths were abandoned, but some were reconstructed (Katlanovo, 1930) and new ones arose (Debar, 1925). However, on the average, geothermal baths decreased in comparison with previous periods. It's necessary to stress that this was also the period when the first serious geological and geothermal investigations were conducted by F. Kosmat (1924); K. Osswald, A. Burm, and Ermansdorfer (1924-25); Cripp (1921); and others.

Federative Republic of Yugoslavia (1945-1963)

After World War II when the Federative Republic of Yugoslavia was established, communist support for all matters involving "public" interest assumed principal importance, stimulating the development of thermal baths for medical and tourism purposes.

Katlanovo Spa, Negorci Spa, Kezhovica, and Debar Spa were modernized, and the first geothermal greenhouse in Bansko (1976) was put into operation. Intensive geological and geothermal investigations resulted in the identification of most of the geothermal fields known today. Vidakovich, Popovich, and Sibinovich (1953); S. Miholich (1955); T. Mitrov, D. Gjuzelkovski, N. Izmailov, and K. Jenko (1955-56); A. Kekich (1961); V. Cirich (1971); G. Kotevski (1971-79); and others started developing these resources in the Republic of Macedonia.

Socialist Republic of Yugoslavia (1963-1991)

A change in the economy after 1963 heavily influenced the development of geothermal resources. The first really modern spa was built in Banja Bansko (Strumica), equipped with geothermally heated rooms, halls, and a swimming pool, as well as facilities for medical treatments with geothermal water. The energy crisis of the 1970s stimulated development of geothermally heated greenhouses and, during that time, Macedonia led the world with this type of geothermal application.

Also during this period, the Negorci Spa was equipped with geothermally heated rooms, and development began of the Kotchany geothermal district heating system. Scientists and engineers like Kotevski, Andrejevski, Micevski, Tcherepnalkovski, Neunov, Gashteovski, Dimitrov, and Popovski are firmly connected with the "flowering" of geothermal development in Macedonia. In 1989, Popovski established the International Summer School on Direct Application of Geothermal Energy, in collaboration with colleagues from Italy (Fanelli, Dickson), Greece (the Martzopoulouses), Macedonia (Dimitrov), and later on from France (Jaudin), Bulgaria, and Romania.

Republic of Macedonia (1992-present)

The period of recent development has been very short. After the end of the "energy crisis" and the decline of Yugoslavia, the Greek embargo, and problems of political and economic transition, it stopped completely. The geothermal project in Vinica has been abandoned and those in Gevgelia, Bansko, and Kotchany are stagnant. Some small activities by the enthusiasts in Kotchany and Gevgelia are apparent, but those cannot change the picture of total stagnation. This situation can change only after the period of privatization and economic transition ends, since geothermal resource application is not a principal interest of a country faced with very complex political and economic situations.

Application Technologies

As previously noted, there is no information about geothermal use during ancient Macedonian times. If one assumes that the use itself came with the Greeks, open air natural pools or pools made by partitioning natural warm water flows can be a realistic supposition.

For the Romans, however, a realistic picture can be reconstructed. Open air pools were augmented with stone or brick side walls and bottoms in the first phase of development. Then, a bathhouse was added that later became a pleasant place for conversing after a bath, from benches lining the inside walls. The last development was the introduction of floor heating.

Effluent water from the pool, except that directly released to the environment, flowed below the floor surface through a system of ceramic tubes or brick channels made of stone cubes or bricks. In that way, the bathhouse, originally intended as a cloak room, became a social gathering spot. Many windows ensured continual fresh air flow, which was very pleasant during the hot summers. The hot floor allowed the room temperature to be increased in colder weather. In the best Mediterranean tradition, it was normal that massage, refreshments, and news were available, and that was the reason why the thermae were among the most popular public places in town.

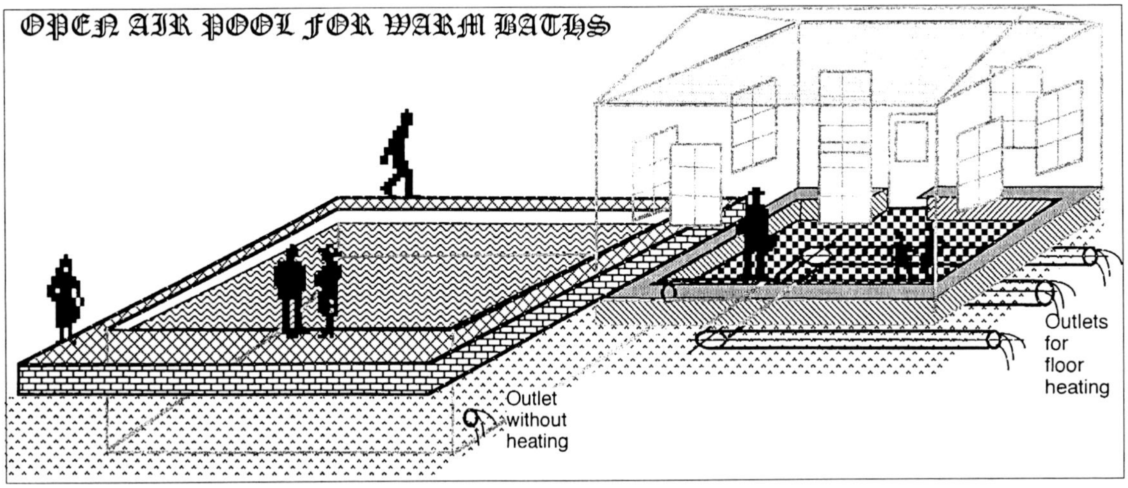

Roman baths in Macedonia include a geothermally heated dressing room.

Unfortunately, the social aspect of the Roman baths was also the reason for their downfall. Very soon after the introduction of Christianity, thermae were identified as decadent. Naked or half naked men and women together in a free atmosphere was opposite from the way of life condoned by the new religion, and slowly, thermal baths were abandoned during the end of the Roman era and the beginning of the Byzantine.

During the Byzantine era, another use for mineral and thermo-mineral waters became popular. Many of them were considered magic, falling under the protection of a saint because they helped to cure many diseases. The Orthodox Church supported this development by establishing monasteries and hospitals near such "miracle" sources. Traces of such treatments can be found even today. For instance, the name of the spring *Sabota voda* means that the water is curative only on Saturdays.

The situation changed when Turks came to Macedonia, bringing another use for water in general and hot water in particular. In addition to continuing the Roman tradition of taking baths in warm and hot water, they also cleansed the body by steaming. That was the background of the characteristic construction and composition of the Turkish baths: a shallow pool with warm or hot water in a completely closed room, full of steam, with very small openings on the top of the oval roof. Hot water served as both the source of the steam and the heating system. A bathhouse, positioned beside the pool, was used for clothing changes, and the pool room itself became the social room. For a small amount of money, one could be washed by an attendant, be offered sweats and cold water, and have a pleasant chat with friends, all at the same time. It's not strange that every town in Macedonia had a Turkish bath, and, where thermal water was available, they were using it. Beautiful examples of Turkish baths in Skopje and other towns reflect the original organization, shape, and construction details.

Baths from the first half of the 20th century have added nothing new to the technology of geothermal water application. They are typically shallow pools, in a closed room, made of cement covered with ceramic plates. Hot water is also used for special medical treatments (baths in mud or bathtubs). Older spas are not equipped with central heating; they use just the heat of the warm water, supplemented by wood stoves for colder days.

Old Turkish geothermal bath.

The first complete geothermal project is the Hotel Tsar Samuil in Bansko, Strumica, which provides nearly all of the hotel's heating needs. The heating system consists of radiators in the rooms and air heaters in the hallways. Heat from geothermal water is extracted with a plate heat exchanger, and a second heat exchanger supplies heat for sanitary warm water and hot water for the kitchen. Used water from both heat exchangers goes to the swimming pool and for medical treatments. Finally, the effluent water from the system is used by farmers to heat greenhouses. On very cold days, a pickup boiler augments the geothermal heat. This system is part of a larger geothermal system, which also heats a greenhouse complex.

Two applications of this system are of particular importance: first, small farmers knew immediately how to use the hotel's effluent water to heat the floors of small greenhouses; secondly, the first totally geothermally heated greenhouse in the world was completed here. The first application illustrates the knowledge of old technology (Roman and Turkish baths), and the second represents a pioneering approach to geothermal resources as "normal" energy resources. That is why the Bansko integrated geothermal system has historical significance, not only for Macedonia but beyond.

Finally, and probably the most important illustration of technological "know-how" for direct geothermal application in Macedonia, is the integrated geothermal district heating system in Kotchany. It is still under development but can be used for a variety of purposes: agricultural (heating greenhouses and drying rice), heating dwellings and preparing sanitary warm water, and industrial (the paper industry). Some of the applied technical solutions are very sophisticated (geothermal-water distribution system without compensators for thermal expansion and contraction over a 6 km length), some are very old fashioned (heating systems in greenhouses), and some are simplistic (projects with existing heating installations connected to the geothermal district heating system). But altogether, it's one of the largest geothermal systems in the world.

Conclusions

An old country like Macedonia has an old history of geothermal applications. However, little concrete evidence exists for us to know exactly when and how geothermal resources were developed. The reason is that virtually no serious investigations have been conducted to clarify this part of Macedonian history.

Archaeological sites and local oral traditions indicate a rich but very changeable history and development of this resource. The history is characterized by flourishing periods lasting for centuries but also deep recessions lasting for many, many years. Different civilizations passed through the country, and each one left traces of a specific approach to this "unimportant" part of life. Ancient Macedonians, under the influence of the Greeks and Romans, adored life and the human body, and it was natural that they discovered and used the benefits of natural warm and hot water resources, enabling pleasant baths and relaxation. The rigors of early Christianity ran contrary to ancient practices and nearly destroyed previous social customs.

However, history is like a rolling ball. The Turks succeeded the Byzantine Empire, and they again reaped the benefits of the Earth's fire. The civilization of today is both built on and enriching the technology of the past.

Selected References

Kotevski, G. 1987. *Hydrogeology of the mineral, thermal and thermo-mineral waters of the territory of the Socialist Republic of Macedonia, self-governing practice.* Skopje. (In Macedonian).

Popovski, K., ed. 1995. *Geothermal energy: State-of-the-art and perspectives in the Republic of Macedonia.* (In Macedonian).

Popovski, K. and T. Novkovski. 1987. The Republic of Macedonia: An area of high potential for low-enthalpy geothermal energy introduction in agriculture and other general uses. *Geothermie Actualite,* 4, no. 4. Orleans.

Velenis, G., E. Vori, E. Gavra, K. Giftopulu, K. Thermos. 1998. Historical materials for interesting thermal spas and monuments in Greece. *Symposium on Thermo-mineral Waters,* 484-500. Thessaloniki. (In Greek).

The Authors:

Kiril Popovski
St. Clement of Ohrid University—Bitola
Faculty of Technical Sciences
Ivo Ribar Lola, bb
97000 Bitola, Republic of Macedonia
Telephone/Fax: 389.91.119.686

Konstantin Dimitrov
Sts. Cyril & Methodius University—Skopje
Faculty of Mechanical Engineering
Naselba Karpos 11, bb
91 000 Skopje, Republic of Macedonia
Telephone: 389.91.363.566
Fax: 389.91.362.298

Hydrothermal minerals (kaolin, sulfur, alunite, and iron oxides) are found on the islands of the Aeolian archipelago, Italy. Many manifestations of this type formed in Southern and Central Italy during recent geological times. Their products were used in prehistoric epochs for a number of purposes. ***Prof. A. Sbrana, University of Pisa, Department of Earth Sciences***

8. Human Rapport with Hypogeal Waters and Geothermal Manifestations in Italian Prehistory

by
Renata Grifoni Cremonesi

Hypogeal waters: cavern waters flowing, dripping, or in lakes; some held ritual appeal for prehistoric peoples.

Abstract: Humans established relationships with hypogeal waters and geothermal manifestations during the Neolithic and Metal Ages in Italy, including cults and therapeutic uses. Some geothermal by-products were used as well: obsidian for tools and weapons, volcanic rocks for millstones, pumice and fine-grained pyroclastics for tempering and grinding materials for ceramic finishes, and sulfur for pigment and ointment mixtures.

A General Look at the Problem

The study of the relationships formed in Italy between prehistoric people and various types of underground waters, particularly thermal waters, depends on the availability of data. In certain instances, significant evidence can be found on cultic practices related to lake, river, and superficial spring waters, as well as to hypogeal waters: cavern waters flowing, dripping, or in lakes. In other instances, solid evidence is lacking, and one must hypothesize about the existence of cultic practices related to hypogeal waters and about the use of thermal waters for therapeutic purposes. Localities with natural manifestations may have direct or indirect indications to suggest that these manifestations were frequented by people for some purpose; however, it is hard to establish if people visited these sites for cultic and/ or therapeutic practices. It is much easier, of course, to identify and study volcanic and hydrothermal products used by the early Italian populations to form artifacts and tools for their daily activities. These products are discussed in the last section of the chapter.

Concerning cults, an intrinsic difficulty lies in defining the concept and in the meaning that cultic phenomena may have held for prehistoric and ancient peoples in different regions of the Mediterranean. Too many times such a delicate subject has been approached with different methodologies, and the term "cultic object" itself has been applied to any type of prehistoric finding or archaeological complex that cannot be explained in terms of functional use (Grifoni Cremonesi, 1994; Di Fraia and Grifoni Cremonesi, 1996).

In Italy, prehistoric artifacts and ruins have been found that are certainly related to the spiritual sphere and the ritual practices of the people, especially for the Neolithic and the Metal Ages: they are mainly circular pits dug on the floors of some caverns and stone circles laid down on the floors of other caverns. Vases or other votive objects were methodically deposited inside the pits and circles. These circular features and the objects inside them probably represent symbols for funerary rituals.

Upside-down vases (perhaps the position represented a way people could connect with the dead and with the chthonian world of dead and underground divinities), pieces of broken vases ("ritual fragmentation") accurately piled up, or piles of vases were also found in some caves; these findings, too, are interpreted in terms of funerary rituals (Radmilli, 1975; Cremonesi, 1976 and 1985-90; Grifoni Cremonesi, 1986 and 1994; Di Fraia and Grifoni Cremonesi, 1996).

Less frequent are Neolithic remnants of true monuments, among which two "altars" of stone drywalls in Apulia, Southeastern Italy, are worthy of mention. For the Copper and Bronze Ages, more evolved and complex remnants of hypogeal and megalithic monuments exist. Many of these are difficult to interpret: they are generally related to agricultural rituals, but they could also refer to both agricultural and funerary rituals. In this latter case, the cult probably aimed at linking the most significant moments of human life: birth, growth, and death.

A different type of cultic phenomenon is related to water and water circulation: cold or warm springs emerging in underground cavities frequented by people, natural manifestations with fluids at more or less anomalous temperatures, hot and ambient-temperature springs outflowing at the surface. Water-related cults did not involve just water; they also involved some of the effects produced by water circulation, such as "solid water" (stalactites and stalagmites), bubbles of gas released in certain water pools, waterfalls and cascades inside caves, and pools or pits formed at the bases of waterfalls (Whitehouse, 1990 and 1992; Di Fraia and Grifoni Cremonesi, 1996). Other phenomena associated with water circulation, such as those resulting from hydrothermal circulation (travertine, chalcedony, and sulfur), relate to functional uses.

The symbolic value of water has been universally recognized and permeates on a global scale the cultural background of religions and popular traditions. Water is considered the origin and blood of any form of life, a symbol of rebirth and regeneration, a fertility element, a magical fluid, and the therapeutic substance *par excellence* (Eliade, 1966). Cults devoted to springs, rivers, lakes, and thermal sources are almost universal, and their persistence is reflected in the traditions and cultures of most peoples on Earth. Writing on this subject is substantial. Concerning Christianity, forms of cults for saints directly or indirectly related to water date back many centuries and still exist today.

Prehistoric Cults and Ritual Practices Related to Hypogeal Waters

Without being too detailed, this section briefly reviews the information available for Italy on prehistoric cults. Regarding the Paleolithic, Bahn (1978) has suggested a connection between some artistic expressions of the Upper Paleolithic and the presence of water in caves. However, no concrete evidence as yet exists for this period.

Evidence for the Neolithic

Abundant and reliable data are available for the Neolithic, beginning in Italy about 5200 B.C. This period is characterized by the arrival, from the Near Middle East to the Italian peninsula, of animal husbandry and agriculture that small communities in permanent settlements began to practice. New life styles and ways of thinking, also originating in the Near Middle East, started taking root among the Italian peoples in the same period. According to opinions of scholars in this field, organized forms of religion began, and cultic rituals, mostly related to agrarian activities, started gradually to take shape and consolidate during the Neolithic. These religions and practices focused on the figure of a female divinity called the "Mother Goddess" (Cauvin, 1994).

All evidence gathered comes from caves or branches of caves, which now, and presumably during the Neolithic, are difficult to access. Unsuitable for frequent visits, these caves probably were used as places for ritual and/or funerary practices or as cultic sites. One must remember that during the Neolithic, caves were no longer regarded as temporary shelters for individuals or small communities, as they were during the Paleolithic and Mesolithic. It is no wonder, then, that the most inaccessible caves were frequented only occasionally for ritual, cultic, and/or funerary practices.

The function of caves as sacred and mysterious places has been extensively debated to try to understand their use in prehistoric, ancient, and even recent times. Relationships have been proved or hypothesized with underworld entities, with the goddess "Mother Earth," with esoteric and initiation rituals, and with incubation practices. For relatively recent times, psycho-analytic implications have been suggested for people frequenting underground or obscure recesses.

All of these relationships are basically the result of the mysterious atmosphere and the sense of fear aroused in people's imaginations by deep caves and lack of light, even today. Mention should also be made of cults for saints, such as St. Michael, related somehow to the under-ground world, cults for "miraculous" waters flowing from natural cavities, and some popular beliefs regarding imaginary beings who guarded fabulous treasures in underground recesses (Di Fraia and Grifoni Cremonesi, 1996; Grifoni Cremonesi, 1996).

Italian caves with Neolithic evidence of cultic practices related to hypogeal waters are Grotta Scaloria Bassa, Apulia; Grotta dei Meri and Pozzi della Piana, Latium; Grotta Zinzulusa, Apulia; and, probably, Grotta Verde, Sardinia.

Grotta Scaloria Bassa. Discovered in 1967, this cave is the lower chamber of the Grotta Scaloria proper. It is a wide, concretion-rich chamber, with a small lake at the bottom. Inside the chamber, Neolithic people placed vases around stalagmites and upon the trunks of stalag-mites intentionally truncated; they also dug a rectangular pit (90 cm x 50 cm) to a depth of 15 cm on the calcareous floor of this chamber. Near the pit, a human skull and other human bones were found in a small lateral branch of the chamber, and near this branch was a hearth with charred animal bones.

Some 60 vases attributed to a culture named "painted pottery with red strips," the predominant culture in Southern Italy during the first half of the 4th millennium B.C., were also discovered in

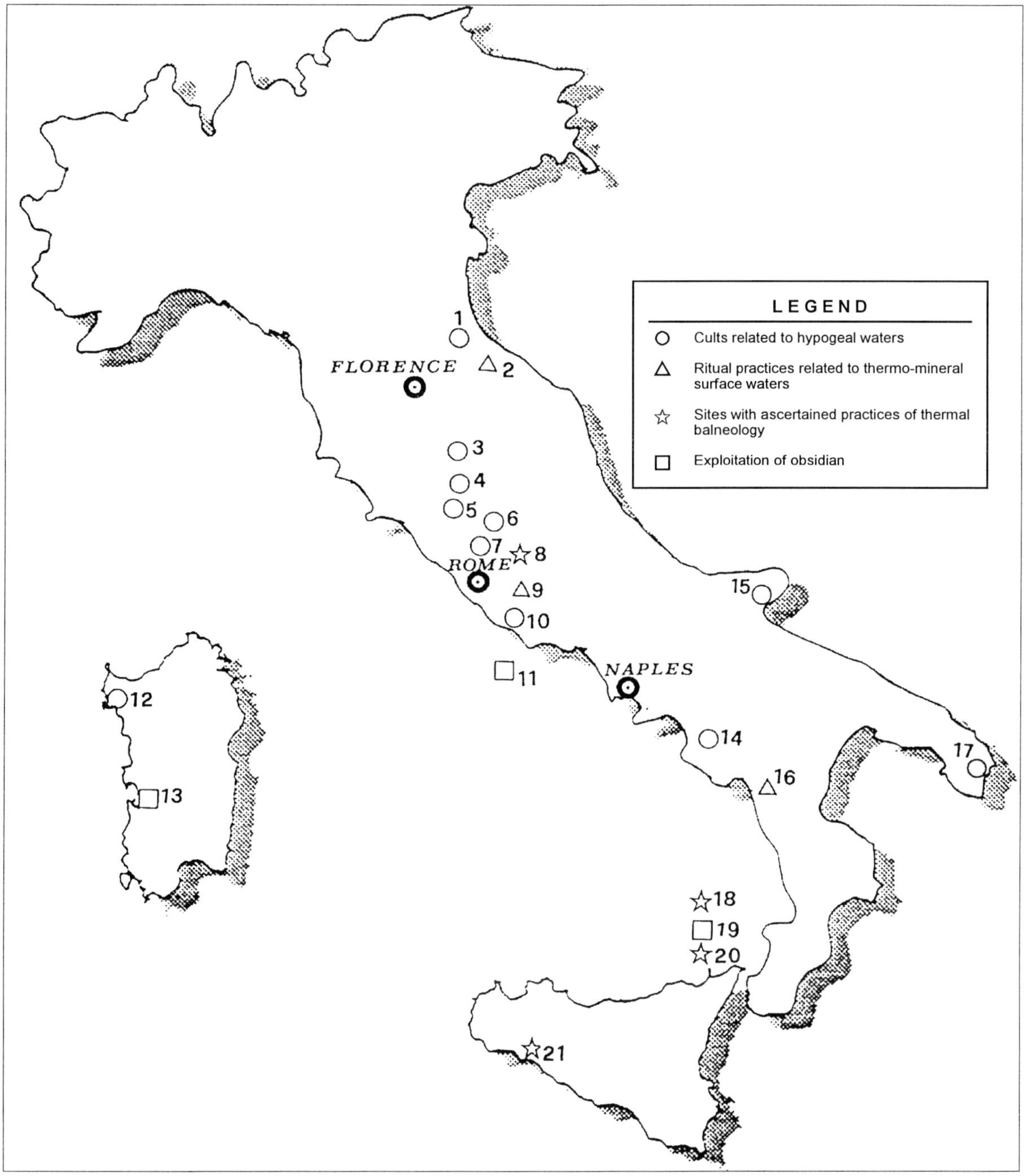

Map of main archaeological sites with ascertained relations between prehistoric people and hypogeal waters or natural manifestations.

(1) Grotta di Re Tiberio, Iron Age; (2) Pozzo della Panighina, Copper Age; (3) Buca del Rospo, Copper Age, and Grotta dell'Orso, Bronze Age; (4) Pozzi della Piana, Neolithic; (5) Grotta Nuova and Grotta Misa, Bronze Age; (6) Caverna dell'Acqua, Bronze Age to Roman times; (7) Grotta dei Meri, Neolithic; (8) Grotta dello Sventatoio, Bronze Age;(9) Lago delle Colonnelle, Bronze Age to Roman times; (10) Lago del Monsignore, Bronze Age to Roman times; (11) Palmarola, Neolithic; (12) Grotta Verde, Neolithic; (13) Monte Arci, Neolithic; (14) Grotta Pertosa, Bronze Age; (15) Grotta Scaloria, Neolithic; (16) Grotte di Latronico, Bronze Age; (17) Grotta Zinzulusa, Neolithic; (18) Panarea, Bronze Age; (19) Lipari, Neolithic; (20) Tholos di San Calogero, Bronze Age; (21) Stufe di San Calogero, Copper Age.

this cave. All vases are painted with large red stripes, edged with narrow brown stripes, and made with a particular decoration technique called "in negative." They are especially valuable compared to the pottery normally produced in Southern Italy at that time. The vases are mainly shaped as bowls, cups, jars of different sizes, jugs, and pots, with subordinate flasks; they were all used to collect dripping water (Tiné, 1975; Tiné and Isetti, 1975-1980).

However, since most of the vases have an open shape, and considering the cave's very difficult access, it is hard to imagine that they were used to transport water outside of the cave; therefore, the water must have been collected for use inside the cave. Furthermore, the locations of vases on the truncated stalagmites suggest that the true interest of the Neolithic people lay in collecting the water dripping from the ceiling or from the stalactites. The fragments of the truncated stalagmites also may have had a cultic function. The subject of the cultic function of mineral and calcareous concretions is widely debated. Important examples of the practice were found by Mellaart (1967) in the Neolithic village of Çatal Hüyük in Western Anatolia, Turkey. Here in an area labeled VI.B.44, fragments of stalactites have been found together with stone statuettes and offerings of cereals. A concretion has been unearthed in another area with one extremity carved in the shape of a human head.

While it is impossible to establish with certainty whether the water in Grotta Scaloria Bassa was collected for cultic, magical, and/or therapeutic purposes, many European caves have numerous stalactites and carbonate concretions from which $CaCO_3$-saturated, opalescent water slowly drips and forms mammary-shaped concretions. Since the color of the whitish waterdrops resembles milk, these concretions are often referred to as *pocce lattaie* ("nursing mammaries"). Because of the resemblance to milk, milkless women used to drink it, frequently in the past and still today in some rural areas (Corrain et al., 1967). Indeed, until a few decades ago, devotees of the cult of St. Michael gathered the milky waters from caves because the waters were believed to have therapeutic properties and were reputed to cure some stomach disorders.

Considering that these popular beliefs and practices have persisted until the present, it is very easy to argue that such waters attracted the interest of prehistoric people for cultic and/or magical and therapeutic applications. Significantly, Neolithic people did not place water-gathering vases by the small lake at the bottom of the Scaloria Bassa cave. Many such vases were put only at the stalactitic-stalagmitic complex. Apparently, Neolithic people visited the cave exclusively to collect the water dripping from the stalactites or the ceiling and not for the water from the small lake.

Grotta dei Meri and Pozzi della Piana. Vases in the Grotta dei Meri and the Pozzi della Piana date from the same period of those of Scaloria Bassa, but they belong to a different culture, known as "pottery with engraved lines," which developed in Latium and Tuscany. Both caverns are large, characterized by very difficult accesses and long labyrinthine passages through winding areas.

In the Grotta dei Meri, a jar was found near the final part of the cave in a place where several narrow passages merge. The jar was located beneath a continual slow drip, which is rather difficult to reach, suggesting that the water was not collected for practical uses.

In the Pozzi della Piana cave, which develops over three levels at different elevations, Passeri (1970) has documented a more complex situation. Stone and pottery objects were located on the various chamber floors, inside small cavities formed around fracture planes, and on some of the rock wall overhangs. In particular, a number of inverted cups 5-7 cm high were deposited carefully inside fracture cavities around a natural, relatively large water pool surrounded by stalagmites and stalactites. Flask-shaped vases were also left in a small natural pool formed under concretionary overhangs. The location and position of these artifacts indicate a ritual or cultic purpose.

Other evidence of cultic practices includes a circular pit (40 cm in diameter and 50 cm deep) dug in the calcareous floor of the cave and filled with ochre clods. A few small millstones were also found in the cave. The presence of these millstones—some left on the ground, others deposited inside pits of caves used for burials—has been highlighted by Di Fraia and Grifoni Cremonesi (1996).

Grotta Zinzulusa. The finds at Grotta Zinzulusa date back to the Late Apulian Neolithic, about the end of the 4th/beginning of the 3rd millennium B. C. This concretionary-rich cave yielded 11 vases, including bowls and flasks; they were all placed along the shores of a small lake located near the most distant and darkest part of the cavern, many hundreds of meters from the entrance (Zezza, 1984). The development of the cave is controlled by a set of subvertical faults, but its path is winding, with an up-and-down labyrinthine course, starting from a rather difficult access now located a few meters above sea level.

Grotta Verde. The access for Grotta Verde is at 75 m, but the cavity quickly deepens down a 45° slope to a small lake of brackish water at sea level. Near the lake, a hole 6 meters deep leads to a small stalagmitic-rich site where human bones were found. From here, a very narrow passage leads to a large flooded chamber at 10 meters below sea level. Marine water is at the bottom, with a 2-6 m thick layer of fresh water floating on it. Many vases, "impressed" pottery, were deposited in the access passage and in the chamber; most are flasks dating to the Lower Neolithic (6th millennium B.C.). Cave visitations continued throughout the Neolithic (Tanda, 1978; Lo Schiavo, 1987).

It is hard to determine whether or not the vases were used to collect drinking water or for some ritual practice; however, considering the extremely difficult access to the chamber, the latter is the most reliable hypothesis.

During the Neolithic, the deposition of vases close to water pools or below dripping water occurred in Italy solely in caves with very difficult access and a labyrinthine development. In many instances, the vases were particularly valuable. All this indicates that the water was not collected simply for daily use but for ritual practices. However, no human remains have been found nor offerings of cereals or other objects, as occurs with ritual forms known for Neolithic sites outside caves, with the exception of Pozzi della Piana. Here, vases filled with wheat were placed along the shores of a small underground stream and the situation resembles that of the Trou Arnaud cave in France, where Late Neolithic vases were found filled with wheat (Blanc, 1976).

Evidence for the Metal Ages

Gathering dripping water in caverns is a custom which continued in Italy until the end of the Copper/beginning of the Bronze Age, as is indicated by the findings in the Buca del Rospo ("Toad's Hole") cave, located on the slopes of Mt. Cetona in Tuscany. Here, two vases were discovered in an area of sustained dripping water, partially embedded inside a stalagmitic concretion (Zanini, 1988). The site is at the very end of the cave, beyond a deep hole that makes reaching it extremely difficult. This situation differs greatly from that of French caves where the vases deposited during the Copper Age were mostly large with large orifices, thought to have been used to collect fresh water for daily use.

In the Bronze Age during the 2nd millennium B.C., the tradition of placing vases below dripping water in Italian caves persisted, but new forms of vase deposition began, with a tendency to favor running waters in caves with easy access and waters at the surface in rivers and lakes.

During the Middle Bronze Age (16th-14th centuries B.C.), vase deposition became frequent along underground streams or in the vicinity of subterranean springs. These vases often contained cereals, seeds, and fruits; other times vegetables were left in small heaps on the ground. In some locations, human bones were placed near the vases. The following are some significant examples.

Grotta Nuova. In the Grotta Nuova cave, Latium, vases were deposited inside the small stream flowing at its bottom. Some of the vases contained wheat and flat beans, and others were placed in an overturned position (Poggiani Keller, 1978; Negroni Catacchio et al., 1989-90).

Grotta Misa. Located in the same area as Grotta Nuova, Grotta Misa has a stream at its bottom, along which many vases and some ravaged burial sites were found. A hearth was also preserved with ashes accurately positioned to form a ring, inside which small heaps of wheat, millet, beans, and flour were carefully arranged (Negroni Catacchio et al., 1989-90).

Grotta dell'Orso. In the Grotta dell'Orso ("Bear Cave"), Tuscany, in the same territory of Buca del Rospo, a number of Bronze Age vases with wheat offerings were found in various chambers of the large cave (Grifoni, 1967; Cremonesi, 1968).

Caverna dell'Acqua. Inside the Caverna dell'Acqua (also called "Cave of the Votive Offerings"), Latium, a spring exists that was used for cultic purposes until Roman times. The most ancient findings in this cave date to the Middle Bronze Age (Rellini, 1920; Guidi, 1991-92).

Grotta Pertosa. The cave with the most abundant and important findings of the Bronze Age is Grotta Pertosa, Campania. It is very large, rich in concretions, and crossed by a stream originating from springs and waterfalls located at the very end of the cave. During the Middle Bronze Age, two pile-dwellings were constructed inside, and many vases were deposited along the shores of the subterranean watercourse. In addition, 324 miniaturized vases were deposited, methodically arranged, and piled up in a narrow branch of the cave near its innermost part; they are interpreted as important votive offerings (Carucci, 1907; Trucco, 1991-92). Another smaller votive offering of the same period was left near the entrance of the cave, on a small overhang of the rock wall on the left riverbank.

Grotta di Re Tiberio. More than 600 Iron Age II miniaturized vases were found in the cave Grotta di Re Tiberio, Romagna, left near a spring flowing out from the inner portion. Miniature vases used for cultic purposes date to Neolithic times, but the practice spread during the Bronze Age, continued to persist for centuries, and assumed a special form of veneration during the Iron Age. For a review of Italian caves where cultic practices occurred, see the papers by Peroni (1989), Guidi (1991-1992), Bernabei and Grifoni Cremonesi (1995-1996), and Di Fraia and Grifoni Cremonesi (1996).

It is clear that a notable increase occurred in the depositions of votive vases on the shores of hypogeal waters during the Middle Bronze Age. However, the offering of vases in this period was not related solely to cultic practices, as it was during the Neolithic, but mainly to agrarian rituals and sometimes to funerary rituals. Vase deposition practices in the Middle Bronze Age shifted from dripping waters and small lakes inside labyrinthine caves with very difficult access to underground watercourses in relatively large caves with easy access.

The custom of vase dedication during the Bronze Age also involved open-air waters (rivers and lakes), as is the case with Monsignore Lake in Latium. Here, the practice of placing votive vases near the spring tributaries of the lake started in about the 15th century B.C. and lasted until early Roman times in the 6th century B.C. (Guidi, 1980).

All of these examples refer to caves and localities in Central and Southern Italy. In Northern Italy, different forms of rituals took place during the Middle and Late Bronze Ages when bronze objects—mostly swords—were offered in rivers, streams, and lakes.

Cults, Ritual Practices, and Use of Thermal and Mineral Waters

Cults and Rites

Due to their peculiarities, the external manifestations of terrestrial heat—thermal and mineral springs, fumaroles, volcanic eruptions, sulfurous springs, and gas releases—could hardly pass unnoticed by prehistoric people. In addition to water-related cults in caverns with hypogeal waters, other forms of cults and rituals related to these manifestations started to appear in about the 3rd millennium B.C.

In the Stufe di San Calogero, Sicily, big vases of the Malpasso culture, dating to the Copper Age (3rd millennium B.C.), were found in natural tunnels from which sulfurous vapor flows out at 39° C. Burials were also found inside these tunnels, and the contemporaneous presence of vases and burials is a puzzle. These finds suggest that the vases were offerings to an underworld divinity that had manifested its presence suddenly, through terrifying phenomena. This may have been an earthquake causing fractures to open through which sulfurous vapors flowed near the end of the Copper Age. Now tunnels used since the Lower Neolithic could be entered no longer (Guidi et al., 1979). However, available data are not sufficient to prove this hypothesis (Maggi, 1976-77; Tiné, 1979).

A. Guidi (1991-92) has provided interesting information about the Grotta dello Sventatoio near Latium. This is a deep cave, 30 m below sea level, consisting of four chambers interconnected by narrow passages. It is classified as a "thermal cave" because warm air continuously flows from the cave and condenses outside, especially during winter. Fragments of 400 Lower and Middle Bronze Age vases associated with food remnants (an entire flat loaf was found near one of them!) were found in this cave, and many vases contained wheat, barley, and flat beans. Pig, sheep, and cattle bones were found in different places, as well as three charred children's skulls.

Since the cave is all but suitable for habitation, the presence of so many artifacts suggests that the "vapor" flowing from the cave may have attracted the attention of people, resulting eventually in customary offerings of vases, food, animals— and perhaps also children— to divinities of the underworld.

Calcara di Panarea is a group of pits located near the shores of an island in the Aeolian archipelago north of Sicily. Archaeologists attribute them to the Capo Graziano culture, dating back to the Lower Bronze Age.

Because of very recent volcanic activity, strong fumaroles spread over a large area around the pits, and even the nearby marine belt displays many releases of hot fluids rising from below the sea floor. People probably did not settle at this site or in its proximity (Bernabò Brea, 1968). Panarea, however, is a small island inhabited since Neolithic times, so the site must have been known to and occasionally visited by the island people. Archaeological evidence indicates human presence at the site since the Late Neolithic.

The pits were dug during the Bronze Age, probably in a period of relatively low fumarolic activity. They were lined with volcanic pebbles and gravels, bound together by a mortar of mud and products of hydrothermal alteration, such as kaolin, alunite, and sulfur obtained from nearby

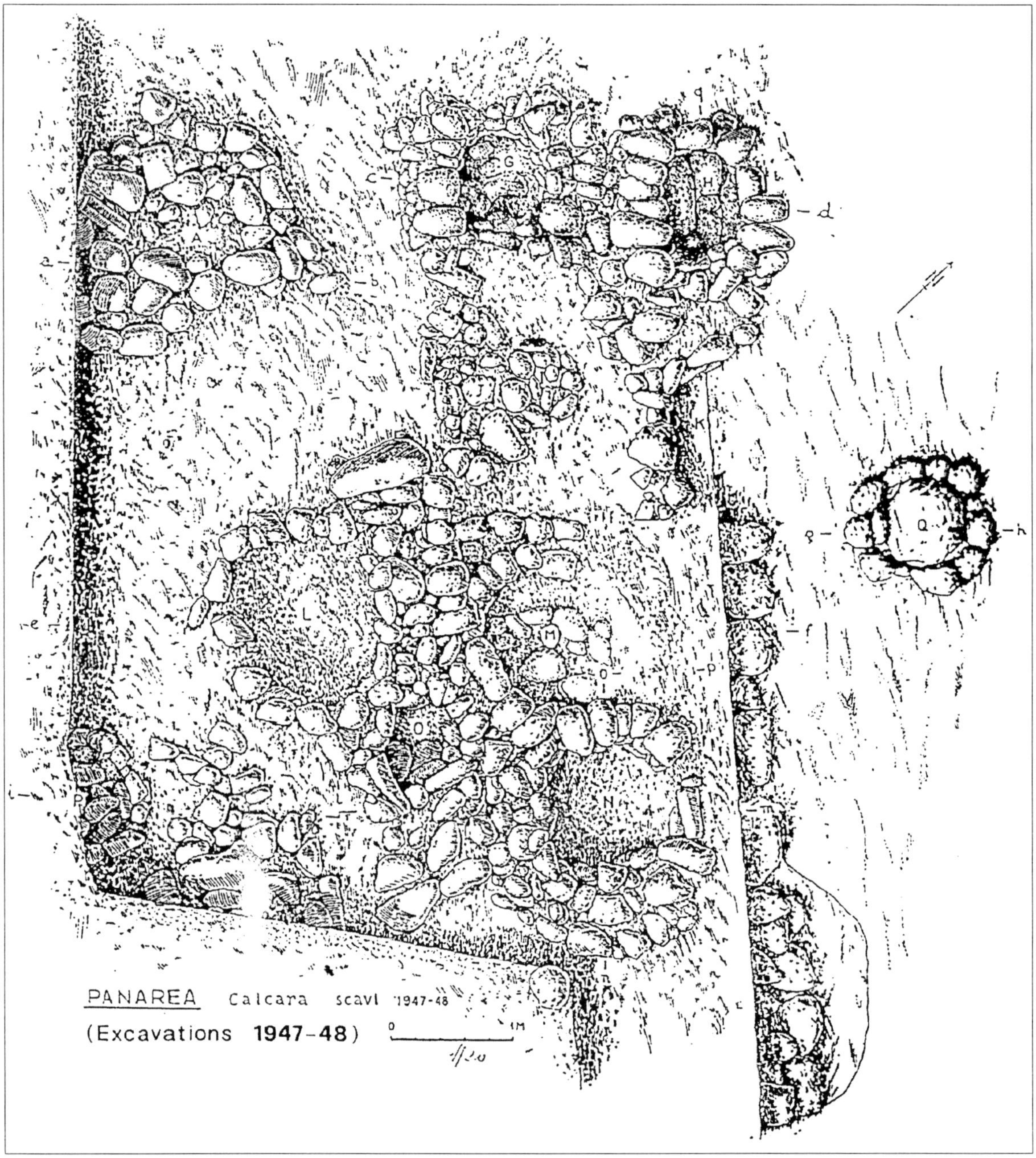

Sketch of the Calcara di Panarea, Lower Bronze Age, as seen after the excavations of 1947-1948. *Bernabò Brea and Cavalier, 1968, with permission*

manifestations. No significant objects were left inside or near the pits, and due to the high temperatures, no functional use can be imagined for these constructions. Perhaps their excavation was motivated by a form of cultic or ritualistic practice dedicated to the chthonian world.

The geological and geothermal conditions in the area of the small Lago delle Colonnelle, Latium, differ from those of the Calcara di Panarea. The sulfur content of the hypothermal water of this small lake is probably related to a secondary branch of the hydrothermal circuit that, in recent geological times, caused the thick travertine deposit to form near Tivoli (the renowned *lapis tiburtinus* of the ancient Romans). Because of a haze and the bad smell rising from the small lake, no people settled in its vicinity. Nonetheless, the site was probably frequented for cultic or ritualistic purposes during the entire Bronze Age, as numerous vases were found on the lake shores (Guidi, 1986).

Grotte di Latronico, Lucania, is located at a distance from sulfurous thermal springs. The cave was frequented during the Bronze Age, as confirmed by archaeological evidence from an expedition conducted almost a century ago (Rellini, 1916): vases full of cereals and fruits were unearthed, and Rellini suggested that they had been left in the cave as votive offerings to "healing waters." Modern excavations of the 1970s and 1980s confirm that the cave was used mostly as permanent shelter by people from the Mesolithic (7000-5500 B.C.) through the Bronze Age (Cremonesi, 1978; Bianco, 1984). Unfortunately, the first excavation did not consider the stratigraphic position and the location of the artifacts, so that a possible connection between the deposition of vases and the thermal springs of the area cannot be established.

Use of Thermal and Mineral Waters

THE USE OF THERMAL AND MINERAL WATERS FOR BALNEOLOGY AND FOR CURING STOMACH disorders probably started in Italy in Neolithic times, or possibly even before. Concrete evidence of these uses dates only from the Bronze Age onwards.

A significant example is the San Calogero *tholos* on the island of Lipari north of Sicily, which was constructed by Mycenaean colonists during the Bronze Age. This *tholos* probably represents the oldest and best documented case of an organized use of hot waters for therapeutic purposes in the Western Mediterranean. It is a building constructed of rows of large jutting blocks enclosing a room saturated with hot vapor (*sudatorium*). The water was gathered from a high-temperature spring in the vicinity and channeled through a stone-lined canal to a pond by the *tholos*, which was probably used for thermal bathing. From this pool, the water was channeled into the *tholos* by a smaller canal through the main entrance, which is finished with a robust architrave. The *tholos* was constructed during the period of the Capo Graziano culture, before 1430 B.C. (Bernabò Brea, 1991-92; Bernabò Brea and Cavalier, 1996). Thus the islands of the Aeolian volcanic archipelago were the sites of both therapeutic and religious practices related to geothermal manifestations (see section on Calcara di Panarea).

The *tholos* di San Calogero, Bronze Age, used as *sudatorium:* (above) entrance to the *tholos*; (below) pond by the *tholos*, probably used as a thermal pool. *Bernabò Brea and Cavalier, 1968, with permission*

The Phlegraean Fields near Naples is another area rich in geothermal manifestations, including fumaroles, hot springs, and hydrothermal activity. For this area, however, no significant archaeological findings exist to indicate therapeutic uses of the thermo-mineral waters in prehistoric epochs. This is because intense volcanic activity affected the area during recent geological times, resulting in the destruction or burial of Neolithic settlements and making the area periodically unsuitable for human habitation (Albore Livadie, 1986).

Ancient authors provide stimulating information on Sicily and Sardinia, the two largest islands of Italy, particularly when describing situations and life styles of the Sican and Sicilian peoples of Sicily and the Protosardinians of Sardinia. These works mention some popular customs of visiting thermal and mineral springs for cures and other purposes. These customs probably took root in prehistoric epochs and persisted for many centuries (or, perhaps, for several millennia) until ancient times (Lilliu, 1988).

For the Nuraghic period in Sardinia (2nd millennium B.C.), cases are reported of sacred places near thermal springs with healing properties. These are the *fonte de los malavidos* ("spring of the ill people") and *funtana de is dolus* ("spring of the pains"), whose waters were particularly sought for the treatment of rheumatism and other maladies. Near these springs during the Middle Bronze Age, people left vases in gratitude for their cures. At Santa Lucia di Bonorva (Sassari), a wide enclosure dating from the Nuraghic period can be seen surrounding an area with many releases of effervescent water. Most commonly, however, water-related sacred places in Sardinia were wells near sanctuaries. Indeed, all types of waters were considered a divine gift and were thus objects of cults (Lilliu, 1988).

For the Protosardinians, ancient writers mentioned a kind of ordeal (popular trial) celebrated at the site of some springs with warm caustic water; the divinities of these waters established the guilt or innocence of people accused of some fault (Lilliu, 1988). Macrobius, grammarian and philosopher of the 5th century A.D., reported similar activity in Sicily. Ordeals were held at the sites of special thermal springs issuing acidic water and gas; Macrobius wrote that the punishment for the people judged guilty consisted of exposing them at length to the noxious blasts of the springs, until they were blind.

Monticchio sul Vulture, on the slopes of the Vulture volcano in Lucania, has some caustic springs near which a *stipis* ("votive offering") was found at the end of the 19th century. It was a pile of miniaturized Iron Age vases and is interpreted as a sign of gratitude for the healing waters (Rellini, 1916).

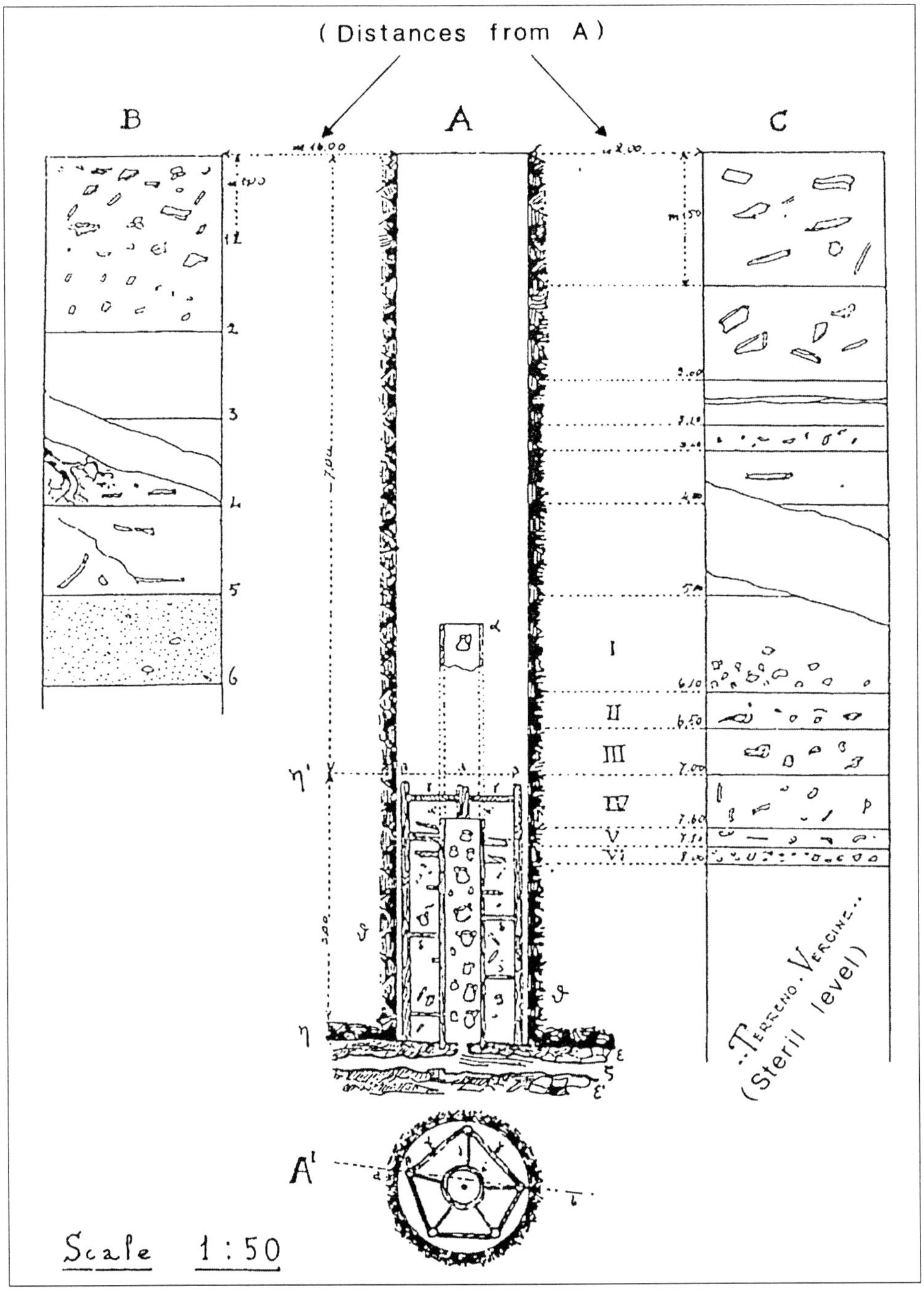

The Pozzo della Panighina: (A) vertical section; (A[1]) horizontal section; (B) and (C) check excavations near the well, made by Ugolini in 1924. *Ugolini, 1924, with permission*

Another example of a functional use of the water is a well, Pozzo della Panighina in Romagna. It was excavated toward the end of the Copper Age and was equipped with wooden pipes. Many flasks and two-handled vases were found inside this well, together with the remains of ropes that prove that the vases were lowered into the well to collect the chloride water and exploit its therapeutic properties (Santarelli, 1902; Pigorini, 1908; Morico, 1988). Well use continued through the Bronze Age. A number of vases were also found at the surface around the well mouth, suggesting a cult of curative waters.

Use of Geothermal By-Products: An Outline

The exploitation of volcanic and hydrothermal products is an extremely interesting branch of prehistorical research, which recently has attracted increased interest by archaeologists, historians, geologists, geographers, and scholars of the history of science and technology. Many works dealing with interdisciplinary investigations on this subject have enriched recent literature. However, I shall limit myself to a few references from Central and Southern Italy.

One of the most desired by-products of geothermal energy in prehistoric times was obsidian. Its systematic use in Italy dates to the Lower Neolithic. Italian obsidian is found on the two volcanic islands of the Tyrrhenian Sea—Lipari in the Aeolian archipelago and Palmarola in the Pontine archipelago—and at Monte Arci, an extinct volcano in the northwestern sector of Sardinia's Campidano graben (an area of subsided land). Obsidian was traded as a raw material on the Italian mainland, Corsica, and the Mediterranean coasts of France. In Northeastern Italy, obsidian also was imported from the Carpathians (Ferrari and Pessina, 1996). In recent years, obsidian has become an important tool to reconstruct the prehistoric cultural relationships among various areas of Southern Europe.

As for hydrothermal products, prehistoric people at Panarea used kaolin, sulfur, and associated compounds of fumarolic activity to prepare a mortar binding pebbles and gravels together to strengthen the walls of the ritual pits. Very likely, kaolin was also used as a component in pottery making.

Data are not yet available regarding the use of native sulfur to prepare ointments and as a component of pigments in prehistoric epochs, but usage can be inferred because this mineral outcropped in a number of areas with fumarolic activity in Tuscany, the Phlegraean Fields, and the Aeolian archipelago. It could hardly pass unnoticed by early Italian peoples. The remains of a sulfur-melting furnace in a Lower Bronze Age settlement was discovered a few years ago in the Monte Grande area, near Agrigento in Sicily (Castellana, 1996). The furnace suggests that the use of hydrothermal sulfur was an established custom by the beginning of the 2nd millennium

B.C., although the use of sulfur might have started much earlier. The size of the furnace suggests that its production capacity exceeded the demands of the local market, and it perhaps served external markets as well, in particular that of ancient Greece. If so, the furnace at Monte Grande would have provided one of the most important trading materials between Southern Italy and ancient Greece during the 16th and 15th centuries B.C., the Late Elladic period I-II.

Among the hydrothermal minerals, alunite and iron oxides, often associated with kaolin and sulfur, were probably used locally in areas of Central and Southern Italy to process textiles and/or to prepare pigments during prehistoric times predating the Bronze Age; however, specific evidence of this use is not yet available.

Volcanic materials, such as trachyte and other hard igneous rocks, were used frequently in prehistory for millstones. Volcanic products, such as fine-grained pyroclastics and pumice, were used for tempering, grinding, and degreasing pottery. Lipari Island had large deposits of pumice, and its exploitation was probably important for the local economy, at least since the Bronze Age.

In addition to obsidian, all of these raw volcanic materials have become important archaeological tools in the last few decades, helping researchers retrace and highlight cultural relations established in prehistory between different and often distant European populations.

Conclusions

The data available so far, even though scanty or unclear in some cases, are sufficient to highlight the interest that prehistoric Italians gave hypogeal waters, thermal springs, high-temperature geothermal manifestations, and by-products of the Earth's energy, beginning several thousand years before the Christian era. From that time, this interest resulted in the formation of cults, rituals, and customs related to the presence of water in caves, water springs of different natures and temperatures, and the emergence of effluents from the subsoil.

All of these elements were involved in the spiritual lives and behaviors of prehistoric people; they evolved and strengthened with time, continued to persist in ancient epochs, and, in a number of cases, influenced present-day cults, rituals, and popular beliefs.

Parallel to the magical and religious aspects, other pragmatic aspects related to the use of fresh water and thermo-mineral springs, high-temperature geothermal manifestations, and by-products of the Earth's energy took root in prehistory. They were gradually refined and enriched through hundreds of generations into a valuable heritage of ancient and modern peoples.

Together, all of these elements contributed significantly to the development of aspects of Italian civilization in prehistoric and ancient times.

Cited and Selected References

Albore Livadie, C., ed. 1986. *Tremblements de terre, éruptions volcaniques et vie des hommes dans la Campanie antique, vol 7.* Publ. Centre Jean Berand, Naples.

Bahn, P. G. 1978. Water mythology and the distribution of Paleolithic parietal art. *Proceedings Prehistoric Society,* no. 44, 125-134.

Bernabei, N. and R. Grifoni Cremonesi. 1995-96. I Culti delle acque nella preistoria dell'Italia peninsulare. *Rivista di Scienze Preistoriche,* 47, 331-366.

Bernabò Brea, L. 1991-92. La Sicilia e le Isole Eolie. Proceedings Congress L'età del Bronzo in Italia nei secoli dal XVI al XIV a.C. *Rassegna di Archeologia,* no. 10, 105-121.

Bernabò Brea, L. and M. Cavalier. 1968. *Meligunìs Lipàra, vol. 3.* Flaccovio Edit., Palermo.

Bernabò Brea, L. and M. Cavalier. 1996. La stufa termale di San Calogero (Lipari, Messina) in Sicilia Orientale, e le Isole Eolie. *Guide Archeologiche del XIII Congresso Internazionale delle Scienze Preistoriche e Protostoriche,* Forlì, no. 12, 11-113.

Bianco, S., ed. 1984. *Testimonianze archeologiche nel territorio di Latronico.* Congedo Edit., Galatina.

Bietti Sestieri, A. M., A. Greco Pontrandolfo, and N. Parise, eds. 1987. Archeologia e antropologia. *Quaderni Dialoghi di Archeologia,* no. 2.

Blanc, A. 1976. Le Trou Arnaud Saint Nazare—Le Désert (Drôme), Livret guide A9. *Proceedings IX Congrès UISPP,* Nice, 41-42.

Carucci, P. 1907. *La grotta preistorica di Pertosa (Salerno).* Stabil. Tipogr. Di Gennaro e Morano Edit, Naples.

Castellana, G. 1996. *La stipe votiva del Ciavolaro nel Bronzo Antico siciliano.* T. Sarento Edit., Agrigento.

Cauvin, J. 1993. *Naissance des divinités. Naissance de l'agriculture.* CNRS Edit., Paris.

Corrain, C., F. Rittatore, and P. Zampini. 1967. Fonti e grotte lattaie nell'Europa occidentale. *Etnoiatria,* 1, no. 2.

Cremonesi, G. 1968. La Grotta dell'Orso di Sarteano: I livelli dell'età del Bronzo. *Origini,* 2, no. 2, 247-331.

Cremonesi, G. 1976. *La Grotta dei Piccioni di Bolognano nel quadro delle culture dal Neolitico all'età del Bronzo in Abruzzo.* Giardini Edit., Pisa.

Cremonesi, G. 1978. *Gli scavi della Grotta n. 3 di Latronico (nota preliminare).* Atti XX Riunione Scientifica Istituto Italiano di Preistoria e Protostoria, Firenze, 177-197.

Cremonesi, G. 1985-90. Su alcune manifestazioni di culto del Neolitico abruzzese. *Abruzzo,* 22-23, 465-474.

Di Fraia, T. and R. Grifoni Cremonesi. 1996. *La Grotta Sant'Angelo sulla Montagna dei Fiori e il problema delle frequentazioni cultuali in grotta.* Istit. Poligraf. Intern., Pisa.

Eliade, M. 1966. *Trattato di storia delle religioni.* Einaudi Edit., Turin.

Ferrari, A. and A. Pessina. 1996. *Sammardenchia e i primi agricoltori del Friuli.* Udine.

Grifoni, R. 1967. La Grotta dell'Orso di Sarteano: I livelli neolitici. *Origini*, 1, 53-115.

Grifoni Cremonesi, R. 1986. Alcuni dati relativi a fenomeni funerari con implicazioni cultuali nella preistoria e problemi di interpretazione. *Dialoghi di Archeologia*, no. 2, 265-269.

Grifoni Cremonesi, R. 1987. Témoignages de cultes dans le néolithique ancien des Abruzzes. *Actes Colloque International Prémières Communautés Paysannes de la Méditerranée Occidentale*, Paris, 413-418.

Grifoni Cremonesi, R. 1994. Observations on the problems related to certain cult phenomena during the Neolithic in the Italian peninsula. *Journal of European Archaeology*, 2, no. 2, 179-197.

Grifoni Cremonesi, R. 1996. Le grotte e la loro funzione. Premessa metodologica. In *L' Antica Età del Bronzo in Italia.* Octavo Edit., Florence.

Guidi, A. 1980. Luoghi di culto dell'Età del Bronzo finale e della prima Età del Ferro nel Lazio meridionale. *Archeologia Laziale*, 3, 148-155.

Guidi, A. 1986. Gli insediamenti perilacustri di riva d'età protostorica nel Lazio centro-meridionale. *Quaderni di Protostoria*, 1, 239-247.

Guidi, A. 1989-90. Alcune osservazioni sulla problematica delle offerte nella protostoria dell'Italia centrale. *Scienze dell'Antichità*, 3, no. 4, 403-414.

Guidi, A. 1991-92. Recenti ritrovamenti in grotta nel Lazio: Un riesame critico del problema dell'utilizzazione delle cavità naturali. *Rassegna di Archeologia*, 10, 427-437.

Guidi, P., D. Marini, and G. Perotti. 1979. Le grotte vaporose del Monte Kronio. *Quarant'anni di ricerche a Sciacca; Atti e Memorie della Commissione Grotte E. Boegan*, Trieste, 19, 19-57.

Güller, A. and A. G. Segre. 1948. La stazione enea del Grottone di Val de'Varri nell'Appennino abruzzese. *Rivista di Antropologia*, 36, 269-281.

Lilliu, G. 1988. *La civiltà dei Sardi.* Nuova ERI Edit., Turin.

Lo Schiavo, F. 1987. Grotta Verde 1979: Un contributo sul Neolitico antico della Sardegna. *Atti XXVI Riunione Scientifica Istituto Italiano Preistoria e Protostoria*, Firenze, 845-857.

Macrobius, Ambrosius Theodosius. c. 410 A.D. *Saturnalia.* Reprint 1937, H. Borneque and F. Richard, trans. Paris.

Maggi, R. (1976-77). Gli scavi nelle Stufe di San Calogero sul Monte Kronio (Sciacca) e i rapporti fra Sicilia e Malta durante il Neolitico. *ΚΩΚΑLOS*, 22-23, 510-524.

Martin, L., A. Nourrit, A. Durand Toullou, and G. B. Arnal. 1964. Les Grottes Citernes des Causses: Le vase à eau et son utilisation. *Gallia Préhistoire*, no. 7, 107-177.

Mellaart, J. 1967. Çatal Hüyük: A Neolithic town in Anatolia. In *New aspects of antiquity.* M. Wheeler, ed. Thames and Hudson Ltd. and The Camelot Press, London and Southampton.

Morico, G. 1988. La Panighina di Bertinoro (Forlì). La fase dell'età del Rame. *Rassegna di Archeologia*, no. 7, 610-611.

Negroni Catacchio, N. 1981. Grotta Nuova. In *Sorgenti della Nova.* CNR Edit., Rome.

Negroni Catacchio, N., L. Domanico, and M. Miari. 1989-90. Offerte votive in grotta e in abitato nelle valli del Fiora e dell'Albegna nel corso dell'età del Bronzo: Indizi e prospettive. *Scienze dell'Antichità*, 3, no. 4, 579-598.

Passeri, L. 1970. Ritrovamenti preistorici nei Pozzi della Piana (Umbria). *Rivista di Scienze Preistoriche*, no. 25, 225 -251.

Peroni, R. 1989. Protostoria dell'Italia continentale: La Penisola italiana nell'età del Bronzo e del Ferro. *Biblioteca di Storia Patria*, 9.

Pigorini, L. 1908. Uso delle acque salutari nell'Età del Bronzo. *Bullettino di Paletnologia Italiana*, 34, 168-191.

Poggiani Keller, R. 1978. La collezione di Grotta Nuova. *Rivista di Scienze Preistoriche*, 33, 225-236.

Radmilli, A. M. 1975. Culti di fertilità della terra testimoniati in alcuni gicimenti neolitici italiani. *Atti Colloquio Internazionale, Les religions de la Préhistoire: Valcamonica Symposium*, Capodiponte, 175-184.

Rellini, U. 1916. La Caverna di Latronico e il culto delle acque salutari nell'Età del Bronzo. *Monumenti Antichi Lincei*, 24, 557-622.

Rellini, U. 1920. Cavernette e ripari preistorici nell'Agro Falisco. *Monumenti Antichi Lincei*, 26, 6-174.

Santarelli, A. 1902. *Bertinoro: Scoperta preistorica nel Fondo Panighina.* Notizie degli Scavi, 541-553.

Segre, A. G. 1951-52. Orcio rinvenuto al Monte Soratte presso Roma. *Bullettino di Paletnologia Italiana*, 6, 136-139.

Tanda, G. 1978. Il Neolitico antico e medio della Grotta Verde, Alghero. *Atti XXII Riunione Scientifica Istituto Italiano Preistoria e Protostoria*, Florence, 45-94.

Tiné, S. 1975. Culto neolitico delle acque nella Grotta Scaloria. *Atti Colloquio Internazionale "Les religions de la Préhistoire: Valcamonica Symposium,"* Capodiponte, 175-184.

Tiné, S. 1979. Stufe di San Calogero: Interpretazione storico-archeologica. *Atti e Memorie della Commissione Grotte E. Boegan,* 19, 58-62.

Tiné, S. and E. Isetti. 1975-80. Culto neolitico delle acque e recenti scavi nella Grotta Scaloria. *Bullettino di Paletnologia Italiana,* 82, 31-66

Trucco, F. 1991-92. Revisione dei materiali di Grotta Pertosa. *Rassegna di Archeologia,* 10, 471-479.

Ugolini, L. M. 1924. La Panighina: Fonte sacra preistorica. *Monumenti Antichi Lincei,* 29, 18-37.

Whitehouse, R. 1990. Caves and cults in Neolithic Southern Italy. *The Accordia Research Papers,* 1, 19-37.

Whitehouse, R. 1992. *Underground religion: Cults and cultures in prehistoric Italy.* Accordia Research Center, London.

Zanini, A. 1988. Sarteano: La Buca del Rospo. Testimonianze della prima età dei metalli. In *Archeologia in Valdichiana.* Multigrafica Edit., Rome.

Zezza, M. G. 1983. *Vasi eneolitici della Grotta Zinzulusa di Castro (Lecce).* Edizioni Scientifiche Quaderni Museo Comunale di Paleontologia, Maglie.

The Author:

Renata Grifoni Cremonesi
University of Pisa
Dipartimento di Scienze Archeologiche
53, Via Santa Maria
56126 Pisa, Italy
Telephone: 39.050.847.303
Fax: 39.050.911.665
E-mail: grifoni@arch.unipi.it

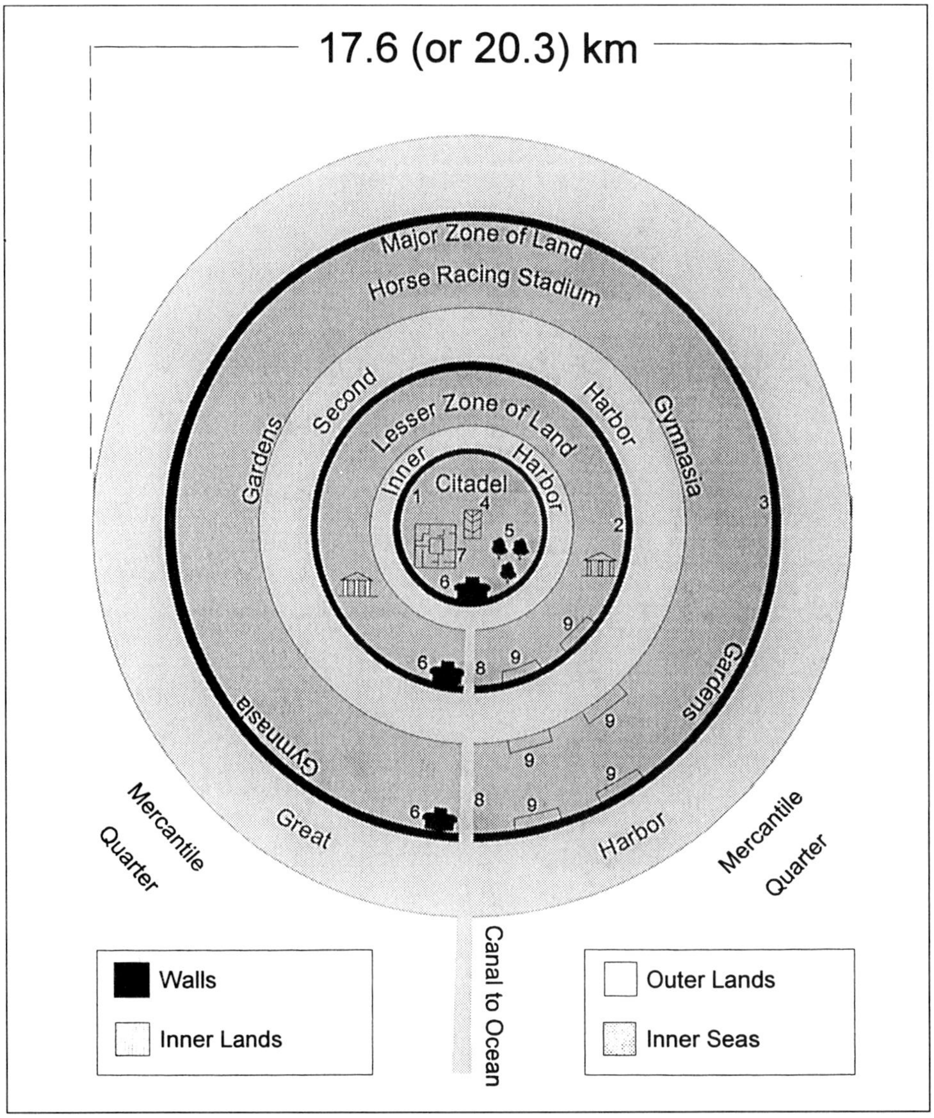

1. orichalc wall; 2. wall plated with tin; 3. wall plated with bronze; 4. Temple of Poseidon and Cleito, surrounded by a golden enclosure; 5. Grove of Poseidon; 6. guard houses; 7. ancient royal palace; 8. underpass for triremes; 9. underground docks.

The metropolis of Atlantis, as imagined by Plato in his *Dialogue, Critias,* c. 350-348 B.C. Plato described in detail the configuration of this fantastic city-state of the imaginary continent with the same name—the famous "lost continent." Among many other features, Plato recounted that within the inner citadel thermal springs existed, supplying the royal palaces with hot water. However fictitious Plato's account may be, it highlights the importance the ancient Greeks attached to lands with thermal springs.

9. From the Mystery of Atlantis to the First Integrated Uses of Geothermal Energy

by
Raffaele Cataldi
Pier Domenico Burgassi

Abstract: About 3,600 years ago, a paroxysmal phenomenon of the Earth's heat struck the volcanic island of Santorini in the Southern Aegean. This probably inspired Plato to write his account of the "lost continent" of Atlantis. The event and the use and trade of geothermal by-products by Cretans, Mycenaeans, Phoenicians, and other ancient peoples of the Eastern Mediterranean are discussed in the first part of the chapter.

The second part illustrates the influence of the Etruscans in mining, processing, and commercializing hydrothermal minerals, using other geothermal by-products, and popularizing thermal balneology. Because these activities were developed so extensively by the Etruscans, we call them the fathers of industrial geothermal development.

Revised English translation of Chapter 3, La geotermia en el período Precolombino en las áreas Mediterránea y Mesoamericana, by R. Cataldi, P. D. Burgassi, and M. C. Suárez Arriaga (1992). Reprinted, with permission, from *Geotermia, Revista Mexicana de Geoenergía*, 8, no. 3, pp. 251-264.

Not knowing what happened before us
is like forever remaining children.
—Cicero

The First Millennia

WHO KNOWS WHAT ANCIENT MEMORIES WERE LOST ON THAT mysterious continent called Atlantis, whose existence and sudden sinking to the bottom of an unknown ocean was recounted by Plato, c. 360 B.C. Let us envision a block of land suddenly disappearing, engulfed in the abysses of the sea, perhaps from a cataclysmic event caused by the Earth's energy some 3,600 years ago in the Southeastern Mediterranean. Here for about two millennia the most ancient European civilization, that of the Minoans, had flourished on Crete and in the Cyclades. The reasons are unclear for the sudden disappearance of the civilization at its highest developmental peak, but modern scholars all agree that the end of the Minoan civilization came from an external event, striking Crete in the 15th century B.C.

What was the cause? A Mycenaean invasion, a deadly plague, or a phenomenon related to terrestrial heat? Some scholars have suggested a catastrophic earthquake followed by a devastating tsunami (Marinatos, 1939; Galanopoulos, 1958; Nikovich and Heezen, 1965; and Luce, 1969). They could have occurred after a sudden rupture in the tectonic equilibrium at the collision margins of the northeastern sector of the African plate as it was subducted beneath the eastern edge of the European plate, eventually forming in the Mediterranean the Tyrrhenian volcanic arc (the Eolian archipelago) to the west and the Aegean volcanic arc

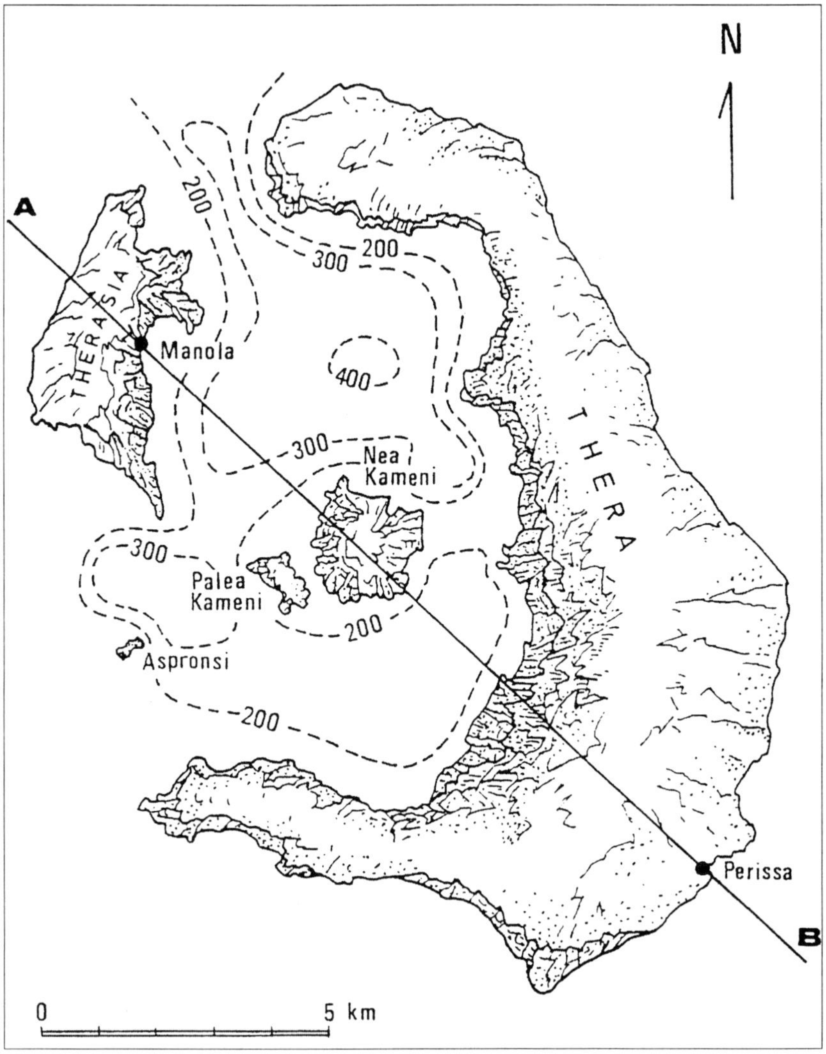

Present morphology and bathymetry of the archipelago of Thera (present-day Santorini), Greece. *Drawing by R. Cataldi*

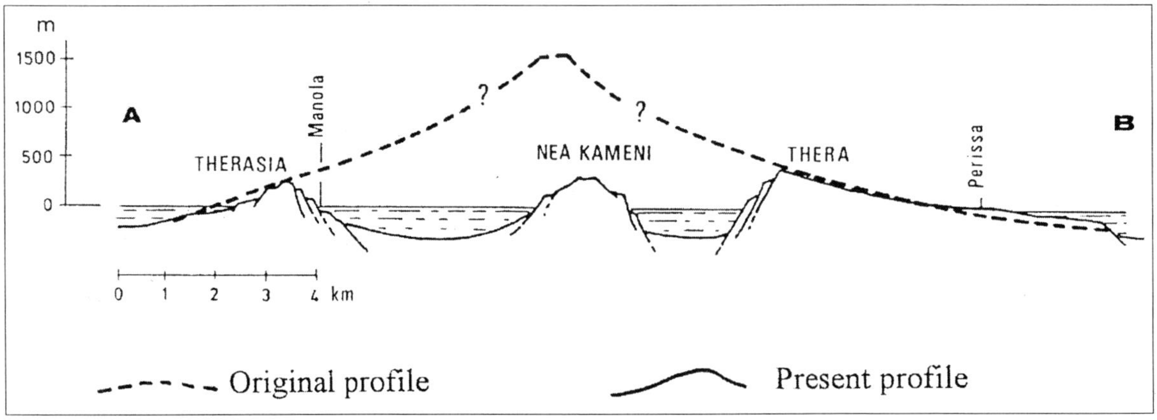

Hypothetical reconstruction of cross section A-B, a morphological profile of the island of Thera before the huge explosion around 1630 B.C. *Drawing by R. Cataldi*

(with islands including Aegina, Milos, Santorini, and Nisyros and others) to the east. The rupture probably allowed sea water to enter a magma chamber below the island of Thera (the present-day Santorini), which caused a terrifying explosion that eviscerated the central section of the old volcanic apparatus forming the original island, as the two drawings illustrate.

If so, the mysteriously quick end of Minoan civilization would be contemporaneous with the Minoan eruption that, until a few decades ago, was dated around 1500 B.C. However, according to the most recent absolute dating, it seems certain that the Minoan eruption occurred some 3,630 years ago (Hammer et al., 1987; Baillie, 1990; Hammer and Clausen, 1990; Vougioukalakis et al., 1994; Cioni et al., 1997; Vougioukalakis, 1997; Fytikas et al., 1999). If, on one hand, these dates seem to disprove the cause and effect relationship between the Minoan eruption and the sudden disappearance of the Minoan civilization, on the other hand they do not exclude the fact that Plato could have drawn from the huge volcanic explosion occurring at Thera around 1630 B.C. the idea for his wonderful account of a lost continent.

Regardless of the reasons why Cretan civilization collapsed, the fact remains that other Mediterranean peoples came to the fore in that period: Achaeans, Dorics, Aeolians, Ionians, and Phoenicians. The Phoenicians in particular, as clever traders and expert sailors, took advantage of the fall of Crete, establishing predominance over the sea routes of the Southern Mediterranean from about 1500 to 500 B.C. Well before Ulysses, they managed to pass the Pillars of Hercules and, sailing near the western coast of Europe, pushed into the North Atlantic as far as the Cassiterides Islands (now the Isles of Scilly) for tin and to the Scandinavian peninsula for amber. However, the economic power of the Phoenicians derived only in small part from exporting fabrics and transferring knowledge about processing and using Tyrian purple to dye wool and linen, at which they were masters. Their power, instead, came mostly from maritime traffic and trade in raw materials, minerals, and manufactured articles of all kinds.

Therefore, it is natural to think that these raw materials included not only metal-bearing ores, such as silver, copper, lead, tin, and zinc, but also many hydrothermal minerals, such as alum, borates, kaolin, iron oxides, silica, and sulfur that were used to manufacture ceramics and glazes, tint glass, prepare ointments and medicines, bleach wool, and treat textiles. All these hydrothermal minerals are associated with the active and fossil geothermal manifestations on many Mediterranean islands, such as Lesbos, Nisyros, Santorini, Milos, Aegina, Pantelleria, Lipari, Vulcano, and Ischia. The same minerals also were found at similar geothermal manifestations on the mainland in Western Anatolia, Southern Peloponnesus, Phlegraean Fields, Alban Hills, Volsini Mountains, and Southern Tuscany.

Besides the Phoenicians, other ancient peoples of the Mediterranean area knew the sites of these manifestations quite well. This probably helped increase the trade of hydrothermal products and expand their uses. It follows that commercialization of obsidian and certain hydrothermal minerals, already existing on a local scale since Neolithic times in Southwestern Anatolia (Mellaart, 1967), expanded regionally and was quite developed during the second millennium before the Christian era (Fytikas et al., 1999).

The Etruscans, Historical Fathers of Geothermal Industries

Toward the start of the 12th century B.C., a new people lived in Tuscany, calling themselves *Rasenna*, or "men." Others called them Tyrrhenians, and we know them as the Etruscans.

Shortly after they appeared in Tuscany, the Etruscans reached a high level of civilization in farming, draining swamps, shipbuilding, construction (including the vault and arch), trade, mining, metalworking, handicrafts, and extracting and processing hydrothermal minerals, such as alum, borates, kaolin, iron oxides, and sulfur. These minerals, found in many thermal localities in Italy, were especially abundant in Etruscan territories, particularly the Boraciferous region in Tuscany (in a triangle formed by the cities of Pisa, Siena, and Grosseto) where many natural manifestations existed, such as hot springs, pools of boiling mud and water called *lagoni*, steam jets known as *soffioni*, and hydrothermal encrustations.

Historians say the Etruscans were creative and pragmatic, capable of enjoying life. This helps us understand why one highly developed Etruscan area was a zone of Tuscany known as the Metalliferous Hills, which includes the Boraciferous region. Here the Etruscans discovered numerous deposits of metal-bearing ores, such as silver, copper, iron, manganese, lead, and zinc, as well as alabaster and hydrothermal minerals (Squarzina, 1965; Moscati, 1987). One can also understand why the natural manifestations of the Boraciferous region, aside from the commercial value of their hydrothermal products, stimulated the Etruscans to develop balneological and therapeutic practices. Therefore, they developed large-scale, integrated uses of geothermal resources. In 1987, Burgassi wrote:

> "The settling of the Etruscan civilization...was to bring something new in the use of geothermal energy, not only in terms of utilization of the manifestations for therapeutic purposes, but also with regard to exploitation of the salt products deposited near the edges of the *lagoni*.

> "From the first point of view...numerous vestiges and handmade objects exist that attest to widespread therapeutic use among the Etruscans of the hot springs, saline encrustations, and thermo-mineral muds near Larderello. From the second point of view, it must be borne in mind that traces of boric salts have been found in the glazes of Etruscan plates and crockery (Fiumi, 1943), a fact testifying to how these people, many centuries before Christ, had already developed a high degree of skill in the grinding and chemical treatment of the borates, and also in mixing these products with the other substances that composed their fine pottery...."

So much development in the Boraciferous region helps explain why the Etruscans also stimulated the practice of balneotherapy in other areas near geothermal manifestations in Southwestern Italy, where they had founded a number of colonies. In some of these zones the Etruscans also extracted and processed the hydrothermal products associated with the manifestations. This probably occurred between the 8th and 5th centuries B.C. at Populonia, Roselle, and Saturnia in Tuscany; Bolsena, Tarquinia, Cerveteri, and Veio in Latium; and in the Phlegraean Fields in Campania. Major Etruscan centers or colonial settlements are documented in all these areas, which have important hydrothermal manifestations. Especially in Roselle, Saturnia, Bolsena, and Veio, the establishment of Etruscan towns and villages was probably encouraged by the presence of hot springs and/or products of hydrothermal alteration with practical applications. Though probably less enterprising than the Phoenicians and other ancient Mediterranean peoples in long-distance transport and marketing of hydrothermal products, the Etruscans were by far the most active in extracting, processing, and using the products themselves. For this reason we consider the Etruscans fathers of industrial development for geothermal resources.

To say that the decline of the Etruscans in the 3rd century B.C. was caused solely by the conflict of interest arising between them and other Mediterranean peoples to control production and trade of the raw materials and hydrothermal products in which Etruria abounded would be an historical exaggeration. However, it seems plausible to hypothesize that Etruscan control over the areas where major Italian geothermal manifestations were located aroused the interest and then an ambition for possession by an emerging people like the Romans. This hypothesis is supported by the fact noted on the following map—that many natural manifestations were in strategic positions along main Etruscan communication routes, which were later retraced by some of the major consular roads.

A summary follows of the locales and the uses of geothermal by-products in the Mediterranean area from the 6th millennium B.C. through the 3rd century B.C.

Use of geothermal by-products in the Mediterranean area from the 6th millennium B.C. to the 3rd century B.C.

Principal by-products (in alphabetical order)

Alum	Obsidian
Borates	Silica
Iron oxides	Thermo-mineral muds
Kaolin	Travertine
Native sulfur	

Main areas from which by-products originated (from west to east)

Continental areas		Mediterranean islands	
Italy	Greece and Turkey	Tyrrhenian	Aegean
Southern Tuscany	Southern Peloponnesus	Ponza	Lesbos
Volsini Mountains	Western Anatolia	Ischia	Aegina
Sabatini Mountains		Lipari	Milos
Alban Hills		Vulcano	Thera (Santorini)
Phlegraean Fields		Pantelleria	Nisyros
			Yialí

Principal uses of by-products

Manufacturing stone objects	Masonry
Manufacturing ceramics	Preparing ointments & other medications
Manufacturing enamels	Bleaching wool
Manufacturing and coloring glass	Treating cloth

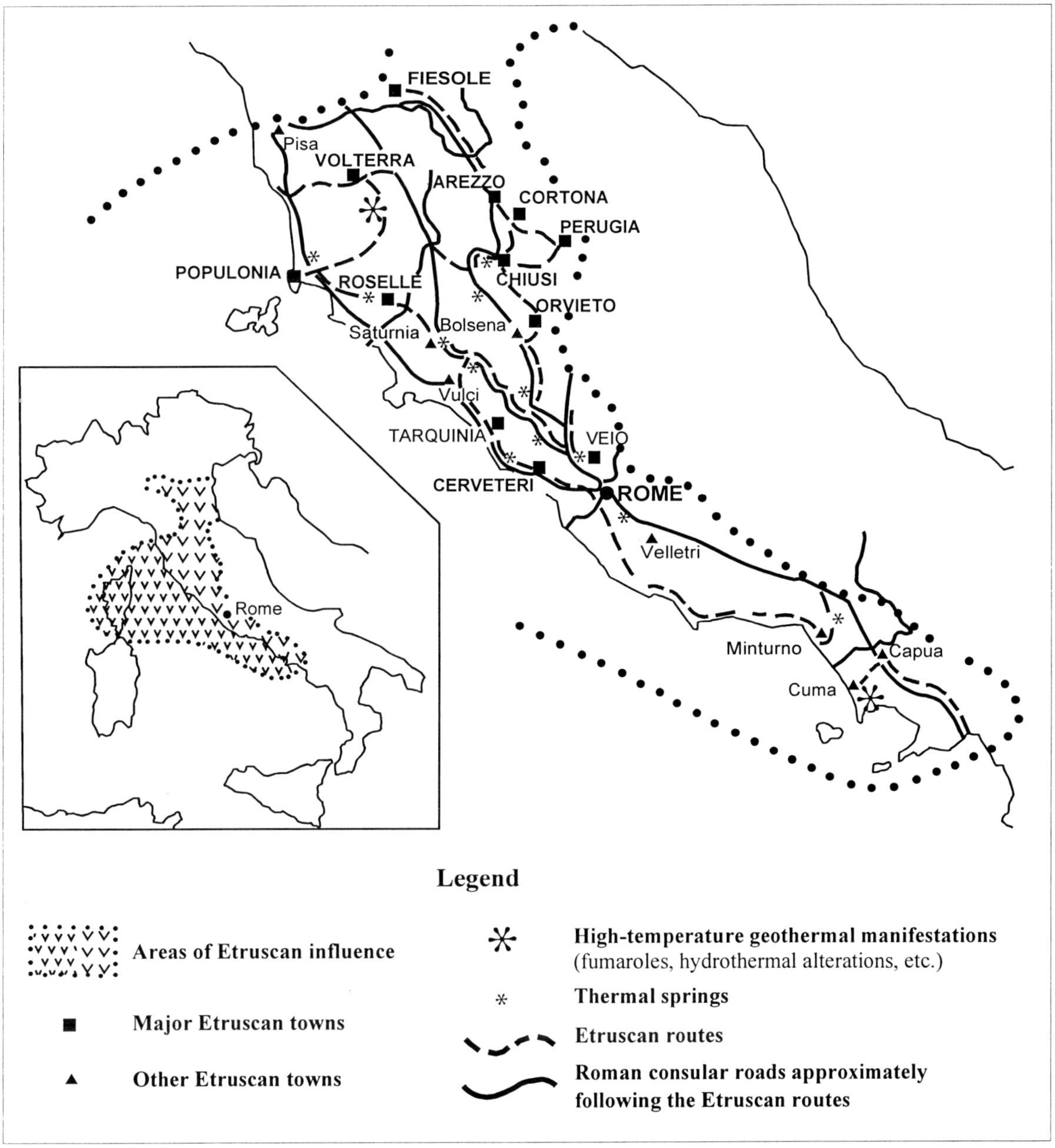

Location of Etruscan towns and main Etruscan and Roman routes linking the sites of principal geothermal manifestations in Central and Southern Italy. *Drawing by R. Cataldi, drafting by T. Boyd*

Cited and Selected References

Baillie, M. 1990. Irish tree rings and an event in 1628 B.C. In *Thera and the Aegean world,* III. B.d. 3. D. A. Hardy, ed. The Thera Foundation, London.

Burgassi, P. D. 1987. Historical outline of geothermal technology in the Larderello region. *Geothermal Resources Council Bulletin,* 16, no. 3, 3-18. (See Chapter 13 in this volume).

Cioni, R., L. Gurioli, A. Sbrana, and G. Vougioukalakis. 1997. Precursory phenomena and destructive events related to the 1628 B.C. Minoan and 79 A.D. (Vesuvius) Plinian eruptions. Inferences from the stratigraphy in the archaeological areas. In *Volcanoes, earthquakes and archaeology.* Geological Society of London (in press).

Dionysius of Halicarnassus. c. 20 B.C. Roman antiquities, vol. 1. In *Quaderni urbinati di cultura classica,* no. 10. Reprint 1970. D. Musti.

Fiumi, E. 1943. *L'utilizzazione dei lagoni boraciferi della Toscana nell'industria medievale.* University of Florence, CYA Edit., Florence.

Fytikas, M., G. Margomenou Leonidopoulou, and R. Cataldi. 1999. *Geothermal energy in ancient Greece: From mythology to late antiquity (3rd century A.D.).* (See Chapter 6 in this volume).

Galanopoulos, A.G. 1958. On the location and the size of Atlantis. *Praktika Akad. Athin.,* no. 35, 401-418. (In Greek, with an English abstract).

Hammer, C. U. and H. B. Clausen. 1990. The precision of ice-core dating. In *Thera and the Aegean world,* III. B.d. 3. D. A. Hardy, ed. The Thera Foundation, London.

Hammer, C. U., H. B. Clausen, V. L. Friedrich, and H. Tauber. 1987. The Minoan eruption of Santorin in Greece, dated to 1645 B.C.? *Nature Bd.,* no. 328/6130, 517-519.

Herodotus c. 440 B.C. The histories, vol. 1. In *Herodotus, vol 1.* Reprint 1921, A. D. Godley, trans. Harvard University Press and W. Heinemann Ltd., London and New York.

Luce, J. V. 1969. *The end of Atlantis.* Thames and Hudson Ltd., London.

Marinatos, S. 1939. The volcanic destruction of Minoan Crete. *Antiquity,* no. 13, 425-439.

Mellaart, J. 1967. Çatal Hüyük: A Neolithic town in Anatolia. *In new aspects of antiquity.* M. Wheeler, ed. Thames and Hudson Ltd. and The Camelot Press, London and Southampton.

Moscati, S. 1987. *L'Italia prima di Roma.* Electa Edit., Milan.

Nikovich, D. and B. C. Heezen. 1965. Santorini tephra. *In Submarine geology and geophysics. Closton Papers,* 17.

Pallottino, M. 1973. *Etruscologia.* 6th ed. Hoepli Edit., Milan.

Plato. c. 360 B.C. Timaeus. In *Plato, vol. 7.* Reprint 1926, R. G. Bury, trans. Harvard University Press and W. Heinemann Ltd., London and New York.

Squarzina, F. 1965. *Italia mineraria.* A.M.I. Edit., Rome.

Vougioukalakis, G. 1997. The volcano of Santorin. *Korfes,* no. 124, 32-38. (In Greek).

Vougioukalakis, G., D. Mitropoulos, C. Perissoratis, A. Andricopoulos, and M. Fytikas. 1994. The submarine volcanic centre of Kolumbos, Santorin. *Proceedings of the 7th Congress of the Geological Society of Greece,* 30, no. 3, 351-360. (In Greek).

The Authors:

Raffaele Cataldi
55, Via del Borghetto
56124 Pisa, Italy
Telephone/Fax: 39.050.59.81.07
E-mail: rafcat@tin.it

Pier Domenico Burgassi
2, Via G. Galilei
56044 Larderello (Pisa), Italy
Telephone: 39.058.86.74.89

Quanta Roma Fuit, Ipsae Ruinae Docent
("Even ruins speak and teach us the greatness of Rome")

Both the Flavian Amphitheater (Colosseum, upper right) and the Baths of Caracalla (the patio is depicted) were significant monuments of Imperial Rome. Hundreds of public buildings aimed at entertainment and thermal bathing were erected in the Roman Empire between the 1st and the 4th centuries A.D. Many, including the above, were made or faced with travertine, an elegant building stone with tiny holes—still used today, and a very important hydrothermal product. ***Photo, 1963, and reconstruction of the Baths of Caracalla patio by R. Cataldi***

10. Flowering and Decline of Thermal Bathing and Other Uses of Natural Heat in the Mediterranean Area, from the Birth of Rome to the End of the First Millennium

by
Raffaele Cataldi
Pier Domenico Burgassi

Abstract: Thermal balneology was developed systematically by the Romans from the 2nd century B.C. From that period on, the use of spas spread throughout the territory under Roman domination. Thermal springs were used as sites for spas whenever possible; however, where none existed, especially in Rome, artificially heated baths were built. At the same time, the Romans developed uses for numerous geothermal byproducts.

The boom for both activities peaked in the Mediterranean area in the 1st through 3rd centuries A.D., declined from the second half of the 4th century, continued to fall rapidly in the 5th and 6th centuries, and, particularly in Italy, entered a long inactive period extending beyond the end of the first millennium.

Revised, English version of Chapter 4, La Geotermia en el período Precolombino en las áreas Mediterránea y Mesoamericana, by R. Cataldi, P. D. Burgassi, and M. C. Suárez Arriaga (1992). *Geotermia, Revista Mexicana de Geoenergía*, 8, no. 3, pp. 265-288.

Thermal Bathing: *Ab Urbe Condita* to the Fall of the Roman Empire (7th Century B.C.-5th Century A.D.)

Even with the many thermal localities around Rome, between 750 and 500 B.C. the early Romans were not yet in the socioeconomic condition needed to profit from the experiences of their Etruscan neighbors by extracting and using hydrothermal products at these sites and practicing thermal bathing. However, such practices began to influence the Romans after the end of the monarchal period, when, starting in the 5th century B.C., they exerted political and military pressures on their neighbors, first to consolidate their young republic and later to extend Roman rule over the entire Italian peninsula.

In the 5th and 4th centuries B.C., the Romans may have used some imported hydrothermal products for pottery, such as kaolin and iron oxides, and Roman soldiers on military expeditions may have made occasional visits to hot springs for thermal bathing. Clearly, the energies of those Romans were not directed toward developing glass- and pottery-making technologies that included processed hydrothermal products or toward thermal bathing, by then widespread in Etruscan territories, especially those north of Rome.

However, during this time, the Romans understood the commercial importance of Etruria's mineral resources, including the hydrothermal products of the Boraciferous region (Cataldi and Burgassi, 1992a; Cataldi and Chiellini, 1995). The Romans probably had also grasped the strategic importance of Etruscan roads that linked many geothermal localities (see map in Chapter 9). In this light one can understand why several Etruscan centers located near hot springs, such as Cerveteri, Veio, Tarquinia, Populonia, and Saturnia, were later chosen by the Romans as sites for military encampments (*castra*).

The slight importance attributed by the Romans to hydrothermal products and thermal bathing would change rapidly after the annihilation of the Etruscans in the 3rd century B.C., especially after the defeat of Hannibal at Zama in 202 B.C. and the destruction of Carthage in 140 B.C., which allowed Rome to rule the entire Mediterranean. Although balneotherapy became more widespread among the Romans between the late 3rd and early 2nd centuries B.C., it still was practiced sporadically, mostly by the elite class at localities with natural hot springs.

From the second half of the 2nd century B.C., Roman attitudes toward thermal bathing changed radically. The republic was now consolidated, the conquered territories were vast, and trade and commercial dealings were intense. Thus, Romans became more familiar with the life styles and technologies of other peoples, the state was wealthy, and people had the time and economic resources to care for their bodies and practice balneotherapy. Now the first public baths (*balnea* and *thermae*) appeared.

In cities such as Rome with no natural hot springs, artificial thermal baths were built and fed with hot water heated in centralized boilers by wood fires. Such a heating system, without pumps to move the hot water, proved inefficient and expensive. Thus the common people of Rome did not practice thermal balneology until several decades later at the beginning of the 1st century B.C.

Then an ingenious Roman citizen of Neapolitan origins, Caius Sergius Orata, inspired by features constructed in previous centuries in natural Greek spas, designed a system of rooms artificially heated to different temperatures by hot air circulation. Fed with steam from a boiler,

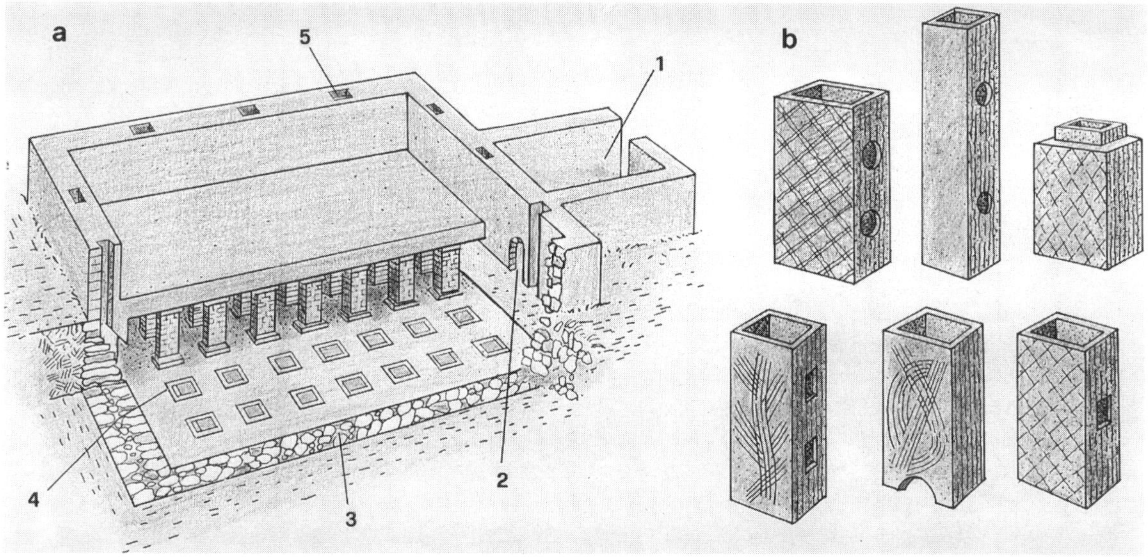

Figure 1. Artificial heating system using hot-air circulation for thermal baths in ancient Rome. a. Cut-away view showing heating system with *hypocaustum* and *alveola*: 1. boiler room; 2. hot air flow channel; 3. spa foundations; 4. underfloor interspace for hot air circulation (*hypocaustum*); 5. channels for the ascent of hot air (*alveola*). b. Sketches showing different types of *alveola*.

the system conveyed hot, humid air through a series of interstices (*hypocaustum*) located beneath the floors of the baths and into a series of channels (*alveola*) embedded in the walls (figure 1a). This system greatly reduced water consumption and lowered the operating costs of thermal establishments. It also made it possible to control the heat in individual rooms and keep each at its desired temperature.

The overall design and technical features—such as regulating the draft by using *alveola* of different diameters (figures 1b, 2), the positions of hot air outlets, external grooves on the *alveola* to firmly hold in place plaster and stucco (figure 1b), chimneys to expel smoke and spent hot air—help us to understand why the basic concepts of this advanced heat engineering system are still used today. Using *hypocaustum* and *alveola* to control temperatures, three types of heated bathing rooms were constructed in each artificial spa: the *laconicum* ("steam room"), the *calidarium* ("hot water pool"), and the *tepidarium* ("warm water pool"). Finally, a *frigidarium*, a "cold water pool," was used to tone the body after a steam or hot water bath (figure 3).

Thus the practice of thermal bathing grew rapidly, peaking in the first three centuries of the Christian era. Besides the numerous private baths installed in practically every villa and many apartment buildings, there were over 1,000 public baths in Rome at the height of the Empire: about one for every 1,000 inhabitants (Montanelli, 1969). Two especially luxurious and spacious thermal complexes were the Baths of Titus and Trajan (80-110) and the Baths of

Figure 2. Different types of *alveola* in the ruins of Roman mansions at Roselle, Tuscany. *Photos by the authors, c. 1967*

Caracalla (217). The latter, enlarged by Aurelian in 275, remained in continual use until the early 6th century (see patio in initial illustration).

During the imperial period, thermal bathing became so widespread and deeply rooted at all levels of society that the *thermae* were a daily meeting place, a reference for every aspect of civic life. Physical exercise was undertaken in attached gymnasiums, reading and study in internal libraries, and there were areas for beauty treatments (massage, shaving, hairdressing, and depilating), recreation (ball games and entertainment), conversation and political discussions,

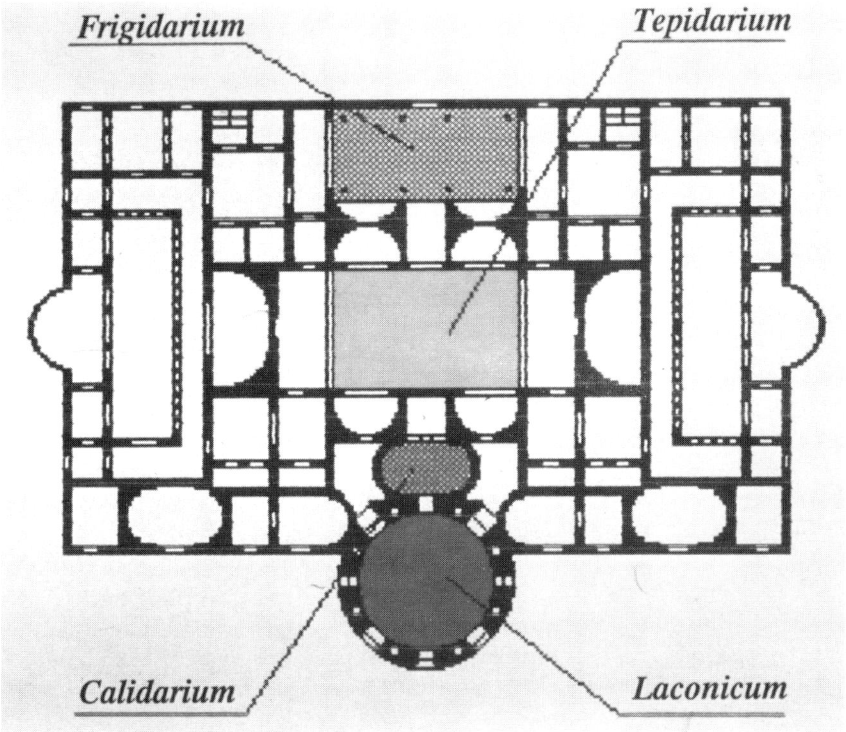

Figure 3. Layout of a typical Roman spa, showing positions of the main bathing rooms. Many other rooms existed for physiotherapy and other purposes.

business dealings, snacks, and formal meals. The baths became institutions for hygiene and recreation and, perhaps most of all, cultural centers where ideas were circulated, public opinion was formed, and basic political choices were prepared.

The boom in thermal balneology during the imperial period was concentrated mostly in Rome and at Baia, near Naples, but it spread rapidly throughout the Empire to the most remote villages. Outside Rome, a number of private mansions had artificially heated thermal pools, and many bathing complexes were built with the heating system devised by Caius Sergius Orata (figure 4). Where possible, the complexes were built near thermal

Figure 4. The patio in the Casale mansion, Piazza Armerina, Sicily, dating to Roman times, is an example of a private thermal pool, heated artificially. *P. D. Burgassi, c. 1967*

manifestations, taking advantage of the natural heat and the therapeutic properties of hydrothermal salts and muds.

The liveliness at the baths in the 1st century A.D. and the bustle dominating each thermal complex inspired Seneca (c. 60) to write this witty passage about Baia:

> "I live just above the public baths; imagine a clamor, a shouting in every tone, that makes you long to be deaf; I hear the moaning of those who are taking exercise: they let out hisses and breathe with difficulty. If someone stays still long enough to have a massage, I hear the slapping of hands on shoulders and it is a different sound depending on whether the slap is given with a flat or a curved hand. When someone arrives wanting to play ball but is unable to do so, if he doesn't shout and begin to count the strikes aloud, it means that the game is finished. There are also the mischief-maker, the robber caught red-handed, the chatterer who, when he speaks, is fascinated by the sound of his own voice; and then there are those who dive into the pool to swim, noisily splashing the water everywhere. But these at least use their own voices! Think of the depilator, who does not keep quiet but, when he relieves someone of their hairs, screams together with the person being depilated. Not to mention the cries of the vendors of drinks, of sausages, of cakes, and of the inn attendants who wander around offering their goods, each one with his own modulation of voice..." (translated by the authors from Paoli's Italian version, 1977).

The main thermal complexes in the territory of the Roman Empire toward the end of the 3rd century are shown on the map (figure 5). However, many of them existed long before the onset of the Roman rule. Inside or very near every thermal complex were statues, inscribed pillars, and temples honoring the gods, including Jupiter, Juno, Minerva, Mars, and Apollo (Panessa, 1983). In many thermal localities, cults of health-giving divinities were documented, most notably those of Asclepius (Musial, 1990), Hygieia (Daremberg and Saglio, 1899), and Sulis Minerva (Salway, 1981). The existence of cults is not surprising if one considers that in Antiquity, springs in general and hot springs in particular were thought of as fonts of life and health—divine gifts and therefore sacred. This explains why the cult of the waters was widespread in Antiquity throughout the whole Mediterranean area.

The recurring presence of monuments to the gods and votive pillars in or near all bathing complexes suggests that thermal springs may have been a catalyst for strengthening religious sentiments and their gradual spread among people of every social class and culture. Such sentiments had begun to form instinctively in prehistoric times (Cataldi, 1992), evolving by historical Antiquity into a mature body of faith. In certain cases, however, the decision to build temples, oracles, or altars near thermal springs may have met a need to justify visiting them for bathing and recreation.

	NAME OF THE SPAS	
	ANCIENT (*)	PRESENT
1	*Aquae Arnemetiae*	(Buxton)
2	*Aquae Sulis*	(Bath)
3	*Aquae Granni*	(Achen)
4	*Aquae Mattiacae*	(Wiesbaden)
5	*Aquae Aureliae*	(Baden Baden)
6	*Luxovium*	(Luxeuil)
7	*Aquae Neri*	(Nens les Bains)
8	*Aquae Elveticae*	(Baden -CH-)
9	*Aquae Pannonicae*	(Baden -A-)
10	*Aquae Calidae*	(Vichy)
11	*Aquincum*	(Budapest)
12	(?)	(Le Mont Dore)
13	*Aquae Segetae*	(Seceaux du Gatinais)
14	*Santicum*	(Villach)
15	*Aquae S.*	(Ilidza)
16	*Germisara*	(Geoagiu)
17	*Aquae Tarbellicae*	(Dax)
18	*Aquae Flaviae*	(Chaves)
19	*Allegre*	(Les Fumades)
20	*Fons Aponi-Aquae Patavinae*	(Abano Terme)
21	*Aquae*	(near Hunedoara)
22	*Aquae Statiellae*	(Aqui Terme)
23	(?)	(Cauterets)
24	*Ad Mediam*	(Baile Herculane)
25	*Aquae Sextiae Salluviorum*	(Aix en Provence)
26	(?)	(Amelie les Bains)
27	*Sarsina*	(Sarsina)
28	*Aquae Iasae*	(Varazdinske Toplice)
29	*Aquae Volaterranae*	(Monterotondo Marittimo)
30	*Aquae Populoniae*	(Campiglia Marittima)
31	*Fortes Clusini*	(Chianciano)
32	*Aurinia*	(Saturnia)
33	*Augusta Emerita*	(Alange)
34	*Centumcellae*	(Civitavecchia)
35	*Aquae Albule*	(Bagno di Tivoli)
36	*Pautalia*	(Kyustendil)
37	*Thermai Phazemoniton*	(Havza)
38	*Baiae*	(Baia)
39	*Thessalonica*	(Salonika)
40	*Neapolis*	(Sardara)
41	*Litus Leuternum*	(Santa Cesarea Terme)
42	*Aidepsos*	(Loutra)
43	*Thermopylai*	(Thrmopylai)
44	*Lipara*	(Lipari)
45	*Ephesos*	(Ephesos)
46	*Hierapolis*	(Pamukkale)
47	*Thermai Himerai*	(Termini Imerese)
48	*Korinthos*	(Korinthos)
49	*Methana*	(Methana)
50	*Aquae Dacicae*	(Sidi Moulay Yakoub)
51	*Therme Selinuntinae*	(Sciacca)
52	*Aquae Calidae Carpitanae*	(Korbous)
53	*Ad Sava Municipium*	(Hammam Guergour)
54	*Novar [-----]*	(Beni Fouda)
55	*Kos*	(Kos)
56	*Bab Khaled*	(Bab Khaled)
57	*Aquae Septimianae*	(Henchir el Hammam)

(*) Referring to the 3rd Century A.D.

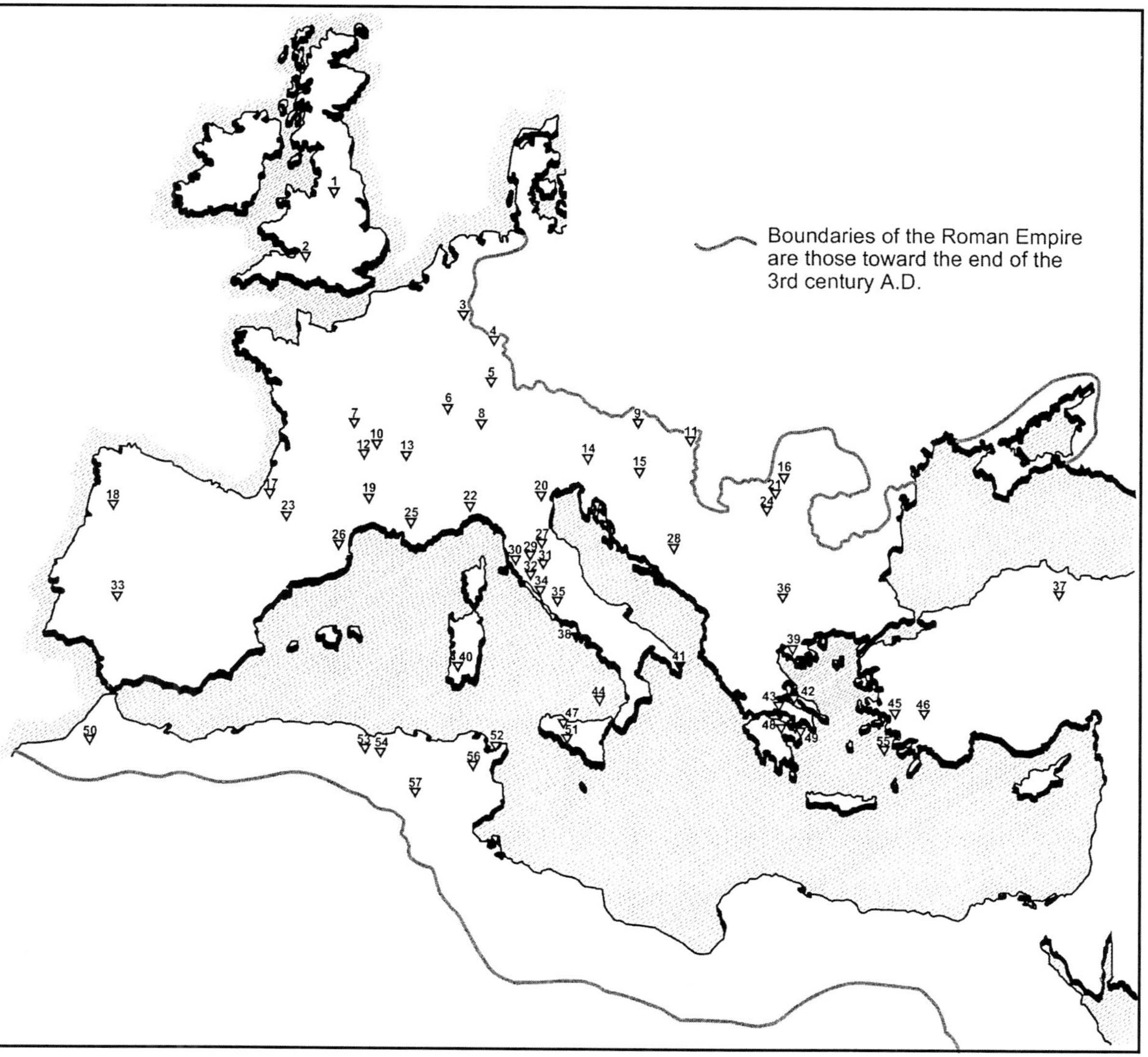

Figure 5. Location of the main natural baths at the height of the Roman Empire.

At any rate, the fact that thermal balneology was practiced by many peoples favored the development of local markets, trade centers, popular festivals, and athletic games at or near important thermal sites. In the latter case, athletes frequently trained at thermal sites to reach top form (*valetudo*) before the games began, and winners built thanksgiving pillars to healing divinities or demigods such as Athena (Minerva), Ares (Mars), and Hercules. The thermal springs of Methana (Peloponnesus, Greece) and the athletic games established at Corinth (near the thermal locality of Loutraki, not far from Methana) are significant in this regard (see Chapter 6 in this volume).

The widespread development of thermal balneology, together with the multiple functions undertaken at thermal sites before and after the start of the Christian era, provided a cohesive element among social classes and contributed significantly to the progress of civilization and cultural aggregation of Mediterranean peoples. Sometimes Roman rulers might have built thermal establishments with the aim of gratifying the masses and easing social tensions, thus curbing (at least in the more remote provinces) attempts at rebellion.

Between the 5th century B.C. and the 3rd century A.D., many Greek and Latin writers mentioned natural manifestations and dealt with aspects of geothermal phenomena. Citing all the works that refer to hot springs or other geothermal phenomena would be impossible, but some important authors are Anaximenes, Apuleius, Aristophanes, Aristotle, Caesar, Catullus, Democritus, Dio Cassius, Herodotus, Juvenal, Hippocrates of Kos, Lycophron, Ovid, Pausanias, Pliny the Elder, Poseidonius, Seneca, Strabo, Suetonius, Tibullus, Virgil, and Vitruvius.

Major thermal establishments are prominent on a world-famous cartographic document known as the *Tabula Itineraria Peutingeriana*, the official geographic map for the final period of the Roman Empire (a portion is in figure 6). This parchment was discovered in 1507 in a library in Alsace, France, and published by Konrad Peutinger, from whom the *Tabula* took its name. It is actually the work of an unknown cartographer who lived between the late 3rd and early 4th centuries and who relied on numerous *itineraria scripta* ("written itineraries") and *itineraria picta* ("mapped itineraries") to construct the map. The cartographer also used a relief model of the imperial territory, which was initially conceived by Caesar, later approved by Augustus, and eventually constructed by Agrippa in 20 A.D.

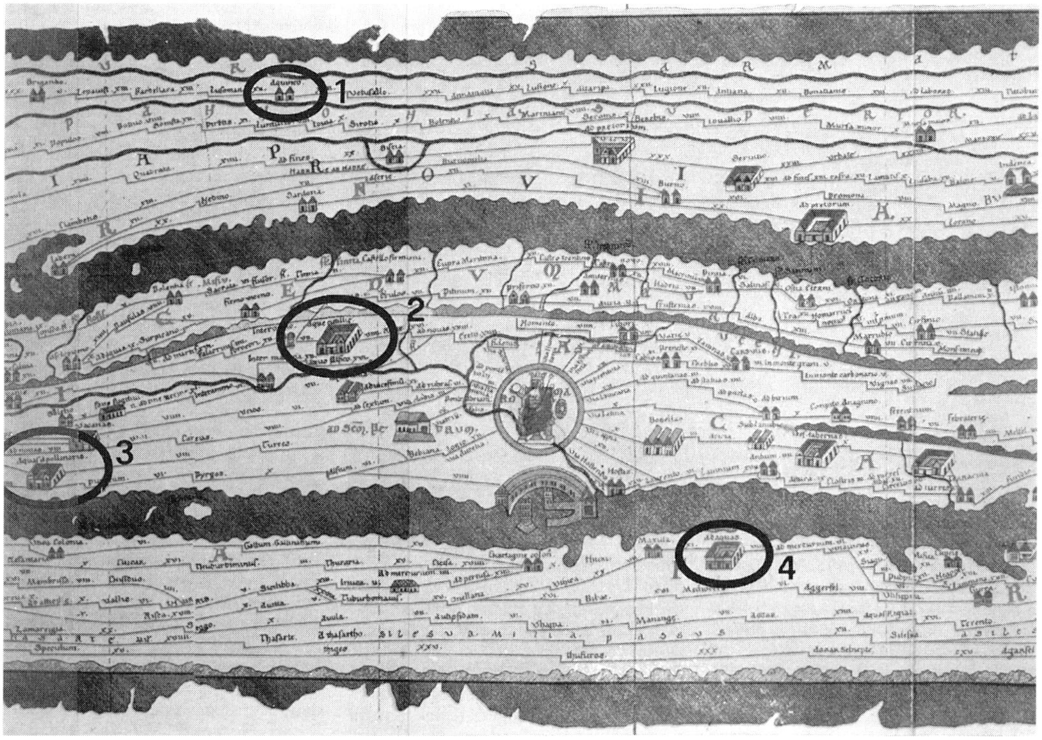

Figure 6. *Tabula Itineraria Peutingeriana*: the area corresponding to Central Italy with a sector from East-central Europe in the upper area and Northern Africa below. Some main baths operating during the 3rd century A.D. are highlighted: 1. Aquincum, Pest, Hungary; 2. Aquae Cutiliae, Palombara Sabina, Central Italy; 3. Aquae Apollinaris, Bracciano, Central Italy; 4. Ad Aquas/ Aquae Carpitanae, Hamman Lif, Tunisia. *P. Burgassi, from a copy of the* Tabula *in the archives of ENEL SpA/Geothermal Department at Larderello, with highlights added by the authors. Courtesy of ENEL SpA/Geothermal Department*

The parchment was composed of a single sheet over 7 m long and just 34 cm high, showing from west to east nearly all of Europe, much of Africa, and all of Western Asia as far as the Ganges River Valley. After its discovery, the original document was subdivided into 12 sheets, the first of which (covering Britain and Spain and bearing the author's name) was unfortunately lost.

The need to represent an enormous expanse of territory on a single sheet required that the author flatten all the land forms and compress the sea areas to thin ribbons. Therefore, the map must be read with the help of the place names and knowledge of the distances between the localities, which are given in leagues (2220 m) for Gaul, miles (1480 m) for the rest of Roman territory, parasangs (6000 m) for ancient Persia, and Indian miles (7000 m) for India (Brancati, 1986).

Only a few of the thermal sites in the Empire are on the *Tabula*, probably those baths that were strategically located, well developed, and famous for balneotherapy and commerce (Burgassi,

1987). Otherwise, why did the cartographer use graphic symbols that emphasized the thermal sites instead of choosing the less prominent graphic symbols selected for most other localities?

The following table summarizes the development of thermal bathing in Roman times from the 8th century B.C. to the 5th century A.D.

Chronological development of thermal bathing in the Mediterranean area, from the founding of Rome to the fall of the Roman Empire

Century	**Event**
8th-6th B.C.	Systematic spread in areas under Etruscan influence Limited development in other Mediterranean areas No interest by the Romans
5th-4th B.C.	Greatest development in Etruscan territories Moderate development in other Mediterranean areas Initial interest by the Romans at some thermal localities on the Italian mainland, primarily for strategic purposes. Commercial purposes were secondary.
3rd-2nd B.C.	Annihilation of the Etruscans (3rd century B.C.) and temporary decline of all geothermal uses in former Etruscan territories Moderate thermal bathing in Mediterranean areas First Roman visits to baths (elite enjoy bathing at thermal springs)
1st B.C.-3rd A.D.	Invention of a heating system with hot air circulation in artificial spas Rapid spread of natural and artificial baths Peak balneotherapy development over all of the Empire Over 1,000 public baths and hundreds of private baths (all artificial) in Rome Hundreds of thermal stations outside of Rome, nearly all natural Multiple functions of baths: • Balneotherapy (water, steam, muds) at different temperatures • Aesthetic practices (massage, shaving, skin care, hair styling, depilation) • Recreation (ball games, parties, and other social activities) • Food services (formal restaurants, snacks) • Centers for public information • Forums for political initiatives • Centers for business negotiations • Magnets for social aggregation and cultural events • Centers for cults • Local markets *Only in localities with natural baths* • Sites for popular festivals • Sites for athletic games
4th-5th A.D.	General decline of baths throughout Europe Thermal bathing practiced locally and at reduced levels Onset of the twilight period

Using Geothermal By-Products between the 3rd Century B.C. and the 5th Century A.D.

The Romans began taking an active interest in using and trading the by-products of geothermal energy in the 3rd or perhaps the 2nd century B.C., particularly after the victory over Carthage in 140 B.C. With trade routes now secured, the Romans dedicated themselves to public works projects, constructing temples and residential villas, and to crafts, arts, aesthetic practices, and health care.

All the uses of geothermal by-products developed by the Etruscans and other Mediterranean peoples before the 3rd century B.C. (see Chapter 9 in this volume) were adopted anew and improved in Roman times, and the Romans developed the following new uses:

- Hydrothermal compounds and pyroclastic products (kaolin, bentonite, perlite, pozzolan, and tuffs) were used as basic components or additives in preparing mixtures for pottery, cement slurries, construction materials, and bleaching solutions. Most compounds came from geothermal localities in Italy's pre-Apennine belt (Tuscany, Latium, and Campania), a few islands in the Tyrrhenian Sea (Ponza, Lipari, and Pantelleria), the Southern Aegean (Milos, Santorini, Nisyros, and Lesbos), and to a lesser extent from other volcanic islands and inland localities in the Southern and Eastern Mediterranean.

- *Palle da cane*, prepared in particular areas with natural manifestations, was widely traded throughout Roman territories. These were cakes of smectic clay, rich in hydrothermal salts (borates and aluminum silicates) taken from mud on the bottom of the pools of high-temperature manifestations in the Boraciferous region and perhaps also from other localities in Southern Italy such as the Phlegraean Fields, Ischia, Lipari, and Vulcano. From the start of the Christian era, the cakes were used widely to cure skin disorders and probably to degrease and tone the skin. The curious name literally means "balls for dogs," perhaps because the cakes, before they were used on humans, were tested at length on domestic animals, especially dogs, to ensure their efficacy at drying and healing sores and curing skin diseases, such as mange (Burgassi, 1987).

- Borax products from geothermal manifestations in the Boraciferous region and iron oxides from the same zone and/or other geothermal localities were used to prepare the glazes needed to paint fine pottery, and the glazes have proved to contain boric

Figure 7. Interpretation of a typical Roman vase, of the type called *vasa sigillata. Drawing by V. Svalova from Cataldi and Burgassi, 1992b*

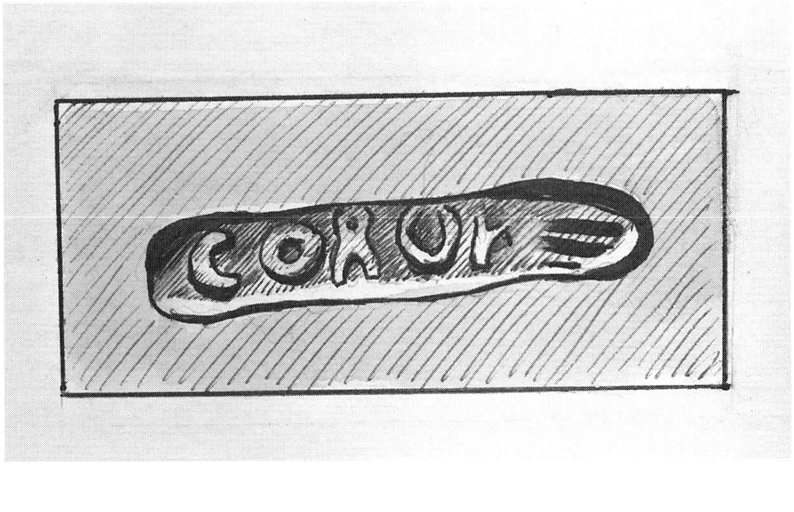

Figure 8. Interpretation of a typical trademark (*sigillum*) used by Roman factories as a guarantee of fine pottery. *Drawing by V. Svalova from Cataldi and Burgassi, 1992b*

compounds (Fiumi, 1943). Reddish-glazed ceramics from the 1st century B.C., called *vasa sigillata,* owe their name to the bas-relief figurines they bear, probably corresponding to the trademark (*sigillum*) of the ancient pottery workshop (figures 7, 8). Examples are displayed in the archaeological museum of Arezzo in Tuscany.

Figure 9. Panoramic view and detail of the travertine deposits of Pamukkale in Western Anatolia, Turkey. *R. Cataldi, 1981*

- Travertine (*lapis tiburtinus*, which means "stone from Tibur") derives its name from the ancient city of Tibur, present-day Tivoli, where a large deposit exists. This typical by-product of geothermal energy was cut to make building blocks or facing stones for buildings and monuments throughout the Empire, including the Roman Colosseum and the Baths of Caracalla, the most famous. Among the many quarries exploited intensively in Roman times, the travertine deposits of Tibur (near Rome) and Pamukkale in Western Anatolia must be cited (figure 9). The latter is famous for its characteristic carbonate configurations of gracefully cascading cups, decorated with organ-pipe encrustations around the outer edges.

The following table summarizes the use of geothermal by-products in Roman times, from the 3rd century B.C. to the fall of the Roman Empire.

Uses of geothermal by-products in the Mediterranean area, from the 3rd century B.C. to the 5th century A.D.

Products	**Uses**
Hydrothermal compounds and fumarolized pyroclastic products (kaolin, bentonite, perlite, pozzolan, and others)	Basic components or additives for • pottery-making • cement and other building materials • bleaching solutions
Thermo-mineral muds	• balneotherapy • care of skin diseases • degreasing and toning skin
Borates and iron oxides	• glazes for painting fine pottery
Travertine	• facing slabs and building blocks for important buildings

A Thousand Fertile Years for Geothermal Resources (5th Century B.C. - 4th Century A.D.)

For nearly a millennium, human relationships with geothermal resources were strong enough to be considered a propelling factor of civilization in the Mediterranean. We have discussed at length the influence of the Romans between the 2nd century B.C. and the 5th century A.D., but this does not mean that other Mediterranean peoples were not offering their own contributions to the development of thermal balneotherapy and the use of geothermal by-products. In reality, all Mediterranean peoples, including the Cretans, Phoenicians, Etruscans,

Gauls, Romans, Greeks, and Turks of Asia Minor, practiced and enjoyed balneotherapy in natural or artificial hot baths and found ways to use by-products of geothermal energy until the collapse of the Roman Empire. If the Etruscans deserve recognition as the historical fathers of the geothermal industry (Cataldi and Burgassi, 1992a), the development of natural heat during this millennium was a cultural heritage common to all the peoples of the Mediterranean and neighboring lands.

Twilight and the Age of Darkness (5th to 10th century A.D.)

After the *Tabula Itineraria Peutingeriana* was drawn, probably in the first decades of the 4th century A.D., thermal sites ceased to develop in the Mediterranean area. From the second half of the 4th century A.D. after the first sack of Rome, the practice of thermal bathing declined rapidly over the entire Empire, starting in the least important localities farthest from the capital. Although balneotherapy continued in some cases (the Baths of Caracalla in Rome, for instance, remained active at a reduced level until the early 6th century), thermal bathing gradually lost the custom of development, tradition, diffusion, and continuous renewal that had made it an international phenomenon for at least a millennium.

The decline continued and became more marked in the 5th and 6th centuries; it then proceeded apace with the dissembling process of the Roman Empire when Western Europe, Northern Africa, and the Middle East entered into a period of unrest, invasions, wars, and territorial partitions, forming the historical backdrop of the Dark Ages (Montanelli and Gervaso, 1965). The decline was not stopped by a precarious aggregation of people and lands established in some sectors of the former Roman Empire in the period between the 6th and 9th centuries when new geopolitical entities were formed in Europe (such as the Western and Eastern Roman, the Carolingian, and the Byzantine Empires).

A contribution to the decline of balneotherapy and the use of hydrothermal products may have come from a widespread belief in the last part of that period that the world would end by the year 1000. This event was insistently foretold to crowds by certain radical preachers for the last two centuries of the first millennium (Montanelli and Gervaso, 1965), and it discouraged visits to thermal baths for therapeutic and, especially, for recreational purposes. Within this framework of geopolitical, economic, and cultural gloom that affected a large part of Europe in the second half of the first millennium, one can hardly be surprised by the flourishing of many place names that associated hot thermal manifestations with the infernal world, such as *Valle dell'Inferno* ("Hell Valley"), *Buca del Diavolo* ("Devil's Hole"), and *Monte Cerberi* ("Mount Cerberus"). All these names were links to legends and fables that attributed the formation of the manifestations to works of the devil; all postulated the existence of a subterranean world peopled with souls condemned to eternal flames. Thus for development of geothermal resources, the last centuries of the first millennium were for much of Europe, and especially for Italy, a period of Dark Ages.

Cited References

Brancati, A. 1986. *Civiltà confronto, vol. 2.* La Nuova Italia Edit., Scandicci (Florence).

Burgassi, P. D. 1987. Historical outline of geothermal technology in the Larderello region. *Geothermal Resources Council Bulletin,* 16, no. 3, 3-18. (See Chapter 13 in this volume).

Cataldi, R. 1992. El año cero de la geotermia. In *La geotermia en el período Precolombino en las áreas Mediterránea y Mesoamericana. Geotermia, Revista Mexicana de Geoenergía,* 8, no. 2, 155-175. (See Chapter 2 in this volume).

Cataldi, R. and P. D. Burgassi. 1992a. Del misterio de la Atlantide a los primeros usos integrados de la energía geotermica. In *La geotermia en el período Precolombino en las áreas Mediterránea y Mesoamericana. Geotermia, Revista Mexicana de Geoenergía,* 8, no. 3, 251-264. (See Chapter 9 in this volume).

Cataldi, R. and P. D. Burgassi. 1992b. Florecimiento y decadencia del termalismo y de los otros usos directos del calor de la Tierra en el área Mediterránea desde el nacimiento de Roma hasta el fin del primer milenio. In *La geotermia en el período Precolombino en las áreas Mediterránea y Mesoamericana. Geotermia, Revista Mexicana de Geoenergía,* 8, no. 3, 265-288.

Cataldi, R. and P. Chiellini. 1995. Geothermal energy in the Mediterranean area before the Middle Ages. *Proceedings of the World Geothermal Congress,* 373-380. (See Chapter 11 in this volume).

Daremberg, C. and E. Saglio. 1899. *Dictionnaire des antiquites grecques et romaines, vol. 3.* Paris.

Fiumi, E. 1943. *L'utilizzazione dei lagoni boraciferi della Toscana nell'industria medievale.* University of Florence, CYA Edit., Florence.

Montanelli, I. 1969. Storia di Roma: 753 B.C.-476 A.D. In *Storia d'Italia.* Rizzoli Edit., Milan.

Montanelli, I. and R. Gervaso. 1965. L'Italia dei secoli bui. In *Storia d'Italia.* Rizzoli Edit., Milan.

Musial, D. 1990. Sur le culte d'Esculape en Rome et en Italie. *Dialogues d'historie ancienne,* 16, 231-238.

Panessa, G. 1983. Le risorse idriche dei santuari greci nei loro aspetti giuridici ed economici. *Annali della Scuola Normale di Pisa,* 13, 359-387.

Paoli, U. E. 1977. *Vita romana: Usi, costumi, istituzioni, tradizioni.* Mondadori Edit., Milan.

Salway, P. 1981. *Roman Britain.* Oxford Clarendon Press, Oxford.

Seneca. c. 60 A.D. Ad Lucilium; Epistulae morales. In *Seneca: Epistles, vol. 1.* Reprint 1917, R. M. Gummere, trans. Harvard University Press and W. Heinemann Ltd., Cambridge and London.

The Authors:

Raffaele Cataldi
55, Via del Borghetto
56124 Pisa, Italy
Telephone/Fax: 39.050.59.81.07
E-mail: rafcat@tin.it

Pier Domenico Burgassi
2, Via G. Galilei
56044 Larderello (Pisa), Italy
Telephone: 39.058.86.74.89

Statue of Hippocrates, famed medical doctor and scientist of ancient Greece, 5th century B.C. Considered the father of medicine, he was the first scientist to describe surface waters systematically and to classify them on the basis of chemical characteristics and temperatures. ***Eng. Demetrios Mylonas; kindly authorized by the Direction of the Hippocrates Museum in Kos, Greece***

11. Geothermal Energy in the Mediterranean before the Middle Ages, A Review

by
Raffaele Cataldi
Paolo Chiellini

Abstract: Archaeological finds dating from the Lower and Middle Neolithic and considerations of refinement of customs, myths, and cults linked to manifestations of terrestrial heat several millennia before the Christian era indicate that the human-geothermal energy relationship can be traced at least to the Upper Paleolithic in the Mediterranean.

After outlining the first human contacts with geothermal manifestations in prehistoric periods and the earliest Mediterranean civilizations, this chapter explains how the Etruscans developed hydrothermal products and how the Romans spread thermal bathing. Major Greek and Latin authors are also mentioned who, from the 6th century B.C. to the 4th century A.D., described many aspects of geothermal manifestations or speculated on their geneses, and who thus may be considered the historical fathers of scientific thought on geothermal heat.

Reprinted, with revisions, from the *Proceedings of the World Geothermal Congress*, 1995, pp. 373-380.

Early Contacts of Mediterranean People with the Earth's Heat

The oldest known artifact reflecting Mediterranean peoples' interest in geothermal phenomena dates from the 7th millennium B.C. (Mellaart, 1967). It is a painting from Çatal Hüyük in Western Anatolia that shows an erupting volcano (probably Hasan Dağı) behind a Neolithic settlement with a highly evolved dwelling structure (see Chapter 2 in this volume). The painting's details illustrate not just morphological features of the volcano but also particulars of the eruption itself, such as the explosion plume and volcanic bombs. Thus by the Lower Neolithic, some Mediterranean peoples already had developed a keen sense of observation and an advanced ability to depict natural phenomena. These peoples must have possessed, collectively, much empirical knowledge on manifestations of the Earth's energy, such as hot springs, fumaroles, and volcanic eruptions (Cataldi et al., 1992-93; Cataldi, 1993a).

Such knowledge probably matured on the individual level over many thousands of years (perhaps even tens of thousands) as direct contacts with manifestations continued, yielding a complex relationship between prehistoric people and geothermal phenomena. Indeed, it is likely that people chose to settle in geothermal localities with "peaceful" manifestations and/or geothermal by-products, places ideal for cooking, bathing, and finding useful materials such as hydrothermal encrustations, obsidian, and other volcanic products suitable for many practical purposes. Other benefits indirectly resulting from the Earth's heat, important at the end of the nomadic period, were the fertility of volcanic soils, the security of some lake basins molded by caldera rims, and the availability of freshwater springs on volcanic outcroppings.

In other localities, it seems certain that the occurrence of volcanic eruptions and earthquakes had the opposite effect, discouraging early peoples from frequenting them. Ever since the earliest days of prehistory, first on an individual level and then collectively, people developed a pragmatic relationship with different types of geothermal phenomena: attraction to some features and avoidance of others (Cataldi, 1992 and 1993b).

Although still rooted in Paleolithic times, this relationship must have evolved gradually and widened in scope, affecting people's spiritual sphere and resulting in the formation of religious sentiments toward underground entities located in geothermal areas (Cataldi et al., 1992-93). Neolithic artifacts found near natural manifestations in the Mediterranean, such as stone statuettes known as *veneri votive*, bone amulets, food offerings, and other objects, clearly relate to religious beliefs in underground entities (Cataldi, 1993a; Grifoni Cremonesi, 1999), as do the megalithic constructions known as "divine stones," found near fumaroles on the volcanic island of Pantelleria in the Western Mediterranean and dating from the 3rd millennium B.C. (Burgassi et al., 1992).

Indirect proof of geothermal influence on prehistoric peoples' spiritual sphere comes from later evidence of customs, cults, and sophisticated myths clearly inspired by natural manifestations and volcanoes. Some examples are the cults of the Great Mother in Western Anatolia, Hephaestus in Greece, and Mephitis in Southern Italy; the myth of Prometheus' theft of fire; and evidence of reverence paid to the nymphs of the thermal waters and the health divinities, such as Asclepius and Hygieia. These cults, myths, and customs could not have appeared suddenly in passage from the Stone Age to the Metal Age. Most likely they represent refinements of religious sentiments born in much earlier times, perhaps stretching back to the

Late Mousterian (35,000 to 40,000 years ago) when, with the onset of burial practices, the possibility of a "dynamic subterranean world, inhabited by endogenous forces capable of giving rise to external manifestations of heat, some benevolent and peaceful, others violent and destructive...entered the mind of mankind" (Cataldi, 1992).

A similar process must have occurred in the evolution of a rational rapport between prehistoric people and geothermal phenomena. Probably thousands of years before the Neolithic, Mediterranean peoples observed volcanoes, natural manifestations, and other geothermal phenomena with curiosity and intellectual interest, eventually thinking that they all could be controlled by an underground cause. That this cause may have been a subterranean entity toward which embryonic religious sentiments were expressed is not important here. What is important is that, at a moment of the Upper Paleolithic, which Cataldi (1992) places between 10,000 and 30,000 years ago, humans began to reason about the Earth's heat in terms of cause and effect. Mediterranean peoples started to differentiate, by at least the Middle Paleolithic, between the "peaceful" and "violent" types of geothermal phenomena.

Thus a complex relationship—with utilitarian, religious, and intellectual aspects—was established between prehistoric people and geothermal heat in the Mediterranean. The relationship must have begun in the Paleolithic, probably over 50,000 years ago, and gradually developed and enriched itself until the end of the Neolithic over that long, long span of time that Cataldi (1992) has called the *year zero* of geothermics.

Geothermal Heat in the Oldest Mediterranean Civilizations

On several important nonvolcanic islands in the Eastern Mediterranean, such as Cyprus, Rhodes, and Crete, excavations have revealed obsidian and lava tools dating from the 5th to the 3rd millennium B.C. They indicate that volcanic materials and other geothermal by-products were used not only in the site of origin but were bartered among the first tribes of people that a millennium or two later would give rise to the oldest civilizations of the Mediterranean area: the Hattis, Hittites, and Phoenicians in Anatolia; Minoans in Crete and the Southern Cyclades; and the Mycenaeans and others on the Aegean coast.

Many of these materials and by-products came from Western Anatolia where, in addition to Çatal Hüyük, other important Neolithic settlements had formed (Mellaart, 1967). Their development probably was helped by the abundance of geothermal by-products in the area, including sulfur and cinnabar, probably first used for pigments.

During the Neolithic, however, and even earlier in protohistoric times, the volcanic islands of the Aegean, such as Lesbos, Limnos, Chios, Nisyros, Yialí, Santorini, and Milos—and other geothermal localities along the Eastern Mediterranean coast—became important for mining and exporting obsidian and other plentiful geothermal by-products (Fytikas et al., 1999), such as alum, bentonite, borates, iron oxides, kaolin, lapilli, perlite, pozzolan, silica, and sulfur. The extraction is documented from the 3rd or 2nd millennium B.C. Neolithic artisans started using some of these compounds and their by-products to make pottery and pigments, tan hides, and color glass. Cooking, thermal bathing, and curing skin diseases with thermo-mineral muds probably became common on a local scale (Cataldi, 1992, 1993b; Fytikas et al., 1999).

At the same time, numerous myths, legends, cults, and beliefs inspired by geothermal phenomena flourished and spread among the Mediterranean peoples, a maturing of the first rational thoughts and religious sentiments growing in the mind and spirit of humans since the darkness of prehistory (Cataldi, 1992). One of many legends is that of Atlantis, which may have been suggested to Plato by an account of the huge volcanic explosion that almost completely obliterated the island of Thera (the present-day Santorini) in about 1630 B.C. (Fytikas et al., 1998; Luce, 1969).

A rising tide of colonization and increased land and sea traffic took place in the Mediterranean area thanks chiefly to the Cretans, Phoenicians, and Mycenaeans from the second half of the 2nd millennium B.C. It extended to many geothermal localities in Italy, such as the Phlegraean Fields, Western Latium, Central-southern Tuscany, the Euganean sector, and islands such as Pantelleria, the Aeolian and Pontine archipelagos, and Ischia, rich in hot springs, active manifestations, hydrothermal minerals, and other by-products of terrestrial heat. All of these localities gained importance for mining and exporting hydrothermal compounds, as did some Aegean islands and other geothermal localities along the Eastern Mediterranean coast.

Cults and legends grew as well, inspired by the Italian natural manifestations. Some were new, and some were grafted onto prior beliefs from other Mediterranean areas (Burgassi et al., 1992). One cult is that of the caretaker gods for the thermal waters of *Fons Aponi* (present-day Abano in the Euganean sector of Italy). Abundant votive offerings date this cult back to the early Venetic period, beyond 1000 B.C.

Among the many Italian legends formed in Antiquity is one where the Phlegraean Fields (Naples) were the scene of a struggle between Hercules and Typhoeus, the giant with 100 fire-spewing heads, who vainly contested Zeus' dominion over the world. Hercules' victory signified the victory of the order instituted by Zeus over the disorder from the "rebel" forces of nature (in other words, a political parable advocating the legitimacy of Mycenaean expansion in Southern Italy). The legend includes a sophisticated attempt by ancient Mediterranean peoples to explain the numerous craters, domes, fumaroles, jets of boiling water, eruption cones, and steaming grounds that existed in the Phlegraean Fields and on the volcanic islands of the Tyrrhenian Sea. The legend points to people's mature desire, at least 1,000 years before the Christian era, to explain the formation of high-temperature manifestations and to find a cause-and-effect relationship for some geothermal phenomena.

THE ETRUSCANS: HISTORICAL FATHERS OF THE GEOTHERMAL INDUSTRY

MYCENAEANS AND PHOENICIANS FREQUENTED THE TYRRHENIAN ISLANDS AND COASTS BETWEEN the 15th and 10th centuries B.C., coinciding with the development of the Italian "Villanovan culture," during which ancient Italic populations grouped into fairly large villages. This characteristic was prominent in the Tyrrhenian sector of Central Italy, where the Etruscans established an organized community starting in about the 12th century B.C. Just a few centuries later, they attained one of the highest degrees of civilization in all activities then practiced in the Mediterranean area, including mining metallic and nonmetallic ores. The rapid rise of the Etruscans was favored by their commercial contacts with Mycenaean and Phoenician navigators, who carried Etruria's riches to foreign markets:

- Metallic minerals: copper, iron, manganese, lead, silver, and zinc
- Evaporitic minerals: alabaster, gypsum, and sodium chloride
- Hydrothermal minerals: alum, borates, hydrated silica, iron oxides, kaolin, sulfur, and travertine

Hydrothermal deposits were mined in the abundant Etrurian areas of active and fossil geothermal manifestations, particularly in the present-day Larderello region (Cataldi and Burgassi, 1992a). This allowed the Etruscans to offer the Mediterranean market a wide variety of products useful for the crafts and industries of the day. Not only did the Etruscans mine many hydrothermal minerals, they sorted, processed, and sold finished products in distant markets. The Etruscans also refined techniques for grinding, processing, and mixing the products, using them to make pottery, paints, glazes, and pigments; prepare ointments and salves; fuse and color glass; bleach wool; and treat fabrics (Burgassi, 1987; Cataldi and Burgassi, 1992a). Numerous archaeological finds prove this, including some very fine ceramics with glazes and paints that have traces of boron compounds (Fiumi, 1943). This list does not exhaust the range of Etruscan activities in the geothermal sector but merely outlines the skill of this people in mining and using hydrothermal minerals. In addition, the Etruscans valued balneotherapy, as recorded by a number of Greek and Latin authors (Burgassi, 1987; Cataldi and Burgassi, 1992a and 1992b).

The relationship between the Etruscans and active geothermal manifestations included religious aspects, such as divination based on the shapes of the steam plumes from fumaroles, a unique belief in the afterlife, and the formation of cults for subterranean divinities. The worship of Velchans (the equivalent of the Greek god Hephaestus, keeper of fire and volcanoes) seems a practice initially borrowed from the Hellenic world. However, the Etruscans elaborated on it and brought the cult first to Etruria proper and afterwards to Southwestern Italy, site of Etruscan colonies and trading outposts, such as Capua and Cuma.

The presence of many spectacular geothermal manifestations, such as steaming grounds and steam jets, deeply influenced the Etruscan concept of the netherworld. They thought it a sad, dark, inhospitable place that the ghosts of the dead were obliged to share with the subterranean gods, who were anything but kind (Pallottino, 1984). This concept is similar (but not identical) to the ancient Greek notion of Hades.

Many important Etruscan centers—Populonia, Saturnia, Bolsena, Tarquinia, Cerveteri, Veio, and Cuma—grew up next to thermal manifestations, a fact that hardly seems coincidental (see Chapter 9 in this volume). We may conclude that the Etruscans fully grasped the importance of geothermal manifestations and exploited their full potential: using hot waters and thermo-mineral muds for therapeutic and recreational purposes; extracting, processing, selling, and using a wide variety of hydrothermal products; and developing urban communities near thermal springs. For these reasons, they have been dubbed the historical fathers of the geothermal industry (Cataldi and Burgassi, 1992a).

Thermal Bathing and Other Uses of Geothermal Resources in the Roman Period

Roman interest in geothermal resources began emerging in the early 4th century B.C., as they increased political and military pressures on nearby peoples to expand Roman rule over the whole Italian peninsula. These pressures were also directed at Etruria, because of territorial contiguity and the region's rich metal-bearing ores and hydrothermal deposits (Cataldi and Burgassi, 1992a, 1992b). The Romans understood the strategic importance of Etruscan roads, which often linked towns near thermal localities: Cerveteri, Veio, Tarquinia, Saturnia, and Massa Marittima (see Chapter 9 in this volume). Initial Roman interest in geothermal localities appears to be based on political and economic concerns, not a need to frequent hot springs for therapeutic or recreational purposes (Cataldi, 1993b). However as early as the 4th century B.C., Roman soldiers and travelers undoubtedly began assimilating many Etruscan traditions, including spa visits for cures and rest. From this early contact, Roman balneological customs grew. Although Rome itself had no thermal springs, the upper-class Romans compensated by bathing in artificial spas built in all luxurious private residences. Balneology in artificial spas made inroads among the Roman middle classes after the Etruscans were crushed in the 3rd century B.C. However it did not become widespread until the early 1st century B.C. when Roman rule over the Mediterranean was more secure and the Romans had started to build large public spas in the capital.

Before the Empire was established in 29 B.C., however, strong support for the spread of thermal bathing among Romans came from interaction with Greeks in Southern Italy, Greece, and the Eastern Mediterranean. Here thermal balneology was established as a local tradition several centuries before the Romans arrived, although it was practiced only in natural baths. Moreover, numerous cults had formed outside Rome with protective divinities of the thermal waters, the most prominent of which were those of Asclepius and Hygieia (Cataldi and Burgassi, 1992b; Fytikas et al., 1999).

In some natural thermal spas under Greek influence, three main types of baths were used: the sweat bath in a hot dry cavern (*laconicum*), a hot-water bath, and a steam bath. The steam bath was made by channeling steam from natural fumaroles into artificial caverns dug in the ground (*hypocausts*). Baia, north of Naples, was one of the most important spas where the three types of rooms were used before Roman times. Baia was developed further by the Romans in the 1st century B.C. and reached its height during the imperial period (Cataldi and Burgassi, 1992b; Pasquinucci, 1987).

It was precisely the arrangement of Baia's natural spa that gave Caius Sergius Orata, an ingenious Baian citizen in the early 1st century B.C., the idea of reproducing it in the artificial public thermae of Rome by building a series of adjoining bathing rooms with three different temperature levels (Cataldi and Burgassi, 1992b). To do so, he heated water in an ordinary wood-burning boiler and used air as a carrier for the heat (see Chapter 10 in this volume). Thus, the public baths in Rome had three main thermal rooms: the *laconicum* with steam for sweating, the *calidarium* with hot water for bathing, and the *tepidarium* with warm water for bathing. Afterwards, the *frigidarium*, a pool at ambient temperature, was used to tone the body. Other rooms were used for physiotherapy. This orderly thermal scheme used heat in cascade, going from hot to cold. A similar process is used in many projects today, and it all started in Mediterranean geothermal localities, especially Baia.

For the next three to four centuries, this scheme was used for public and private artificial baths, not only in Rome and important localities of the Empire, such as the baths of Curium in Cyprus far from active geothermal manifestations, but also in places with natural hot water or steam. The balneological scheme developed by Caius Sergius Orata was thus fundamental to the rapid growth of thermal bathing in every social class in the Roman Empire until the early 4th century A.D. In the 3rd century in Rome alone, 1,000 public baths existed—about one spa for every 1,000 inhabitants (Montanelli, 1969).

In the first three centuries of the Christian era, thermal bathing became deeply rooted in the Mediterranean: the baths were a daily meeting place and a main site for civic life. In the last decades of the republican period and even more so during the imperial period, the thermae were no longer just baths. They were also beauty institutes (massages, depilating, and hair styling), gymnasiums (ball games and physical exercise), libraries, reading centers, meeting places, and complexes equipped with all kinds of services, such as

Hypocaust of the ancient thermae of Kourion (Curium), West of Limassol, Cyprus, dated to the imperial period of Rome. *R. Cataldi*

restaurants and shops—open to men and women of all ages and social classes (Pasquinucci, 1987; Cataldi and Burgassi, 1992b). Everywhere the baths became socialization sites and centers where public opinion was formed, capable of influencing political decisions on all levels.

Spas located near natural manifestations were places for worship and religious practices. Temples, inscribed pillars, statues, altars, and other votive artifacts in honor of the gods or the protective divinities of hot waters have been found in or near each thermal complex. In Roman times, betrothed young ladies used to go through a cycle of ritual baths at natural spas immediately preceding their marriages. According to popular beliefs, this cycle was needed to appease the gods of fertility through purification of the female body. It was believed that, if performed in thermal waters with particular therapeutic properties, these ablutions enabled deflowered young ladies to recover their chastity. This practice seems to have occurred in the San Casciano dei Bagni thermal springs. Such ritual practices recall the legend of the goddess Hera (Juno) who, before retiring into the bridal room with her demanding spouse Zeus, would fly every evening to the sacred thermal spring of Kanathos, near ancient Nauplia, taking a warm bath to restore her virginity (Pausanias, 155-170 A.D.).

These practices show the importance of thermal bathing in the first four centuries of the Christian era until the fall of the Roman Empire in the 5th century A.D. They are described by many poets, philosophers, geographers, and historians, including treatises on medicine, natural sciences, and engineering by Greek and Latin authors (Cataldi and Burgassi, 1992b; Fytikas et al., 1999). The importance of some thermal sites in the imperial period is also illustrated in the *Tabula Itineraria Peutingeriana*, from the early 4th century, the official geographic map of the Roman Empire during its final period (see Chapter 10 in this volume). This map shows several

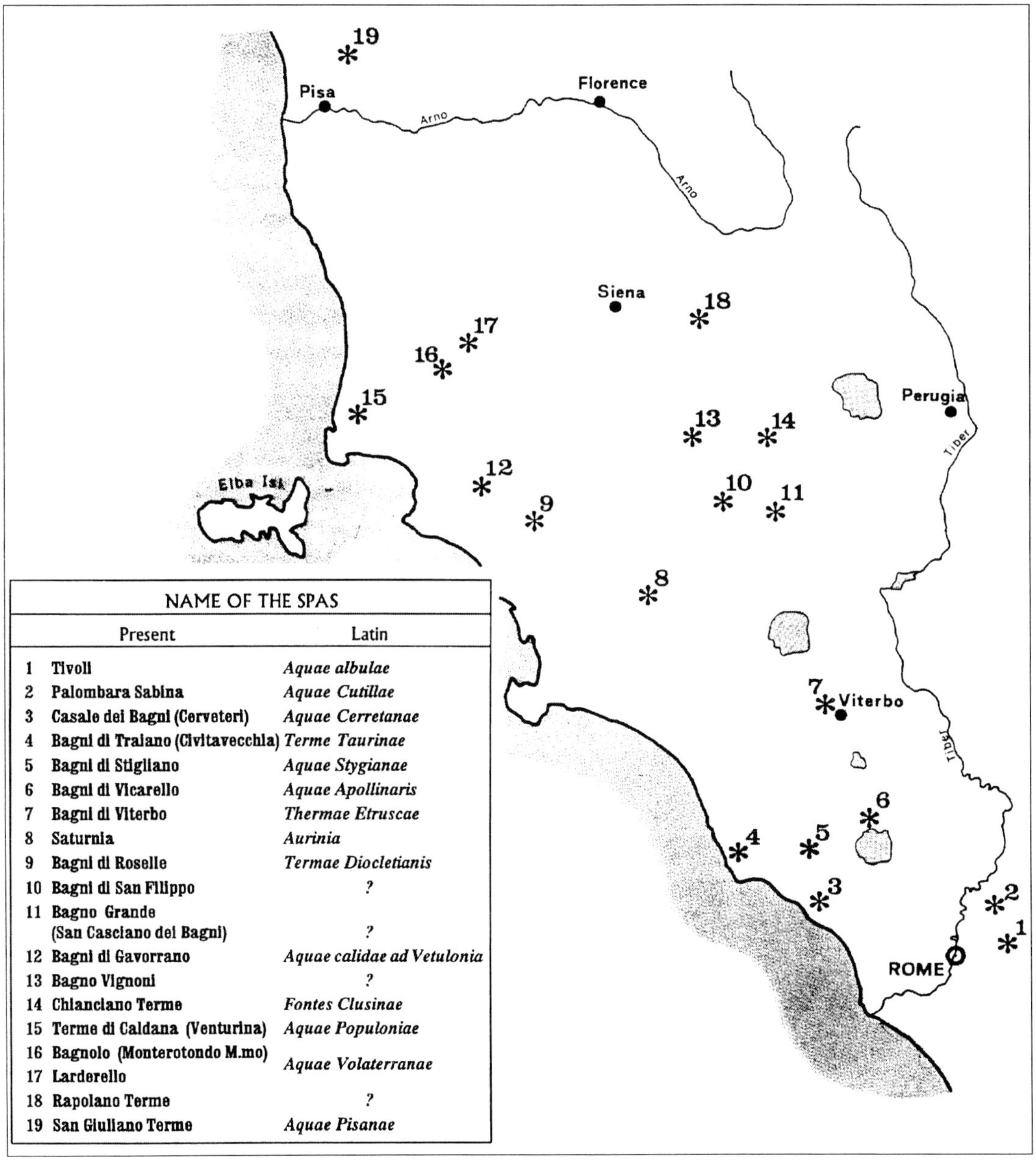

NAME OF THE SPAS		
	Present	Latin
1	**Tivoli**	*Aquae albulae*
2	**Palombara Sabina**	*Aquae Cutillae*
3	**Casale dei Bagni (Cerveteri)**	*Aquae Cerretanae*
4	**Bagni di Traiano (Civitavecchia)**	*Terme Taurinae*
5	**Bagni di Stigliano**	*Aquae Stygianae*
6	**Bagni di Vicarello**	*Aquae Apollinaris*
7	**Bagni di Viterbo**	*Thermae Etruscae*
8	**Saturnia**	*Aurinia*
9	**Bagni di Roselle**	*Termae Diocletianis*
10	**Bagni di San Filippo**	?
11	**Bagno Grande (San Casciano dei Bagni)**	?
12	**Bagni di Gavorrano**	*Aquae calidae ad Vetulonia*
13	**Bagno Vignoni**	?
14	**Chianciano Terme**	*Fontes Clusinae*
15	**Terme di Caldana (Venturina)**	*Aquae Populoniae*
16	**Bagnolo (Monterotondo M.mo)**	*Aquae Volaterranae*
17	**Larderello**	
18	**Rapolano Terme**	?
19	**San Giuliano Terme**	*Aquae Pisanae*

Main natural thermae in operation during Etruscan and Roman times between the Tiber and Arno Valleys.

thermal stations, indicating their great importance not just as spas but as commercial and strategic centers (Cataldi and Burgassi, 1992b). These thermal stations, however, are only a few of those operating at the end of the 3rd century; many hundreds of thermae were spread throughout the territory of the Roman Empire. For example, in only the western sector of Central Italy, some 20 thermal stations could be found in a territory of about 20,000 km^2, most of them dating to Etruscan times (7th-4th century B.C.).

Beginning near the end of the 2nd century B.C., the Romans started systematically developing industrial and artisanal activities everywhere they governed. They began a vast construction program that included public buildings, temples, monuments, stadiums, and villas. At the same time, they greatly expanded the use of hydrothermal minerals and other by-products of terrestrial heat.

Over five centuries, the Romans developed the following uses of geothermal by-products (Cataldi and Burgassi, 1992b):

- Pyroclastic and fumarolized rocks, and hydrothermal minerals, such as bentonite, perlite, and pozzolan, to prepare cement slurries

- Travertine and volcanic materials, such as lavas, tuffs, and lapilli, for building

- Kaolin to make fine ceramics and bleaching solutions for the textile industry

- Boron compounds and iron oxides to prepare glazes and enamels for painting fine pottery

- Preparation and wide commercialization of *palle da cani* ("balls for dogs"), soap-like masses of smectic clays rich in hydrothermal minerals used to cure skin diseases (Burgassi, 1987)

The importance of these uses and of ones previously mentioned, including thermal bathing, began to decline in the second half of the 4th century. In the next two centuries, they almost disappeared in many areas of Europe. Especially in Italy, thermal balneology and the use of hydrothermal by-products started to fade in the second half of the 5th

Chronological development of ancient scientific thought on geothermal heat (6th century B.C.-4th century A.D.) Revised from Cataldi, 1993b

Century	Author	Work	Subject Treated	Speculation or Theory
6th B.C.	**Anaximenes**	See Aristotle, *Meteorology*	Earthquakes	Origin of earthquakes due to deformation of the Earth's crust, caused by "drying up of the soil" during arid periods, or by "swelling up of the soil" during rainy periods.
5th B.C.	**Herodotus**	*The Nine History Books*	Thermal manifestations	Description of many Greek thermal manifestations. Some of them would form after volcanic eruptions, accompanied by "great clouds of fire."
5th B.C.	**Hippocrates**	*Air, Water, Land*	Organoleptic and physio-chemical characteristics	Classification of natural waters as stagnant, springs, and thermal. Salinity of the thermal waters is a function of temperature.
5th-4th B.C.	**Democritus**	See Aristotle, *Meteorology*	Earthquakes	Origin of earthquakes due to subterranean water flow, with formation of hydrostatic overpressure, or pressure drops, in underground cavities.
4th B.C.	**Aristotle**	*Meteorology*	Earthquakes	Origin of earthquakes due to "gas streams" caused by "thermal expansion" arising from the Earth's "internal fire," or from the sun's heat.
3rd B.C.	**Lycophron**	*Alexandra* (or *Cassandra*)	Physio-chemical characteristics of the waters	Description of the healing properties of boric waters from the main geothermal area of Tuscany (Larderello).
2nd-1st B.C.	**Poseidonius**	See Strabo, *Geography*	Volcanism and seismo-volcanic phenomena	Description of changes in the Earth's surface due to various natural phenomena (floods, erosion, etc.), especially volcanic phenomena.
1st B.C.	**Strabo**	*Geography*	Volcanism, volcanic phenomena, and thermal manifestations	Description of the main geothermal manifestations in the Mediterranean area. Detailed description of the volcanic explosion on Thera at the beginning of the 2nd century B.C., and of the encrusting properties of thermal waters.

1st B.C.	**Pompeius Trogus**	See Justin, *Epitome of Phil. Hist.* by P. Trogus	Volcanism and seismo-volcanic phenomena	Description of earthquakes in the southeastern sector of the Aegean Sea and the volcanic explosion at Thera in 197(?) B.C.
1st B.C.	**Tibullus**	*Elegies*	Thermal manifestations	Comparison between the natural manifestations of Tuscany and the area near Naples.
1st B.C.	**Ovid**	*Metamorphoses*	Earthquakes and explosions	Description of earthquakes, volcanic eruptions, and hot springs. In particular, description of the explosion at Methana (Greece) in the 2nd century B.C. (probably a phreatic explosion).
1st B.C.	**Vitruvius**	*Architecture*	Volcanism and thermal manifestations	Genetic interpretation of the manifestations of the Phlegraean Fields, due to decomposition of minerals in the subsoil. The author probably had in mind a heat-formation process now called "exothermal reactions."
1st B.C.-4th A.D.	**Various Greek and Latin authors**	See A. Corretti, *Baia*	Volcanism, thermal manifestations, and physio-chemical characteristics of the waters	Description and/or comparison of various geothermal localities in the Mediterranean area. According to Strabo, the manifestations of the Phlegraean Fields can all be attributed to the same volcanic phenomenon.
1st A.D.	**Pliny the Elder**	*Natural History*	Volcanism and thermal manifestations	Description of most main thermal manifestations in Southern Europe and the Mediterranean area. Description of the eruption of Vesuvius in 79 A.D.
2nd A.D.	**Pausanias**	*Description of Greece*	Earthquakes, natural manifestations, and physio-chemical characteristics of thermal waters	Inventory of most thermal springs of Greece and Southern Italy. Description of the warning signs of earthquakes.
1st-4th A.D.	**Apuleius and others**	See T. Ritti, *Hierapolis: Excavations and Research*	Thermal manifestations	Description of the phenomenon of gas stratification.

century, marking a long period of recession that would extend beyond the threshold of the second millennium (Cataldi and Burgassi, 1992b).

The First Nucleus of Scientific Thought on Geothermal Heat

A depiction of how volcanic eruptions were seen by early prehistoric peoples in the Mediterranean area is in a painting from the Early Neolithic settlement of Çatal Hüyük in Western Anatolia (see Chapter 2 in this volume). Ancient myths and legends illustrate indirectly the desire of protohistoric Mediterranean peoples to find a cause and effect relationship for manifestations of the Earth's heat. Such elements formed a foundation for later scientific study of geothermal phenomena.

The first direct description of events caused by the Earth's heat dates to the 6th century B.C., when Egyptian priests gave Solon a detailed oral account of a natural catastrophe that had occurred about a thousand years earlier. The account, handed down orally by Egyptian priests from generation to generation, most likely referred to the huge volcanic explosion on Thera around 1630 B.C. (Fytikas et al., 1999).

After the 6th century B.C., scientific information about geothermal phenomena is found in the works of numerous Greek and Latin thinkers. The main authors, their works, and the geothermal aspects treated in them are summarized in the table. The geothermal phenomena described in some of these works are often accompanied by speculation on their causes, such as the existence of a heat source (the "internal fire" of the Earth). Other works deal with the classification of spring waters on the basis of temperature and chemical characteristics, the encrusting properties of water, the stratification of CO_2 as a function of density, some underground chemical reactions, and the surface characteristics of geothermal areas. The need to explain geothermal phenomena concerned Greek and Latin thinkers, who, of course, developed their interpretation using the logic of the philosophical theories of their time. Together, these concepts form the first scientific body of thought on geothermal energy, knowledge that gradually evolved and enriched itself with time, remaining an indispensable basis for understanding the development of scientific approaches to this matter. For this reason, these authors should be considered the historical fathers of geothermal science.

Summary and Concluding Remarks

The relationship of humans and the Earth's energy in the Mediterranean has roots in the night of time, in a moment of the Paleolithic that cannot be specified but which dates back many tens of thousands, perhaps even hundreds of thousands, of years. From that moment on, this relationship must have slowly matured over that long, long period called the *year zero* of geothermics (Cataldi, 1992). During this period, people's contact with the Earth's heat was probably confined to a few localities and limited to cooking and occasional thermal baths. Gradually, however, it included the use of geothermal by-products: obsidian to make stone tools; thermo-mineral muds for therapeutic applications, such as drying sores and hemostasis; and hydrothermal compounds for pigments, such as cinnabar, iron oxides, and sulfur.

Toward the end of the *year zero*, at the start of the Metal Age about 5,000 years ago, all of these uses began to take root in other geothermal localities on the Mediterranean mainland and the major volcanic islands. Thus, indigenous centers formed naturally, harnessing natural heat and its by-products, but they remained isolated for many centuries until systematic contacts began among distant populations and trade developed in a number of raw materials, including obsidian and hydrothermal compounds such as alum, borates, and kaolin. At first, these materials were bartered and the market was limited. A few production centers existed in some places, the most important of which were probably located in Western Anatolia.

In historical times, 2,000-1,500 years before the Christian era, the market for hydrothermal products gradually expanded to the Mediterranean coastal areas under the influence of Cretan and Phoenician navigators. As a result, the popularity of hydrothermal compounds increased, and they were used systematically for making pottery and pigments, coloring glass, preparing ointments and salves, and tanning hides. Near the end of the 2nd millennium B.C., these uses began to proliferate, taking root in the geothermal and nongeothermal areas of the Mediterranean coast.

Starting around the 7th century B.C., thanks chiefly to the Etruscans, methods for extracting and processing hydrothermal compounds improved remarkably, resulting in wide commercialization throughout the Mediterranean. An important industry and a true international market based on geothermal by-products formed between the 6th and the 3rd centuries B.C. This market boomed between the 1st century B.C. and the 2nd century

A.D., spreading to all territories under Roman rule and growing richer still at the height of the Roman Empire when new mines opened in Italy and other Mediterranean geothermal localities. During the imperial period, systematic use was made of all hydrothermal compounds and of many other geothermal by-products, such as travertine, bentonite, pozzolan, perlite, lavas, pyroclastics, and tuffs, which were important in the building industry (Cataldi and Burgassi, 1992b). Nearly all geothermal by-products held great importance in Antiquity.

At the same time, the practice of thermal balneology grew gradually in the early and ancient Mediterranean civilizations. Even before the Roman period, thermal bathing was popular in Western Anatolia, Eastern and Southern Greece, and peninsular Italy (especially in the Phlegraean Fields and Etruria). The boom in thermal balneology between the 1st century B.C. and the early 4th century A.D. came from the Romans, who built countless spas in localities with and without hot springs. Thermal bathing was so popular in the imperial period that it became a daily custom for several million people of every social class, ethnic group, and gender. The thermae became complete health service centers, places for meetings, debates, entertainment, reading, business dealings, sports, and cultural events—all catalysts for many social activities. Considering that each ancient Mediterranean people made its own contribution to thermal bathing, its development in Antiquity must be considered a common sociocultural heritage of them all (Cataldi and Burgassi, 1992b).

In addition, geothermal manifestations in Antiquity also influenced the formation of cults to worship the protective divinities of hot waters and the creation of many myths and legends. Many Greek and Latin thinkers sought to explain geothermal events (Cataldi, 1993b), resulting in the formation of the first nucleus of systematic ideas and theories on the genesis of the Earth's heat and related phenomena.

From at least the Upper Neolithic until the end of the Roman Empire, the development of geothermal energy was an important facet of Mediterranean civilization.

Cited and Selected References

Bibliographical note: Relatively few citations are included in the text. Interested readers can refer to the bibliographies contained in the works cited, as well as those from Nenci-Vallet (1984), Pallottino (1984), Panessa (1991), and Ritti (1985).

Burgassi, P. D. 1987. Historical outline of geothermal technology in the Larderello region. *Geothermal Resources Council Bulletin,* 16, no. 3, 3-18. (See Chapter 14 in this volume).

Burgassi, P. D., R. Cataldi, J. L. Hernandez Galán, M. Moggi, R. Rubinovich, and J. Saldaña. 1992. Primera contribución al conocimiento historiográfico de la energía geotérmica en el area Mediterránea y en la America Latina. Contribution to *Simposio sobre la história de la geotermia,* Ciudad de Mexico. (This work has been published as a limited edition in the "grey literature"; it is available for consultation at selected places. Information on locations can be obtained from the first three authors of the publication).

Cataldi, R. 1992. El año cero de la geotermia. In *La geotermia en el período Precolombino en las áreas Mediterránea y Mesoamericana.* In *Geotermia, Revista Mexicana de Geoenergía,* 8, no. 2, 155-175. (See Chapter 2 in this volume).

Cataldi, R. 1993a. Review of historiographic aspects of geothermal energy in the Mediterranean and Mesoamerican areas prior to the modern age. *Geo-Heat Center Quarterly Bulletin,* 15, no. 1, 13-16.

Cataldi, R. 1993b. Utilizarea energiei geotermale in antichitate: Din vremea Etruscilor pina la finele Evului Mediu timpuriu. In *Analele universitatii din Oradea: Fascicola electrotecnica.* Ministerul Invatamintului/ Universitatea din Oradea Edit., Oradea.

Cataldi, R. and P. D. Burgassi. 1992a. Del misterio de la Atlantida a los primeros usos integrados de la energía geotérmica. In *La geotermia en el período Precolombino en las áreas Mediterránea y Mesoamericana.* In *Geotermia, Revista Mexicana de Geoenergía,* 8, no. 3, 254-264. (See Chapter 9 in this volume).

Cataldi, R. and P. D. Burgassi. 1992b. Florecimiento y decadencia del termalismo y de los otros usos directos del calor de la Tierra en el área Mediterránea, desde el nacimiento de Roma hasta el fin de primer milenio. In *La geotermia en el período Precolombino en las áreas Mediterránea y Mesoamericana. Geotermia, Revista Mexicana de Geoenergía,* 8, no. 3, 265-288. (See Chapter 10 in this volume).

Cataldi, R., P. D. Burgassi, and M. C. Suárez Arriaga. 1992-1993. La geotermia en el período Precolombino en las áreas Mediterránea y Mesoamericana In *Geotermia, Revista Mexicana de Geoenergía,* 8, no. 2, 145-175; 8, no. 3, 251-303; 9, no. 1, 5-58. (See Chapters 9, 10, 12, and 31 in this volume).

Fiumi, E. 1943. *La utilizzazione dei lagoni boraciferi della Toscana nella industria medioevale.* University of Florence, CYA Edit., Florence.

Fytikas, M., G. Margomenon Leonidopoulou, and R. Cataldi. 1999. In *Geothermal energy in ancient Greece: from mythology to late antiquity (3rd century A.D.).* (See Chapter 6 in this volume).

Grifoni Cremenosi, R. 1999. *Human rapport with hypogeal waters and geothermal manifestations in Italian prehistory.* (See Chapter 8 in this volume).

Luce, J. V. 1969. *The end of Atlantis.* Thames and Hudson Ltd., London.

Mellaart, J. 1967. Çatal Hüyük: A Neolithic town in Anatolia. In *New aspects of antiquity.* M. Wheeler, ed. Thames and Hudson Ltd. and The Camelot Press, London and Southampton.

Montanelli, I. 1969. Storia di Roma: 753 B.C.-476 A.D. In *Storia d'Italia.* Rizzoli Edit., Milan.

Nenci, G. and G. Vallet. 1984. *Bibliografia topografica della colonizzazione greca in Italia e nelle isole tirreniche.* Scuola Normale Superiore/Ecole Française de Rome, Centre J. Berard Naples Edit., Pisa-Rome.

Pallottino, M. 1984. *Etruscologia.* Hoepli Edit., Milan.

Panessa, G. 1991. *Fonti Greche e Latine per la storia dell'ambiente e del clima nel mondo Greco.* Scuola Normale Superiore Edit., Pisa.

Pasquinucci, M. 1987. *Terme romane e vita quotidiana.* Panini S.p.A. Edit., Modena.

Pausanias. 155-170. Description of Greece. In *Pausanias: Description of Greece, vol. 1.* Reprint 1918, W. H. S. Jones, trans. Harvard University Press and W. Heinemann Ltd., London and New York.

Ritti, T. (1985). Fonti letterarie ed epigrafiche. In *Hierapolis: Scavi e ricerche,* G. Bretschneider, Ed., Rome.

The Authors:

Raffaele Cataldi
55, Via del Borghetto
56124 Pisa, Italy
Telephone/Fax: 39.050.59.81.07
E-mail: rafcat@tin.it

Paolo Chiellini
153, Via 4 Novembre
56026 Cascina (Pisa), Italy
Telephone: 39.050.74.11.14

Portion of the G. Vasari wall painting in the Clement VII (Giulio de'Medici) salon of the Palazzo Vecchio in Florence (c. 1560). The painting depicts the siege of Volterra by the Florentine army during the "War of the *Allumiere*" (1470-1472) between the towns of Volterra and Florence, which were fighting for possession of the Boraciferous region in Southern Tuscany. This region held rich deposits of hydrothermal minerals, locally called *allumiere*, very important for the Italian wool processing industry in the Late Middle Ages. *Courtesy of the Commune of Florence, Department of the Culture-Conservancy of Museums*

12. Utilization of Geothermal Resources in Italy from the 11th through the 16th Centuries

by
Raffaele Cataldi
Pier Domenico Burgassi

Abstract: For about six centuries after the fall of the Roman Empire in the 5th century A.D., the development of natural heat and its by-products progressed in Italy at a very slow pace. Then development resumed, starting from the second half of the 11th century when systematic exploitation of hydrothermal products began in the Boraciferous region in Central Italy for pharmaceutical, industrial, and craft purposes. These products gradually increased in commercial importance over the next five centuries, eventually causing a series of disputes in the Late Middle Ages between some Tuscan communes and the Bishopric of Volterra for ownership of the area where they were found.

Thermal balneology and other uses of natural heat remained slow until the 13th century, but in the next three centuries it flourished once again when over 100 thermal localities were developed in Italy for balneotherapy and recreation.

Revised, translated, and reprinted with permission from *Geotermia, Revista Mexicana de Geoenergía,* 1992, 8, no. 3, pp. 289-303.

The sun did not stop in the opposite half of the sky,
nor did the dawn of the second millennium remain dark.
The ground shook again that day;
volcanoes still erupted streams of lava,
and Earth continued to give mankind
the fertile heat of her restless body.
—R. Cataldi

The Transition Period, between the 1st and the 2nd Millennium

However widespread and sincere the belief may have been that the world would end in the year 1000 (Cataldi and Burgassi, 1992b), the reawakening at the dawn of the 11th century began a period of deep renewal of customs and lifestyles in Italy that in the succeeding centuries would lay the historical groundwork for the Renaissance.

The revival is attributed to many reasons, such as the end of "barbarian" invasions in Central and Southern Europe, rapid population increase, decrease in feudal anarchy, Muslim expansion, birth and development of local autonomies, a maturing religious spirit, conflicts among different religions, and the Crusades to the Holy Land. It was expressed in numerous flourishing commercial enterprises that gradually established a market economy in nearly all fields of producing and processing raw materials. From the first decades of the 11th century, many old mining extraction activities were revitalized, new deposits of metal-bearing ores and nonmetallic minerals were discovered, and systematic exploitation began in a number of areas. These activities became especially intense in the Metalliferous Hills in Tuscany (Central Italy) where they contributed substantially to the development of the regional economy (Bargellini, 1984).

Hydrothermal products, however, were an initial exception. Indeed from the 9th century on, some of them, particularly boron compounds used for preparing pharmaceuticals, were imported from the Middle East by Saracen traders following the Saracenic expansion into Northern Africa and sectors of Southern Europe. The most important compound imported in the 9th and 10th centuries was tincal, most probably a mixture of borax $Na_2[B_4O_5(OH)_4] \cdot 8H_2O$, colemanite $Ca[B_3O_4(OH)_3] \cdot H_2O$, and ulexite $NaCa[B_5O_6(OH)_6] \cdot 5H_2O$ extracted from sedimentary deposits in Iran and volcanic-lacustrine deposits in Western Anatolia. The word "borax" derives from the Arab *buraq,* meaning "white."

In the following centuries, bitter maritime clashes for commercial supremacy in the Mediterranean greatly hindered large-scale trade of hydrothermal products. The conflicts included those between the Saracen fleet and the Italian marine republics of Amalfi and Pisa (9th to the 13th centuries), and of Genoa and Venice (10th through the 16th centuries). However, with the possible exception of some borates, these products did not have the economic importance of many other raw materials and manufactured goods then traded by sea.

Because importing tincal from the Middle East had become impossible, new initiatives were launched in the mid-11th century to exploit the rich hydrothermal deposits of the Boraciferous region in Tuscany. These efforts paved the way for the birth of the geothermal chemical industry at Larderello in the following century. However, thermal bathing remained depressed nearly everywhere until the early 13th century.

Development in the Boraciferous Region

Exploitation of Hydrothermal Compounds

The Boraciferous region is by far the most important geothermal zone in Italy; it is located in Southern Tuscany in the area presently known as Larderello. The systematic development of geothermal resources started here in the second half of the 11th century when the bishops of Volterra, who administered the region in the name of the Holy Roman Empire, authorized local businessmen to mine the rich hydrothermal deposits associated with the high-temperature geothermal manifestations of the area (Volpe, 1923). These manifestations were then called *allumiere* or *lumaie*. *Lumaie,* with its Latin etymology, signifies either a locality with minerals with glittering specks (*lumen*) or a place where *allume* ("alum") is mined; hence, the name *allumiere.* These deposits are now better known in local jargon as *biancane* or *lagonicci.* The term *biancane* ("whitish spots") reflects the dominant white color in areas with geothermal alterations. *Lagonicci* comes from *lagone* ("hot water pool") and refers to the hydrothermal deposits covering ground around high-temperature manifestations.

The following compounds were increasingly produced and traded from the 12th through the 16th centuries:

- Yellow sulfur (S), much in demand for the preparation of pharmaceuticals (ointments and salves), bleaching solutions for the wool industry, and from the 13th century onward, a component of gunpowder

- Alum ($KAl[SO_4]_2 \cdot 12H_2O$), a product used in processing wool

- Green vitriol (ferrous sulfate heptahydrate: $FeSO_4 \cdot 7H_2O$), widely used to prepare disinfectants and hemostatic products, and as a component of dyes for the wool industry and other processes

- Blue vitriol ($CuSO_4 \cdot 5H_2O$), used as a disinfectant, antiparasitic, and coloring component

- *Nitro volterrano*, which, on the basis of manuscripts from 1527, 1531, and the work of Mercati (1717), is now identified as boric acid (H_3BO_3). From very ancient times this product was recommended—and is still used today—for treating eye diseases and as a weak acid for disinfecting sores.

Throughout this period, all of these products gradually increased in commercial importance; consequently, a series of disputes arose over awarding mining contracts. Several bitter struggles occurred from the 12th century onward between the Tuscan communes of Florence, Pisa, and Siena and the Bishopric of Volterra for ownership of all or part of the Boraciferous region. A state of permanent conflict among these communes, and between some of them and the Bishopric of Volterra, continued for three centuries. Finally, after a relentless war between the Commune of Volterra and the Republic of Florence, the whole Boraciferous region was passed to the Signory of the Medicis in 1472 (Targioni Tozzetti, 1769; Nasini, 1930). A painting of the pitched battle during this war is shown in the photograph at the front of this chapter.

Now the Florentine Republic assigned control for mining the *lumaie* to its flourishing wool guild, called *Arte della Lana* ("wool merchants"). With no need to pay duties on the hydrothermal products necessary to process wool and dye fabrics, the guild soon became so prosperous that it gained a near monopoly over the Italian textile industry, a situation that lasted well beyond the end of the Renaissance.

Other Uses

Beginning in the 13th century, almost certainly the *lagoni* of the Boraciferous region were used for small local crafts. This seems especially true for the *Lagòn Cerchiaio*: according to Fazio degli Uberti (c. 1350), the boiling waters of this pool were used to soak barrel staves and to expand the metal hoops used to bind the staves together. Moreover, by very ancient tradition (perhaps dating to Etruscan times), the steam manifestations of the Boraciferous region were used to dry farm produce and cook chestnuts. Until a few decades ago, the latter were a staple of the region's inhabitants. Often the chestnuts were boiled in steam escaping from very hot manifestations. The soil of the fields immediately around the manifestations, made fertile both by natural heat and impregnated potassic salts, was often used to grow vegetables and grapes. Balneotherapy was practiced in some localities with hot springs in the Boraciferous region, beginning around the end of the 12th century. The most important localities were *Bagno a(d) morba*, Bagnolo, and Terme Antica Querciolaia. *Bagno a(d) morba* means "bath of the illnesses," and the persistence of the Latin word *morba* shows the site's ancient fame as a place for treating diseases with steam baths and thermo-mineral muds. Several authors of the Late Middle Ages wrote about this spa, including Dante Alighieri (c. 1300), Ristoro d'Arezzo (1282), M. Savonarola (1523), Ugolino da Montecatini (1553), and Mengo Blanchello Faentino (1553).

The wide variety of uses of hydrothermal minerals in the following list explains why the exploitation of hydrothermal products and the use of natural heat for different purposes contributed considerably to the economic growth of the Boraciferous region during the first five to six centuries of the present millennium.

Uses of geothermal resources in the Boraciferous region from the 11th through the 16th centuries

Hydrothermal products and their uses	
Product	Use
Yellow sulfur (S)	Pharmaceutical compounds (ointments, salves, and others) Bleaching solutions for wool industry Gunpowder, from the 13th century
Alum ($KAl[SO_4]_2 \cdot 12H_2O$)	Processing wool
Green vitriol ($FeSO_4 \cdot 7H_2O$)	Pharmaceutical compounds (disinfectants and hemostatic agents) Coloring agents (wool, textiles, and others)
Blue vitriol ($CuSO_4 \cdot 5H_2O$)	Pharmaceutical compounds (disinfectants) Coloring agents Antiparasitic agents
Nitro volterrano (Boric acid: H_3BO_3)	Pharmaceutical compounds (for eye diseases, disinfectants for sores, and others)
Other uses	
	Soaking staves and thermally expanding barrel hoops Drying farm produce Cooking chestnuts Growing vegetables and grapes Thermal bathing, from the 13th century

Development in Other Areas

Exploitation of Hydrothermal and Pyroclastic Products

Besides those of the Boraciferous region, other geothermal by-products were exploited during the Late Middle Ages in other areas of Italy. Most were located in Western

Latium, Western Campania, and on some islands of the Tyrrhenian Sea, especially Ponza in the Pontine Archipelago and Lipari in the Aeolian Archipelago.

One or more of the following by-products were exploited in these areas: pumice, pozzolan, bentonite, kaolin, sulfur, silica, and travertine. They were used mainly in cement slurries and building materials, and less importantly for crafts and pharmaceuticals. Most were used locally and never attained the economic importance of the hydrothermal compounds in Tuscany's Boraciferous region (Cataldi and Burgassi, 1992a and 1992b).

Thermal Balneology

Following a long period of recession, balneotherapeutic customs began taking root again in many thermal localities from about the second half of the 13th century, after autonomous municipalities (communes) were established and consolidated, especially in regions of Northern and Central Italy. The most important thermal establishments developed from the 13th through the 16th centuries were Acqui Terme, Abano (see engraving), Montegrotto, Bagno di Romagna, Bagni di Lucca, Montecatini, Chianciano, San Casciano dei Bagni, Bagno Vignoni, Viterbo, and Tivoli in Northern and Central Italy, and Pozzuoli, Ischia, Lipari, Santa Cesarea Terme, Sciacca, and Sardara in Southern Italy and on Italian islands.

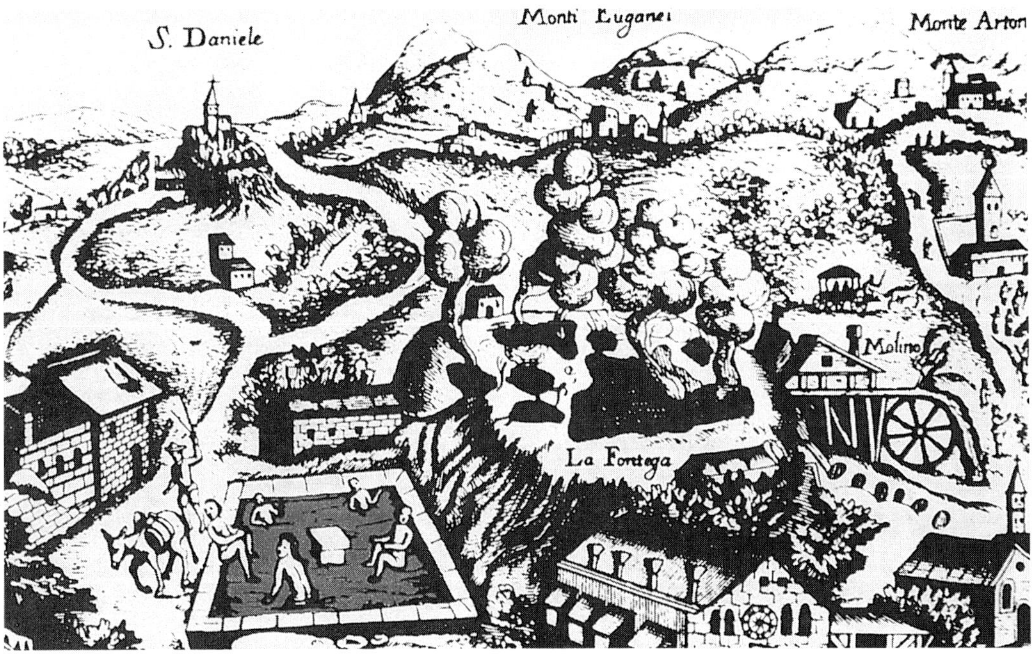

A 16th century engraving of a bathing scene in the thermal pool of Abano in Veneto, Italy. *From J. M. Couder and F. Jaudin, 1989. Courtesy of the authors, with authorization of the BRGM*

Illustrious people visited many of these establishments for cures—sometimes becoming their patrons—and this contributed remarkably to the reputation and fame of the sites. Examples include Bagni di Lucca, where the Holy Roman Emperor Frederick II went for a cure in 1245, the Italian poet L. Ariosto in about 1530, and the French essayist M. de Montaigne in 1581; and Bagno Vignoni, where Pope Pius II Piccolomini went for treatment around 1460 and the Prince Lorenzo de'Medici (the Magnificent) in 1490 (Rosetti and Valenti, 1997).

At least 100 other minor thermal sites were active in Italy during the Late Middle Ages, and many were already in existence before Medieval times. Often these sites had simple facilities and were open during specific periods only to serve local workmen and common people from nearby areas.

Conclusions

The revival of thermal balneology during the Late Middle Ages in Italy is testified to in treatises of natural history, medicine, and hydrology of the epoch, as well as in works by poets, geographers, and scientists of the same period and of the Renaissance. Many passages can be found in these works on the physical and chemical characteristics and the healing properties of thermo-mineral waters and muds at each locality, and on their abilities to cure internal and external diseases. In addition to those quoted for the Boraciferous region, the following authors deserve a special mention here: Bauer, better known as Agricola (1546 and 1556); Van Merle, called Merula (1605); and Fallopio (1606).

Simultaneously with the reawakening in Italy, thermal balneology started to thrive again, as in ancient times, in many Western and Central European countries, such as Austria, Bulgaria, the Czech Republic, France, Germany, Hungary, Iceland, Poland, Portugal, Romania, Slovakia, Spain, Switzerland, and the states of the former Yugoslavia. During the first half of the present millennium, thermal bathing again flourished at Bath in England and began taking root in the areas of Reykjavik, Ingolfafjall, and Hvitarholt in Iceland.

Mining some hydrothermal minerals became rather important during the Late Middle Ages in a few other European countries besides Italy. This is the case with native sulfur in Iceland (Sigurdsson, 1992) and bentonite and other minerals in some Greek islands of the Southern Aegean.

The development of thermal balneology and the use of hot water for district heating and other purposes from the 11th through the 16th centuries are highlighted in other chapters of this volume for other European countries. From these chapters and from this, one can realize that the development of geothermal resources in Europe during the first five to six centuries of the second millennium was a phenomenon of international importance. Harnessing geothermal resources, however, did not then attain the peak of development and popular diffusion it had enjoyed about a millennium earlier in territory under Roman rule at the height of splendor of the Roman Empire.

Cited and Selected References

Agricola (Bauer, G.). 1546. *De natura eorum quae effluent ex terra.* Frobenius Edit., Basel.

Agricola (Bauer, G.). 1556. *De re metallica.* Frobenius Edit., Basel.

Alighieri, D. c. 1300. *La vita nova e il Canzoniere.* Reprint 1868. Le Monnier Edit., Florence.

Bargellini, P. 1984. *La splendida storia di Firenze, vol. I.* Vallecchi Edit., Florence.

Cataldi, R. and P. D. Burgassi. 1992a. Del misterio de la Atlantide a los primeros usos integrados de la energía geotérmica. In *La geotermia en el período Precolombino en las áreas Mediterránea y Mesoamericana. Geotermia, Revista Mexicana de Geoenergía,* 8, no. 3, 254-264. (See Chapter 9 in this volume).

Cataldi, R. and P. D. Burgassi. 1992b. Florecimiento y decadencia del termalismo y de los otros usos directos del calor de la Tierra en el área mediterránea desde el nacimiento de Roma hasta el fin del primer milenio. In *La geotermia en el período Precolombino en las áreas Mediterránea y Mesoamericana. Geotermia, Revista Mexicana de Geoenergía,* 8, no. 3, 265-288. (See Chapter 10 in this volume).

Cohut, I. and M. Arpasi. 1995. Ancient uses of geothermal energy in the Pre-Carpathian area and in the Pannonian Basin. *Proceedings of the World Geothermal Congress,* 381-384. (See Chapter 15 in this volume).

Coudert, J. M. and F. Jaudin. 1989. *La géotermie: Du geyser au radiateur.* BRGM Edit., Orléans.

Fallopio, G. 1560. Opera genuina omnia, vol. 1. In *De thermalibus aquis.* Jo Antonio and Jacopo Francesco Edit. Reprint 1606. Venice.

Fazio degli Uberti. 1350?. *Il dittamondo.* Reprint 1826. G. Silvestri Edit., Milan.

Fendek, M., A. Rebro, and M. Fendekova. 1999. *A spell cast: Historical aspects of thermal spring use in the Western Carpathian region.* (See Chapter 16 in this volume).

Fridleifsson, I. 1995. Historical aspects of geothermal utilization in Iceland. *Proceedings of the World Geothermal Congress,* Florence, 427-432. (See Chapter 19 in this volume).

Gibert, P. and F. Jaudin. 1999. *Using geothermal waters in France: The district heating system of Chaudes-Aigues from the Middle Ages.* (See Chapter 18 in this volume).

Mengo Blanchello Faentino. 1488. De balneo de morba volterrano. In *De balneis omnia quae extant apud graecos, latinos et arabos.* Reprint 1553. Giunta Edit., Venice.

Mercati, M. 1717. *Metalloteca Vaticana. Opus posthumum, Armarium, vol. 4.* Ex Officina Io Mar and Salvioni Romani C. Edit., Rome.

Merula (Van Merle P.). 1605. *Cosmographiae generalis (lib. tres) and Geographiae particularis (lib. quatuor).* Ex Officina Plantiniana Edit., Raphelengie.

Nasini, R. 1930. *I soffioni ed i lagoni della Toscana e l'industria boracifera.* Italia Edit., Rome.

Nasini, R. and G. Grassini. 1932. Sulla acoperta dell'acido borico nei Vasi Sigillati Aretini. In *Studi Etruschi, vol. 6.* Florence.

Ricettario fiorentino. 1567. Stamperia Eredi di B. Giunti, Florence.

Ricettario senese. 1567. Stamperia Eredi di B. Giunti, Florence.

Ristoro d' Arezzo. 1282. *Libro della composizione del mondo.* Reprint 1864. G. Daelli Edit., Milan.

Rosetti, E. and L. Valenti. 1997. *Terme e Sorgenti di Toscana (note, meno note e sconosciute).* Le Lettere Edit., Florence.

Savonarola, M. 1460. De balneis Vulterrae. In *De balneis omnia quae extant apud graecos, latinos et arabos.* Reprint 1553. Giunta Edit., Venice. (See also Nasini, 1930).

Sigurdsson, J. 1992. Namafjall: Fire and brimstone. In *Atlantica. (Iceland Review for Icelandair).*

Sokolowski, J., J. Sokolowska, and B. Kapinski. 1999. *Using geothermal energy in Poland in the second millennium.* (See Chapter 17 in this volume).

Targioni Tozzetti, G. 1769. *Relazione d'alcuni viaggi fatti in diverse parti della Toscana, vol. 3.* Stamperia Granducale, Florence.

Ugolino da Montecatini. 1420. De natura et virtute balnearum Comitatus Vulterrae. In *De balneis omnia quae extant apud graecos, latinos et arabos.* Reprint 1553. Giunta Edit., Venice.

Volpe, G. 1923. *Volterra: Storia di Vescovi Signori, di Istituti Comunali, di rapporti tra Stato e Chiesa nelle città italiane nei secoli XI-XV.* Biblioteca Storica Toscana, Florence.

The Authors:

Raffaele Cataldi
55, Via del Borghetto
56124 Pisa, Italy
Telephone/Fax: 39.050.59.81.07
E-mail: rafcat@tin.it

Pier Domenico Burgassi
2, Via G. Galilei
56044 Larderello (Pisa), Italy
Telephone: 39.058.86.74.89

Lithograph titled "View of one of the volcanic localities of Larderello," showing the different types of uncovered and covered *lagoni* ("boiling pools") and installations and facilities of the geothermal-chemical industry at Larderello, as seen around 1850. In describing Larderello as a volcanic area, the title's author reflects the belief of the time that the natural steam had an igneous origin.

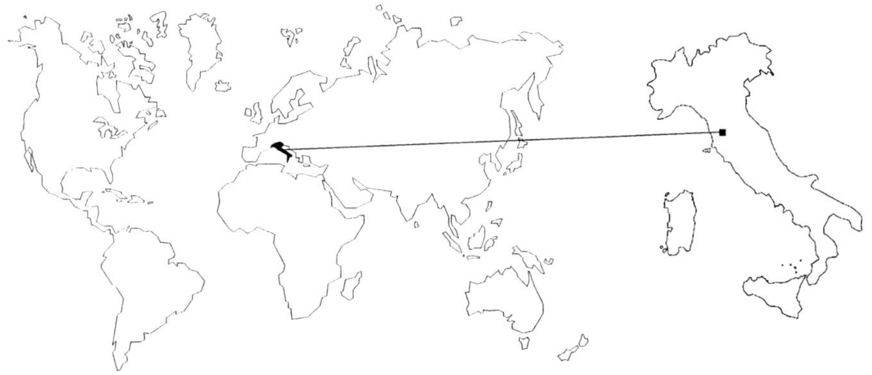

13. Historical Outline of Geothermal Technology in the Larderello Region to the Middle of the 20th Century

by
Pier Domenico Burgassi

Abstract: The chapter discusses the impact of geothermal resources on the inhabitants of the Larderello region in prehistoric, ancient, and recent times, up to 1944.

Industrial processes exploiting boric acid and other boron compounds from geothermal fluids and drilling technologies for high-temperature geothermal resources were developed at Larderello, starting in the early 19th century. An integrated drilling, chemical, and agricultural industry at Larderello paved the way in the 20th century for the birth of the world's first geothermal-electrical industry. The first successful experiments conducted between 1904 and 1910 to generate electricity using hot natural fluids are recounted, and mention is made of the geothermal power plants installed at Larderello up to the beginning of World War II.

Revised and published by permission of the Geothermal Resources Council from the *Geothermal Resources Council Bulletin,* March 1987.

At the Roots of Civilization, between Prehistory and Legend

The geothermal manifestations in the region of Central Italy now known as Larderello are very young from the geological point of view: on the order of 0.5 to 1 million years old (figure 1). No one, however, can say if some of our prehistoric ancestors observed their formation and what feelings they may have had: surprise? curiosity? terror? Whatever the answer, indirect information shows that in the Upper Paleolithic, fumaroles, steam jets, hot springs, *lagoni*, and gas exhalations were already objects of attention by the first inhabitants of the Larderello region. The term *lagoni* (singular, *lagone*) refers to hot natural pools from which gas and saturated steam escape into shallow circulating waters, resulting in the formation of hydrothermal deposits at the surface, especially boron compounds.

Probably as early as the Neolithic, all these manifestations were considered signs of the existence of underground divinities, who

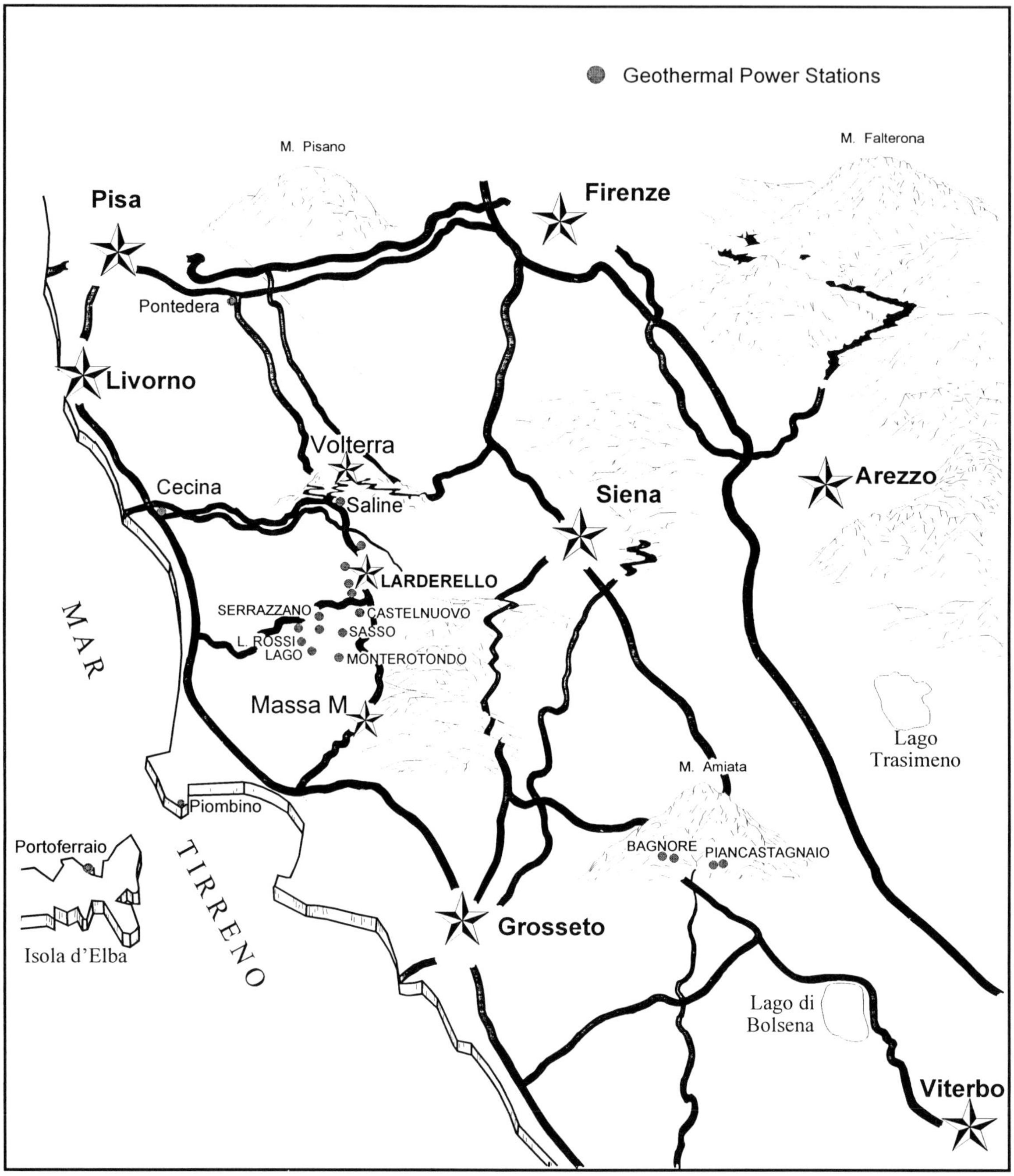

Figure 1. Location map of Larderello and the surrounding region (modified from a 19th century chorographic sketch).

occasionally rose through them up to the surface with hisses, rumbles, and subterranean noises. Perhaps then the belief began to spread in prehistoric epochs that these divinities wanted people to be aware of their intrinsic power, that they demanded signs of respect from whoever dared to tread on the roof of their abode.

Based on legends whose roots fade away into Neolithic times, some believe today that the area of manifestations was considered by the early inhabitants of the region to be the uncontested realm of infernal divinities, but this idea is not sound because the concept of hell certainly came after the Neolithic period. Instead, Neolithic people probably believed not in the devil and hell but in the existence of underground divinities, some powerful and dangerous, others gracious and benevolent. Thus people learned gradually, at their own expense, to respect the former by not arousing their wrath and to have confidence in the latter, enjoying their healing gifts. These included the heat perennially supplied by the gracious manifestations as an antidote to the cold and the therapeutic properties of certain thermal muds and waters, probably perceived as generous gifts from benign deities—certainly not from malevolent angels, as the concept of hell would suggest.

Whichever interpretation one places on the first relationships that prehistoric people established with geothermal manifestations in the Larderello area, whether these were exclusively functional or included a religious aspect, as I believe, the fact remains that the first contacts of the region's early inhabitants with geothermal manifestations have faded away from the roots of civilization into the night of time.

From the Etruscans to the End of the First Millennium

From the Paleolithic through the Neolithic, the development of civilization was long and slow, a fact reflected in the evolving relationships between prehistoric peoples and the natural manifestations of Central Tuscany. Etruscan settlements in this region around 1300 B.C. and their flourishing growth for over a millennium brought something new to the harnessing of geothermal energy and its by-products—the use of thermo-mineral waters and muds for therapeutic purposes, and the exploitation of hydrothermal minerals deposited near the edges of the *lagoni*.

Numerous vestiges and handmade objects found in Tuscany's Metalliferous Hills—the hilly, low mountainous region between Florence to the east and Grosseto and Pisa to the west—attest to widespread therapeutic use among the Etruscans of the hot springs and muds in the Larderello area. Moreover, traces of boron compounds found in the glazes of Etruscan plates and crockery show how these people, many centuries before Christ, had developed a high degree of expertise in grinding and processing borates, mixing these products with other ingredients to manufacture fine pottery.

As time passed, the fame of the therapeutic properties of the thermo-mineral waters and muds of the Larderello area extended to other regions in the Mediterranean basin. The Greek poet Lycophron in the 3rd century B.C. wrote in his poem "Alexandra" (or "Cassandra") that "the river Lynceum, running through the territory of the Etruscans, originates in an area with special manifestations...whose waters are used to treat numerous diseases of the eyes...." The river Lynceum was identified later as the present-day river Cornia, originating near Larderello. Its catchment basin covers a large part of the area with high-temperature manifestations.

The glazes of the *vasa sigillata*, dating from the 1st century B.C. and conserved in the archaeological museum of Arezzo, offer another proof of the ancient use of boron compounds associated with the Larderello manifestations. *Vasa sigillata* means "seal-bearing vases," and this red-glazed pottery often had designs in relief resembling seals, perhaps a sort of trademark for the vases. The glazes for this pottery are a mixture of iron oxides and boron compounds, both found near the manifestations of Monterotondo, a few kilometers south of Larderello. Other origins of these compounds can be ruled out, since the only important deposits of boron minerals in Italy are associated with thermal manifestations in the Larderello region.

The fact that such products were used during the 1st century B.C. is not surprising if one considers that the Etruscans until the 3rd century B.C., and the Romans after them, knew and mined all the mineral resources of Central Italy for more than a millennium. These include hydrothermal deposits in the Larderello area and the rich metal ores of the Metalliferous Hills and the island of Elba, such as copper, iron, zinc, lead, and silver. After the 1st century B.C., many Latin poets and writers, such as Tibullus, Strabo, Pliny the Elder, and Vitruvius, mentioned these uses of hydrothermal minerals and metals along with the balneologic and therapeutic uses of hot springs and muds from Central Italy to the south of Pisa.

The *palle da cani* probably date from the beginning of the Christian era. These were small balls of mud gathered from the edges of natural hot pools, then dried and sold to treat skin diseases. The term itself means "balls for dogs," most probably because the mud balls were tried on the sores of dogs before they were used on people.

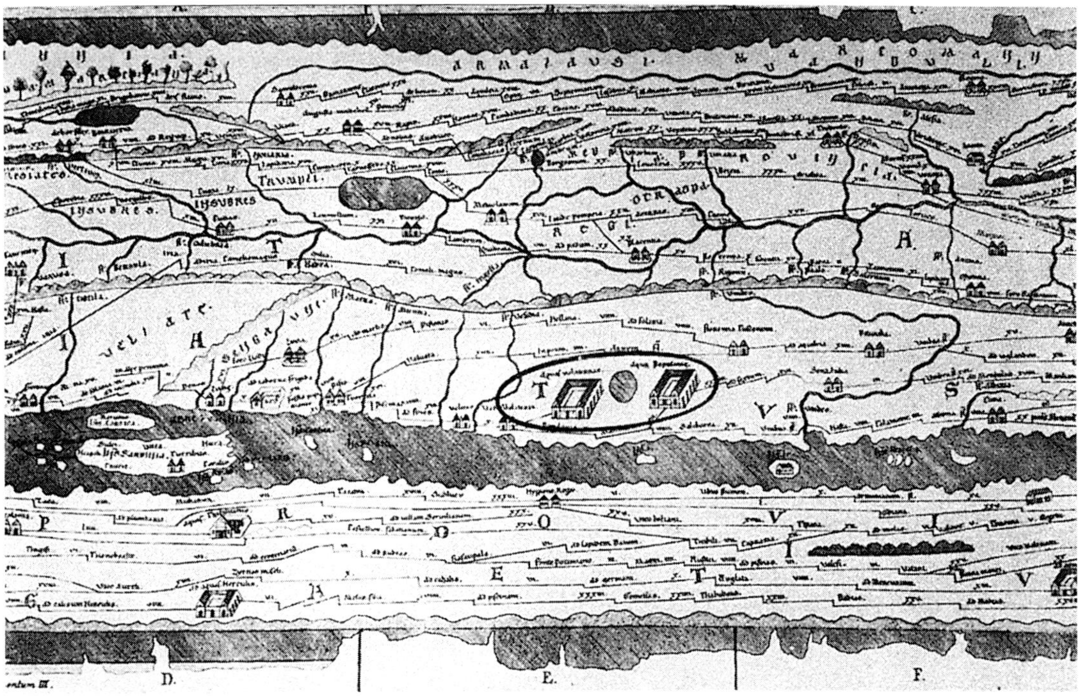

Figure 2. Detail of the *Tabula Itineraria Peutingeriana*, showing the location of Aquae Volaternae and Aquae Populaniae.

The first cartographic document recording two spas near Larderello (Aquae Volaternae and Aquae Populaniae) is the *Tabula Itineraria Peutingeriana* (figure 2). This document, dating from the late 3rd or early 4th century A.D., is a map of the main routes and military roads of the Roman Empire. The fact that these two spas, among the many existing in Central Italy, were singled out in the portion of the map shown in figure 2 means that they had long since reached a stage of advanced development and fame.

Most likely not only were thermal waters and muds used for therapeutic purposes, but hydrothermal minerals associated with natural manifestations were traded since Etruscan times. This trade was well established before the Christian era began and flourished for several centuries in Roman times. This exploitation of hydrothermal products slowed toward the end of the 4th century when the Roman Empire began to decline. The place names of localities such as Montecerboli (from the Latin *Mons Cerberi*), "Devil's Mountain," and *Valle dell'Inferno*, "Hell Valley," probably originated between 800 and 1000, clear indications of people's belief that thermal manifestations had an infernal origin.

In Italy, as in most of Southern Europe, the last two to three centuries of the first millennium were a period of low economic and social growth, generated mainly by historical causes and subordinately by the widespread conviction that the world would come to an end in the year 1000. All these factors strongly reduced balneotherapeutic practices as well as exploitation and use of hydrothermal products in that period.

From 1000 A.D. to the Discovery of Boric Acid

With the beginning of the second millennium, when the expectation of the end of the world had passed, a far-reaching cultural reawakening over all Italy led to relaunching many old enterprises and beginning many new ones. This occurred in all commercial and mining activities, including the exploitation of hydrothermal minerals in the Larderello region. The first contracts between the bishops of Volterra, who had jurisdiction over the region in the name of the Holy Roman Empire, and the lessees entrusted with mining the *lumaie* date from the middle 11th century. The term *lumaie* comes from the Latin word *lumen.* The word may mean "bright spot," alluding to whitish mineral deposits at the surface caused by condensation of thermal fluids, or *lumen* may refer to localities with alum deposits.

The main products extracted and marketed from the *lumaie* are listed below:

- Yellow sulfur, commonly used to prepare pharmaceuticals, such as ointments and salves, and bleaching agents for the wool industry

- Cyprian (green) and Roman (blue) vitriol (heptahydrated ferrous sulfates: $FeSO_4 \cdot 7H_2O$), used by pharmacies to prepare disinfectant and anti-hemorrhagic agents, and in the textile industry to dye wool. It should be pointed out, however, that green vitriol was also produced artificially in the Larderello area by the reaction in sulfurous waters of pyrites and/or bits of iron

- Alum, $KAl(SO_4) \cdot 12H_2O$, a product then indispensable for the wool industry

- *Nitro volterrano*, "nitre from Volterra," a product whose chemical formula is unknown, but on the basis of two 16th century manuscripts known as *Ricetti* ("Books of Medical Prescriptions") of the towns of Florence and Siena (*Ricettario Fiorentino* and *Ricettario Senese*) is thought to be boric acid. We learn from the manuscripts that *nitro volterrano* was prescribed for curing certain eye diseases. Mercati wrote about this product in 1717: *"nitrum nativum scissile ex quo venetis borax conflatur"* ("processed native nitre from which Venetians obtain borax").

These products were of great commercial importance in the Middle Ages. Consequently, bitter disputes and even wars broke out among some Tuscan communes over who would own the areas with *lumaie* in the Boraciferous region. A war between the towns of Florence and Volterra ended with victory for the Florentines; therefore, in 1472 ownership of the *lumaie* passed from Volterra to the Florentine craft guild for wool, known as *Arte della Lana.*

The geothermal manifestations in the region continued to be used for therapeutic purposes as they were commercially exploited for hydrothermal products. Many treatises on natural history and medicine from the Late Middle Ages, the Renaissance, and the Modern Age deal with the beneficial effects that thermal waters, muds, and gaseous exhalations had for treating all kinds of internal and external disorders. The authors who described the uses include Ristoro d'Arezzo (1282), Ugolino da Montecatini (1420), Savonarola (1460), Agricola (1546 and 1556), Fallopio (1560), Merula (1605), Cluverius (1624), and Targioni Tozzetti (1769).

During a period of grave economic crisis that befell Tuscany in the second half of the 18th century, the Grand Duke of Tuscany, Pietro Leopoldo, decided to inventory all of the natural resources in his territory. He asked a number of specialists to make on-site area surveys, with descriptions of the situation, sample collection, and analyses of minerals, thermal and fresh waters, gas, hydrothermal compounds, natural resources, and products of every kind. Targioni Tozzetti in 1769, while describing the situation at Larderello, pointed out the state of obsolescence into which the vitriol production plants, the alum quarries, and the spas had fallen since the end of the 17th century. He recommended that a number of specific analyses be undertaken. It was in analyzing the waters of *lagoni* at Monterotondo and at Castelnuovo in 1777 that Hoefer, director of the grand-ducal pharmacies, identified boric acid, then called "Homberg's sedative salt."

In 1779 Mascagni, noted anatomist, physiologist, and chemist, proved that boric acid was a component characteristic to all thermal fluids in the Larderello region. In one of his papers, Mascagni discussed the possibility of using natural heat to evaporate waters from the *lagoni* by burying metal boilers close to thermal manifestations. In a following paper, Mascagni published further analyses of thermal waters in the area, adding a detailed description for a method he had patented to exploit natural heat with boilers buried near fumarolic manifestations. Each boiler actually amounted to a heat exchanger, and it was the first equipment in the world used to obtain evaporative heat from geothermal fluids.

The Technology of Chemical Production between the End of the 18th Century and the Mid-19th Century

The discovery of boric acid, the studies by Hoefer and Mascagni on this compound and Mascagni's ideas—ingenious for his time—on using heat exchangers, mark the beginning of a period of strong growth for producing chemicals from geothermal fluids.

In the 18th century, the boric acid market was extremely important all over Europe and boric acid was obtained by processing a boron compound (tincal) imported from Persia. Just the cost of transporting tincal from such a distant country makes it easy to understand the worth of boric acid in those days. At that time, all of Europe was characterized by revolutions, national independence movements, and wars; therefore, the situation was poorly suited for launching a new product obtained from the hydrothermal manifestations of the Larderello area, and—perhaps more importantly—bringing a local product into the market would clash with established and powerful business interests that had grown in Europe around boric acid.

Nonetheless, a company was formed in 1812 to attempt industrial exploitation of boron compounds deposited around some manifestations in the Larderello area with the methods proposed by Mascagni. The venture failed for organizational and economic reasons. In 1815-1816, a new company was set up to develop the evaporation process for boric waters. However, instead of using Mascagni's heat exchanger to supply evaporation heat, the process was based on burning wood taken from the thick forests in the area. This time the initiative was successful and in just ten months (June 15, 1817, to April 14, 1818) some 36 tons of boric acid were produced and immediately found a receptive market in France. This explains the interest aroused in that country by boric acid produced in Tuscany.

In 1818, the company Chemin-Prat-La Motte-Larderel was formed and incorporated in Leghorn, where the charter members, all French, were living in exile. The company leased the *lagoni* of Montecerboli for a period of six years, with an option to renew, and named one of the partners, Francesco Larderel, as the executive manager for technical activities. The evaporation process used the same technology just described (evaporating the waters of the *lagoni* in metal boilers heated with wood fires) and immediately yielded very positive results. Indeed, thanks to the exploitation both of natural *lagoni* and artificial *lagoni* dug next to some "dry" manifestations, an average of about 50 tons per year of boric acid was produced for about 10 years.

However, as time passed, the increased use of firewood began depleting the forests. The resulting wood shortage slashed the company's profits and pointed out the precariousness of the technology used to evaporate the boric waters. Thus the French company came into trouble, but Francesco Larderel refused to abandon the project. He became a naturalized citizen of the Grand Duchy of Tuscany, bought out his three partners, and in 1827 began to apply some innovative ideas derived from his past experiences.

Above each artificial *lagone,* he had a brick dome built with a riveted metal pipe inserted at the top to redirect the steam accumulating inside. This construction, called a *lagone coperto* ("covered lagoon"), enabled a substantial amount of steam to be gathered and used to evaporate boron-bearing waters drawn off at the bottom of the dome (figure 3, center). The *lagone coperto* was essentially a low-pressure separator, while the steam collection pipes and the boilers they fed constituted a refined version of Mascagni's heat exchanger. Both Mascagni's

Figure 3. Lithograph of the 19th century showing a "covered lagoon" in the center and the related facilities, as seen around 1860.

boiler and Francesco Larderel's *lagone coperto* were inventions embodying remarkably advanced technology for their times.

Another innovative technology, introduced by Larderel as a follow-up of an idea conceived by his then technical advisor, the French chemist Payen, was to connect several natural and artificial *lagoni* in cascade (figure 4). In this configuration, the boric waters, which had at the source a boron content of 5 to 10 g/l, could be enriched gradually by natural evaporation before they were channeled into decantation tanks and evaporation boilers, where they became boric brines.

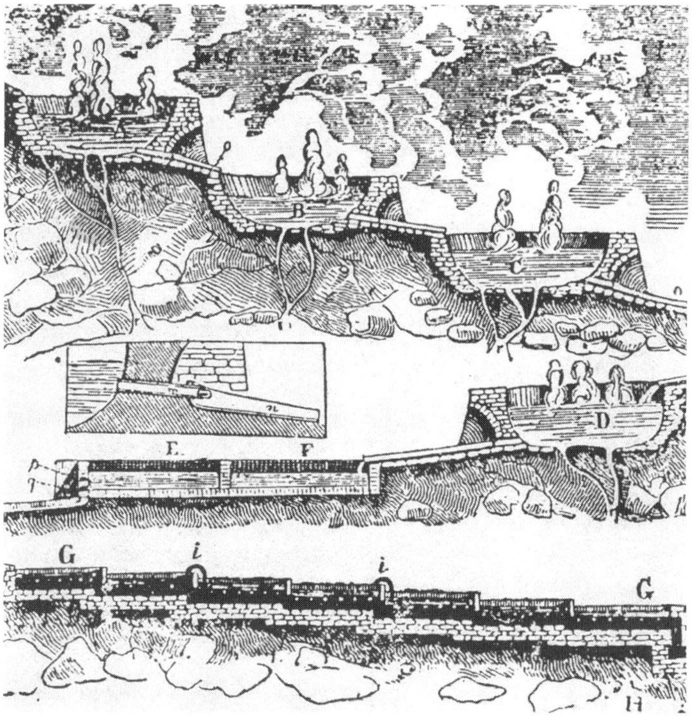
Figure 4. Print from 1841, showing a system of cascading lagoons (*lagoni*). *From Payen, 1841*

For the drying process, Larderel drew off part of the steam collected at the top of the covered lagoons and circulated it through a network of brick conduits beneath the floor of the processing plants. This operation concluded the multiple use cycle of the geothermal fluids: production of chemicals, evaporation, and drying. With these technological innovations in place, no less than 125 tons of boric acid were produced in 1829 and sold across Europe.

The ready availability in Western Europe of boric acid from Larderello had a great impact on the market, which had been dominated by English and Dutch companies importing *tincal* from Persia and other Middle Eastern countries. One offered to buy Larderel's company, but he had the farsightedness to refuse.

From 1840 on, Larderel's sons began collaborating in the company: Federigo eventually succeeded his father as the general manager and Adriano later became technical director. Both prepared for their jobs by studying industrial chemistry at the University of Pisa. They first improved the processing cycle—the drainage of boric waters, vaporization, crystallization, and drying—with newly invented equipment. For the vaporization process, a new type of boiler was designed, named the *caldaia adriana* ("adrian boiler") after its inventor and designer, Adriano Larderel. This boiler consisted of a series of brick conduits internally lined with lead (figure 5). Boron-rich waters were circulated above the conduits in counterflow to the steam stream. In this way, the heat content of the steam was fully exploited, achieving a much higher efficiency than the previous boilers at considerably lower operating costs. It will not go unnoticed that the *caldaia adriana* is a forerunner of modern counterflow heat exchangers. Thanks to this advanced process cycle and because of increased availability of boric waters from new drilling technologies discussed below, boric acid production rose from 125 tons in 1829 to over 1,000 (!) tons in 1850.

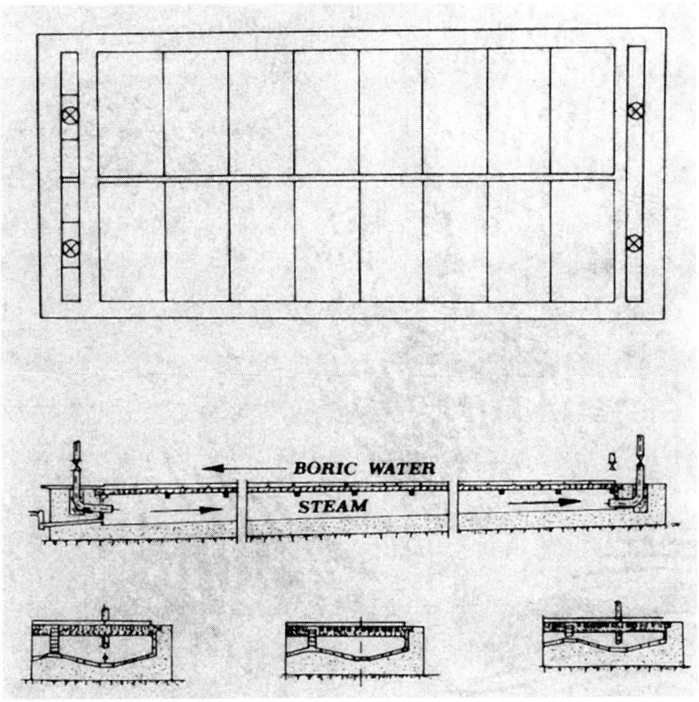

Figure 5. Layout (top), longitudinal section (middle), and cross section (bottom) of the *caldaia adriana* ("adrian boiler").

Another innovation, introduced in 1845 by Federigo and Adriano Larderel, was the use of steam to heat workshops, plants, and civil buildings. At first, simple copper tubes were used, but later, lead-surface heat exchangers were adopted, increasing the heat output.

Geothermal Drilling Technology in the 19th Century

Until the end of the 18th century, the exploitation of geothermal fluids and the extraction of boron minerals were based on simple holes, hand-dug to shallow depths (4 to 5 meters). The first geothermal wells, ordered by Francesco Larderel in 1832, were drilled with hand-operated drills in the immediate vicinity of fumaroles and natural hot pools. Although the wells had small diameters (about 10 centimeters) and were very shallow (6 to 8 meters), in most cases they produced fluids in greater amounts and at higher temperatures, pressures, and boron concentrations than fluids obtained directly at the surface from natural or artificial lagoons. However, after the first experiments were made, technical adviser Payen convinced Larderel to suspend all attempts to drill wells because of the possibility of uncontrolled blowouts.

A few years later in 1838, the Florentine scientist and engineer Gazzeri wrote an article discussing the possibility of increasing boric acid production by drilling "deep wells." Gazzeri did not specify the depth of the wells, but considering the drilling technology then available, he probably was thinking of initial depths on the order of 12 to 15 meters. In the same article, the author contended that fluids containing boric acid did not exist in small accumulations found only at the bottom of each surface manifestation but were supplied, instead, by a single deep, and presumably large, deposit: "The most elementary notions of geology persuade us that the mineral, whose decomposition gives off the boric acid contained in the ardent vapor of the steam jets, is not distributed in isolated masses at the base of each steam jet or *lagone*, but forms a single and huge deposit, the extent of which is unknown. If we were able to reach the deposit by means of drilling, a spontaneous blowout of natural fluids should follow, similar to that occurring in artesian wells. Alternatively [if a spontaneous blowout does not occur], we could obtain a stimulated production [of the well] by judicious injection of [fresh] water or by other methods...." Thus, Gazzeri believed that it was necessary to reach the deep reservoir by drilling, not only to tap the larger amount of fluid produced spontaneously by the wells, but also to pump fresh water through the wells into the reservoir for more sustained leaching of boric acid.

In another paper in 1841, Gazzeri developed these concepts further and announced that the engineer Manteri, stimulated by the theory and operative suggestions presented in Gazzeri's article of 1838, had built new drilling equipment that could be used without a winch and could drill holes about four inches in diameter. Using his drill, Manteri had bored three wells 6 to 7 meters deep that he sited over 100 meters away from natural manifestations at Lago.

The production from these three wells, located for those days at such great distances from the manifestations, seemed to prove Gazzeri's theory—and for some aspects concerning the large size of the geothermal reservoir, it actually did. Gazzeri's hypotheses on the size of the geothermal reservoir and on drilling wells far from manifestations were important contributions to geothermal knowledge and development technology.

Also in 1841, a small company (not Larderel's) successfully drilled a few "deep" geothermal wells using a rig called the *verga artesiana* ("artesian bar"), illustrated by A in figure 6.

Figure 6. Old print showing (A) the *verga artesiana* ("artesian bar") and (B) a square-based drilling rig with winch.

Therefore, in consideration of the encouraging results by this company and by Manteri, and having noted that wells could be kept under control without risking the blowouts Payen had predicted, Larderel and his sons decided to resume drilling wells using the *verga artesiana*.

In a few years, the boost to drilling by the Larderels brought considerable technological improvement: four-legged drilling rigs (B in figure 6), a winch to enhance percussion efficiency, various types of drilling bits (some with a flat cross section, others with a helicoidal shape), and bucket-type tools, called bailers, to remove cuttings from the well bottom. Furthermore, to obtain a hole with a round cross section, a method of manually rotating the drill stem was developed that allowed lining the first 6 to 8 meters of the hole with an anchor pipe (surface casing) of riveted metal plate.

By 1842, all these technological improvements enabled drilling wells at a depth of 25 to 30 meters. Moreover, between 1841 and 1855, a large pit (5 to 6 meters deep) was usually dug at the well site and drilling actually began at the bottom of the pit. In this way, well depths in the order of 30 to 35 meters were reached around the mid-19th century. However, the function of this pit was not so much to increase the well depth but to hold the liquid fraction of the steam-water mixture produced by the well. In this manner, an artificial *lagone* was created at each drilling site. The pit was covered with the characteristic brick dome (described earlier and illustrated in figure 3), resulting in an artificially covered lagoon.

Figure 7. Sketch of a drilling rig in 1862, showing (1) walking beam; (2) work floor; (3) *gomito* ("elbow"); (4) winch; (5) protective shield; (6) ropes for percussion drilling.

The year 1856 marked the beginning of a new stage in geothermal drilling technology, one that in a few years would enable depths to be reached of between 150 and 200 meters. Innovations from 1856 to 1870 were the first rig with an elevated working floor (2 meters high), walking beams to operate the drill stem and increase the percussion efficiency of the bit, a protective shield for winch operators, and a wellhead with a side outlet called a *gomito* ("elbow"), which constituted the first blowout control device ("blowout preventer") for geothermal wells. The device had two main parts, the *chiusino* ("lead plug") and the *bisesta* ("arm with keyhole"). See figure 7.

The practice of casing wells as drilling proceeds comes from this period. This was done with pipe joints of riveted plate in standard two-meter lengths. The bottom joint was equipped with a toothed ring of hardened metal that reamed and enlarged the wellbore. The joints were coupled every two meters with a riveted metal band, and the casing string was given a jerky, rotating motion so that the toothed ring on the bottom joint acted as a hole reamer. In this manner, drilling could proceed under reasonably safe conditions.

Another drilling technique developed at Larderello between 1856 and 1870 was the stimulation of well production. Since wells were gradually increasing in depth (up to 200 meters), they often encountered shallow, cold aquifers that in many cases prevented them from producing spontaneously because of the cooling and hydrostatic pressure by fresh water on the reservoir fluid. The Larderels and their drilling personnel understood the importance of this phenomenon and managed to find a way to lighten the hydrostatic load on the producing layers. For this, a special tool was devised, a series of coupled metal disks with a slightly smaller diameter than the casing.

Sandwiched between the two disks of each couple were hemp or rags, and lengths of chain were used to connect the various couples of disks, one below the other. For its length (some 10 meters) and flexibility, the tool was called a "serpent." With the help of a lead weight, it was lowered into the well—25 to 30 meters below water level—to piston out the water. To do this, a team of workmen manually extracted the serpent from the well as fast as possible, which lowered the water level in the well, making it produce. Sometimes the operation was repeated several times before stable production was achieved.

By the end of the 19th century, the combination of all these new drilling techniques and the new equipment, mostly developed autonomously at Larderello and continuously refined over the years, made it possible to drill wells 250 to 300 meters in depth.

Before concluding this outline of drilling technology at Larderello, it is interesting to recall a few organizational aspects of this work toward the end of the 19th century. As rig capacity increased, clear-cut rules for drilling were laid down by the Larderels: each rig was equipped with a metal winch when the well depth was greater than 130 meters, and with two winches when the well was expected to exceed this depth. The work was carried out in one 10-hour shift, with two teams alternating every 10 minutes to rest. The number of workmen on each team was five down to 30 meters, six from 30 to 50 meters, and one additional worker for every 10 meters of depth thereafter. The gradually increasing well depths also required standardization of equipment, allowing for interchangeability between drilling rigs operated simultaneously at different sites.

This, along with the fact that Larderello was far from other towns and had to develop as an autonomous community, made it necessary to set up workshops to build and maintain all equipment, tools, and materials (figure 8). The shops not only supported the drilling activity but provided services for the chemical production plants and for every other need of the chemical industry.

The First Modernization Period: 1860 to 1900

As the number and depth of the wells, and therefore the fluid production, increased over the second half of the 19th century, rapid technological improvement continued to occur in systems for gathering fluid and processing boron compounds. These improvements

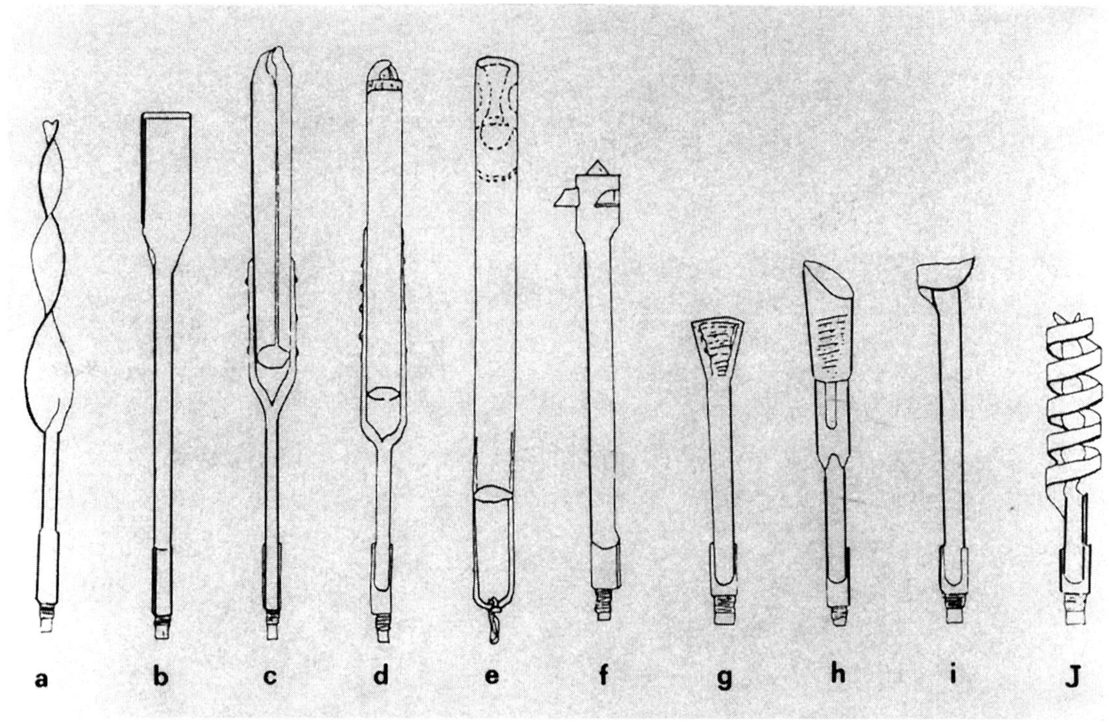

Figure 8. Sketch of various types of percussion bits (a and b), shell augers (c, d, and e), and recovery tools (f, g, h, i, and j) made and used at Larderello from 1860 to 1900.

resulted in increased boric acid production, which rose from 1,000 tons in 1850 to 2,000 (!) in 1860. For the next 10 to 15 years, Larderello's boric acid production grew so fast in the European market that it became a near monopoly.

However, in about 1875, when boric acid produced from borates of sedimentary origin, such as ulexite, colemanite, and rasorite, began penetrating the market, especially those from the United States, Larderello's chemical industry suffered a sudden setback that threatened to nullify its technological achievements and experience in the use of geothermal fluids for chemical production. The owners and workers of Larderello reacted by fostering further technological innovations, introducing organizational improvements in geothermal fluid extraction and processing, and diversifying chemical production as much as possible. This resulted in the production of a wide range of borate compounds and the development of a philosophy for geothermal fluid utilization that we today call "multiple use." Among the technological innovations made in this period, the following are worth mentioning:

- Development of steam-driven pumps and their systematic use for well stimulation and for pumping borate solutions to processing plants (actually, the first experiments on using the mechanical energy of natural steam to drive pumps date back to 1859. However, owing to the rather low pressure of the steam produced by the wells between 1850 and 1870, this use remained at the experimental stage until 1875)

- Modernization of the gathering system for steam and boric brine solutions. This was obtained by replacing the old wooden or brick channels with riveted pipes, insulated to reduce heat dissipation

- Use of multi-tubular boilers as heat exchangers in all production processes requiring pure steam or hot water of nongeothermal origin

As to the diversification of chemical products, the principal achievements of the period 1880 to 1900 were threefold: production of refined, highly pure boric acid, use of ammonia present in residual boric brines for the production of ammonium sulfate, and production of borax by the reaction between boric acid and sodium carbonate.

Organizational aspects and scientific advances in the second half of the 19th century included the following:

- Establishment in 1854, and expansion in 1880, of a chemical laboratory at Larderello to monitor the characteristics of geothermal fluids and optimize the quality of chemical products

- Execution of many geoscientific studies aimed at better understanding the conditions of the Larderello geothermal system. As a result of all these studies, toward the beginning of the 20th century, an eminent geologist, Bernardino Lotti, formulated the "theory on the juvenile origin of endogenous steam"

- Execution of lab analyses and on-site experiments in a continuous modernization of the chemical processing cycles. Among other achievements in this field, in 1895 a group of scientists from the Universities of Pisa and Padua, headed by the renowned chemist Nasini, discovered helium and argon in the gas contained in the steam at Larderello. This discovery was followed in a few years by the detection of radon and the quantitative assessment of the radioactivity of geothermal fluid

THE START OF GEOTHERMAL POWER PRODUCTION: 1904 TO 1913

NEAR THE BEGINNING OF THE 19TH CENTURY, THE LARDERELLO CHEMICAL INDUSTRY CAME under the direction of Prince Piero Ginori Conti, son-in-law of Florestano Larderel. The most outstanding achievements of Ginori Conti's commitment to innovation are the studies and experiments on the use of geothermal steam as an energy source for electrical generation. This work was carried out at a fast pace, and in only a couple of years it was rewarded with success: on July 15, 1904, the first five light bulbs were lighted at Larderello using geothermal power. The success of the experiment is immortalized in a picture familiar to geothermal specialists the world over (figure 9).

Figure 9. Photo showing the first experiment in the world of electrical generation from geothermal steam, carried out by P. Ginori Conti at Larderello in 1904.

This experiment used a piston engine coupled with a 10 kilowatt dynamo; the engine was driven by pure steam produced in a small heat exchanger fed with wet steam from a well near Larderello. Therefore, the first use of geothermal steam for power generation was carried out with a transformation cycle that we today call an "indirect cycle." Although Ginori Conti and his technical staff had also studied the possibility of feeding the piston engine directly with natural steam, the indirect cycle was chosen because, aiming to relaunch his chemical company, Ginori

Conti wanted to lay the technological basis for the simultaneous use of geothermal fluids for power generation and chemical production.

The encouraging results of this experiment quickly led to the first prototype of a geothermal power plant, which went into operation in 1905. The plant was formed by a Cail reciprocating engine connected to a 20 kilowatt dynamo. This first small power station, along with another installed in 1908 composed of a Neville reciprocating engine coupled to a second 20 kilowatt dynamo, enabled electrification of Larderello's most important industrial plants and the main residential buildings.

In 1913, after absorbing all the small local firms that produced boric acid with obsolete technology, Ginori Conti's company assured itself of the possibility of extending exploitation of geothermal fluids to the whole Boraciferous region, an area of several hundred square kilometers. In the same year, the first commercial geothermal power plant went into operation. The energy it produced was fed into a network serving all the chemical production plants and the main buildings of Larderello and the other villages of the region. This power plant, named Larderello 1, was equipped with a turbine generating 250 kilowatts of electricity. It was designed and built by the Tosi Electromechanical Company to operate with wellhead fluid pressures of up to three atmospheres. The turbine was driven with pure steam obtained from a heat exchanger supplied by two wells.

Achievements from 1900 to 1944

Although Ginori Conti initiated a further step in modernizing the chemical plants in 1903, it was not until 1908 (when electricity from geothermal power stations became available) that he was able to start a radical renewal of almost all activities of his company and launch new initiatives. From the beginning of this century to the end of World War II, the principal achievements can be summarized as follows:

Chemical processing of geothermal fluids. The processes were further expanded to obtain new products, such as sodium perborate, ammonium carbonate, carbonic acid, talcum powder, and other compounds widely used in the pharmaceutical industry. Very wide multiple use of geothermal fluids for chemical production was thus achieved.

Drilling. The activities were organized as a completely autonomous service sector of the company. Its main operational characteristics were

- 1906: the use of mechanical winches driven by geothermal steam. These were replaced with electrically-powered winches in 1914

- 1907: oxyacetylene welding, enabling radical change in well casing configurations

- 1923: drilling partly converted from the percussion to the rotary system. The new "H" and "BCF" rigs, built by Termomeccanica Co., were equipped with water circulation systems. The rotary system was only used to drill impermeable formations. Once the first producing fracture was encountered (usually at the top of the carbonate formations forming the principal reservoir), the rotary rig was replaced with a percussion rig until it reached the main producing layers. Using two different rigs to drill a single well, however, was abandoned after a few years when more versatile drilling equipment became available

- 1929: introduction of Trauzl rigs, equipped with high-powered pumps suitable for both rotary and percussion drilling

- 1938-1940: percussion techniques abandoned definitively in favor of rotary drilling

Scientific achievements. A few words will suffice to emphasize the intensity and advanced nature of the scientific effort that accompanied geothermal development at Larderello during this period: geologic and structural surveys, hydrogeologic investigations, paleontological analyses, chemical and physical analyses in company and university laboratories, geoelectric prospecting, well logging, seismic recording in special stations to single out possible relationships between well production rates and earthquakes, measurements of production characteristics and reservoir studies, lab tests and field experiments with pilot plants to improve chemical processing cycles of geothermal by-products, and studies on thermodynamic cycles for power generation. The results of all studies carried out until about 1928 are included in a voluminous monograph published by Nasini in 1930, prepared in collaboration with several other Italian scientists and top technical staff of the Larderello company. The monograph described the situation up to the publication date, but commitment to science and technology actively continued in the following decade, before the outbreak of World War II.

In particular, based on results of the geoscientific nature, the criteria for selecting drilling sites were gradually improved after 1925 and wells were located at ever greater distances away from localities where natural manifestations were seen. Although it would be incorrect to say that Lotti's "theory on the juvenile origin of endogenous steam" was outdated after 1925, it is worth mentioning that the technical archives at Larderello contain unpublished data and reports from the period that discuss the limits of this theory and hypothesize on the meteoric origin of geothermal steam. A special contribution came from the results of the well Soffionissimo 1 ("a very powerful well"), shown in figure 10, that initially produced 230 t/h of steam at 4.5 abs. atm. and 205° C.

Figure 10. The well "Soffionissimo 1" during the initial stage of production in 1931.

Space heating. This use was introduced around 1845, but it was not until 1910 that it began systematically taking root. By 1940, all establishments, plants, facilities, workshops, and residential buildings, plus a number of greenhouses and agricultural installations created by the Larderello company—not only in the main village but also in many other localities of the Boraciferous region—were warmed with the residual heat of low-pressure steam.

Electrical generation. This is the most significant achievement of geothermal development at Larderello in this period. The first 250 kW_e commercial power plant, dating from 1913, was followed in 1916 by two 3.5 MW_e turboalternator units manufactured by Tosi Co. These units used the indirect cycle for power generation and were supplied with pure steam obtained in heat exchangers. The size of these two units, although very small by today's standards, equaled that of most of the world's installed hydroelectric and thermal power plants of the time.

The first pilot turbine fed directly with natural steam produced by the wells ("direct cycle") had a capacity of 23 kW_e and was installed at Serrazzano in 1923. It remained in operation without any significant problems for about two years, when it was assigned for use in training technical personnel at the Larderello company school. After this pilot turbine, other direct cycle turbines were installed at Castelnuovo (a 600 kW_e unit in 1926 and an 800 kW_e unit in 1927) and Larderello (a 3.5 MW_e unit in 1930). This raised the total installed capacity in the Boraciferous region, as of December 1930, to 12,150 kW_e, of which 7,250 used the indirect cycle and 4,900 the direct cycle.

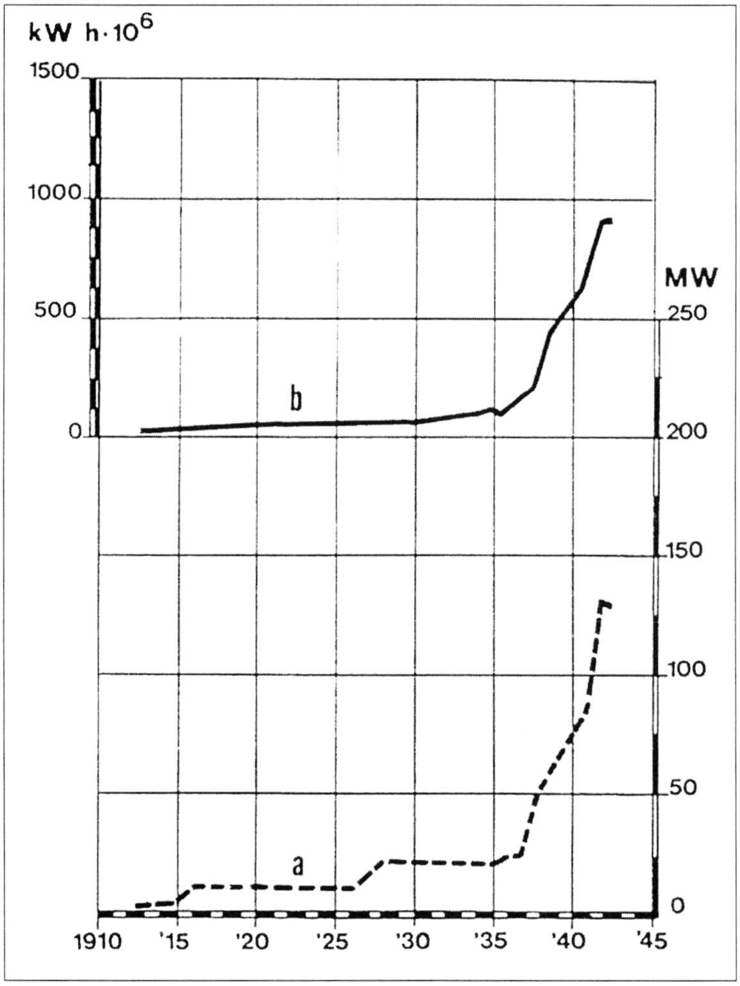

Figure 11. Installed capacity (a) and electricity generated (b) at Larderello from 1913 to 1943.

The first large geothermal power plant, called Larderello 2, was completed in 1939 and included six 10 MW_e units fed with pure steam in an indirect cycle. Adoption of this cycle was dictated by the need to use most of the natural steam produced by the wells at Larderello to extract valuable chemical by-products.

Between 1940 and mid-1943, other units went into operation: four 10 MW_e units at Castelnuovo (similar to those at Larderello), one 3.5 MW_e direct cycle unit at Sasso Pisano, and two units for a total of 8.5 MW_e at Serrazzano, again based on the direct cycle.

At the end of 1943, the total installed capacity in the

Boraciferous region was 132 MW_e, of which 107 MW_e was fed by pure steam, the indirect cycle (figure 11). The others were exhausting-to-atmosphere units, the direct cycle.

At the start of 1944, the Larderello region was dramatically involved in the events of World War II. The Larderello power plants were strategically important because they were fed from an indigenous energy source and were supplying electricity to the whole railway network of Central Italy. In the spring of 1944, not far from Larderello, the retreating armies then in Italy organized the "Gothic Line." This was the strategic line separating the belligerent armies of the two opposing forces.

All the geothermal power stations and chemical plants in the zone were heavily bombed and destroyed, and almost all the production wells were blown up by charges placed at the base of the master valve. A desolate scene of noise and ruin spread over the region. Thus, the modern phase of geothermal development in Italy ended tragically in early 1944.

Among the ruins of war, miraculously still intact, was the 23 kW_e exhausting-to-atmosphere turbine used at the company school to train technical personnel since 1925 (figure 12). Thanks to it and to the determination of the people of Larderello, the curtain rose in the summer 1944 on the present act of Italian geothermal history.

Figure 12. First exhausting-to-atmosphere geothermal turbine (23 kW_e), installed at Serrazzano in 1923.

Acknowledgments

The author is indebted to ENEL SpA/Geothermal Department for authorization to consult its historical archives and permission to use all illustrations included in this chapter. Thanks are also due to Dr. R. Cataldi for encouragement to write the chapter, for his critical review of the manuscript, and for editing the original and the revised versions of the work.

Cited and Selected References

Agricola (Bauer, G.). 1546. *De natura eorum quae effluent ex terra.* Frobenius Edit., Basel.

Agricola (Bauer, G.). 1556. *De re metallica.* Frobenius Edit., Basel.

Belli, E. 1891. *Della trivella: Appunti d'un lagonaio.* Museum of Larderello. (Unpublished report).

Burgassi, P. D. 1983. Energia geotermica nelle Colline Metallifere. *Mem. Soc. Geol. It.,* 25, 135-140.

Burgassi, P. D. 1985. I de Larderel e la perforazione geotermica: Studi, esperienze e lessico tra XIX e XX secolo. *Ricerche Storiche,* 15, no. 3, 631-648.

Burgassi, R., and P. D. Burgassi. 1981. I soffioni boraciferi della Toscana e le originali industrie a cui hanno dato luogo. *Riv. Ind. Min.,* Serie III, 2, no. 2, 13-23.

Cluverius, F. 1624. *Italia antiqua. Lugduni Batavorum, vol. 1.* Ex Officina Elseviriana Universitaria Edit., Pisa.

Fallopio, G. 1560. *Opera de thermis.* Reprint 1606. Giunta Edit., Venice.

Fiumi, E. 1943. *L'utilizzazione dei lagoni boraciferi della Toscana nell'industria medioevale.* University of Florence, CYA Edit., Florence.

Gazzeri, G. 1838. Lettera al giornale del commercio, arti e manifatture. *Dec. II,* 1, no. 31.

Gazzeri, G. 1841. Induzione ora verificata di ottenere nuovi soffioni di acido borico per mezzo della trivellazione del terreno. *Atti Accademia dei Georgofili,* 19.

Ginori Conti, G. 1931. *La perforazione a Larderello e il soffionissimo.* Tip. G. Cencetti, Florence.

Hoefer, U. 1777. *Memoria sopra il sale sedativo di Homberg.* Stamperia Granducale, Florence.

Lotti, B., and L. Cresti. 1879. Studi sopra i soffioni boraciferi della Toscana. *Regia Accademia de' Lincei,* Ser. 3, no. 4, 33-34.

Lotti, B. 1907. Sulla provenienza dell'acido borico nei soffioni della Toscana. *Rass. Mineraria ed Industria Chimica,* 26, no. 14, 1-5.

Lotti, B. 1928. I soffioni boraciferi della Toscana. *Boll. Soc. Geol. Italiana,* 47, no. 2, 1-10.

Manteri, V. 1841. Prove di perforazione artesiana sulle rive del Lago di Vecchienne. *Atti Accademia dei Georgofili.*

Mascagni, P. 1779. *Dei lagoni del Senese e del Volterrano.* Stamperia Poggini Carlo & Figli, Siena.

Mercati, M. 1717. *Metalloteca Vaticana. Opus posthumum, Armarium, vol. 4.* Ex Officina Io Mar and Salvioni Romani C. Edit., Rome.

Merula (Van Merle, P.). 1605. *Cosmographiae generalis (lib. tres) et geographiae particularis (lib. quatuor).* Ex Officina Plantiniana Edit., Raphelengie.

Micheletti, V. 1956. *Storia intorno ai soffioni.* Rassegna Larderello, Larderello (Pisa).

Nasini, R. 1930. *I soffioni e i lagoni della Toscana e l'industria boracifera.* Italia Edit., Rome.

Nasini, R., and G. Grassini. 1932. Sulla scoperta dell'acido borico nei Vasi Sigillati Aretini. In *Studi Etruschi, vol. 6.* Florence.

Payen, A. 1847. Acide borique des soffioni de la Toscane. *Annal. Chim. Phys.,* 1, 247.

Ricettario fiorentino. 1567. Stamperia Eredi di B. Giunti, Florence.

Ricettario senese. 1567. Stamperia Eredi di B. Giunti, Florence.

Ristoro d' Arezzo. 1282. *Libro della composizione del mondo.* Reprint 1864. G. Daelli Edit., Milan.

Savonarola, M. 1460. De balneis Vulterrae. In *De balneis omnia quae extant apud graecos, latinos et arabos.* Reprint 1553. Giunta Edit., Venice.

Targioni Tozzetti, G. 1769. *Relazione d'alcuni viaggi fatti in diverse parti della Toscana, vol. 3.* Stamperia Granducale, Florence.

Ugolino da Montecatini. 1420. De natura et virtute balneorum Comitatus Vulterrae. In *De balneis omnia quae extant apud graecos, latinos et arabos.* Reprint 1553. Giunta Edit., Venice.

The Author:

Pier Domenico Burgassi
55 Via G. Galilei
56044 Larderello (Pisa), Italy
Telephone: 39.0588.67.489
Fax: 39.0588.22.910

Extracting boron compounds by evaporating and concentrating boron-rich geothermal waters from natural boiling pools (*lagoni*) near Larderello, about 1540. A: natural boiling pool (*lagone*) lined with wooden boards and bricks; B: pots; C: ladles; D: wooden boxes and iron pans; E: tongs. Earthenware pots were placed in the *lagoni* and ladles of hot saline water were added, filling the pots half full. The water was evaporated from the pots until a thickened brine remained. Then the pots were clasped with tongs and the brine was poured into iron pans standing on four heavy legs near the borders of the *lagoni*. As nearly boiling hot water flowed freely around the pans, the saline brines condensed even further until the last of the water was evaporated. The mixture of boron salts was then shoveled out of the pans and the process begun anew. From *De re metallica* by Agricola, Italian translation by M. F. Fiorentino, H. Frobenio, and N. Episcopo. Basilea, 1563. *Courtesy of ENEL SpA/Geothermal Department*

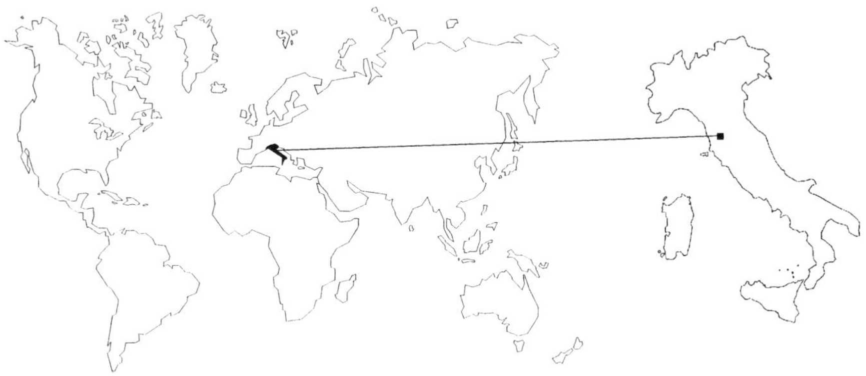

14. Four Centuries of Scientific Investigation at Larderello, from the 16th Century to the 1st International Geothermal Congress in 1928

by
Pier Domenico Burgassi
Raffaele Cataldi
Claudio Donati

Abstract: The chapter describes geoscientific studies and laboratory analyses carried out in the Larderello area of Central Italy, after the Middle Ages until the first International Geothermal Congress in 1928. It examines hypotheses and theories advanced by numerous Italian and foreign scientists explaining the origin of hot fluids, the nature of hydrothermal compounds, and the formation of geothermal fields. During these four centuries, a number of new hydrothermal minerals and some rare gases were identified at Larderello, and many other research activities were undertaken. All of these studies reflect the high level of scientific knowledge, technological advancement, and development reached at Larderello by the beginning of the 20th century.

Historical Background

THE DESCRIPTION OF THE PRINCIPAL GEOTHERMAL MANIFESTAtions and the formulation of the first hypotheses explaining their occurrence in the Mediterranean area predate the Christian era by many centuries. Hot springs, fumaroles, active volcanoes, earthquakes, and other manifestations of the Earth's heat aroused the interest of many Greek and Latin writers, who described or interpreted them in light of the cosmogonal theories of their time. Thus nearly 1,000 years before the beginning of the Middle Ages, a first and very important nucleus of knowledge was formed concerning geothermal phenomena and the ways that terrestrial heat manifests itself at the surface.

The uses of natural hot water for balneotherapy and the exploitation of hydrothermal products for a wide range of practical applications increased remarkably during the millennium preceding the Christian era and eventually extended over all

Roman territory, reaching an apex during the period of maximum splendor of the Roman Empire, the 3rd century A.D. After the fall of Rome in the 6th century, these uses declined quickly throughout Southern Europe and entered a long period of disuse lasting until the beginning of the second millennium (Cataldi and Burgassi, 1992b).

In a gap of some seven centuries between the collapse of the Roman Empire and the middle of the 12th century, few works can be found referring to geothermal phenomena or describing the uses of the Earth's heat and its by-products. However, the geothermal knowledge gained in Antiquity was not lost and became the basis of a more advanced understanding of terrestrial energy that started in the first centuries of the second millennium (Cataldi and Chiellini, 1995). A reawakened interest in hydrothermal minerals started at Larderello in Central Italy during the first decades of the 12th century. The existence of conspicuous hydrothermal deposits around natural manifestations began to be a matter of contention among some Tuscan communes, lasting about three centuries (Burgassi, 1987; Cataldi and Burgassi, 1992c).

A new literary flowering inspired by geothermal manifestations began in the 13th century. Ristoro d'Arezzo is worthy of mention for the scientific importance that his description of the geothermal phenomena in the Larderello area has had in understanding the evolution through the centuries of the Larderello Geothermal Field. In 1282, he described in detail a huge explosive event that took place in a *lagone* (a natural pool of boiling water) in the southern part of the Larderello area, probably a phreatic explosion. Numerous other authors from Italy and abroad started writing about the natural manifestations: some interested in religious considerations, others in the phenomena and interpretative hypotheses, and others in the therapeutic properties of thermal waters and thermo-mineral muds (see Chapters 11 and 12 in this volume).

16th and 17th Centuries

In the first half of the 16th century, the first important author who dealt systematically with the natural manifestations and hydrothermal products of the Larderello area was Georg Bauer (1494-1555), a German naturalist and medical doctor who translated his family name into Latin and is better known as Agricola. Considered the "father of mineralogy," Agricola wrote detailed descriptions of the manifestations, gases, and hydrothermal compounds at Larderello, explaining them scientifically and comparing them with similar ones in other Italian geothermal areas. He described techniques for extracting and processing hydrothermal minerals (see the initial illustration). The results of his studies are in three general works: *De veteribus et novis metallis* (*Old and New Metals*), 1546; *De natura eorum quae effluent ex terra* (*Nature of Fluids Escaping from the Earth's Surface*), 1546; and *De re metallica* (*Minerals and Mining*), published posthumously in 1556.

Contemporary with Agricola's work is the treatise *De balneis omnia quae extant apud graecos, latinos et arabos* (*All Thermal Stations in Greek, Latin, and Arabian Territories*, 1553). Each monograph in this treatise covers a major geothermal area from these territories and includes a geographic description of the site and detailed information on the therapeutic properties of the thermal waters and thermo-mineral muds. Monographs for the Larderello area include those by Mengo Blanchello Faentino, Franciotti, Savonarola, Bartolomeus Taurinensis, and Ugolino da Montecatini.

The work of naturalist Baccio in 1571 followed shortly afterward. This author was concerned with the fundamental causes of natural manifestations, attributing their formation to the presence of a "subterranean fire." He discussed the origin of sulfurous encrustations associated with fumarolic manifestations.

In 1575, Aldrovandi dealt with natural manifestations and hydrothermal compounds at Larderello. He described the presence of a mineral that he called *nitrum nativum* ("native nitre"). From his description and from later sources (Mercati, 1717; Hoefer, 1777), we know that *nitrum nativum* was boric acid.

In 1606, anatomist and medical doctor Fallopio published an important work describing the characteristics of the manifestations at Larderello. He paid particular attention to the therapeutic properties of different thermal waters and their abilities to enact medical cures. He agreed with Baccio, attributing the formation of the manifestations to the presence of a "subterranean fire."

A little later, the German geographer Phillip Clüver (1580-1622), who took the Latin name of Cluverius and is considered the founder of historical geography, described the Larderello region in detail and prepared the first geographic map of the area with all of the major natural manifestations. His work, with the map, was published posthumously in 1624.

In the second half of the 17th century, the Larderello region attracted the attention of the Danish physician and geologist Nils Steensen, who is better known as Nicolaus Steno (1638-1686).

During his stay in Florence as physician for Grand Duke Ferdinand II, he studied the geological characteristics of many Tuscan areas, including Larderello, which he described in detail because of its geological peculiarities. In his work *De solido intra solidum naturaliter contento dissertationis prodromus* (*Preliminary Dissertation on Natural Solids inside Rocks*), published in 1669, he discussed the nature of volcanoes, maintaining that the natural manifestations at Larderello are "exhalations" of cryptovolcanic activity.

18th Century

Toward the beginning of the 18th century, the mineralogist Mercati began to study systematically the hydrothermal deposits associated with geothermal manifestations in the Larderello region. In a 1717 work, Mercati described many hydrothermal minerals, including what he called *nitrum volaterranum* ("nitre from Volterra"). Based on his indirect description of this mineral *"nitrum nativum scissile ex quo venetis borax conflatur,"* which means "processed native nitre from which Venetians obtain borax," and from Hoefer (1777), we know that he was speaking of boric acid.

A few decades later, the scientist Targioni Tozzetti was appointed by the Grand Duke of Tuscany to evaluate potential development of the agricultural and mineral resources of the Grand Duchy. In 1742, he began 25 years of detailed field surveys in many areas, the results of which were published in 1769. The author did not stop with a simple description of natural manifestations and *lagoni* or with merely listing hydrothermal minerals amenable to extraction and commercialization, such as alum, *nitrum volaterranum*, sulfur, and vitriol (heptahydrated ferrous sulfate); he also discussed critically Larderello's geothermal phenomena in light of theories already proposed. Regarding the natural manifestations and the *lagoni* of the region, Targioni Tozzetti wrote,

> "In all of them, I have identified alum, salt, sulfur and vitriol. Baccio says that a 'subterranean fire' is the cause of heat and boiling; however, if he means 'real fire,' as in the case with volcanoes, he is wrong simply because there is no fire escaping [from those manifestations]. At most, there is a continuous hot fermentation, like certain fermentations that occur chemically with two or more cold liquids....
>
> "Not only the *lagoni*, but all the mineral waters and *mofettes* I have observed, spring out from fractures of primordial mountains, but never from gently-sloped terrains. This means that the minerals formed within rock mountains. As time passed, some small portions of the veins in which the inflammable material was trapped came to outcrop; therefore, this material could come into

contact with outside air and a small amount of [surface] water started to percolate into the veins. This water may be sufficient to trigger off the fermentation process, drive out the air trapped within fractures, release heat, and cause the water to boil. For sure, large masses of inflammable material are scattered within rock mountains, much larger than we might suppose, and one day in the future will come to outcrop and start to ignite [by fermentation]....

Sketch of a *lagone* in the area near Larderello, as seen in 1750. From Targioni Tozzetti, 1769. *Courtesy ENEL SpA/Geothermal Department*

"In other *lagoni*...deep, round-shaped holes are to be seen, like those made by the tarantula; from this type of hole, nothing but very hot air blows out. If you hold a silver coin against the air stream, a veil of condensation forms, much as it happens if you breathe on a mirror. This is an extremely fine coat of water, tasteless but sulfur-smelling; and in less than a minute the coin turns black, as if it were made of iron. On the upper surfaces of many of these holes a substance is deposited similar to pumice. I doubt very much, however, if this substance is that which Aldrovandi calls '*alumen volaterranum...facultatis astringentis*' ['alum from Volterra...with astringent properties']."

Targioni Tozzetti even recognized the effects of the self-sealing process; however, he could not fully understand the causes of this phenomenon. He wrote,

"The *lagoni* continually undergo an expansion process, and tend to move toward the mountain top. Areas with fields, which until a few years ago were richly fertile, now yield nothing because they have been crossed by the tortuous fissures created by the spreading *lagoni*. This expansion and encroachment on once fertile fields indicates that this type of *lagoni* eventually reaches and settles in correspondence to large veins of the original material, and this continues until the whole material has been consumed.... However, there is no doubt that the expansion is a process that becomes weaker day by day, and that in certain places it halted a long time ago. The many dried-up and cold, round-shaped lacustrine depressions, which can be seen here and there in the valleys, are evidence of this. These appear as extremely barren places and look like craters, the bottoms of which are filled with collapsed blocks of *alberese* ['marly limestone']; thus, these are to be considered [old *lagoni*] similar to the presently active *lagoni*, the only difference being the fact that no steam is released...."

In 1777, a few years after Targioni Tozzetti concluded his Larderello surveys, a very important scientific finding was made by Hoefer, chief chemist for the pharmacies of the Grand Duke of

Tuscany. Hoefer analytically documented the presence of boric acid in the water of two *lagoni* of the Larderello area, near Monterotondo and Castelnuovo.

At that time, boric acid (then called "Homberg's sedative salt") had not yet been recognized as one of the compounds in the hydrothermal deposits of the Larderello area, nor it had been separated chemically. The confirmation of its presence in the area and the success of Hoefer's chemical separation process opened the way for a flourishing geothermal industry in Southern Tuscany, and the area took on the name of the Boraciferous region.

Boric acid, used widely in the pharmaceutical industry at that time, was obtained from minerals imported overland at great expense from Persia. Hoefer's discovery therefore resulted in immediate plans for industrial production of boric acid at Larderello from hydrothermal minerals. With the support of the Grand Duke of Tuscany, two complementary groups of activities soon began at Larderello: drilling for hot fluids (used for extracting boric acid and as a primary energy source to drive mechanical engines) and systematic scientific research to improve knowledge of geothermal phenomena.

In this regard, we must highlight the scientific contributions made by Mascagni, a professor of chemistry, physiology, and anatomy at several Tuscan universities. In 1779, he published a detailed mineralogical description of hydrothermal products in the Boraciferous region, maintaining in particular that certain "white concretions" that Targioni Tozzetti had thought to be selenite were actually boric acid. In the years that followed, trying to find ways to exploit large quantities of boric acid for industrial production, Mascagni made many more chemical analyses of hydrothermal minerals and conducted site-specific studies on the extension of the manifestations. He proposed that natural heat, rather than wood fire, could be used to evaporate boric brines and designed two different types of metal boilers for this purpose, one to be embedded in hot ground on the edges of the natural manifestations and the other to be partially sunk into the *lagoni* (Burgassi, 1987).

19th Century

After the discovery of boric acid and along with—and because of—the rapid exploitation of hydrothermal deposits beginning around 1820, scientific interest in geothermal phenomena at Larderello grew quickly. Many new geological investigations and mineralogical and chemical analyses were undertaken by Italian and foreign scientists. The Frenchman Dumas stated in 1828 that boric acid reaches the surface after interaction between meteoric water and deeply-rooted large masses of hydrothermal minerals, including boric acid and sulfur. In this way, he explained the occurrence of hydrogen sulfide in gas escaping from the *lagoni* and steam jets.

Important geological works were published in 1833 and 1863 by Savi, professor of natural sciences at the University of Pisa. They describe rock formations in the Larderello region and surrounding areas, analyze the characteristics of the solfataras and *lagoni*, and speculate on the patterns of hydrothermal circuits that, rising to the surface, can alter outcropping terrain. Savi recognized the existence of folding and faulting structures in the mountains and hills throughout the Metalliferous Chain, also known as the Metalliferous Hills, due to their rich mineral deposits extending over the whole Boraciferous region and the area to the west (Cataldi and Burgassi, 1992a, 1992c). Savi wrote that a number of anticlines had been formed by the emplacement of intrusive bodies at relatively shallow depth. These could be inferred from outcrops of trachytes and selagites (lamprophyric differentiations from quartz-latitic, alkali-bearing magma) found in areas of Central and Southern Tuscany.

From 1832, several proposals were presented to develop drilling technologies suitable for conditions at Larderello. The one made by Gazzeri (1838 and 1841) is worth mentioning for the geoscientific background on which it was based. Gazzeri dedicated himself to developing drilling methods for Larderello after he had studied the geological situation and the depositional characteristics of hydrothermal minerals in the area. He maintained that the carrier fluids for boric acid were originally contained in one huge, deep, single reservoir. He said that all natural manifestations in the area were nothing but local discharges of deep fluids, "The most elementary notions of geology persuade us that the mineral, whose decomposition gives off the boric acid contained in the ardent vapor of the steam jets, is not distributed in isolated masses at the base of each steam jet or *lagone*, but forms a single and huge deposit, the extension of which is unknown. If we were able to reach the original deposit by means of drilling, a spontaneous blowout of natural fluids should follow, similar to that occurring in artesian wells. Alternatively, [if such a blowout does not occur] we could obtain stimulated production by a judicious injection of [fresh] water or by other methods...." There is no need to stress how modern these concepts are, considering the present knowledge of the geothermal field at Larderello.

In 1838 the French chemist Payen, technical adviser of the Larderello chemical company, began to study the origin of boric acid in the field, first determining the concentration of the compound in the *lagoni* and steam jets as a function of temperatures and flow rates. After some years, noting that both of these parameters and the concentrations of boric acid were decreasing significantly, he wrote that the declines stemmed from the ever-growing extraction of boron-bearing waters from the *lagoni* for chemical production. He thus began seriously to consider Gazzeri's theory of a single, large geothermal reservoir at depth but strongly rejected Gazzeri's idea to drill "deep" wells out of concern for the possibility of uncontrolled blowouts.

To slow the fall of temperatures and boron concentrations of natural *lagoni* and other nearby hot manifestations, Payen developed an ingenious system of connected, cascading, artificial *lagoni* in 1841 (see figure 4 in Chapter 13). The system created a more efficient extraction cycle for boric acid (Burgassi, 1987). Payen's theoretical studies and experimental results on the subject are in an extensive monograph, published in 1847.

At the same time, the Tuscan Boraciferous region was studied by the geologist Pilla. He wrote in 1845 that high-temperature heat is responsible in some areas for forming coal and natural manifestations: "Where partial outlets of igneous rocks occur, the vegetable material that transforms itself into lignite seems elsewhere, [in places it] is baked into lithantrax because of the exceptionally high underground temperature. The famous *lagoni* in Tuscany can be considered the legacy of these underground conditions. One of the characteristics of these is the very high temperatures of the terrain around...." Pilla also discussed the formation of the geothermal field. He was first to maintain that steam jets in the Larderello area result from a regional thermal anomaly caused by recent granitic intrusions and effusions of porphyritic and trachytic rocks.

Pilla claimed that boric acid originates from the decomposition of tourmaline: "The tourmaline found in granitic and porphyritic formations outcropping in the region indicates that the steam jets are related to these.... It is a well-known fact that boric acid is a component of tourmaline, and this clearly points to a connection between the presence of boric acid in the steam jets and the formations just mentioned...."

The geological peculiarities of the manifestations at Larderello and the fame the region gradually acquired after 1830 from the production of boric acid caught the attention of many foreign geologists. Among them were Hamilton, Murchison, and Warington from England, whose works were published in 1844, 1850, and 1854, respectively.

In 1851, the work of Murchison was translated into Italian by Savi and Meneghini, who were studying the geology of Larderello. They wrote an appendix to Murchison's work, with the following comment: "As concerns the numerous steam jets of the area, it should be noted that they seem to spring out from fissures running along northwest-southeast trends.... These regular, parallel, eruptive fissures are clearly the rejuvenation of older fissures that had been formed in preceding eras as a result of violent underground movements caused by the high-temperature regime...."

Murchison visited the Larderello region in 1852 and concluded that the steam jets were not a "phenomenon apart." He maintained that in the framework of the regional geological structure of Tuscany, the steam jets were late manifestations of the same volcanic process that once had affected several areas of Central Italy. Moreover, in agreement with Savi and Meneghini, Murchison pointed out that steam jets and other natural manifestations were all situated along northwest-southeast alignments, corresponding to the main tectonic trend of Tuscany and the whole Italian peninsula.

1818 print showing an alignment of steam jets along a northwest-southeast fault scarp in the southern part of the Larderello region. *Courtesy ENEL SpA/Geothermal Department*

In 1858 and 1878, the mineralogist Bechi stated that boric acid originates from the decomposition of boron silicates contained in the serpentine rocks that widely outcrop in the Larderello region and are many hundreds of meters thick. He maintained that the decomposition of these rocks is caused by carbon dioxide streams rising from deeper formations. In 1858, Bechi identified a new hydrothermal mineral, which would be called *larderellite,* describing it with the formula $(NH_4)_2 \cdot B_8O_{13} \cdot 4H_2O$. The formula was perfected by D'Achiardi in 1878.

Others, such as Deville from France (1857) and Wagner from England (1875), maintained that boric acid resulted from interactions between hot water and deeply-rooted masses of boron azides. This is how they justified the presence of ammonia in the waters of *lagoni* and thermal springs.

In 1867, Meneghini published a paper dealing with the production of boric acid. He discussed the formation and depositional characteristics of all hydrothermal minerals of the Boraciferous region in light of prevailing ideas in his time on the magmatic origin of geothermal fluids.

Meneghini's paper became a fundamental reference for the Abbot Stoppani, noted naturalist and geographer, who published a notable work in 1874 with an ample dissertation on the geothermal phenomena at Larderello. Using the ideas of his time and many of his own, Stoppani discussed the characteristics of the geothermal field and of the area, such as tectonic control of manifestation alignments, migration of steam jets through time, and thermal anomalies determined by igneous processes, fitting them into a geological framework for all of Tuscany. Not only did he understand the mechanism that controls the occurrence of phreatic explosions, but also the cause of the phenomenon that we call "self-sealing." A passage of Stoppani's work reads,

> "As I have already pointed out, the steam of the manifestations carries not only boric acid but also other compounds, such as ammonia, sulfates, iron, manganese, magnesia, and soda. These crystallize when the steam cools and then they encrust their subterranean passages to the point of closing them completely. Boric acid, for instance, causes the scaling of artificial boreholes.... During drilling, large masses [of hydrothermal minerals] are found, which prevent the steam rising to the surface.... There is nothing more natural than a steam jet closing in by itself, finding no outlet. As a consequence, the closed steam jet gathers all its power before breaching its prison walls and springing out again to the surface like an unshackled prisoner who has just escaped. Sometimes steam jets blow out spontaneously, just like a bomb; the soil is tossed into the air, and a steam cloud blows out.... The steam is accompanied sometimes by a jet of boiling water, which brings to mind a geyser...."

In years to come, several Italian and foreign scientists continued speculating on the origin of boric acid. Dieulafait, from France, proposed in 1877 that boric acid is carried by the steam and that the steam forms by vaporization of meteoric waters when they can percolate through fissures at great depth, reaching very hot formations. The Italian mineralogist D'Achiardi did not support Dieulafait's theory. In 1878, he maintained that the steam escaping from natural manifestations originates in much older rock formations that underlie the Miocene evaporitic sediments and have no intercalated saline layers. In 1888, Dieulafait's theory was again supported by Stohaman and Schwarzerberg from Germany.

In 1897, a few years later, geologist De Stefani strongly opposed Dieulafait's theory on the origin of boric acid from Miocene evaporitic formations. After studying the lithostratigraphic sequence and the geologic setting of all geothermal areas in Tuscany, as well as the hydrogeological conditions and the hydrothermal species occurring in all active and fossil manifestations in these areas, De Stefani stated that boric acid could only form at great depth:

> "The considerable surface area of the territory where these [manifestations] occur, the fact that natural steam carries boric acid everywhere it escapes, the presence of other acids and gases, the high temperature and energy of the steam, are all facts proving that both steam and [boric] acid originate at depth, and that they have nothing to do with shallow sediments and [Miocene] saline deposits.... Since it is widely accepted that temperature increases with depth at an average rate [of some 3° C/100 m], steam with a temperature of about 175° C must come from a depth of some 5 km....
>
> "The geothermal waters have nothing to do with sea water; therefore, they should be of atmospheric origin and percolate underground through the numerous cavities until they reach the deepest recesses of the Earth....
>
> "The presence of boric acid in the steam jets can only be justified if one accepts that its origin is in the depths of the Earth, and that the steam jets are the last legacy of volcanic activity.... This occurred in Tuscany quite recently, and even today the final signs of such activity are quite evident. The steam jets also carry carbonic acid and, as steam rises from underground, it comes into contact with very old borate-bearing rocks; thus, boric acid is drawn from the ensuing decomposition [of the borates].... The volcanic origin of the boric acid contained in the [waters of the] *lagoni* is indirectly confirmed by the fact that the same compound is contained in the steam plumes of active volcanoes and solfatara craters. They, too, just like the *lagoni*, bear witness to the presence of ammoniac substances, sulfurous gases, and carbonic acid...."

Throughout the second half of the 19th century, parallel with geological research and studies on the origin of boric acid, laboratory analyses were made systematically for all hydrothermal minerals from Larderello. Besides confirming the occurrence of alum, boric acid, and sulfur, previously unknown sulfates and borates were discovered as well, such as cerbolite and larderellite. Cerbolite was identified in 1872 by Popp from England and described with the formula $(NH_4)_2 SO_4 \bullet MgSO_4 \bullet 6H_2O$. Larderellite had been identified in 1858 by Bechi, but the formula he wrote for the mineral was incorrect. The formula was perfected by D'Achiardi in 1878 as $(NH_4)_2 \bullet B_{10}O_{16} \bullet 5H_2O$.

In 1895, the chemist Nasini began a systematic study of the gases contained in the steam and dissolved in the hot waters of geothermal manifestations and wells. His series of analyses was repeated for many years. He discovered the presence of helium and argon in geothermal gases at Larderello, quantifying the average amounts in wells and field areas, and the results were published in 1930.

From 1900 to 1930

In the period between the end of the 19th and the beginning of the 20th centuries, other scientists began studying the Larderello region. In 1904, Perrone published an extensive work on the hydrographic conditions of Southern Tuscany. In illustrating the situation of the hydrologic basins in the region around Larderello, he also dealt with the geological characteristics of the geothermal field and presented his own ideas on the genesis of hot fluids. Concerning the origin of boric acid, he again took up the hypothesis advanced by Pilla some 60 years earlier, discussing it in light of recent well data. Considering that the greater drilling depths reached between the end of the 19th and the beginning of the 20th centuries had produced fluids with temperatures above 200° C (not on the order of 175° C, as stated by De Stefani in 1897), Perrone claimed that the original temperature of the primary steam should be at least 250° C; thus, the steam had to come from depths of between 5 and 8 kilometers. At these depths, hot granitic masses containing tourmaline species were likely to exist, as could be inferred by the presence of Late Miocene-Pliocene granitic outcrops in several areas west of Larderello. Perrone concluded that with such temperatures and geological conditions, boric acid could only originate from the decomposition of tourmaline.

Geologist Lotti started in-depth investigations in the Larderello region and wrote many papers about the geothermal field, as well as several general papers on the geology of Tuscany. His most important works were published in 1900, 1907, and 1928. In 1907, Lotti categorically refuted Pilla and Perroni's arguments on the origin of boric acid from tourmaline:

> "Boron, fluorine, and chlorine, along with superheated water, constitute the most energetic dissolvents that contribute to the crystallization of eruptive rocks. Thus, it can be argued that

> boric acid is not derived from tourmaline-bearing granites, but forms before, and not after [the eruption of igneous rocks]. Rather, the deep circulation pattern of boron-rich waters must be taken into account because on the one hand, [while percolating downwards, the shallow waters] may reach the granitic stock, interact with it, and facilitate the formation of tourmaline; and on the other hand, [while rising up, the deep fluids] can nourish the steam jets and the fluid produced through drilling...."

From the structural point of view, Lotti recognized that the whole Metalliferous Chain of Tuscany, which includes the Larderello region, has two tectonic styles: one characterized by folding and the other by faulting. Taking up Savi's hypotheses of 1863 and adding many new arguments, Lotti attributed the folded structures to a prolonged phase of igneous activity that began in the Miocene with the sequential intrusion of granitic bodies at the Tyrrhenian islands of Elba, Giglio, and Montecristo, and inland at Gavorrano and Campiglia in Southern Tuscany. During the Quaternary, the granitization process moved upward from great depths until it reached the first shallow formations, resulting in the eruption of trachytic melts in several Tuscan localities, such as Capraia, Montecatini, Orciatico, Roccastrada, and Mt. Amiata. More or less regular anticlines could form in this way, following the uplift of rising igneous bodies: some intrusive, others extrusive.

At Larderello, since no igneous rocks outcrop there or in the surrounding region, Lotti argued that a granitic intrusion, rooted in an active magma chamber, is likely to exist at a relatively shallow depth beneath the field area. He concluded that the cooling magmatic body would have to supply the geothermal system with both heat and steam. In this framework, steam jets and other natural manifestations at Larderello should be regarded as the final stage of an active magmatic process. Lotti speculated that a similar explanation should be given for high-temperature fluids at any other geothermal field in the world, with or without occurrence of recent volcanic activity.

On this basis, he formulated a general theory in 1928 on the formation of geothermal fields, now called "Lotti's theory of the juvenile origin of geothermal fluids," which was accepted worldwide until the end of the 1950s. While making field surveys, Lotti compiled a detailed geologic map of the Larderello area. It was used as a guide for field excursions during the First International Geothermal Congress, held at Larderello in 1928 under the auspices of the Italian Geological Society.

In 1926, the mineralogist D'Achiardi joined Lotti in rejecting Pilla and Perroni's theory on the origin of boric acid. He stated that, even though granitic masses were likely to exist at great depth beneath the geothermal field, their tourmaline content could in no way be sufficient to form the large quantities of boric acid produced at Larderello.

Through the first three decades of the 20th century, besides continuing to study the occurrence and nature of geothermal gases, Nasini and his team of industrial chemists from the Universities of Pisa and Padua conducted a series of research projects, lab tests, and on-site examinations aimed at improving the production processes of all chemical compounds of the Larderello company. These processes resulted from many years of theoretical investigations into the chemistry and thermodynamics of geothermal fluids. In 1930, Nasini's work culminated with the publication of a voluminous monograph that remained a benchmark in geothermal literature until after the 1950s.

Concluding Outlook

Due to geological characteristics, the area around the village presently known as Larderello has attracted the attention of indigenous populations since prehistoric epochs. From Antiquity through the Late Middle Ages, it has aroused the scientific interest of many writers for the same reasons. From the 1470s, this interest grew more quickly when the region came under the rule of Florence and intensive use began of the hydrothermal minerals deposited around the natural manifestations in the area (Cataldi and Burgassi, 1992c). For about three centuries scientific research remained at the level of speculation and was aimed more at understanding geothermal processes than at finding ways to improve development of the rich hydrothermal deposits.

Following the discovery of boric acid at Larderello in 1777-1779 and after new techniques allowed large quantities of this compound to be produced there, applied and scientific research increased, starting in 1830. From then on, advances in scientific research paralleled achievements in the technological sector (see Chapter 13), and the two become interdependent. To better evaluate the former, it is necessary to have in mind the latter and vice versa.

In summary, the principal advances in scientific research at Larderello from the 16th century to the early 20th century include the following:

- Identification and characterization of hydrothermal minerals
- Igneous and hydrothermal processes

- Structural control of manifestations
- Presence of a strong thermal anomaly as a result of relatively young igneous processes
- Formation of anticlines and creation of rock fractures by granitic intrusions or magmatic extrusions
- Percolation of meteoric water along fractures and/or through outcrops of permeable formations
- Presence of reservoir rocks
- At-depth interaction of meteoric water with reservoir rocks and hydrothermal minerals
- Self-sealing processes and evolution of natural manifestations over time
- Underground circulation patterns and interaction of meteoric water with deep, hot fluids
- Origin of noncondensable gases associated with geothermal steam and waters
- Accumulation at shallow depth of over-pressured hot fluids, resulting in phreatic explosions and/or steam jets

Knowledge of these aspects accrued gradually, evolving slowly over 400 years and leading to the first structured theories on the formation of Larderello Geothermal Field between 1890 and 1930. If we consider all of the above in light of the practical achievements—the technological innovations, the chemical production of many compounds, the use of natural heat for space heating, the first experiment of geothermal-electrical generation in 1904, and the industrial production of electrical power from 1913 on—then we can realize why the knowledge of Larderello Geothermal Field and ways to exploit its resources for many uses became familiar to many Italian and foreign scientists and engineers by the first decades of the 20th century. These technological achievements and scientific discoveries formed the basis for future development of the Earth's heat, not only at Larderello and other geothermal areas in Italy but in many other geothermal fields around the world.

Cited References

Agricola (Bauer, G.). 1546. *De veteribus et novis metallis, vol. 2.* Frobenius Edit., Basel.

Agricola (Bauer, G.). 1546. *De natura eorum quae effluent ex terra, vol. 3.* Frobenius Edit., Basel.

Agricola (Bauer, G.). 1556. *De re metallica, vol. 12.* Frobenius Edit., Basel.

Aldovrandi, U. 1575?. In *Museum metallicum.* B. Ferroni Edit., Bologna.

Baccio, A. 1571. *Opera de thermis.* Reprint 1606. Giunta Edit., Venice.

Bartolomeus Taurinensis. 1553. De balneis Vulterrae. In *De balneis omnia quae extant apud graecos, latinos et arabos.* Giunta Edit., Venice.

Bechi, E. 1858. Studi sulla formazione dei soffioni boraciferi. *Atti Accademia dei Georgofili,* Florence.

Bechi, E. 1878. Teoria sui soffioni boraciferi della Toscana. *Atti Regia Accademia dei Lincei,* 2, ser. 3, 1-8.

Burgassi, P. D. 1987. Historical outline of geothermal technology in the Larderello region. *Geothermal Resources Council Bulletin,* 16, no. 3, 3-18. (See Chapter 13 in this volume).

Cataldi, R. and P. D. Burgassi. 1992a. Del misterio de la Atlantide a los primeros usos integrados de la energia geotérmica. In *La geotermia en el período Precolombino en las áreas Mediterránea y Mesoamericana. Geotermia, Revista Mexicana de Geoenergía,* 8, no. 3, 251-264. (See Chapter 9 in this volume).

Cataldi, R. and P. D. Burgassi. 1992b. Florecimiento y decadencia del termalismo y de los otros usos directos del calor de la Tierra en el área Mediterránea desde el nacimiento de Roma hasta el fin del primer milenio. In *La geotermia en el período Precolombino en las áreas Mediterránea y Mesoamericana. Geotermia, Revista Mexicana de Geoenergía,* 8, no. 3, 265-288. (See Chapter 10 in this volume).

Cataldi, R. and P. D. Burgassi. 1992c. Hacia el renacimiento en Europa. *La geotermia en el período Precolombino en las áreas Mediterránea y Mesoamericana.* In *Geotermia, Revista Mexicana de Geoenergía,* 8, no. 3, 289-303. (See Chapter 12 in this volume).

Cataldi, R. and P. Chiellini. 1995. Geothermal energy in the Mediterranean area before the Middle Ages. *Proceedings of the World Geothermal Congress,* 373-380. (See Chapter 11 in this volume).

Clüver (Cluverius), F. 1624. *Italia antiqua. Lugduni Batavorum, vol. 1.* Ex Officina Elseviriana Universitaria Edit., Pisa.

D'Achiardi, A. 1878. Sull'origine dell'acido borico e dei borati. *Atti Società Toscana Scienze Naturali,* 1, 7-8.

D'Achiardi, G. 1926. Il ritrovamento dell'anidride nella regione dei soffioni boraciferi a Castelnuovo Val di Cecina (Pisa). *Atti Soc. Tosc. Sc. Nat.,* 35, 25-30.

De Stefani, C. 1897. I soffioni boraciferi della Toscana. *Memorie Società Geografica Italiana,* 6, 410-435.

Deville, C. 1857. Recherches nouvelles sur le bore et ses affinités, et en particulier son affinité pour l'azote. *Compt-rend. de la Royale Acad. de Science,* 45.

Dieulafait, L. 1877. L'acide borique, méthodes de recherche, origine et mode de formation. *Ann. de Chim. Phys.,* 12, 318-354.

Dumas, A. 1828. *Traité de Chimie appliquée aux arts.* Paris.

Fallopio, G. 1560. Opera de thermis. Reprint 1606. Giunta Edit., Venice.

Franciotti, G. 1553. De balneis Vulterrae. In *De balneis omnia quae extant apud graecos, latinos et arabos.* Giunta Edit., Venice.

Gazzeri, G. 1838. *Lettera al giornale del commercio, arti e manifatture.* Dec. II, Anno I, no. 31. Florence.

Gazzeri, G. 1841. Induzione ora verificata di ottenere nuovi soffioni di acido borico per mezzo della trivellazione del terreno. *Atti Accademia dei Georgofili,* 19.

Giunta, T., ed. 1553. *De balneis omnia quae extant apud graecos, latinos et arabos.* Venice.

Hamilton, W. 1844. Observations of the geology of some parts of Tuscany. *Quarterly Journal of the Geological Society,* 1.

Hoefer, U. 1777. *Memoria sopra il sale sedativo di Homberg.* Stamperia Granducale Edit., Florence.

Larderel, F. 1831. Memoria sopra l'acido borico scoperto in Toscana e sulla sua applicazione. *Atti Accademia dei Georgofili,* Florence.

Lotti, B. 1900. I soffioni boraciferi della Toscana. *Rass. Mineraria,* 12, no. 13, 27.

Lotti, B. 1907. Sulla provenienza dell'acido borico nei soffioni della Toscana. Rass. *Mineraria ed Industria Chimica,* 26, no. 14, 1-5.

Lotti, B. 1928. I soffioni boraciferi della Toscana. *Boll. Soc. Geol. Italiana,* 47, no. 2, 1-10.

Mascagni, P. 1779. *Dei lagoni del Senese e del Volterrano.* Stamperia Poggini Carlo & Figli, Siena.

Meneghini, G. 1867. *Sulla produzione dell'acido borico dei Conti De Larderel.* Nistri Edit., Pisa.

Mengo Blanchello Faentino. 1553. De balneo de Morba Volterrano. In *De balneis omnia quae extant apud graecos, latinos et arabos.* Giunta Edit., Venice.

Mercati, M. 1717. *Metalloteca Vaticana. Opus postumum armarium, vol. 4.* Ex Officina Io Mar and Salvioni Romani C. Edit., Rome.

Murchison, R. 1851. *Memoria sulla struttura geologica delle Alpi, degli Appennini e dei Carpazi.* P. Savi and G. Meneghini, trans. (Includes an Appendix on the geology of Tuscany by Savi and Meneghini.) Stamperia Granducale, Edit. Florence.

Murchison, R. 1852. On the vents of hot vapour in Tuscany. *Quarterly Journal of Geological Society of London,* 6.

Nasini, R. 1930. *I soffioni e i lagoni della Toscana e l'industria boracifera.* Italia Edit., Rome.

Payen, A. 1847. Acide borique des soffioni de la Toscana. *Annales de Chimiq. Physiq.,* 1, 247.

Perrone, E. 1904. *Carta idrografica d'Italia: Fiora, Cecina, Cornia, ecc.* Ministero Agricoltura, Industria e Commercio, Rome.

Pilla, L. 1845. *Breve cenno sopra la ricchezza minerale della Toscana.* R. Vannucchi Edit., Pisa.

Popp, O. 1872. *Ann. Chem. Pharm.*, 8, supplement 1.

Ristoro d'Arezzo. 1282. *Libro della composizione del mondo.* Reprint 1864. G. Daelli Edit., Milan.

Savi, P. 1833. *Memorie per servire allo studio della costituzione fisica della Toscana.* Nistri Edit., Pisa.

Savi, P. 1863. *Saggio sulla costituzione geologica della provincia di Pisa.* Statistica della Provincia di Pisa.

Savi, P. and G. Meneghini. 1851. *Considerazioni sulla geologia della Toscana.* Stamperia Granducale, Florence.

Savonarola, M. 1460. De balneis Vulterrae. In *De balneis omnia quae extant apud graecos, latinos et arabos.* Reprint 1553. Giunta Edit., Venice.

Steensen (Steno), N. 1669. *De solido intra solidum naturaliter contento dissertationis prodromus.* Stella Edit., Florence.

Stohaman, G. and P. Schwarzerberg. 1888. *Enzyclopädisches handbuch der technischen chemie.* Tat Bor, Braunschweig.

Stoppani, A. 1874. *Il Bel Paese. Conversazioni sulle bellezze naturali, la geologia e la geografia fisica dell'Italia.* C. Agnelli Edit., Milan.

Targioni Tozzetti, G. 1769. *Relazione d'alcuni viaggi fatti in diverse parti della Toscana, vol. 3.* Stamperia Granducale, Florence.

Ugolino da Montecatini. 1420. De natura et virtute balneorum Comitatus Vulterrae. In *De balneis omnia quae extant apud graecos, latinos et arabos.* Reprint 1553. Giunta Edit., Venice.

Wagner, R. 1875. *Nuovo trattato di chimica industriale, vol. 1.* A. Cossa, trans.

Warington, R. 1854. *Chemical Gazette,* 419.

The Authors:

Pier Domenico Burgassi
2, Via G. Galile
56044 Larderello (Pisa), Italy
Telephone: 39.0588.86.74.89

Raffaele Cataldi
55, Via del Borghetto
56124 Pisa, Italy
Telephone/Fax: 39.050.59.81.07
E-mail: rafcat@tin.it

Claudio Donati
33, Pod. Bufalona-Zona Belvedere
58024 Massa M.ma (Grosseto), Italy
Telephone: 39.0566.90.33.89

The indoor pool with geothermal waters, designed around 1900 for the Hotel Gellert in Budapest. ***J. Lund***

15. Ancient Uses of Geothermal Waters in the Precarpathian Area of Romania and the Pannonian Basin of Hungary

by
Ioan Cohut
Miklós Árpási

Abstract: The first uses of geothermal waters in the Precarpathian area, Romania, and in the Pannonian Basin, Hungary, are lost in the darkness of prehistory. However, many Neolithic peoples settled near thermal springs. Heat from natural manifestations became the first therapeutic relief for the early inhabitants who came to consider the thermal waters as gifts from the gods.

In historic times, the use of these waters and thermo-mineral muds became more frequent and systematic. During the Late Middle Ages, these natural sanatoria were visited by local people. During the Late Middle Renaissance, the thermal localities were rediscovered, growing especially famous by the second half of the 19th century.

Reprinted by permission of the International Geothermal Association, from the *Proceedings of the World Geothermal Congress,* 1995, pp. 381-384.

INTRODUCTION

IN THE AREA OF THE LARGE INTER-CARPATHIAN AND Pannonian Basins of Romania and Hungary, thermal springs constitute the only manifestation of geothermal heat. Archaeological finds, dating from the Neolithic period, indicate that mankind settled in these friendly areas very early on. From prehistoric times to the present, the human community has continued to live near and develop a variety of geothermal areas: Budapest and Héviz in Hungary; Baden in Austria; Felix, Herculane, Geoagiu, and Călan spas in Romania.

Since prehistoric times, hot springs have been thought of as fonts of life and health, divine and sacred gifts. Before the Roman conquest, the practice of balneotherapy had been an indigenous custom among the Celtic and Dacian peoples with autonomous and developed civilizations. Celts dominated much of Western and Central Europe in the 1st century B.C. The word "Celt" is derived from *Keltoi*, the name given to these people by Herodotus and other Greek writers. To the Romans, the European Celts were "Gauls" and those in Britain were "Britanni." Dacia was an ancient kingdom of Eastern Europe inhabited by Thracian descendants and roughly equivalent to modern Romania. Unfortunately, these people, whose main occupations were agriculture, hunting, and sheep breeding and whose principal building material was wood, did not survive for more than 2,000 years. They left no writings describing their traditions of thermal bathing.

However, the oral tradition recounts a legend from the southwestern part of Romania about Ioan Iorgovan. He was a local Hercules who, using hot water redolent with hydrogen sulfide fumes, killed a seven-headed dragon from a cave near the present Herculane spa. This legend is comparable to similar legends from the Mediterranean area, for instance, the victories of Hercules against the giant Laestrygonians at the Santa Cesarea spa (Southern Apulia, Italy) and of Hercules over Typhoeus at the Phlegraean Fields (Campania, Italy).

Use of Thermal Springs during the Roman Period

The true importance of localities with thermal springs in the Precarpathian area and the Pannonian Basin is demonstrated by the fact that the springs were all included in the network of baths developed by the Roman Empire during the conquest and Romanization of provinces in Pannonia in the 1st century and in Dacia at the beginning of the 2nd century A.D. (The map illustrates this region.) For more than three centuries, the northern limit of the Roman Empire generally followed the Danube River and included the middle Danube (the provinces of Pannonia Superior, Pannonia Inferior, and Moesia Superior) and the lower Danube (the province of Moesia Inferior) up to Pontus Euxinus (the Black Sea).

Roman behavior towards thermal bathing was influenced significantly by interactions with several peoples in the Mediterranean area, such as the Etruscans and Greeks, from the 5th through the 1st centuries B.C. (Cataldi and Chiellini, 1995). In the 1st century B.C., when the Roman Republic was consolidated, and after 29 B.C., when the Empire was formed, the increasingly vast conquered territories expanded their commercial dealings and exchanges with each other. This, combined with benefits derived from knowing the lifestyles and technologies of other peoples and readily available state funds, left the Romans with time to care for their bodies—for balneotherapy (Cataldi, 1993).

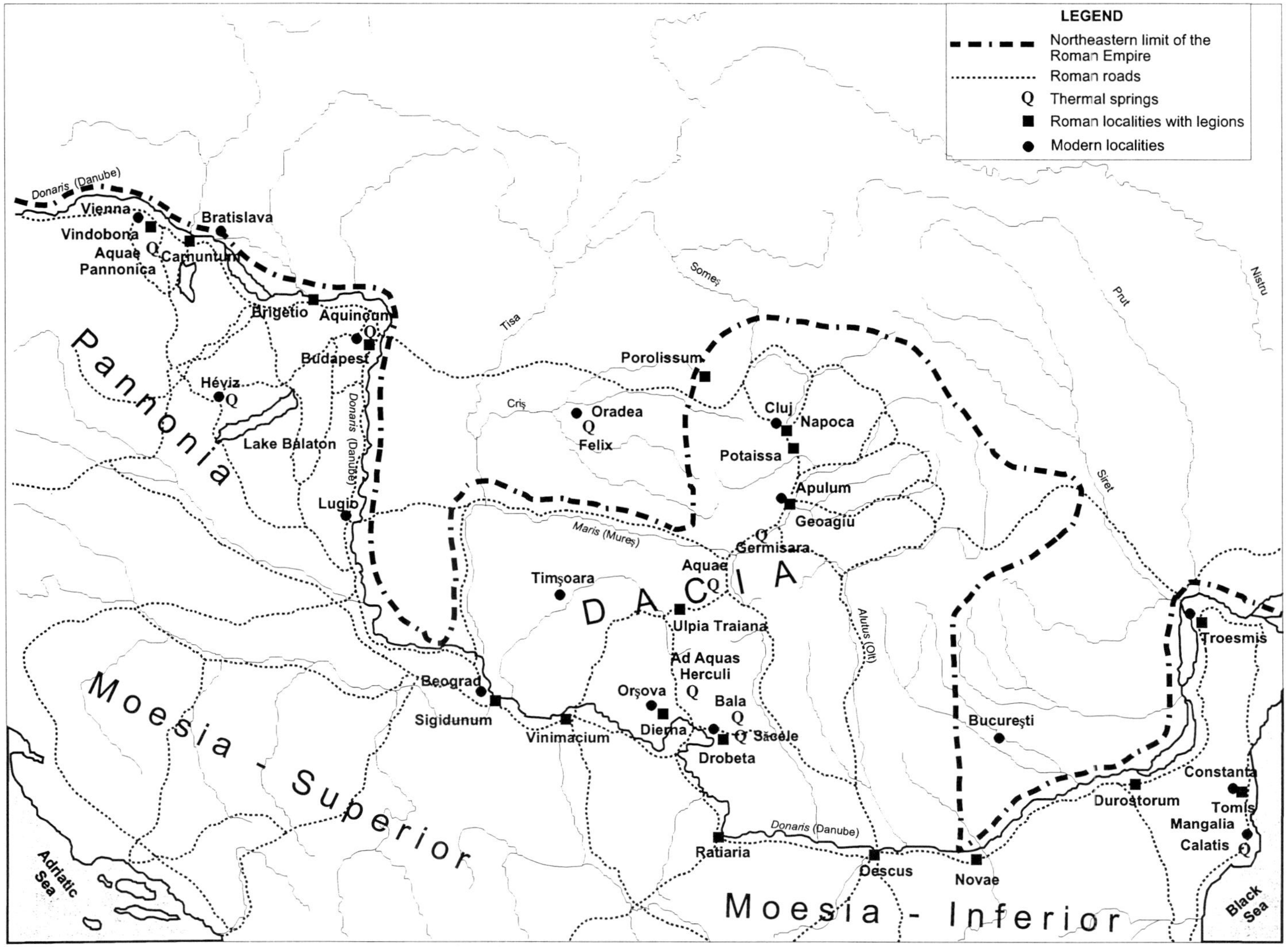

Map with locations of thermal sites in the Precarpathian area and the Pannonian Basin in the 3rd century.

The Roman Emperor Trajan conquered rich and powerful Dacia at the beginning of the 2nd century A.D., and the Romans ruled Dacia for almost two centuries. During this period, they constructed spa complexes in all localities with thermal manifestations in order to use the natural heat and the therapeutic properties of the salts and muds; moreover, some of the localities became commercially and strategically important (Cataldi, 1993). In fact, the Romans apparently chose several Celtic and Dacian centers (Vindobona, Carnuntum, Brigetio, Aquincum, Apulum, Dierna, and Ad Mediam) for their military encampments precisely because of their proximity to hot springs.

The importance of thermal bathing in Central Europe during this period can further be deduced from a famous atlas, the *Tabula Peutingeriana* (the official map of the Roman Empire in the 4th century A.D.). In the fifth of eight segments, this document shows the principal thermal spas of Pannonia and Dacia and indicates that the spas held great balneological and military importance. The Romans built and fortified Vindobona (Vienna) near Baden (Austria), the capital of Pannonia Superior and the main site of a Roman legion, and Aquincum (Budapest), the capital of Pannonia Inferior. The latter site had a thermal complex near the left bank of the Danube River.

Dacia, with its capital of Ulpia Traiana Sarmizegetusa, had many thermal springs within its borders. Two large ones northeast of the capital were Aquae (Călan, Hunedoara) and Germisara (Băile Geoagiu). The center of the 13th Roman legion Gemina at Apulum (today Alba Iulia) was located near the thermal complex of Germisara. The fortress of Drobeta was built near Ad Aquas Herculi, the most important bath in this part of the Roman Empire. Moreover, Dierna (on the left bank of the Danube), Ad Mediam, and Tibisum were located along the imperial route to a sector of Dacia. This arrangement suggests that the layout of the main Roman routes was purposely established to connect, as much possible, the different thermal localities with each other, additional evidence of their importance.

During the 2nd and 3rd centuries A.D., within the territory of Pannonia and Dacia under Roman rule, thermal bathing became a way of life so rooted at all levels of society that each spa became not only a place for cures (sanatorium, or *valetudinarium*) but for meeting and, therefore, a reference point for every aspect of civic life. The elevated civic importance of thermal bathing occurred in all thermal localities in the Roman Empire (Cataldi, 1993).

Thermal balneology, however, was a local tradition long before the arrival of the Romans. Very early, many cults formed with protective divinities for thermal waters. It was from the Dacians, whose religion was monotheistic, that the Romans adopted their cult of local divinities. They dedicated numerous votive offerings in honor of the hot springs themselves (*fontes calidae*), in honor of the sites where they were found (*genus loci*), and in honor of the divinities of water (*numines aquarum* or *nimphae salutiferae*). However, they dedicated the greatest number of statues, inscribed pillars, altars, and temples in honor of the health-giving divinities: Hercules, Asclepius, and Hygieia (Macrea, 1969).

Another proof of the symbiosis of Roman and Dacic cultures was the merging of the Roman divinity Dis Pater (the underground god Pluto) with the Dacic divinity Terra Mater, as can be seen on a bronze plate found at Ad Mediam and on an altar table found at Germisara.

Herculane Spa

After the 106 A.D. conquest of Dacia, the Romans started to frequent the curative thermal springs in the narrow canyon of the Cerna River (Southwestern Romania, near the Danube), which they named Ad Aquas Herculi Sacras (also called Ad Mediam, among other names). It is certain, however, that these springs were used by Dacians long before the Romans arrived. The archaeological discovery of a settlement in Outlaws Cave, inside the spa area, indicates that people lived there throughout the Neolithic period. In the area of these baths, Roman coins from 93-92 B.C. prove the existence of an active population dealing in handicrafts and trade, exchanging goods with peoples from the South-Danubian territory of Moesia, which was under Roman occupation (Macrea, 1969).

Bronze statue of Hercules at the Herculane spa. This is a copy made in 1852 of a marble statue dating from the 3rd century. *I. Cohut, Romania Geothermal Association*

The proximity of Roman roads coming from the Danube, as well as the mild climate, fresh air, and picturesque natural surroundings, facilitated Roman development of thermal springs. Buildings were constructed for cures, pools, and aqueducts. The Hercules thermae was famous for a very long period of time. After the Emperor Aurelianus left Dacia in 273, interest in the spa dropped as pressures grew from incursions of "barbaric" peoples. Eventually, the complex fell into decay. The thermal springs, however, continued to be used by the local population and by the occasional traveler.

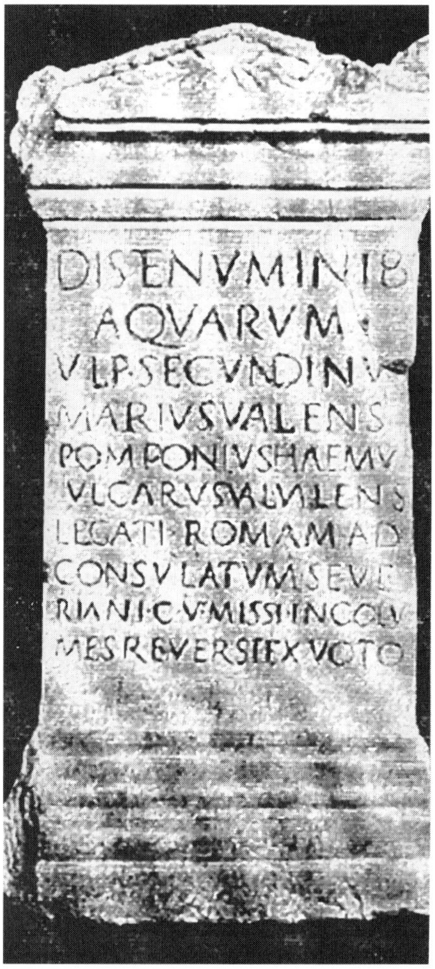

Votive *tabula* at Germisara spa dedicated to the "Gods of Waters" in 153 A.D., perhaps placed in honor of Marius Valens, an envoy sent on consultation to Rome who returned unharmed in the reign of Severus. *Translation by Peggy Lodeesen*

In 535, the Emperor Justinianus of Constantinople set up the bishopric Ad Aquas of Dacia Ripensis (Gogâltan, 1980) at the Herculane spa. Between the 6th and 15th centuries, despite many stormy invasions by a number of peoples, the Herculane spa remained active. Spa attendance was noticeably affected by many wars, especially those related to the conflict between the Ottoman and Austrian Empires in the 16th and 17th centuries. (During two centuries of Ottoman occupation, the Turks partly rebuilt the old Roman spas.) After armies under the command of Prince Eugene of Savoy ousted the Turks from the area, the Herculane spa was rediscovered.

Many Roman artifacts are found at the spa whenever the baths are modernized: statues, sarcophagi, altars, and votive tablets *(tabulae)*. The most frequently represented votive offerings are to Hercules Salutiferous ("the healer"); there are also numerous votive offerings dedicated to Hygieia and Asclepius. Unfortunately, most of the statues and other priceless objects were either lost (a ship carrying archaeological treasures sank in 1755 in front of the city of Pest) or sent to the Court of Vienna to decorate halls in the imperial library (Cristescu, 1978). The 18th and 19th centuries brought the spa international fame after Austrian modernization.

Germisara Spa

The Germisara spa (Geoagiu Spa, Alba County) was almost continuously active for more than 20 centuries, enjoying great fame and growth in the 2nd and 3rd centuries A.D. The ruins of this spa, which date back to the Roman period, include remnants of imposing stone and brick buildings, water tanks and pools dug in massive rock, and a complex network of pipelines for both hot and cold water. All this attests to intensive balneological activity in antiquity.

In addition to these structures, numerous remnants have been found of monuments, statues, bas-reliefs, sanctuaries, inscribed pillars, oracular altars, and craters. They contain dedications to gods, demigods, and *nymphae*, such as Sulis Minerva, Diana Augusta, Fortuna, Hercules Invictus, Asclepius, Hygieia, and *nimphae sanctissimae et salutiferae*. The dedications are signed by persons of the ruling class, wealthy people, veterans of the Roman army, and also by people of modest circumstances, confirming that balneology was practiced widely by people of all social levels.

Aquae

Aquae, at Călan, Hunedoara County, is mentioned in Ptolemy's *Geographia* (2nd century B.C.) under its Greek name of *Hydata* ("thermal springs") and is shown in the *Tabula Peutingeriana* as Ad Aquae. The springs are in the vicinity of the political, administrative, and religious metropolis of the Dacic capital, Colonia Ulpia Traiana Sarmizegetusa. During the Roman period, this thermal locality was famous for a travertine quarry.

The ruins of two other Roman *thermae* were discovered in southern Dacia at Săcele (the remnants of health bathing facilities and a sanctuary dedicated to Hygieia and Asclepius) and at Bala de Jos near Drobeta. At both localities, oak tubs for bathing (*alvae*) and offerings of Roman coins from the 2nd and 3rd centuries have been found (Tudor, 1958).

Felix Spa

Balneological practice at the Felix Spa, near Oradea, and in the thermal stream of Peţa, which is fed by copious hot springs under the lake bed, is documented only from the 12th century, after Transylvania became a province of the Hungarian Kingdom. However, some archaeological finds in the Salca-Oradea area prove that on the banks of a 12 km segment of the Peţa stream from Felix spa to Oradea, human settlements were established in the lower Neolithic period and developed mostly in the Bronze and Iron Ages. Existence of the Dacic civilization is documented by abundant archaeological evidence from the Hallstadt and La Tene cultures. Traces of Roman domination are absent, however, as the Oradea area was beyond the limits of the Roman Empire.

Mary Rowley stands by the Felix spa, 1996. The original spring is in the center of the spa. *J. Lund*

In the 11th century, the Bishopric of Oradea was established. The bishopric took possession of the hot springs near Felix and founded the Bishop's spa, which operated for over nine centuries. Paleontological evidence attests to the existence of the thermal springs in this area, at least since the end of the Tertiary period. In the warm waters of the Peţa stream, two fossils of that age have been preserved: the aquatic plant *nimphae lotus thermalis* and the gastropod *melanopsis parreisi*.

The Felix spa resort began to be known and visited after 1720; however, its true development started in 1885 when the first geothermal well in Romania was drilled at this site. The well was 51 m deep, with a flow rate of 195 liters per second and a wellhead temperature of 49° C.

Aquincum

THE AQUINCUM, A ROMAN SPA COMPLEX INITIALLY KNOWN AS AQUAE CALIDAE SUPERIORES ET Inferiores, is located on the right bank of the Danube River in the urban area of present-day Budapest. The copious thermal springs feeding the spa have been in constant use for balneotherapeutic purposes since the time of the Emperor Claudius (260-268). It was during

Roman rule that the thermae were given the Illyric name of *Aquincum*, which means "hot spring with high flow rate." Aquincum became very famous in this part of the Roman Empire.

After the fall of the Roman Empire in the 5th century, this spa lost international importance. However, it continued to be visited by the local people until 890, when Aquincum started to assume a new strategic importance, especially after the Hungarian leader of that time, Árpád, constructed a fortified encampment near the thermal springs. Once again, these thermal waters became known internationally.

Turkish baths at Aquincum in Budapest, Hungary. *M. Árpási, Hungarian Geothermal Association*

In the 13th century, thermal waters from the springs were transported from Aquincum to the Fortress of Buda through a wooden pipe 10 km long (Fekete, 1974). The water temperature along the pipe fell only about 10° C.

In the following period, during Turkish rule, the thermal station was expanded by constructing numerous *ilidse* ("Turkish baths"), which enabled thousands of people to enjoy hygienic and therapeutic treatment each day. Among the most important *ilidse* were those built by Mustapha Sokoli-Pasha in 1566 and those built by Veli Bey in the l7th century. The importance of the latter *ilidse* is stressed in the work of the Turkish illuminist Evlia Tcelebi.

After driving out the Turks in 1686, the Austrians modernized and expanded the whole thermal complex of Aquincum, which by the 18th century again enjoyed international recognition and attendance (Grove, 1977). In the 19th century, the need for more thermal water at the complex led to drilling two wells, in 1867 and 1878. The additional water supplied many indoor and outdoor thermal pools. By the late 19th century, the Aquincum was one of the largest therapeutic and recreational centers in Europe.

Héviz Spa

At Héviz spa, near Lake Balaton, copious under-lake springs (36° C) have flowed continuously since the end of the Tertiary period, as evidenced by the travertine and mud deposits. They have created the largest thermal lake in Europe. Local people have used the thermal waters and mud since prehistoric times, and archaeological sites date back to the Bronze Age. After the Roman conquest, a major balneological center (*valetudinarium*) was developed near the geothermal lake at Héviz. Many bas-reliefs and inscribed pillars have been found with dedications to the gods, such as Zeus and Sulis Minerva, and to the *nimphae salutiferae*.

View of the geothermal lake at Héviz, the largest thermal lake in Europe. *M. Árpási, Hungarian Geothermal Association*

References

Caryophilus, P. 1743. *De usu et praestantia thermarum Herculanarum quae nuper in Dacia Traiani detectae sunt.* Reprint 1924. Bucharest. (In Romanian).

Cataldi, R. 1993. *Utilizarea energiei geotermale în antichitate: de la Etrusci la sfârşitul Evului Mediu Timpuriu.* Special volume dedicated to the 30th Anniversary of the University of Oradea, Romania.

Cataldi, R. and P. Chiellini. 1995. Geothermal energy in the Mediterranean area before the Middle Ages. *Proceedings of the World Geothermal Congress, vol.1,* 373-8.

Cristescu, I. 1978. *The thesaurus of Cerna (Tezaurul Cernei).* Sport-Tourism Publishing, Bucharest. (In Romanian).

Fekete, G. 1974. *Magyarország.* Budapest.

Gogâltan, I. and D. Gogâltan. 1980. *Herculane Spa.* Sport-Tourism Publishing, Bucharest.

Grove, D. 1977. *Hungary's unrivalled leisure resource.* UNDP-HUN (71) 511 Project, Budapest.

Macrea, M. 1969. *Life in Roman Dacia (Viaţa în Dacia Romană).* Publishing House of The Romanian Academy. (In Romanian).

Tudor, D. 1958. *Oltenia Romană.* 2nd ed. Scientific Publishing, Bucharest.

The Authors:

Ioan Cohut
University of Oradea
5 Armatei Române St.
Oradea 3700
Romania

Miklós Árpási
MOL Inc.
H-1117 Budapest
Október huszonharmadika u. 18
Hungary

Surge et Ambula ("rise and walk").
The man breaking his crutch is the symbol of Slovakia's most famous spa, Piešt'any.

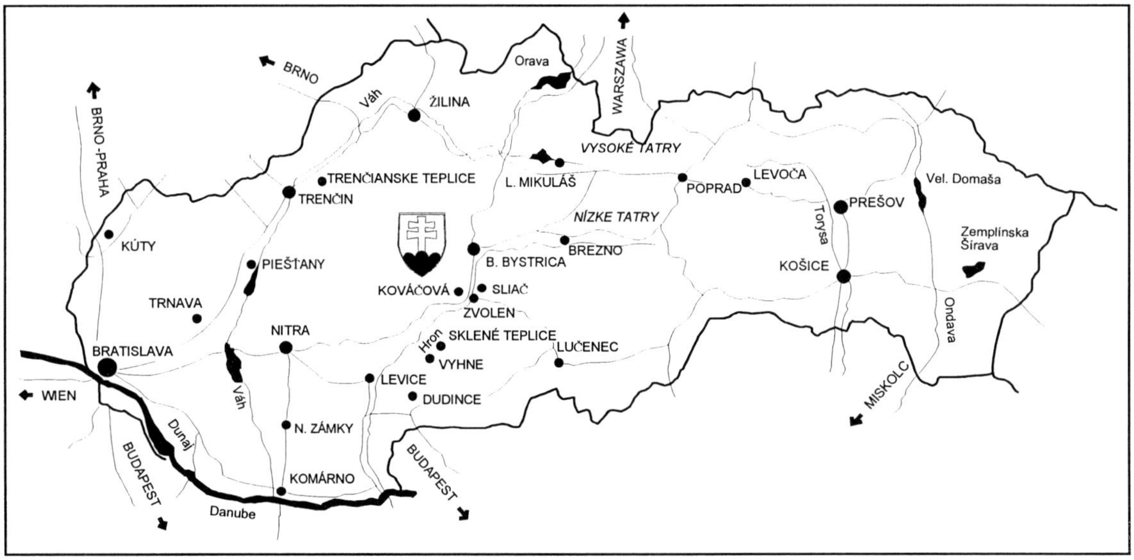

Western Carpathian localities, including spas.

16. A Spell Cast: Historical Aspects of Thermal Spring Use in the Western Carpathian Region

by
Marián Fendek
Augustín Rebro
Miriam Fendeková

Abstract: The history of geothermal resource use in the Western Carpathian region is mainly a history of the use of thermal springs. Thermal balneology established itself as a local tradition in prehistoric times; attention was paid to mineral and thermal waters not only by common people but by travelers, foreign visitors, and scientists. The richness of these waters is reflected in the geographical names of villages, towns, and springs. These names persist today, even in places where thermal waters are not presently found. The age of humanism gave rise to more precise investigation of these waters, their qualities and singularities, as well as possibilities for their technical and curative use.

Western Carpathian Geology

The Western Carpathians, which occupy the territory of Slovakia, consist of an Alpine folded mountain system and Tertiary basins. The mountain system is divided lengthwise by the Klippen Belt into the Outer (Flysh Belt-Paleogene Sediments) and Inner Western Carpathians. The Inner Western Carpathians are characterized by the following (Fendek, 1995):

- Abundant pre-Upper Carboniferous crystalline schists
- Variscan granitoids
- Late Paleozoic sediments and volcanics
- Largely carbonate Mesozoic
- Pre-Senonian nappe structure
- Alpine metamorphism
- Formation of granitoids
- Post-Cretaceous vertical movements that modified basins of deposition

- Tectonics that gave rise to morphostructural elevations (mountains) and depressions (basins) with widespread post-nappe Paleogene and Neogene sedimentary and volcanic formations

The geological setting is favorable for the occurrence of thermal springs with temperatures higher than 25°C only to the south of the Klippen Belt, in the Inner Western Carpathians. These thermal springs are largely associated with Triassic dolomites and limestones of the Krížna and Choč nappes (Fatricum and Hronicum), less frequently with Neogene sands, sandstones, conglomerates, Neogene andesites, and related pyroclastics. Many archaeological finds at the site and nearby thermal springs indicate that people settled in these friendly areas. In historic times, the use of these waters and their thermo-mineral muds became more frequent and systematic, so that many thermal localities acquired recreational and curative importance.

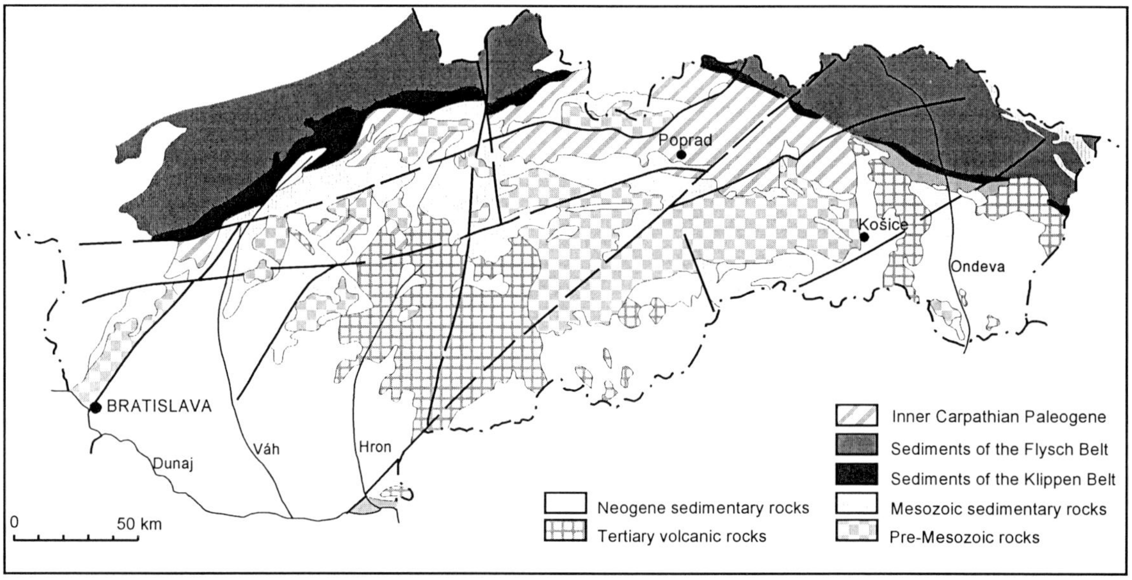

Geological map of the Western Carpathians of Slovakia, by J. Nemčok, 1984. *Reprinted with permission of the Geological Survey of the Slovak Republic*

Thermal Springs from the 16th to the 19th Centuries

Thermal springs in the Western Carpathians were known as early as the Middle Ages, and in the 15th and particularly the 16th centuries. Georg Wernher, a humanist scholar, brought them to the attention of the public in his work *De admirandis hungariae aquis hypomnemation* (Basilae, 1549). Although small in scope, his works surprised Europe and complemented the mining issues attracting the attention of scientists. His studies are cited by all known specialists and writers dealing with mineral and thermal waters. Of particular interest were the "cement" waters in Smolník and in Špania Dolina, which were used as cheap tools to extract high quality copper.

In the 17th century, alongside the odes written in praise of Piešt'any's thermal waters (Trajan, 1642) and mineral waters in Svätý Jur (Peucker, 1679), a scientific work appeared, the first hydrogeological study of this region. It was the eighth dissertation of Martin Szentiványi's *Miscellaneas* (1689). Szentiványi, a professor at Trnava University, explained and interpreted surface and ground water movement and the various mysterious powers and properties of lakes, springs, rivers, and other waters, not only in this region but also in other parts of the world (Rebro, 1983).

The treatise had a hydrogeological character. However, his interpretations were not original; they were taken from authorities such as A. Kircher and R. Descartes. Primarily, he accepted the ideas of Aristotle. He gathered and mediated his own scientific view from the domain of hydrology, even hydrogeology. Neither he nor other scientists, such as M. Zeiler, D. Fröhlich, K. Reyger, Sr., and J. Tollius, could let this unique part of nature go unnoticed (Rebro, 1996).

The 18th century witnessed an explosion of interest in mineral and thermal waters. Scientists, as well as political leaders, tried to apply to economics and medicine what had been advantageously exploited abroad.

In the first half of the 18th century, Matej Bel, scientist and historian, was the leading personality in this field. He was the third influential personality who devoted exceptional energy to the study of mineral and thermal waters. His literary legacy was extraordinary, and he published books as well as manuscripts in Latin, German, Hungarian, and Biblical Czech. Matej Bel introduced a team approach into his scientific work, and his most important books were the results of this collaboration (Rebro, 1983).

His most important work was *Notitia Hungariae novae historico geographica* (*Historical and Geographical Knowledge of Contemporary Hungary*). However, only four volumes were published in Vienna in 1735, 1736, 1737, and 1742. They contained data mainly from the western districts of Slovakia. The other districts of Slovakia were mentioned only in draft manuscripts. Matej Bel, together with his colleagues, assembled the known facts about mineral and thermal springs and their practical applications. He wrote not only of contemporary knowledge but often spoke of their history, of his own experiences with springs of mineral and thermal waters in Slovakia, and of their use. At those times, thanks to K. O. Moller, head of the medical preparatory school in Banská Bystrica, several physicians were already engaged in thorough, profound analyses and descriptions of the most important thermal waters (A. Hermann, 1726, Trenčianske Teplice; J. J. Torkos, 1745, Piešt'any and Pezinok). Thanks to Bel and his cohorts, 137 places with mineral and thermal waters received much publicity.

Maria Theresa (Queen of the Austro-Hungarian Empire) and her enlightened advisors were probably influenced by Bel's treatise and decided to investigate medical and social uses for domestic mineral and thermal waters, thereby preventing the outflow of wealth abroad where nobility traveled for entertainment and health. The first official registration of mineral and thermal springs, which had been documented by chemical analyses and therapeutic indication instructions, occurred between 1763 and 1769. The official doctors were assigned to undertake the registration. The data from the registration attracted a wider public, who were interested not only in learning about mineral and thermal waters but also in their professional promotion via monographs, geographic publications, and balneographies.

One of the first balneographies was by H. J. N. Crantz (1777), a professor at a Viennese university. It recorded 158 places with mineral and thermal springs in the Western Carpathians. Later he would personally analyze many of those waters. Most admired were the well-known spas: Piešt'any, Trenčianske Teplice, Rajecké Teplice, Turčianske Teplice, Sklené Teplice, Vyhne, and Sliač, but also Svätý Jur, Pezinok, Lipovce, and Rudno. The latter mostly had their own scientists and promoters.

A tireless analyzer of mineral waters, Professor Pavel Kitaibel from Budapest began his work in the late 18th and early 19th centuries. His posthumously published 1829 work *Hydrographica Hungariae* (*Hungarian Hydrography*) showed a depth of knowledge about many significant localities and included botanical information, particularly his monographs about mineral and thermal waters and spas in Slatina, Turčianske Teplice, and Bardejov. His descriptions included details on the discovery of new springs, their composition, and their other qualities.

Professor L. Tognio, an Italian born in Trenčín, was another enthusiastic expert in mineral waters. He visited many places in Slovakia, analyzing and evaluating mineral waters. He also encouraged medical graduates, such as D. Szüsz, J. Calas Nagy, K. Emresz, and A. Zsigmondy, to write dissertations on mineral waters.

Although several balneographies appeared in the mid-19th century, written mostly by physicians, it was David Wachtl's 1859 work *Ungarns kurorte und mineralquellen* (*Hungarian Spas and Mineral Springs*) that was most significant and instructive. The author not only assembled topographic, analytical, and medical data, but he also briefly categorized the springs' geological position.

Using Thermal Springs through the Ages

Piešt'any

The most attractive sulfur thermal springs in the Western Carpathians are at Piešt'any. Various forms of this name are found in Hungarian, German, and Latin: *Pestien*, *Pestyen*, *Postyeny*, *Pistinien*. The temperature of the thermal springs, 68° C, is the highest in the Western Carpathians. People have utilized the Piešt'any thermal springs since the oldest times. The extraordinary density of Stone Age settlements here was apparently closely connected to the occurrence of thermal waters; the area was also frequented by game, such as mammoth, reindeer, and red deer. Archaeological finds dating from this period include stone tools, animal bones, shell necklaces, and sculptures of women, particularly Venus. The excavations also show that the Celtic tribes of the Cotins settled in this region, as did the later German tribes of the Quadi (Šípoš and Kovačevič, 1972).

An old tale says that a big and beautiful lame peacock discovered the medicinal waters in Piešt'any. The peacock dragged itself to the Váh River below Piešt'any, and after a time it was completely cured. According to another tale, the medicinal effects of the local waters were discovered by Roman legions marching to the north in the time of Marcus Aurelius. Their presence in Trenčín *(Laugaricio)* is noted by the well-known inscription on the Trenčín castle rock, and excavated Roman artifacts include objects made of bronze, iron, silver, and gold. Once, after a long march, the Roman horses were so tired that they could not walk, and so they rested for a few days in a marshy meadow by the Váh River near Piešt'any. After this rest, to the surprise of the soldiers, they continued their way north, quite fit and fresh. People heard of the medicinal benefits and started to use the Piešt'any waters to cure sore feet and joints. They dug hollows by the Váh River where hot water flowed in and they treated their sore limbs. In a short time, they enjoyed improvements in their health, and in this way people underwent treatments for many centuries.

In the 7th and 8th centuries, Piešt'any and its surroundings were already settled by Slavic inhabitants, as indicated by important artifacts dating from the Great Moravian Empire: the remains of a fortified redoubt and merchant center at Pobedim (Hunka, 1988) and the vast spaces of the fortified Great Moravian magnate seat on the Kostolec near Ducové. The magnate seat is located on a strategically important hill that overlooks the whole of the middle Váh Basin. There are stone foundations of a rotunda and other buildings. During reconstruction, the whole site was enclosed by a high palisade that could be seen from afar. The distance from Piešt'any is approximately 8 km.

The first written mention of Piešt'any is found in the *Zobor Chart* of 1113, which demarcated and listed the boundaries of monastery properties on King Koloman's order. Other written documents on life in Piešt'any's springs up to the end of the 15th century primarily are private correspondence. A letter written by Andreas Mesterházy to Palatine Thomas Nádasdy in Budapest is important: he praised the healing power of Piešt'any's waters. In addition to these letters, there are many other references by travelers, physicians, and poets describing the bustle at the springs, healing effects, water treatment, and bathing.

Wernher wrote that Piešt'any spa was well-known for its hot springs and medicinal benefits, surpassing all other hot springs in Hungary (1549). Its speciality was that it had no permanent well, that the thermal springs incessantly changed their locations. The bathers were therefore obliged to dig new hollows on the Váh riverbank. Water treatment was simple in those times. In places where thermal water gushed forth, people dug holes and covered them with dead branches. They then let the warm water cool in the holes, or they cooled it with water from the river. The sick were submerged in the holes and thus recovered their health.

Physician Johann Crato de Craftheim, in his 1571 work *Five Books of Medical Recommendations* (Bottan, 1996), suggested a period of 30 days to treat ischias. However, not everybody consulted a doctor, and many people came to Piešt'any to treat all diseases. They believed in the superstition that the holy water would show within three days who could be cured and who would die.

In the 16th century, Habans came to Western Slovakia. Some of them worked in the simple spas of those times. They took care of patients, carried out sanitary measures, and performed minor medical services such as extracting teeth, blood taking, blood letting, and clyster administration. They applied tin bottles filled with hot thermal water to aching parts of the patients' bodies (Mulík, 1982).

During the 17th century, Piešt'any was frequently visited by sick people. Therefore, Emperor Leopold I issued a charter in 1682, *Salva Guardia*, by which he took Piešt'any and Banka villages under his protection. It meant that both settlements were protected against military requisitions and compulsory accommodation of soldiers. The most beautiful 17th century image of the healing springs of Piešt'any is that of Czech exile Adam Trajan of Benešov who settled down as a preacher in Drahovce near Piešt'any. His ode, "Saluberimae Pistiniensis Thermae" ("The Healing Springs of Piešt'any"), was published in 1642. He described the baths of the sick in the hollows, gave advice and recommendations for taking baths, but he also praised nature and the springs. For this reason, the main thermal spring of the spa was named after him.

In 1745, the doctor and physician of Bratislava Torkos built a dike in Piešt'any to prevent the destruction of thermal springs by floods and to enable the construction of more permanent buildings for patients bathing near the springs. He also recommended connecting the springs to wells by drilling. Professor Johan Crantz was interested in Piešt'any's thermal waters and was critical that the thermal water sources were not captured in a workmanlike manner. In 1771, he recommended the construction of permanent buildings. Nine years later, Count Johann Nepomuk Erdödy constructed the first wooden spa building at a site where patients already bathed in wooden bathtubs. Two other buildings were added: a second building for baths—in a mixed basin with a board roof—and a walled inn on the right side of the Váh River. Several farmhouses for the patients also were built there. Thus the colony of Teplice arose by the Váh River, about 2 km from the village of Piešt'any. Life in the settlement of Teplice was described in a comical poem by Count Jozef Gvadáni: "Clouds of dust or seas of puddles welcomed the guests arriving on wagons. It was a procession of a small defeated army. First wagons full of furniture, chair, beds, clothes, kitchen utensils, foodstuffs. This advance-guard was followed by horses with rich harnesses drawing the carriage of the spa guest..." (1787). Many other authors described their experiences at Piešt'any spa.

The spa was visited by important personalities: the Hungarian Palatines and Cardinal Peter Pazmán. In 1801, Ludwig van Beethoven underwent treatment. He was a frequent guest in the castle of the Brunswick family at Dolná Krupá (Bottan, 1996). He traveled quite often to Bratislava where his famous musician friends lived and gave piano lessons to Babeta, a gifted girl in the Keglevič family. His letter to an "immortal beloved," written here, was never sent to the intended recipient. Although the reason for not sending this letter will never be known, it is important that he wrote this letter in a spa: it was apparently written in Piešt'any in 1802. Beethoven's ardent love, felt from each line of this letter, inspired him to compose the "Moonlight Sonata."

At the beginning of the 19th century, the spa's proprietor was the county administrator and state minister, Count Jozef Erdödy. This man already saw a possibility for revenues to pay off debts of the Hlohovec dominion, and so he wanted to improve the spa. The authorities sent him officers and soldiers wounded in the Napoleonic war to undergo treatment. The cost was, of course, paid by the army.

Piešt'any was often struck by great catastrophes. In 1813, after a four-day rain, the Váh burst its banks; the water level rose 4 m and flooded the whole valley, an expanse of 9 km. Most houses in old Piešt'any were washed away. The flood destroyed or damaged 355 houses, claimed lives (Šípoš, 1981), and washed away the spa buildings. Other catastrophes included long periods of drought in 1816 and 1817 and the great fire of 1822, which destroyed many farmhouses.

The spa prospered in the 1820s. In Teplice, an extension was built to the Gentlemen's Inn, together with a superstructure and a casino. On the opposite side, the pharmacy house, the house of the county commissioner, and a Jewish dining hall appeared. The aristocracy started to build private villas in Teplice, doctors constructed private sanatoria, and private hotels appeared. Count Erdödy launched a major reconstruction of the spa. He razed the old buildings used for bathtub treatments, replacing them with a classical brick building. The first mirror room, "Napoleon's Bath," was built in 1822. In the same year, the Count developed regulations for the spa, and the first list of patients was compiled. Most patients drove to the spa in carriages and stagecoaches, which also served as mail transports. Letters were collected in a large, book-shaped wooden box.

Following Joseph Erdödy's death, the spa began to deteriorate. Dr. Francis Scherer, the permanent doctor of the spa since the 1820s, rented the spa and started to reconstruct some of the buildings (Mulík, 1984). To his merit, the Military Institution of the Spa was established in 1863. The arrival of Dr. Scherer thus brought about improvements in treatments at the spa.

A speciality of Piešt'any must be mentioned here. Since even seriously ill patients were lodged rather far from the spa, they used to lease *infanterits* to be transported there. The infanterits had special push-carts with a booth in which the patients could be seated comfortably.

A new era for the spa began in 1876, when a railway opened from Trnava to Nové Mesto nad Váhom, and in 1883 it was extended to Žilina. Thus Piešt'any was connected to the rail network of Hungary, a most important event. The patients grew in numbers and romanticism, too, because in front of the railway station many fiacre-drivers waited for guests they could drive to the spa.

In 1889, the firm of Alexander Winter and Sons rented the spa for 100 years from Count Francis Erdödy. The number of facilities and buildings showed that the Winters family was able to fulfill their plans to revitalize the world-famous spa of Piešt'any.

Trenčianske Teplice

The curative qualities of the thermal waters (38°-40° C) and the mineralized sulfur mud have been used for ages at the Trenčianske Teplice spa, as its environs were a hunting ground and hence a suitable locality for prehistoric habitation. Archaeological findings include Paleolithic and Neolithic stone implements and Bronze Age artifacts at a dwelling on Čvirigovec hill and at an urn burial site in the lowland of the Teplička brook, not far from the spa. Later, a Celtic tribe (Cotins) settled there and after that a German tribe of Quadi. The latter fought against the Roman legions. A Slavonic housing estate in the valley of the Teplička brook dates back to the 7th century, the period of the Great Moravian Empire.

The oldest written records of the Trenčianske Teplice thermal springs are from 1242 and 1265, during King Belo IV's reign. As a spa, it was mentioned in a 1488 document (Šípoš and Kovačevič, 1972). In Trenčianske Teplice there were three holes dug at a spring. The holes had no artificial bottoms, only mud. The advantage of the water was its temperature, which did not vary and needed no cooling or heating. The people could not only dangle their feet in the water but also could walk in it. This was the first mention of rehabilitation treatment in the Slovak spas of the 16th century. The pools were built on the springs; they had roofs, and the walls were covered with wood. There was mud nearby. The spa had a spa-master, who introduced the "spa order" according to an Austrian model that was gradually implemented by other Slovak spas. The order had 23 items (Horváth, 1984). For example, patients must use a laxative before the cure. During the cure, patients must follow a diet, and after their baths they could have only a short rest. It was also necessary to avoid negative emotional conditions. Patients must sleep during the night. All parts of the body should be covered with mud. A perspiration cure was advised not only when the tonsils swelled but also for hypochondria and hysteria. As to the length of the treatment, three to four weeks was sufficient for a lighter case. Patients were also advised to drink water early in the morning for 3, 7, and 9 days, day after day.

The first literary reference to the spa is found in G. Wernher's book (1549). Thirty years later, the spa was visited and explored by a learned provincial doctor of the Moravian margrave, Thomas Jordan von Klausenburg, who described it in detail, conducted chemical analyses of its curative waters, and gave remarkable advice on how to rationalize methods of cures. The results of his study were published in 1581 in Czech and in 1586 in Latin. During his visit, the spa consisted of only two pools in the open air and several log cabins (Šípoš and Ondrejkovič, 1989).

The thermal springs and the spa were a part of the Trenčín castle estate from the very beginning. As early as the 16th century, various Hungarian aristocrats were its owners. The spa reached its first real prosperity during the rule of Count Joseph Illésházy, who inherited it in 1724. He built a manor, new houses, and five spas (Mulík, 1981). He improved the whole environment of the spa, extended the parks, and introduced a dazzling social life. His descendants continued improvements, especially the last of kin, Count Štefan Illésházy, who was the owner from 1799 until 1835. By reconstruction and construction, according to a well-considered plan, the Trenčianske Teplice spa became the most popular of Hungarian spas, frequented by Hungarian, Austrian, Moravian, Czech, Silesian, Polish, Russian, and Prussian nobility. In 1835, a Viennese banking magnate, Juraj Sina, bought the spa from Štefan Illésházy and improved and commercialized it (Rebro, 1983). In 1856-1876, his son, Šimon Sina, continued developing the prosperous spa as did his granddaughter, Iphigenia d'Harcourt, from 1876-1909. Among other innovations, she enriched Trenčianske Teplice with a precious architectural jewel: the "Hamman Bath," completed in 1888. In 1909, the spa was bought by the Hungarian Bank and Trade Association. Its aim was to build a worldwide spa center at Trenčianske Teplice, another Pyrmont. But the association's extensive investments were thwarted by World War I.

Sklené Teplice

The Sklené Teplice spa is one of the oldest known health resorts in Hungary. The village was Sklenô, known for glass production. According to information from the 16th century, mud was collected from thermal waters in Sklené Teplice and applied in curing procedures. It was the first instance of spa mud therapy in the Western Carpathians (Horváth, 1984). Mud formed in places where water gushed from the rock and was applied to joints, ulcers, and wounds. Five roofed spas had benches, and there was also a steam cave, *Parenica* ("sweating bath"). In 1671, the English traveler Edward Brown, on a visit to mining towns, described Sklené Teplice in great detail (Rebro, 1984). In 1690, the Dutch scholar Tollius wrote that there were about 30-40 thermal springs there (37°-52° C). Thirty years later, Slovak scholar Matej Bel found fewer springs. He thought that they had disappeared due to human interference. Fourteen of the springs are registered today (Rebro, 1996).

At the beginning of the 18th century, after the Lippay family had died, the spa was forfeited to the state (Šípoš and Kovačevič, 1972). Then it was in the hands of ministerial authorities in Štiavnica, who promoted its further growth. They built the "nobility" bath in 1701. In 1777 Professor Henrich von Cranz mentioned three baths: nobility, common, and poor. In the 19th century, mining administrator Zipser commissioned a bathhouse to be built above the spring, which was named the "Zipser Bath" after him (Rebro, 1983). The local mining administration also provided a permanent spa doctor. Present reality, however, is far from past glory. The Sklené Teplice spa has remained a small resort, set in magnificent natural surroundings with fond vestiges of the past.

Vyhne

Vyhne, a mining town, boasts thermal water (37°-40° C) and a spa dating from the 16th century. Known as Rössler's spa after the owners, it was also called *Hadie* ("Snake spa") since young snakes used to appear in a pool rising from the gallery of a mine. Thermal waters were collected during mining operations (Rebro, 1984), and the spa flourished and declined with the mining enterprise. English traveler Edward Brown came to Vyhne after visiting Sklené Teplice spa in 1671. At that time, there were three baths, used mainly by miners for treatment of movement diseases. Thermal waters were used also in silver processing. Despite efforts to secure higher capacity, balneal business has not been revived at the present time.

Sliač

The health-giving thermal springs of the Sliač spa (33° C), located in the Zvolenská kotlina depression, have been known for centuries. Today, these springs bring relief and comfort to those whose cardiovascular systems have been impaired by the pace of modern life. The earliest written record about the springs dates from 1244, during the reign of King Belo IV. Since the mid-15th century, numerous eminent scholars and writers have mentioned the springs' curative effects. G. Wernher characterized them as waters with deadly effects (1549), but Matej Bel cast doubt on that view and proved their harmlessness. It was just evaporation of carbon dioxide that was deadly (Rebro, 1983).

Sliač has served as a resort for centuries and in the past was among the most frequented spas of former Hungary. It owed its popularity and attractiveness to the curative nature of its thermal waters and to visits from eminent personalities. The spa expanded before the Enlightenment, and it has remained popular until the present, chiefly for the treatment of cardiovascular diseases. Bathing pools are directly over the curative springs, allowing the mineral water, rich in carbon dioxide, to be applied directly in its most natural and effective form. The gas emanating from the spring is also exploited for curative purposes as subcutaneous injections.

Kováčová

THERMAL WATERS IN KOVÁČOVÁ WERE DISCOVERED BY THE VIENNESE FIRM TRAUSL, WHICH mined coal in the Zvolenská kotlina depression. The firm drilled a 405 m borehole in 1899 and hit a free outflow of thermal waters with a temperature of 48° C. A spa with a hotel was located next to this source of thermal water. The spa focuses on the treatment of adolescents and on the rehabilitation of motion organs.

Dudince

THE FIRST MENTION OF DUDINCE (DYUD) DATES FROM 1284 (ŽIGO, 1992). THE DUDINCE SPA IS located in the Štiavnica river valley. In the past, this area was crossed by an important trade road, the "Via Magna," which connected Budapest with Kraków (Poland) and the Baltic Sea. There is also evidence of thermal springs in Dudince in ancient times. More than 30 Roman baths, miniature pools hollowed in travertine, are remnants of the visits of the Roman legions. However, monuments to prove this view are lacking.

The oldest known written document about the use of thermal waters in this region is the work of G. Wernher (1549). D. Wachtel, the first to write about water analyses of the Dudince thermal springs, described the water as iodine, alkaline-saline, hydrogen acidulous (Hyánková and Melioris, 1993). At the same time, he described seven springs in travertine. The thermal water was used for bathing in holes that were dug. Thermal springs (27°-28° C) have long been utilized to treat rheumatism, eye inflammations, skin diseases, and digestive organs. In 1890, specialists from the Geographical Institute in Vienna came to survey the springs, and after them other specialists from the Mining Geological Institute visited the area. Based on research results, a decree was issued in 1894, officially acknowledging the thermal springs of Dudince as healing waters. The area around Gestenec hill was declared a protected zone (Krajčík and Krajčíková,

1995). The mineral water was used later for drinking treatments for stomach acid or pressure disturbances. It was also bottled and sold in stores. Bottled mineral water in special boxes was transported by trains to Komárno and from there by river ships to seaports, where it was sent to America.

A Few Words at the End...

"ALL THINGS CONSIDERED, YOU ARE LIKE A LUCKY PRINCESS IN A FAIRY TALE HAVING IT ALL—youth, beauty, curative art, love of all the people, and at the same time, lots of excellent virtues—even the virtue of taciturnity. Reposing silently in a lap of mountains, mysterious and so shy! However, a single look at you and the spell is cast."

These lyrical words were written more than a century ago, in 1872, by the great Czech writer Jan Neruda about the Mariánské Lázně spa. However, they actually apply to all the spas in the Western Carpathians, spas with old traditions and new approaches to treatment, with reputations achieved through abundant natural thermal resources with excellent curative properties. Like a king's crown sparkling with gems, the Western Carpathian region glistens with the life-giving gifts of nature. And like oases of health, spas have grown for centuries around these gems: settlements and towns with sanatoria, colonnades, parks, cultural and social facilities. It is not only the vast quantity of curative springs—preliminary data in the territory of Slovakia shows 1,382 mineral and thermal springs were registered by 1980 (Rebro, 1979)—but their exceptional quality that makes them noteworthy: their many-faceted chemical compositions, physical properties, and curative effects through time.

Modern Selected References

Bel, M. 1735, 1736, 1737, and 1742. *Notitia Hungariae novae historico geographica, parts I, II, III, IV.* Vienna.

Bottan, R. 1996. *Piešt'an—Information for visitors.* Alfa, Bratislava.

Chropovský, B. 1981. *Vývoj osídlenia a územnej organizácie.* Mapa č. 11. Atlas SSR. SAV, Bratislava.

Cranz, H. J. N. 1777. *Gesundbrunnen der Oesterreichischen monarchie.* Vienna.

Fendek, M., A. Remšík, and M. Král. 1995. Geothermal energy of Slovakia. *Slovak Geological Magazine,* 1-95, 59-64.

Hermann, A. 1726. *De thermis Trentsiniensibus commetariolus historico-medicus.* Lipsiae.

Horváth, G. 1984. Fyzikálna liečba v Slovenských kúpel'och do konca 18. storočia. *Balneologický Spravodajca,* 23, 72-6.

Hunka, J. 1988. Nálezy mincí z okolia Piešt'an a ich historický význam. *Balneologický Spravodajca,* 27, 112-19.

Hyánková, K. and L. Melioris. 1993. The unusual chemism of mineral waters at Dudince. *Geologica Carpathica,* 44, 121-31.

Kitaibel, P. 1829. *Hydrographica Hungarieae I and II.* J. Schuster, Budapest.

Krajčík, J. and B. Krajčíková. 1995. *Dudince.* Alfa konti, Bratislava.

Mulík, J. 1981. *Dejiny kúpel'ov a kúpel'níctva na Slovensku.* Osveta Martin.

Mulík, J. 1982. Piešt'any za feudalizmu. *Balneologický Spravodajca,* 19, 23-45.

Mulík, J. 1984. Piešt'any za uhorského kapitalizmu. *Balneologický Spravodajca,* 23, 25-34.

Peucker, J. 1679. *Die königliche frey stadt sanct Georgen in nider Hungarn.* Wittenberg.

Rebro, A. 1983. *S Matejom Belom o kúpel'och a minerálnych vodách.* Alfa, Bratislava.

Rebro, A. 1984. Zvučný hlas o sláve tekovských kúpel'ov. *Balneologický Spravodajca,* 23, 35-62.

Rebro, A. 1996. *Vzácne a obdivované vody Slovenska. Balneologické múzeum Piešt'any.*

Rebro, A., M. Klago, P. Krahulec, J. Mulík, A. Porubský, and P. Tkáčik. 1979. *Vody uzdravujúce a osviežujúce.* Osveta Martin.

Šípoš, J. 1981. Iniciálny rozvoj kúpel'ov v Piešt'anoch. *Balneologický Spravodajca,* 18, 63-86.

Šípoš, J. and I. J. Kovačevič. 1972. *Slovenské kúpele.* Osveta Martin.

Šípoš, J. and I. Ondrejkovič. 1989. *Trenčianske Teplice.* Obzor, Bratislava.

Szentiványi, M. 1689. *Curiosiora et selectiora variarum scientarum miscellanea.* University of Trnava, Trnava.

Torkos, J. J. 1745. *Schedisma de thermis Pösthenienisibus.* Bratislava.

Trajan, B. A. 1642. *Saluberrimae Pistinienses thermae.* Trenčín.

Wachtel, D. 1859. *Ungarns kurorte und mineralquellen.* Oedenburg.

Wernher, G. 1549. *De admirandis Hungariae aquis hypomnemation.* Basilae.

Žigo, P. 1992. *Z historickej toponymie Dudiniec. V. balneohistorická konferencia.* Dudince, 60-63.

The Authors:

Marián Fendek
Geological Survey of Slovak Republic
Mlynská dolina 1
817 04 Bratislava, Slovakia
Telephone: 421.7.59375.355
Fax: 421.7.54771.648
E-mail: fendek@gssr.sk

Augustín Rebro
A. Hlinku 18/34
921 01 Piešt'any
Slovakia

Miriam Fendeková
Department of Hydrogeology
Faculty of Natural Sciences
Comenius University
Pavilon G, Mlynská dolina
842 15 Bratislava, Slovakia
Telephone/Fax: 421.7.65425.446
E-mail: podzvody@fns.uniba.sk

The entranceway of the "Maria" thermal establishment at Ladek Zdroj in the Sudetes region of Poland, as seen in the late 17th century. This spa was visited for cures by many important persons in the 17th and 18th centuries, including John Quincy Adams, who would later become the sixth President of the United States. At the end of his visit, Adams declared, "I have never seen a spa, the location and appearance of which would be as much favorable to health preservation and restoring as Ladek."

The figure was first published in *Fons salutaris en Ladek et Silesia (Healing springs in Ladek and Silesia)* (1693, Vienna). The original drawing has been granted in recent years to Professor W. Ciekowski, who published it in his work *Ladek Zdroj* (1992), and who has authorized the authors to reproduce it in this chapter.

17. Using Geothermal Energy in Poland in the Second Millennium

by
Julian Sokolowski
Jozefa Sokolowska
Beata Kepinska

Abstract: Due to geological and geothermal conditions, Poland is characterized by low to moderate heat flow values. The only signs of the Earth's heat at the surface are a few dozen geothermal springs with temperatures between 18° and 44° C, located in Central Poland (Polish Lowland), the Sudetes region (Lower Silesia), and in the southern sector of Poland (Carpathian Mountains). Some of these springs were probably known and used since prehistoric times, but the systematic use of Polish geothermal waters for curing purposes began about 1,000 years ago. Since then, the tradition of thermal bathing has increased progressively, suffering temporary recessions as a result of conflicts and political changes that have affected Central Europe, Poland in particular, from the beginning of the second millennium to the mid-20th century.

Introduction

The use of geothermal heat has accompanied the development of people since the beginning of their existence on the planet. Initially, this heat helped people survive cold periods of a changing climate: prehistoric people managed to find shelter in caves and, especially in winter, they harnessed natural heat from geothermal sources whenever possible. Examples of such behavior are apparent in Poland, confirmed by numerous archaeological findings dating back to the Upper Paleolithic-Neolithic and to the Bronze and Iron Ages. Major archaeological sites are shown on the map.

Stone Age artifacts include pottery, flint tools, arrows and spears, fishing tools, and traces of fortified strongholds. Among the most important Neolithic findings are the underground flint mines in the Oronsk, Krzemionki Opatowskie, and Krasnik regions and by the Kraków-Wielun Jurassic caves, which were inhabited from at least 5000 B.C. In addition, numerous

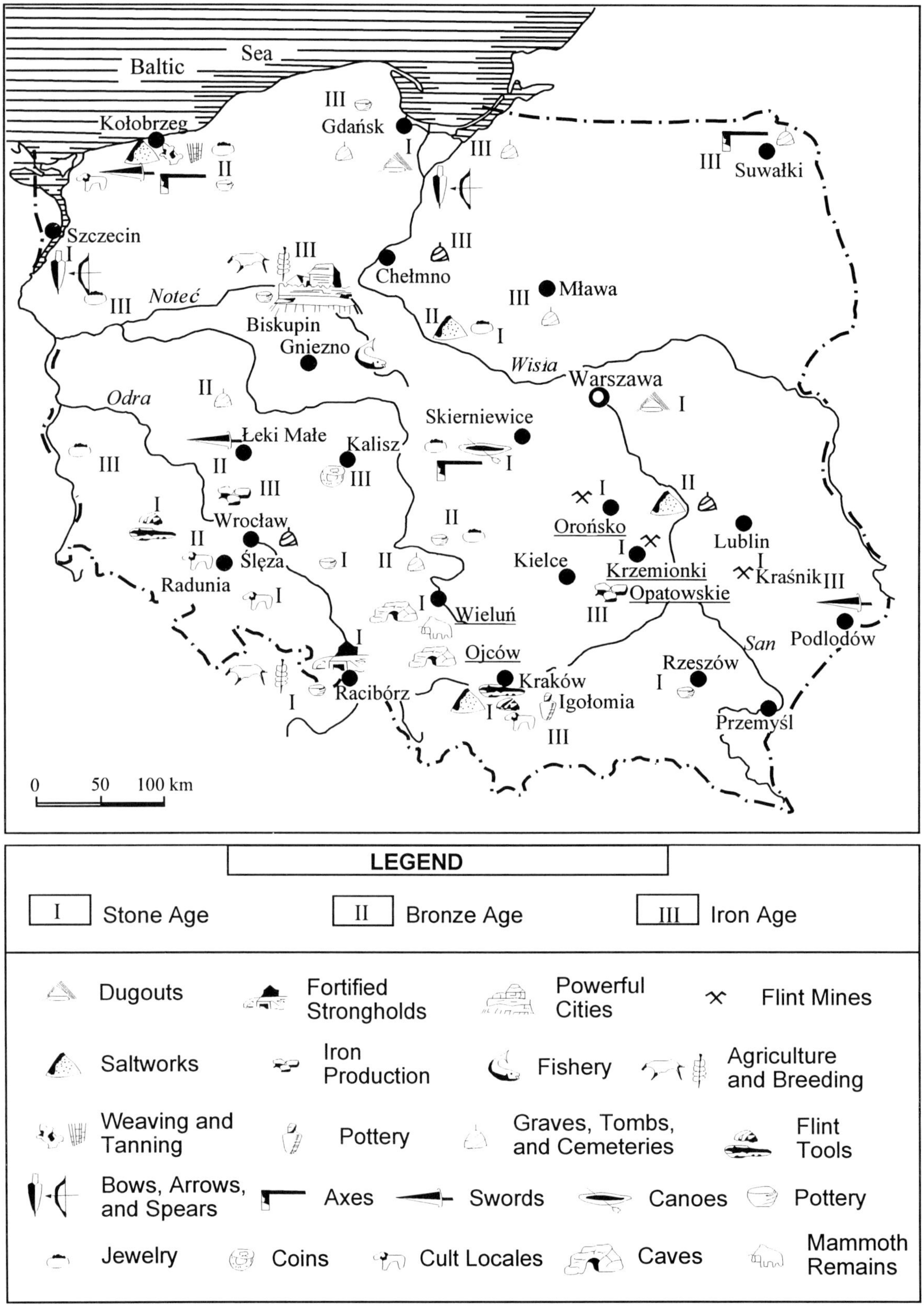

Main archaeological sites and findings in Poland. The underlined localities are inhabited caves and underground mines where ancient people took advantage of natural heat. *Compiled by the authors*

objects and utensils dating back to the Bronze Age have been found in the central and northern parts of the country. The most spectacular Iron Age site is a reconstructed fortified settlement in the Biskupin area. Other Iron Age artifacts include ceramics, pottery, coins, and graves and cemeteries.

Starting probably from the Upper Paleolithic, and certainly from the Middle Neolithic, people settled in Poland and began to take advantage of the relatively warm environment and constant temperatures in natural caverns warmed by the Earth's heat. The heat also made it easier for the people to mine flint and ore veins. In fact, flint tools from Poland have been exported to a number of Northern European regions since ancient times.

Historical Highlights

"Modern" Polish statehood began in 966, sanctioned by the baptism of Mieszko I. Written records about the establishment of governance date to that time, as well as written information concerning the occurrence and use of geothermal springs in certain parts of Poland. From the year 1000, countless documents provide evidence that Poland has often changed its area and property. The present borders, however, approximate those established around 1000.

The frequent fluctuations of Poland's boundaries over the past 10 centuries resulted in localities with warm springs changing their national ownership from time to time. Regardless of whether they belonged to Poland or nearby countries, however, and despite political circumstances, the warm springs of present-day Poland have always attracted people from Central Europe because thermal facilities were relatively unaffected by political downfalls and were often protected by temporary governments and local administrations. For this reason, geothermal spas could develop in Poland over the centuries. The main historical highlights of Poland follow, illustrating the territorial changes within which thermal bathing developed.

The first Polish ruler, Mieszko I (960-992), ruled over two different regions traced back to the mid-10th century: one belonged to the "Polans" and the other to the "Vistulans." At the end of Mieszko I's reign, these two regions united to form the first Polish nation, including the nearby lands of Slavonic tribes; its first bishopric was established in Poznan.

During the time of Boleslaw Chrobry (992-1025), other bishoprics were established in Gniezno, Kolobrzeg, Wroclaw, and Kraków. Records from that time indicate that warm springs were known and had probably long been in use by local people in the region of present-day Jelenia Gora and Cieplice. The latter name means "warm springs."

During the reign of Boleslaw the Skew-Lipped (1107-1138), new bishoprics were established in Lubusza at Oder and at Plock. On his death, Poland was partitioned among his sons.

In 1306-1333, some previous Polish territories were reunited under Wladyslaw the Elbow. During this period, the caves of Ojcow-Wielun, located near Kraków (then the Polish capital), were used temporarily by people as places of refuge during invasions of foreign troops. However, some caves in the Tatra Mountains and in the Swietokrzyskie ("Holy-Cross") Mountains were also used as shelters by bandits.

Wladyslaw the Elbow's successor, Casimir the Great, broadened the Polish borders and strengthened both the defense system and the Polish economy.

Under King Wladyslaw Jagiello (1386-1434), the Polish Kingdom united with the Great Lithuanian Kingdom. The territories of the new political entity extended as far as the Black Sea and included important towns such as Polock, Smolensk, and Kiev. However, the southwestern part of Poland, together with Lower Silesia and such important towns as Wroclaw, Opole, and Jelenia Gora (Cieplice), passed to the rule of the Czech crown.

Under the reigns of Sigismund III (1587-1632) and Wladislaw IV (1632-1648), the united Polish and Lithuanian Kingdoms were first called "Crown," then "Polish Republic I." In the years following the Swedish invasion (1655-1660), under Ian III Sobieski (1674-1696), the Polish territory still extended to the east as far as Kiev.

During 1772-1793, the Polish territory was partitioned among Austria, Prussia, and Russia. However a small area, called the Polish Kingdom with its capital in Warsaw, remained untouched and formed the core of the new Polish nation that emerged in 1918 with new borders, the "Polish Republic II."

This new nation flourished until 1939, when it was again divided as a result of World War II. In 1945 after the end of the war, the new borders of Poland were restored to more or less what they were 1,000 years ago.

Geological and Geographical Background

The Polish territory can be divided into three main geological regions:

- The Polish part of the Middle-European Epiplatform Province: it consists of a fragment of the Pre-Cambrian East-European platform (Northern and Eastern Poland) and a fragment of the Paleozoic West-European platform (Western and Central Poland). The older formations are covered by thick Permian, Mesozoic, and Cenozoic sedimentary deposits (6-12 km total thickness). From a geographical point of view, this region is called the Polish Lowland; it is part of the Middle-European Lowland.

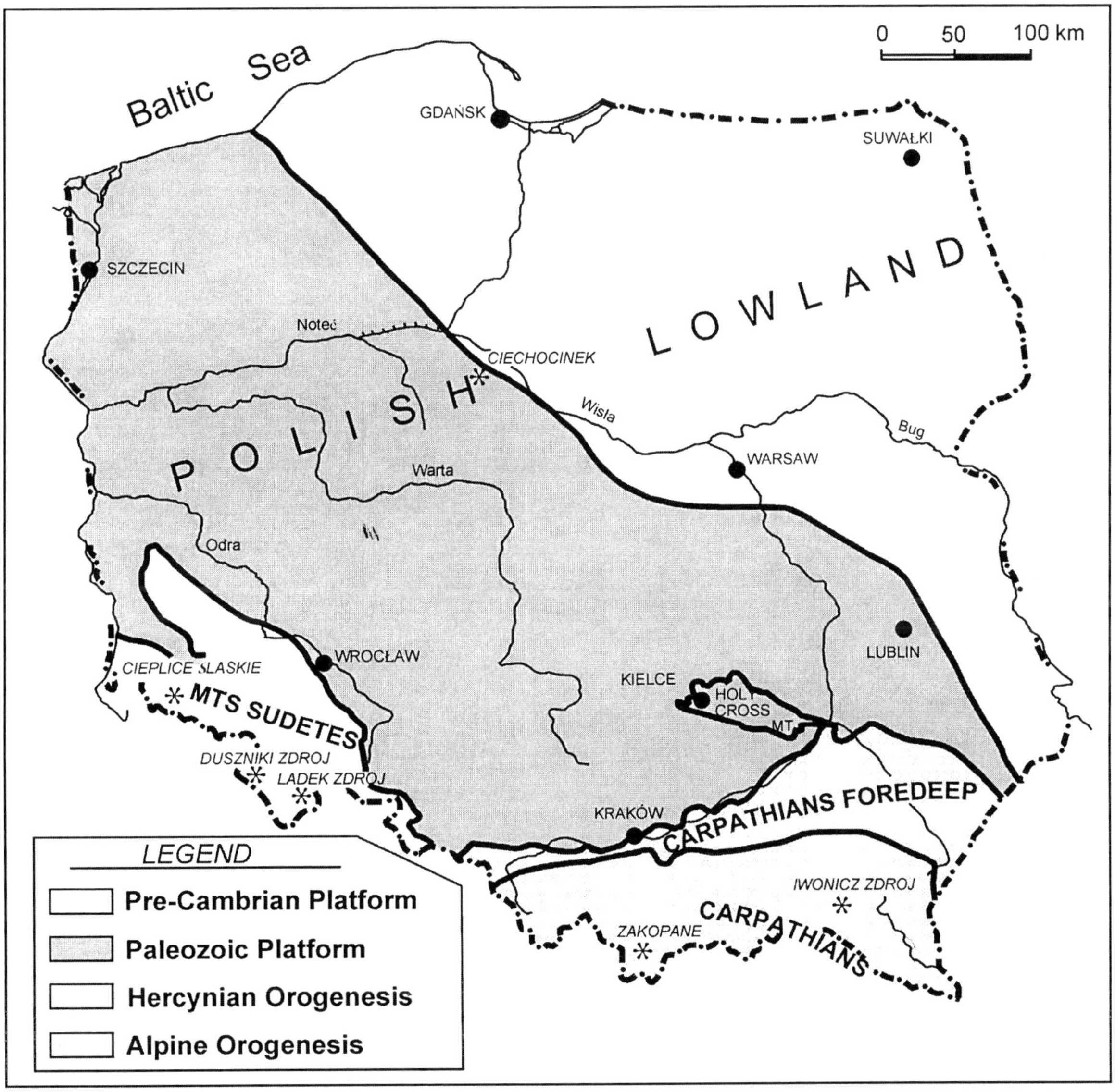

Sketch of the main geological regions of Poland with the locations of health resorts (*) using geothermal springs.

- The Sudetes region, consisting mainly of Pre-Cambrian and Paleozoic formations: it is a sector of the Czech massif, part of which forms the mountain range of the Sudetes, while the rest forms the Fore-Sudetes block. The block is covered by Tertiary and Quaternary formations.

- The Carpathian region, which belongs to the Alpine orogenic system, includes the Outer Carpathians, with Cretaceous-Paleogene flysch formations, and the Inner Carpathians, with Paleozoic crystalline cores and Mesozoic sedimentary formations; the Pieniny Klippen Belt is located between the Inner and Outer Carpathians. This very narrow and elongated ridge, with an extremely complex tectonic structure affected by Mesozoic-Paleogene folding, resulted from the collision of two continental plates. The youngest part of the Carpathians is known as the Carpathian Foredeep, which formed during the Miocene.

Poland is characterized by low to moderate terrestrial heat flow values (30-70 mW/m^2), with temperature gradients of between 2° and 3° C/100 m (Sokolowski, 1995). A few dozen warm springs (18°-44° C) are found in the country, in the hilly and mountainous regions of the Sudetes and the Carpathians.

Geothermal Balneology

Over the past 1,000 years, geothermal heat from natural springs in Poland was used only for bathing and balneology at Ciechocinek in the Polish Lowland; Cieplice Slaskie, Ladek Zdroj, and Duszniki Zdroj in the Sudetes; and Iwonicz Zdroj and Zakopane in the Carpathians. Although no evidence has been found proving similar uses in earlier times, it is likely that some of these thermal localities were known and frequented by people for bathing since prehistoric and ancient periods.

Ciechocinek (Polish Lowland)

This resort used curative brines flowing from natural springs of 10°-13° C. The high content of chemicals in these spring waters comes from salt-rich Mesozoic formations, which exist at shallow depth in the Ciechocinek area. The salt minerals are dissolved and carried to the surface by water of meteoric origin.

Centuries of use and the excellent healing properties of the warm brines resulted in the mid-19th century drilling of wells to increase the springs' yield. This, in turn, brought about the rapid development of this resort. At the same time, to exploit more efficiently the heat and healing properties of the geothermal brines, modern wooden installations were built near the wells to capture the warm water spray. These structures were designed by Stanislaw Staszic, the "pioneer and father" of Polish geology and mining.

Ciechocinek is situated in Central Poland, on the left side of the Vistula River Valley. Development of the town and its neighborhoods began after the partition of Poland among Austria, Prussia, and Russia in 1772, when Poland lost the Wieliczka salt mine. At that time, by the order of Stanislaw August Poniatowski, king of the remaining Polish nation, brine sources for salt extraction were sought in the Ciechocinek area. In 1823, after the government had bought the brine springs area from private owners, Ciechocinek began to develop rapidly: old brine springs known since the 12th century were restored, new wells were drilled, settling systems were improved, and several salt processing facilities were installed.

Ciechocinek. Wooden installations designed by S. Staszic for spraying warm brines used in curative inhalations. From the mid-19th century. *Mieczkowski, 1873*

From 1836, when the saline springs began to be used systematically for healing purposes, a special therapeutic station was established in the spring area. From 1841to1860, the first wells (100- 300 m deep) were drilled. They discharged brines with temperatures of 18° C (Mieczkowski, 1873), about five degrees warmer than the natural spring water. The wooden spraying installations built by Staszic included a number of graduation towers; altogether, these installations were some 2.5 km long (see illustration on prior page). The warm iodine-bromine brines created an ocean-like microclimate, which was especially suitable for natural curative inhalations. At the same time, the infrastructure of the areas around the springs and the wells was improved: embankments protecting the lowland against possible flooding of the Vistula River and a number of other waterworks were built, and a theater was erected (*The Popular Encyclopedia*, 1973).

In 1910, the population was 2,000; however, after Poland regained its independence in 1918 and Ciechocinek was granted civic rights in 1919, the existence of the renowned therapeutic station contributed substantially to the town's development. Thus, Ciechocinek became one of the main Polish resorts: some 60,000 patients were treated there in 1970 and more than 100,000 in 1995.

For present-day treatments, warm iodine-bromine and ferric brines at 30° C are extracted from three wells (1300-1400 m deep), and peat is used for baths. Gynecological diseases, rheumatism, circulation illnesses, problems of the central nervous system, and upper respiratory maladies currently are treated at this resort.

Resorts in the Sudetes

Over the centuries, the Sudetes have become famous throughout Central Europe for beautiful landscapes, a mild climate, and numerous health spas. A number of mineral spring waters have been used for curing purposes, and some have issued thermal waters in use since ancient times. The most important of these springs are Cieplice Slaskie, Ladek Zdroj, and Duszniki.

Cieplice Slaskie. This resort is 6 km southeast of Jelenia Gora, located at about 350 m in the Jelenia Gora Depression, which is surrounded by four mountain ranges: the Karkonoszes to the south (with the highest massif in the Sudetes, Mt. Sniezka, at 1602 m), the Janowickie Rudawys to the east, the Izerys to the west, and the Kaçzawas to the north. In 1785, the population of

Cieplice was 1,391 (*Polish Cities in the Millennium*, 1965), but it grew to 15,500 in 1970, mostly as a result of the geothermal spas.

The Cieplice resort is supplied by warm mineral waters from six springs, which yield 0.2-4 m^3/h of water with temperatures of between 20° and 44° C and flow from the Karkonosze mountain granites. Warm waters are used mostly for therapeutic baths; to a lesser degree, however, they are used to treat orthopedic-traumatic and neurological diseases, nephropathies, and urinary tract diseases.

A legend says that the Cieplice warm springs were discovered in 1175 by a Piast prince, a descendent of the Silesian King Boleslaw I the Tall (1127?-1201), when he was chasing a wounded deer. However, the oldest historical record of these springs is found in a document of March 18, 1281. It concerns the donation of *caldius fons* ("warm springs"), along with 250 fieves and a 20-year release from taxation, made by Prince Bernard from Lwowek to the Silesian monastery of Strzegom, which belonged to the Knights of St. John of Jerusalem. This monastery had undertaken the task of curing ill persons through the use of thermal waters.

In 1288, the first curative house was built, and the Czech king authorized the monastery to erect an inn at Cieplice to serve a growing number of patients. The Slavonic name of this locality (*Cieplowody* or *Chleplevode*, which means "warm waters") can be found in records of Gabriele da Rimini, a papal functionary who visited Silesian villages to collect overdue taxes at the beginning of the 14th century.

In 1381, the resort was purchased by the noble Schaffgotsch family. In 1403, the family founded a prebend of the warm springs on behalf of the Krzeszow Cistercian monastery, established there in 1292. (The monastery received funds from operation of the springs.) After the monastery belongings were liquidated in 1810, the resort was managed directly by the Schaffgotsch family.

The most famous patient who visited Cieplice was Polish Queen Maria d'Arquién Sobieska, in 1687. She came to treat her ailments and to strengthen her poor health and was accompanied by her whole court—some 1,500 people (Kurek and Arczynski, 1970). She was the beloved wife of one of the greatest Polish kings, Ian III Sobieski, whose army heroically helped stop the Turkish invasion in Europe in the famous 1683 battle of Vienna. Two of the warm springs in Cieplice are named after King Sobieski and his wife. A scene of group bathing was painted in

Group bathing in the "Cloister Bath" at the Cieplice Slaskie thermal resort during the 18th century. *Photograph of a picture in the archives of the Cieplice Resort State Company*

the 18th century in the "Cloister Bath," one of the most important thermal spas in Cieplice Slaskie.

The resort flourished in the 19th century, when additional warm springs were channeled towards the intake system and more modern therapeutic methods were introduced. Towards the end of the century, however, Cieplice faced increasing competition from German and Czech resorts and declined in popularity. The local economy of the town and neighboring areas was reoriented to the machine industry. The Cieplice resort was revived again in 1945. With the warmest curative waters in Poland, it became, and still is, one of the most important spas in Central Europe.

Ladek Zdroj. Ladek Zdroj is located at 440 m on the slopes of the Zlote Mountains, the highest peak of which is Mt. Kowadlo (989 m). This mountain range surrounds the Klodzka Valley from the east, and radioactive and sulfide waters flow from fractured Pre-Cambrian gneisses. Depending on the season, the flow rate from five springs is 1-17 m^3/h, with water temperatures between 20° and 30° C. These warm waters are used mainly for treating patients with motor system, vascular, oral, and dermatological diseases.

The first records of the Ladek Zdroj thermal springs date back to 1242. Ladek was granted civic rights by the Czech crown in 1325, and the first bathing house was built near the end of the 15th century; Ladek continued to develop at a slow but steady pace for about three centuries. At the beginning of the 19th century, the spa began to flourish and the number of patients increased from about 850 per year in 1800 to more than 10,000 per year in 1900 (*Polish Cities in the Millennium*, 1965). Among the most important visitors who stayed at Ladek for treatment were the German poet Johann Wolfgang Goethe; John Quincy Adams, sixth President of the United States; and Russian novelist Ivan Turgenev.

Duszniki Zdroj. This spa is located at 550 m, at the western piedmont of the Klodzka Valley, about 40 km west of Ladek Zdroj. It is supplied with iron- and CO_2-rich warm water from one spring with a temperature of 19°-20° C. The first records date back to 1408. Since then, the water has been used for curative drinking and thermal bathing.

In 1823, the Duszniki resort hosted an outstanding composer: Felix Mendelssohn Bartholdy from Germany. In 1826 Frederic Chopin, the great Polish composer and pianist (1810-1849), stayed at Duszniki for healing treatments. Chopin was only 16 years old when, after finishing secondary school in Warsaw, he went to Duszniki with his mother and sister. Since his early childhood, Chopin had delicate health, suffering from pneumonia. The young pianist's letters provide insight into how he passed the time at Duszniki: "it is the second week of my drinking whey and waters...and, they say, I look better.... In the morning, at six at most, all the patients gather together at the spring, and very poor music...accompanies them in their slow walk..." (Kurek and Arczynski, 1970). During his stay, the young artist gave one of his first public concerts, so arousing the admiration of the audience that their negative impression of the music played by the resort orchestra easily was offset. This was one of the first performances which opened the doors to Chopin's artistic career.

In subsequent years, numerous Poles visited Duszniki (under Czech rule at that time) and built a monument to Chopin, as well as a theater bearing his name. To commemorate the artist and his stay at Duszniki, the warm spring was named *Pieniawa Chopina*, and an international music festival is held there each year.

Resorts in the Carpathians

Like the Sudetes, the Carpathians abound with low temperature mineral springs, which enabled the establishment of numerous health resorts. But because of the different geological structure and hydrogeological conditions of the two regions, the Carpathians have relatively fewer geothermal springs, which are located in two areas: Iwonicz and Zakopane (see second map). These springs have been known and used by local people for many centuries.

Iwonicz Zdroj. This resort is situated at about 410 m, on the foothills of the Beskid Niski Highlands, one of the Carpathian chains. Records of warm springs usage date back to 1578 and 1630 (*Polish Cities in the Millennium*, 1965), when they were recognized and described by the royal physicians. However, it wasn't until 1793 that the resort started to flourish, after facilities were built to encourage more people to take advantage of thermal balneology and mineral water cures.

At the beginning of the 19th century, reputable chemists and physicians voiced favorable opinions about the high curative value of these waters. In particular, in 1856, Jozef Dietel, a professor at Jagiellonian University, declared Iwonicz the "prince of iodine waters." Since then, Iwonicz waters have been bottled and widely sold in Central Europe.

To increase the production of the iodine-rich warm waters, some wells (400-600 m deep) were drilled in this area around the start of the 20th century. The warm brine discharged by one of these wells drilled in 1911 is still used for cures today. Due to water production from the wells, however, the natural springs started to dry up. Water production was subsequently moved to the wells only. Today, warm water for curing purposes is taken from nonproducing oil wells.

The interwar period (1918-1939) was a real boom for this spa. Afterwards, following a decline which lasted some 20 years, Iwonicz again flourished in the late 1950s. It thus became, and still is, one of the best known Polish resorts. Twelve wells, with a maximum depth of 1000 m, produce warm water ($T=20^{o}$ C, and TDS = 6-19 g/l) from Eocene sandstones. The water is especially rich in bromine, iodine, boron, and free carbon dioxide. Rheumatism, diseases of the motor, alimentary, and respiratory systems, dermatological diseases, and many other illnesses are treated at this spa. The water is also used for drinking, for the preparation of peat baths (mud or "fango" therapy), and the extraction of salts for the cosmetic industry (Karwan, 1989).

Zakopane. Zakopane is located in Southern Poland, on the northern slopes of the Tatra Mountains, the highest sector of the Carpathians. These are the only Polish mountains involved in the Alpine orogenesis; the highest peak is Mt. Rysy (2499 m). The Tatras region, in particular Zakopane and the nearby Podhale area, constitutes the main centers for tourism and winter sports in Poland. Their popularity started in the second half of the 19th century and continued to grow until the present time. With a population of about 30,000 inhabitants, this region is visited now by 3.5 to 4 million tourists each year.

In addition to the scenic attractions, a restorative summer climate, and winter ski facilities, a significant contribution to tourist development in Zakopane and Podhale is the exploitation of geothermal waters. In the region of Podhale in the town of Zakopane, a large-scale district heating system has been under construction since the end of the 1980s. Started as the first system of its type in Poland, it will be accompanied by other geothermal uses. Thermal bathing, however, began many centuries ago at Jaszczurowka (a borough of Zakopane), where a warm spring (20° C) was used by local highlanders. The spring was in existence until the late 1960s. Due to a salamander species (*salamandra salamandra*) that found its optimum habitat in the wet area around the warm pool, the spring was named *jaszczur*, which means "reptile" (Radwanska-Paryska and Paryski, 1973).

The spring was first described scientifically in 1844 by L. Zejszner, an outstanding geologist, forerunner of the geological investigations in the Tatra Mountains. He wrote: "not a trace of an igneous process will be found in the neighborhood [of the spring].... The spring is located in correspondence to sedimentary rocks...." Therefore, Zejszner concluded, the water derived its temperature from depth. From the hydrogeological point of view, Zejszner correctly stated that the spring was the result of an ascensional water circuit that reached the surface and outflowed through an open segment of the regional fault marking the limit of the northern border of the Tatras ridge.

Other scientists and medical doctors in the mid-19th century studied the spring water's chemical and therapeutic properties. They recognized that the Jaszczurowka warm water was particularly suitable for curing dermatological, rheumatic, and endocrinological diseases, as well as treating patients with contagious diseases, such as leprosy.

This spring was easily noticed because "in winter, the place is constantly steaming, and even during very severe winter days, fresh grass keeps on growing..." (*Geographical Dictionary of the Polish Kingdom and Other Slavonic Countries*, 1882). This same source also reports that long before the 19th century the local highlanders soaked "the diseased parts [of their bodies] in the warm water for hours, six hours or more, without interruption.... It usually sufficed to take one such bath to alleviate ailments...." The first investigators of the Jaszczurowka spring

The Jaszczurowka spa near Zakopane: general view of the bathing place using geothermal water (20° C), with the wooden pier ("sunny beach") and related facilities, as seen at the beginning of the 20th century. *Photo by Z. Sulkowski; reprinted by R. Bukowski from the collection of Tytus Chalubinski Tatra Museum in Zakopane*

also observed that it was one of the few warm springs in the Polish territory to have such efficacious healing properties.

In 1862, the owner of the spring's site constructed a small bathing facility (Zwolinski, 1966). The spring was timbered, with a perforated floor added for water outflow. Next to the pool, modest wooden buildings were constructed as dressing rooms, and an inn was built nearby. Thus, the owner could charge the users for bathing.

Since the warm spring and the bathing facilities were located in a beautiful spot, among fir trees at the foothills of the Tatra Mountains, an increasing number of tourists began visiting the site (Radzikowski, 1870). Towards the end of the 19th century, the Polish sector of the Tatra Mountains and the village of Zakopane were "discovered" by Dr. Tytus Chalubinski, a well known personality in Poland; consequently, many representatives of Polish high society frequented Zakopane.

Exceptional natural beauty and the highlanders' rich folk culture attracted many other people of elite society, as well as artists and scientists. Mountain tourism flourished, and sports and recreational activities became fashionable. Moreover, due to the fresh summer climate and the

The Tatra Mountains with the characteristic peak Giewont (~1900 m) in the background. The Zakopane resort and geothermal bathing facilities are located in this beautiful sector of the Tatras. *Photo by M. Kowalski, 1996*

therapeutic characteristics of the thermal water, treatments for lung and respiratory diseases were especially sought at Zakopane, and mineral water cures became more popular as well. The facilities at Jaszczurowka thus became a curing and social "must" for the many tourists who visited the Zakopane area from 1880-1900.

This is evidenced by numerous written records of that period: "in summer visitors coming to Zakopane are used to enjoying baths..." (*Geographical Dictionary of the Polish Kingdom and Other Slavonic Countries*, 1882), and Walery Eljasz Radzikowski, author of the first tourist guide to the Tatra Mountains, wrote in 1870: "it is hard to get a bath before noon; ...much time must one wait for taking a bath, otherwise, as it often happens, one has to return home without having had any bath...."

The Jaszczurowka spring is recognized in Polish literature and folk culture. A beautiful poetic description of the place can be read in Stanislaw Witkiewicz's novel *At the Mountain Saddle* (1891). The enchantment created by the beauty of the Tatra Mountains and the extraordinary abundance and originality of the highlanders' culture are described eloquently. The author was the father of Stanislaw Ignacy Witkiewicz (Witkacy): a world-renowned playwright, art critic, philosopher, painter, and forerunner of the Theater of the Absurd. In describing the spring, Stanislaw Witkiewicz also quoted some of the tales related to Jaszczurowka. In particular, he recalled "Sabala," the most famous Tatra storyteller, singer, gifted violinist (and poacher as well!), who had described the Jaszczurowka spring in previous years.

A passage from Witkiewicz's novel reads, "By coach or on foot, from the very early morning hours, crowds of people are going to the emerald warm spring at Jaszcurowka. Closed in a pool made of fresh fir timber, the spring resembles light sapphire set in pale gold. The bathing people loom as greenish *jaszczur* ('reptiles'): naturally I am speaking of men only! The spring, surrounded by bushes of marsh-marigold and coltsfoot, flows out at the boundary of a ravine like a little mountain stream.... Jaszczurowka is romantic: not only for its warm, light blue and unique spring, but also for the black-orange spotted and clumsy salamanders living there. After thirty years of peaceful life, the salamanders inflate their bodies enormously and become 'dragons.' Due to this change, the 'reptiles' become equipped with a number of heads and gain a terrible hunger. Bones of these 'dragoned' reptiles can be found, according to 'Sabala,' in many caves and recesses. However, not only for its 'dragons' is Jaszczurowka so romantic; it is for its people who extol the beauty of their spring in a poem, whose refrain says: 'Oh, sweet, bright, and transparent waters; with the mirror wave....'"

Between the world wars (about 1919-1938), Jaszczurowka continued to flourish and reached its peak in balneological attendance. Next to the spring pools, a wooden pier was built (the "beach") where people used to rest and sunbathe. Swimming competitions and water-polo matches with a great number of competitors and fans were organized.

Two pools supplied by the Jaszczurowka spring remained in use after the end of World War II until the 1960s. There were also plans to modernize the curing establishment and organize new balneological and therapeutic treatments. After drilling a well to raise the spring's output, however, the hydrogeologic circuit was drowned and the warm spring disappeared, due to cold water drawn in from the neighboring stream. Consequently, all plans to increase balneotherapeutic activity were deferred temporarily.

However, to revitalize and further develop the geothermal bathing tradition of the Zakopane region while expanding the use of the Earth's heat for other purposes, new initiatives were begun at the end of the 1960s. A modern geothermal bathing establishment was erected in the center of Zakopane (see prior photograph); it is supplied with warm waters (26°-36° C) from two wells and is frequented, especially in summer, not only by Poles but by many foreign tourists. In addition to this spa, a large-scale geothermal district heating project was begun recently in the nearby Podhale area to replace the polluting coal heating system with clean, natural heat.

Concluding Remarks

Poland as a whole is characterized by low to moderate terrestrial heat flow values, and the country has three geological areas with thermal springs at temperatures of between 18° and 44° C. Considering the severe climatic conditions of Polish winters, prehistoric and early historic people very likely took advantage of the natural heat that was responsible for relatively mild and constant temperatures in the caves where they settled in the Wielun-Kraków area and in the Swietokrzyskie ("Holy-Cross") Mountains. While it is also likely that these people knew and used the water from some of these thermal springs, no evidence has been discovered within the Polish territory to support this hypothesis.

The first written records describing the use of geothermal waters for balneotherapy date back to the 12th century and provide evidence that the practice of thermal bathing has old roots in

Poland. Warm springs were used in the Sudetes and the Carpathians, as well as at Ciechocinek in the Polish Lowland. While undergoing flourishing and recessive periods, this practice developed to the point that some of the Polish geothermal sites (e.g., Cieplice, Duszniki, and Iwonicz) became renowned spas in Central Europe, similar to those of Karlove Vary in the Czech Republic. The benefits of these spas were experienced by famous personalities in Polish and international artistic society, such as Frederic Chopin at Duszniki. In a few cases, the occurrence of thermal springs spawned folk legends about the warm water sources.

The tradition of geothermal balneology and bathing has grown in Poland for the past 1,000 years, providing the foundation of recent geothermal resource development. This development, however, is limited to a few locations and is based on warm waters issuing from natural springs, which are limited in quantity and temperature. In most cases, therefore, natural heat will be tapped in the future by wells drilled to 1-3 km deep. This is happening in the Podhale region and in other parts of the Carpathians, as well as in the Polish Lowland. Here geothermal resources will be used for integrated applications: domestic heating, industrial processes, greenhouses, fish farming, and balneology.

Acknowledgments

The authors are indebted to the employees of the Tytus Chalubinski Tatra Museum in Zakopane: to Mrs. A. Liscar and Mr. J. Darowski for their help in preparing the source material related to the Jaszczurowka spring, to Mr. J. Filar from the High Tatras Company, and to Mr. M. Kowalski for photographs of the bathing place in Zakopane. Thanks are also due to Mrs. T. Kepinska and Mr. J. Kepinski for their assistance in archival investigations and preparation of the figures. The authors are also grateful to Mrs. W. Fortuna from the Cieplice Resort and to Professor W. Ciezkowski (Wroclaw University) for providing the figures from the Sudetes region and for their kind permission to publish them in this chapter. The authors express their special gratitude to Dr. R. Cataldi for his valuable contribution to the final content and form of the chapter.

Cited and Selected References

Ciezkowski, W. 1992. *Ladek spa (Ladek Zdroj)*. Oficyna Wydawnicza Sudety, Wroclaw. (In Polish).

Geographical dictionary of the Polish Kingdom and other Slavonic countries (Slownik geograficzny Krolestwa Polskiego i innych krajow slowianskich), vol. 3. 1882. F. Sulimierski, B. Chlebowski, and W. Walewski, eds. Warsaw. (In Polish).

Karwan, K. 1989. *Mineral and therapeutic waters of the Carpathian health resorts (Wody mineralne i lecznicze uzdrowisk karpackich)*. Academy of Mining and Metallurgy Publishers, Kraków. (In Polish).

Kulik, Z. 1986. *Jelenia Gora and vicinity. A guide (Jelenia Góra i okolice)*. Sport and Turystyka Publishers, Warsaw. (In Polish).

Kurek, T. and S. Arczynski. 1970. *Lower Silesia (Dolny Slask)*. Sport and Turystyka Publishers, Warsaw. (In Polish).

Mieczkowski, L. 1873. *The historical and therapeutic aspects of Ciechocinek (Ciechocinek pod wzgledem historycznym i leczniczym)*. Warsaw. (In Polish).

Polish cities in the millennium (Miasta polskie w tysiacleciu), vol. 1. 1965. The Ossolineum Publisher, Wroclaw-Warsaw-Kraków. (In Polish).

The popular encyclopedia (Encyklopedia powszechna). 1973. Polish Scientific Publishers, Warsaw. (In Polish).

Radwanska-Paryska, Z. and W. H. Paryski. 1973. *Encyclopedia of the Tatra Mountains (Encyklopedia tatrzanska)*. Sport and Turystyka Publishers, Warsaw. (In Polish).

Radzikowski, W. E. 1870. *New illustrated guide to the Tatra and Pieniny Mountains (Nowy illustrowany przewodnik do Tatr i Pienin)*. J. K. Zupański, ed. Poznan. (In Polish).

Sokolowski, J., J. Sokolowska, S. Plewa, S. Nagy, M. Krokoszynska, and U. Krzysiek. 1995. *Geothermal provinces and basins in Poland*. MEERC PAS Publishers, Kraków.

Witkiewicz, S. 1891. *At the mountain saddle—Impressions and pictures from the Tatras (Na przeleczy—Wrazenia i obrazy z Tatr)*. Gebethner and Wolf Publishers, Warsaw. (In Polish).

Zejszner, L. 1844. *On the temperature of the Tatras' and adjacent areas' springs (O temperaturze zrodel Tatrowych i pasm przyleglych)*. Biblioteka Warszawska, t.2, Warsaw. (In Polish).

Zwolinski, T. 1966. *The Tatras. A guide (Tatry. Przewodnik)*. 12th ed. Sport i Turystyka Publishers, Warsaw. (In Polish).

The Authors:

Julian Sokolowski
Polish Academy of Sciences
Mineral and Energy Economy Research Institute
Polish Geothermal Association
Ul. J. Wybickiego 7, 31-261 Kraków, Poland
Telephone/Fax: 48.12.632.24.35
E-mail: pga@min.pan.krakow.pl

Jozefa Sokolowska
Polish Geothermal Association
Ul. J. Wybickiego 7, 31-261
Kraków, Poland
Telephone/Fax: 48.12.632.24.35
E-mail: pga@min.pan.krakow.pl

Beata Kepinska
Polish Academy of Sciences
Mineral and Energy Economy Research Institute
Geothermal Laboratory Banska Nizna
Polish Geothermal Association
34-424 Szaflary, Poland
Telephone/Fax: 48.12.73.218
E-mail: labgeo@zakopane.top.pl

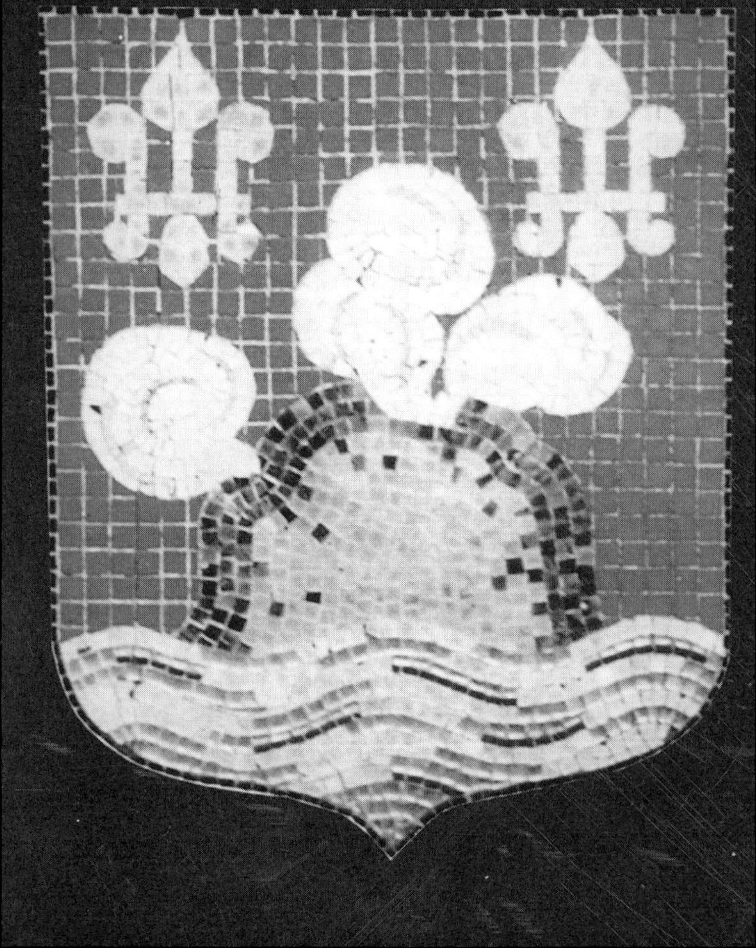

Shield of the town of Chaudes-Aigues, France, in Southern Auvergne. The design in colored mosaic tile depicts a volcano releasing plumes of whitish steam while surrounded by undulating waves of hot water. In the Late Middle Ages, the town began a system of heating private homes and public buildings with geothermal water. The heating system, modernized and expanded, continues today.

18. Using Geothermal Waters in France: The District Heating System of Chaudes-Aigues from the Middle Ages

by
Jean Pierre Gibert
Florence Jaudin

Abstract: This is the story of how Chaudes-Aigues, in Cantal, France, has used its geothermal waters for nearly six centuries. Tax records for supplying natural hot water to dwellings in Chaudes-Aigues date from the 14th century. Water from hot springs gradually came to be used for an increasingly wide variety of purposes. Today, the Chaudes-Aigues geothermal heating network services 150 homes in the winter and, during the season of health cures, supplies spas with water for therapeutic purposes.

Introduction

DEEP IN THE VALLEY OF THE REMONTALOU RIVER, Chaudes-Aigues, at the southern edge of Auvergne in the Central French massif, a region inhabited since prehistoric times, has one of the hottest natural thermo-mineral springs in Europe: 82° C.

General view of the spa town, Chaudes-Aigues.

The main spring, called *le Par*, is one of about 30 gushing springs located in a small area. From mid-October to the end of April, 150 homes are now heated by hot water flowing through a 5 km network of pipes from five of these springs. Houses built on the spring field itself use the hot water directly below their floors for heating. The first writings describing this hot water use date back to the 14th century.

Present view of the old thermal spring, *le Par*, at Chaudes-Aigues.

Thus this small town has a rich geothermal history. In 886, the area around Chaudes-Aigues was evangelized. Lord Bodon gave the "villa du *Par*" (le Par villa) to the chapter of Saint-Julien de Brioude. The name *le Par* dates back more than 10 centuries. According to traditionally accepted etymology, the word "par" comes from *parer les cochons* ("to clean pigs"). Another possible etymology is a word found in several medieval texts: *parare*, which means to prepare wool. *Parare* was also used during the 14th century to mean "care for, treat, cure."

1300-1400

CHAUDES-AIGUES, KNOWN BY VARIOUS NAMES (*CALIDAE AQUAE* OR *AQUAS CALIDADES* IN LOW Latin, or *Caldas Aygues* in *Langue d'Oc*, then used in Southern France), started to develop rapidly. A 1334 *terrier*, a register of real property listing the names of vassals or tenants, with details of their holdings, services, and rents, mentions distribution of hot water in houses three times. One tenant owes his lord five *sous*, the monetary unit used in Southern France at that time, "for the hot water which is taken in the public road between Adrien Brugier's house below the baths and R. Mercier's house and for the pipe...." Two other tenants of *la place* ("the square") each owe five *sous* for bath water, one for his house and the other for his building. This is the earliest recorded reference to Chaudes-Aigues' use of geothermal waters for heating.

1400-1500

CHAUDES-AIGUES, NOW GROWN LARGE ENOUGH TO BE CALLED ONE OF THE "GOOD TOWNS OF the high country of Auvergne" with the right to send a representative to the "Auvergne States," was the subject of an ownership dispute between the Counts of Armagnac to the south and the Dukes of Bourbon to the north.

According to a 1474 *terrier*, the distribution of thermal water was widespread, and about 20 "hot houses" are mentioned: "These concessions are numerous. Almost all of the inhabitants of the *rue du Par* ('Par Street') or of *la place* have the right to draw water at its source, above the pool, in the *futdyras* ('wooden pipes')...."

Innkeepers, barbers, apothecaries, and several leading citizens had similar rights. For instance, a barber named Jean Neuvéglise, who had his shop in the lower sector of *la place*, near the town gate, was allowed to draw hot water from the spring *le Par* "with his ancient servitudes..." above the bath, whereas Jean Roux, despite his position as bailiff, was permitted to "tap hot water for his house on the *rue du Four* only in the dispensary of the apothecary next door, and only holds the rights in a precarious manner; therefore, he has agreed to give them back to the Duke whenever asked to do so..." (Felgères, 1890).

The hot water network was used not only to heat houses; it also had industrial applications: "The inhabitants of Chaudes-Aigues like to work and are very industrious. They profit as much as possible from their hot waters.... They wash wool in the spring water before spinning...and treat textiles with the spring water, as well..." (Berthier, 1810).

Other mentions of hot water usage at Chaudes-Aigues are included in the 1474 *terrier*: "Etienne Bouniol, miller, acknowledges his right to draw hot water in the wall next to the gate of the square and to draw it to his mill in La Vernède...," and "Astruge Ferrande, from Saint-Julien, draws, at the same spot, the hot water that comes from the pool on the square along the wall, and carries it to his mill in order to triturate, card, and finish his textile products..." (Felgères, 1890).

This same *terrier* clearly stipulates the relationships between the inhabitants and the weavers of Chaudes-Aigues: "Therefore, for the convenience of the textile industries, the other users are required to return their bath water to a collector canal which flows from the spring *le Par*, crosses the square and runs along the wall before flowing through a hole in the wall and into the stream gravels. It is here tapped and carried to the various mills by a series of aqueducts, several of which cross the stream..." (Felgères, 1890).

Piercing pine tree trunks to make pipes. Hand-halved pine trunks have been used to channel thermal waters since the 14th century.

Starting in the 14th century, hot water was distributed via pinewood pipes, hollowed out with a long auger and fitted together. Another use of the hot waters at Chaudes-Aigues, which has endured through the ages, was for health treatments. The *terriers* of 1334 and 1474 make numerous references to the *bàynh*, a large public spa fed by the spring *le Par*.

This spa became a popular center for hydrotherapy in Europe, as the following passages from the Bishop of Saint-Flour indicate: "People come in crowds from many parts of the world to regain their health at Chaudes-Aigues..." (Bertrand de Cadoèn, 1426), and "In 1400, the consuls of Saint-Flour paid to send a pauper named Guillaume Girbal there...to cure him from *pelsicadura*..." (Felgères, 1890). *Pelsicadura* is an ancient term indicating a malady of the skin, perhaps psoriasis.

The site became particularly renowned for the treatment of paralytics, similar to its current specialization as a spa to treat rheumatism. In 1440 Gilles Bouvier, first herald of Charles VII, wrote that the most frequented spas of his time were Forges, Barbotan, and Chadesaigues [*sic*]. Moreover, Felgères wrote that "Paralytics and lepers, especially, are taken there for cure...."

1500-1600

THE TOWN OF CHAUDES-AIGUES WAS BURNED IN THIS PERIOD DURING THE RELIGIOUS WARS. The wealth of the ancient barony dwindled, and municipal freedoms were lost.

During the 16th century, the technique of water distribution to the mills was refined further, as was the use of wooden pipes. In 1545, Philander described the distribution methods: "Part of the hot water is drawn off through channels from the baths and the springs and is conveyed out of town by means of half-pipes of wood; hot water is then used to degrease and treat textiles, which are piled in wooden washtubs and struck energetically with hammers and pestles. The rest of the hot water is recuperated at the foot of the mountain and scooped up by the buckets of a spatulated water wheel moved by the cold-water stream that flows nearby...."

Another traditional use for the thermal waters is deeply rooted in the history of Chaudes-Aigues. Down through the centuries, visitors to the town noted that the inhabitants used the hot water for cooking. Philander reported that "the inhabitants used the principal spring to cook eggs, pluck chickens, remove the hair from baby pigs, and from the heads and feet of sheep, and heat up meats until they lose their bad taste...." In 1605, Jean Banc wrote that the waters at Chaudes-Aigues "are so hot that people cook eggs perfectly...."

Artist's impression of a water mill. From the encyclopedia *Diderot-Alembert*.

Philander also mentioned an open-air public bath: "In this town, very hot waters spring out with great force from various spots. The water of the main spring, as soon as it falls in the pool, is drawn off by an underground pipe to a public, open-air bath...." Neither the *terrier* of 1615 nor any other later documents, however, mention this specific pool; apparently, it dried up between 1545 and 1615. Baths were also taken in small covered pools inside buildings.

In addition to the existence of public baths, Philander described methods of thermal bathing: "I saw exudations done in two ways in the town of Chaudes-Aigues in Auvergne.... First of all, hot water is conveyed through open lateral canals to little rooms they fill with very hot steam; afterwards, water flows to the bathing pools. You are free to choose one of these two ways for exudation, but I don't know if you would be able to remain without harm in one of these steam rooms because the heat is suffocating inside them, and there is no exit other than a little door. But this is not the only use they make of this water...."

In addition to the springs used for public or private baths, others supplied water to some private homes located very near the mountain and to nearby inns.

1600-1700

THE 17TH CENTURY USHERED IN AN ERA OF DECLINE IN THE USE OF HOT WATER AT CHAUDES-Aigues. However, using hot water for bathing resisted the passage of time. In *The Renewed Memory of the Marvels of Natural Waters in Favour of French Nymphs* (1605), Jean Banc wrote: "The truth is that the thermal baths [at Chaudes-Aigues] are the baths used for the longest period of uninterrupted time in France; they are very neatly and cleanly equipped in private homes. For the most part, hot water is derived by canals from the nearby springs, as one wishes...."

In 1623, Jean Passenaud, an apothecary living on *la place*, acknowledged two-thirds of the hot water pipes branching off from the main pipe that gathered the water from *le Par* spring "next to the pillory on the square." He was allowed to draw "the water to his bath without being able to convert it, contrary to customary usage, and afterwards [he had to] return it to the main pipe."

The woolen industry, too, started to decline significantly during the 17th century. The *terrier* of 1623 indicates that "the quantity of *cazals* ['houses'] fallen in ruins that one can see in all the neighborhoods attests that here, like elsewhere, civil discord has accomplished its mission of death. The number of mills has dropped from 15 to 4..." (Felgères, 1890).

1700-1789

IN 1787 LEGRAND D'AUSSY WROTE ABOUT THE THERAPEUTIC VALUE OF CHAUDES-AIGUES' HOT water: "Capable of acting on the human body because of their iron, gas and heat content, these waters can provide sick people not only with a drink cure but also with steam baths, hot-water baths and showers. Outside of the area, however, they are unknown...." Soon after, the health spas underwent a progressive decline. Those frequenting the spas imposed such a hardship on the community that in 1790, the mayor of Chaudes-Aigues, Mr. Sauret, declared to the Regional Department, "the town is incapable of feeding all of the disreputable and sick people who flock to the hot springs..." (Felgères, 1890). The last spa, *le Barlier*, closed in 1792.

In 1787, the Municipal Government of Chaudes-Aigues sent a report to the Provincial Assembly in an attempt to obtain authorization and funding to establish a hydropathic center and a charity workshop, which would involve restoring and modernizing the pipe network. However, the pipe network had been repaired and further developed a few years before. The 1775-1778 municipal accounts of the mayor, Jean Devèse, mention repairs on the main pipe from the spring *le Par* and on the Gravier pipe.

In the 1787 report, the Municipal Government proposed the construction of woolen and paper mills to help revitalize the town. In his 1890 treatise, Felgères explained the strategies of such an undertaking: "There might be nowhere else in the realm a town more well suited to such an establishment. The waters are hot enough for all these operations. Several sources of hot water can be seen along the river. It would be easy to increase their number. A little help from the Provincial Assembly would eliminate begging and vagrancy, which have increased in Chaudes-Aigues because of idleness...."

The distribution of hot water to private homes had also been extended to almost all of the town in these years: "A certain volume of water is gathered at *le Par* spring, flows under the roadbeds through wooden pipes, and then through private pipes, and is distributed to the ground floor of each house.... The houses on the other side of the river have their own pipes which cross the river on a bridge..." (Legrand d'Aussy, 1787).

The technique used for domestic heating was very similar, or even identical, to that used now, as described by Legrand d'Aussy: "At the entrance of most houses there is a stonework pipe with a sluice gate, and in the middle [of each of these houses] there is a small pool covered by a removable stone.... The water flows in through the pipe, circulates in the pool, thus heating the tile, and then flows back out to the river. By opening or closing the small sluice gate, the regulation of the hot water flow rate, and thus of the desired temperature, is obtained. When room temperature reaches the desired level, and one no longer wishes to increase it, the gate is closed and the water does not enter the house but flows directly into the river...."

Hot water also made the workers' lot more bearable. Bonnet de la Bragesse wrote in 1782 of the advantages workers experienced by laboring in geothermally heated locations: "The hard-working craftsman should not fear that his hands will be numbed by cold, dropping the shuttle that is his livelihood. It is also in these heated rooms that women and children gather, sitting on the tile, to knit with the rapidity acquired by habit of a skill learned at a young age...."

After 1750, the woolen industry gained new life, and the wool was treated by pressing and pestling by foot or pestles the raw textiles in a washtub of hot water. In 1765, Ballainvilliers wrote of the desirability of Chaudes-Aigues products: "There is a considerable trade in caddis, a light woolen twill manufactured 18 inches and 4 lines wide, which corresponds to five twelfths in Paris. This material, which is well suited for linings, is sold by the length, 32 ells long in Parisian measure. The wool is woven by women and children, both in the town and in the surrounding villages, and then is finished by the weavers. The merchants in Chaudes-Aigues and Rhodez buy the material in its natural state in the fairs and markets of the town and, after having it dyed and dressed, send it to the neighboring provinces, especially to Lyons, where most of the material is used. Some merchants simply buy the treated material and send it off. A large quantity of bulky-knit woolen stockings are also made in the town and in the surrounding villages. They are known for their good quality. These stockings are mostly worn by common people and the *petit bourgeois*. Some of the higher classes, however, also wear these stockings to keep their legs warm...."

The literature of this period also details many accounts of the culinary use of hot water, such as the following:

- "These thermal waters are used to treat the caddis, and to blanch the heads and feet of calves and sheep.... This water is said to preserve the green color of vegetables when they are cooked, and is also used to make very good bread. It is said that the inhabitants of Chaudes-Aigues have taken every possible advantage from the waters..." (Bosc d'Antic, 1771).

- "Women bearing jugs come continually to fetch water and occasionally there is even a crowd. At meal times, the women of the common people come to prepare their food along the canal. In a pot, they put slices of bread and a little butter and salt. They fill up the pot with spring water, put it in the canal as they would for a *bain-marie* [double boiler], and in less than half an hour the soup is ready. People come to cook eggs, remove the hair from pigs, pluck chickens, etc. The hot water is also used to make bread, and for ordinary uses..." (Legrand d'Aussy, 1794).

The years of physical therapy at Chaudes-Aigues culminated eventually in a physical therapy center. In this regard Ballainvilliers (1765) wrote: "This town is renowned for its hot waters. Their temperature is too high for bathing; therefore, paralytics and cripples can use these waters only after they have been left to cool in pools...."

1789 - 1900

WITH THE FRENCH REVOLUTION IN 1789, THE FRENCH REPUBLIC EMERGED AND ANY PREVIOUS privileges were abolished. At Chaudes-Aigues, a local administration replaced the lord and declared that hot water should be used exclusively for purposes of public interest. Consequently, the distribution of hot water was reorganized to enable domestic heating in the town.

The users were given the responsibility of withdrawing water from the hot springs and maintaining the pipes. However, Chevallier noted in 1868 that users were too haphazard: "People drew water in a disorderly fashion and without moderation from *le Par* spring, and carried it to their homes. They then passed it on to several neighbors, asking them to contribute for the expenses. This led to a confusion of rights, quarrels and sometimes also ill-founded shortages of

Pig hide depilation was undertaken with geothermal waters at the beginning of the 20th century.
Chaudes-Aigues Council

hot water...." Forty years earlier, Chevallier had written that from this disordered distribution of hot water, the streets "had become like cesspools, so that they were difficult to cross. From the puddles, putrid vapors emanated which were harmful to the health..." (Chevallier, 1828).

Furthermore, the pipes were not well maintained, as Legrand d'Aussy (1794) explained: "I will not mention the annoyance caused by all of this water that, during the course of the year, makes the town dirty and muddy. A much more serious inconvenience, however, results from the continuous release of steam from pipes, which necessarily makes the air unhealthy...."

In an attempt to clean up what threatened to become a public health menace, the mayor of Chaudes-Aigues determined in 1817 that "the chief district engineer had to draw up a plan establishing the elevation of the spring that should be used. He had then to build a large reservoir at the highest point aimed at collecting all of the water of *le Par* spring, and after this reservoir had been built he had to plan the distribution of hot water to each neighborhood, using pine-wood pipes that should run down both sides of the streets, so that the pipes no longer had to cross under the roadbeds.... In this way, some seven eighths of the approximately 350 houses of the town could be supplied with hot water, and there was no longer any degradation, nor were there cesspools or any other cause of insalubrity that could be attributed to the distribution of the water.... The administration then appointed a fountain-keeper to be in charge of the day-to-day maintenance of the reservoir and of the hot water network. He only took orders from the Municipal Government, and in this way the most perfect operation of the water-distribution service was attained..." (Chevallier, 1868).

Nevertheless, these installations did cause some problems, as reported in the November 9, 1851, minutes of the Municipal Council: "The mayor informed the Council that there had been innumerable complaints and disputes, the object of which were the poorly defined obligations of the fountain-keeper. He [the mayor] then expressed the urgent need to draw up clear-cut regulations aimed at establishing the obligations of the fountain-keeper in maintaining the reservoir, the pipe network, the collectors, the fountains and the sinks, as well as channeling cold and hot water...." Consequently, the Council issued a set of regulations; Article 7 stipulated that "the fountain-keeper is no longer the only person authorized to supply wooden pipes for hot water. Private individuals, too, if informed with fair advance, can supply the wooden pipes...."

However, the gravity-driven distribution network of the hot water and the district heating system was not changed. The hot water continued to flow "under the floors of the houses, where there is a series of small pools that slow down the speed of the flow, and the houses are heated by the heat drawn from the pools. These pools are covered by large slabs, carefully cemented to prevent the humidity of the steam from entering the rooms. By means of faucets or plugs, inhabitants can, at will, increase or decrease the heat depending on the volume of water; they can even shut down completely the water inlet to their homes. The apartments heated in this way are very sought after during the winter. The room temperature may even be rather high, from 22° to 26° C, depending on how well the dwelling is sealed..." (Bouillet, 1834).

Regarding the industrial use of hot water, Berthier (1810) wrote: "The town of Chaudesaygues [*sic*] has long been devoted to the manufacture of textiles and woolen stockings, which are very renowned. Wool is washed in the spring water before spinning and, instead of treating textiles with clay, as is done elsewhere, they use this water. Without knowing its composition, people guessed its alkaline nature, and used it to its best advantage...."

In the early part of the 19th century, numerous scientific and medical studies were conducted on the therapeutic uses of hot water. Following the initial observations of doctors Grassal, Verdier, Bremont, and Mathieux, two medical theses with similar titles (*Dissertation on the Thermal*

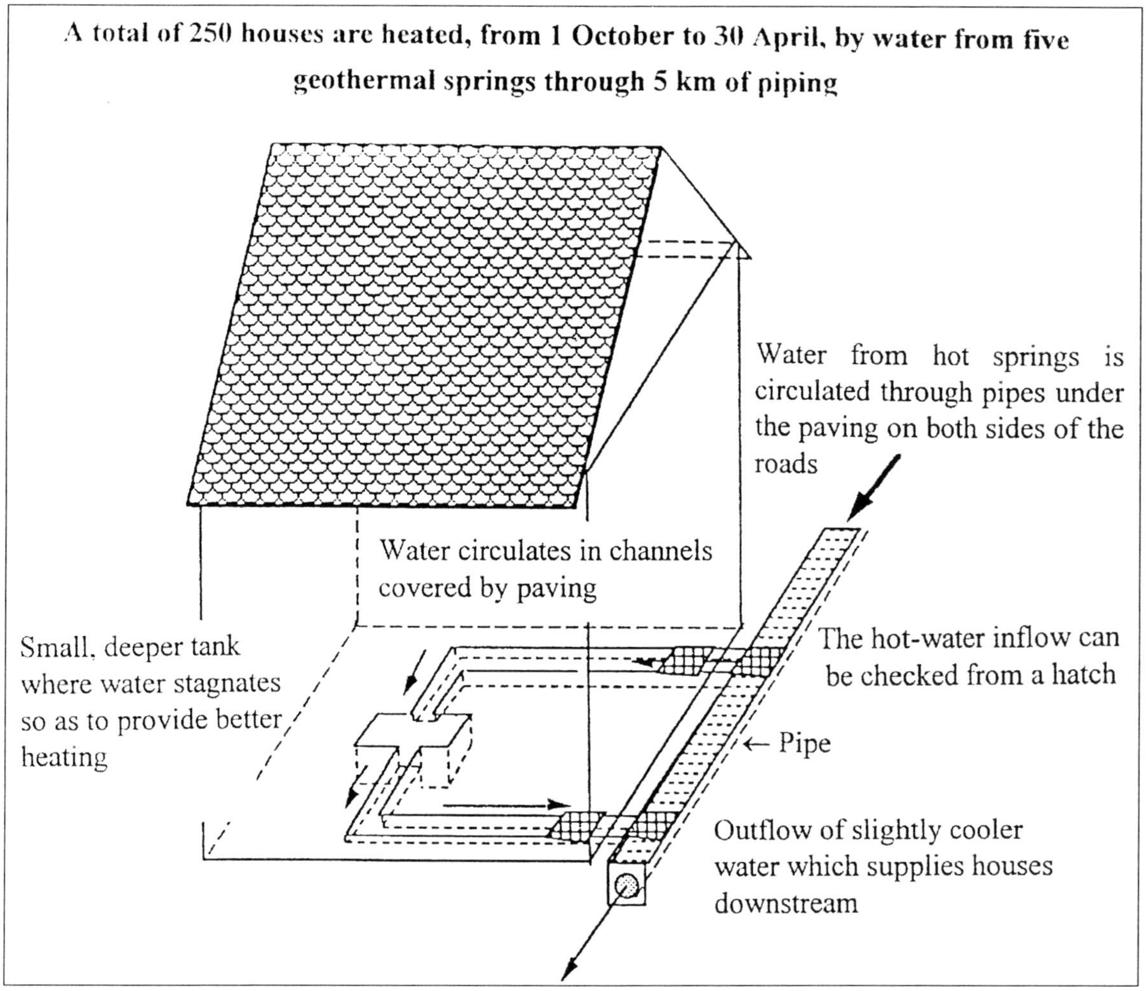

Sketch of the domestic heating system at Chaudes-Aigues in the 19th century.

Waters of Chaudes-Aigues and *Dissertation on the Mineral Waters of Chaudes-Aigues*) were presented on the same day, July 17, 1833, by Mr. Bonniol and Mr. Podevigne.

The water cures of Chaudes-Aigues gained new strength and became increasingly popular in the 19th century. After Felgères' hydropathic establishment, those of Clavières, Voissier, Verdier, and Fabre opened one after the other. The number of people taking the waters for a cure increased from only a few dozen in 1820 to several hundreds in the 1830s, and reached a thousand during the 1850s.

The use of the waters for cooking continued: "The poor come to prepare their meals at the spring itself. Indeed, there are no detrimental effects on health; the waters seem to be healthy for there are very few sick people in the area, the blood is good, and the young girls have a freshness and a sparkle which is rarely seen elsewhere..." (Berthier, 1810).

Recent Developments (1900-1980)

From the beginning of the present century, water distribution techniques stabilized and public administrative and management conditions improved. On November 21, 1920, the Municipal Council established an annual usage fee: "The heated houses are to be divided into three categories depending on their temperature.... An [annual] fee per square meter of heated surface for each of these categories is established: 3 francs for category 1, 2.5 francs for category 2, and 2 francs for category 3. Each house will pay the fee according to the category in which it will be classified, and to the number of square meters of surface heated.... The surfaces heated will be measured in three ways: 1) for houses with M-shaped pipes, or for pipes which crisscross the house several times, the heated surface will be determined by multiplying the length of the heating pipes by the width of the heated areas; 2) for houses with A- or V-shaped pipes, the same calculation will be used, but the surface is to be reduced by one third; 3) for those houses which are crossed only once by the pipe, the calculation will be done by measuring the length of the pipe and multiplying it by 1 meter.... Furthermore, there will be a minimum charge of 10 francs per house; this means that those houses that, according to the category in which they have been placed and their heated surface, would normally pay less than 10 francs, shall nevertheless have a 10 francs fee to pay...." Although the payment system was complex, methods were introduced to help calculate the amount due by each house for the use of hot water.

The usage fee, however, did not resolve legal issues. Since the distribution system, while occasionally improved, had not substantially changed through the centuries, many disputes and lawsuits among houses and neighborhoods occurred, as the following examples illustrate:

- "The construction and maintenance of pipes were the responsibility of the users, but the lord had to supply the necessary wood. To put an end to the complaints and abuses for which this supply was a pretext only, this rule was abolished in the *terrier* of 1615 and was replaced by a reduction of rents... (Felgères, 1890).

- "In 1787, the inhabitants of the neighborhood of Saint-Joseph (below the church) sued those of a neighboring street, accusing them of retaining the hot water, mixing cold water with the water they rendered, obstructing pipes, etc. Verification was done, [but since] the pipes were intact, there had been no manipulation of the water system. The plaintiffs were then ordered to pay the law-suit costs..." (Felgères, 1890).

- "I heard the inhabitants of Chaudes-Aigues complain about the poor maintenance of pipes and distribution of water to the houses..." (Legrand d'Aussy, 1794).

The minutes of the Municipal Council of August 13, 1854, Article 5, specified, "It is expressly forbidden to have even the smallest opening for drawing water inside of the heated houses. The same quantity entering [each house] must be returned with no loss of heat other than that occurring during its passage, and this loss must not exceed 2° C...." Following development work, several neighborhoods still complained of poor distribution. The Municipal Council of February 12, 1855, therefore stated that the Council itself had "attempted to act as a good father, dividing the piece up among his children...." One can hardly be more paternalistic.

Nevertheless, the temptation of withdrawing hot water from the public pipes into private homes did not disappear. Nearly a century later, on November 21, 1920, the Council determined that "inlet gates enabling this withdrawal must be closed before February 5 of the following year...." To enable drawing of hot water into private houses or buildings for uses other than heating, the Municipal Council installed two street fountains to be used as public collection points of hot water. These old fountains still exist today, alongside five newer fountains of the same type.

Until the 1950s, the hot water flowed through pinewood pipes, hollowed out with a long auger and fitted together; the flow rate was regulated in lead basins. The disadvantage of using these pipes for collection and distribution was the rapid clogging by carbonate scaling. Streets were continually dug up to unclog and replace pipes. For this reason, between 1955 and 1960, the Municipal Council of Chaudes-Aigues decided to adopt a new collection system and improved distribution techniques.

The first technique, installing iron pipes, did not yield good results. Due to the severe climatic conditions in the area and the temperature of the hot water, the pipes rusted very rapidly from the outside. Not until PVC pipes were able to resist temperatures of about 70° C was a solution found. The whole hot water distribution network for domestic heating was modernized some 20 years ago using PVC pipes, and hand-excavated holes now allow maintenance personnel to easily reach the shallowly buried pipes and either unplug or replace them, without digging up the streets.

The first central heating system, fed by a heat exchanger, was installed in a primary school in 1957. This was followed in 1959 by a heater installed in a private home. Because this technique allows heat to reach upstairs floors with no scaling problems, it has been used in 10 buildings since 1959. Currently, 150 homes use geothermal heating, but many individual houses are still heated by the old system. Now, as in past centuries, "the only houses without steam rooms are those located above the spring area.... Most of the houses nearest to the spring do not draw their water from the pipes, but have their own private springs for heating..." (Legrand d'Aussy, 1794).

Public geothermal fountain.

Finally, as regards ambient conditioning by natural heat, worthy of mention is the case of a homeowner who wanted to have both a cellar and a floor heating system. He built a copper basin under the ground floor of his home and covered it with a tile floor. The thermal water circulated in the copper basin, heating both the apartment rooms and, in a moderate way, the cellar.

Other notable developments follow:

- In 1902, the first hydropathic establishment, which had remained closed for many decades, was reopened by Mr. Ginisty, an inhabitant of Chaudes-Aigues.

- In 1928, a public wash house fed by a thermal spring was built.

- In 1934 a new therapeutic establishment, the *Thermes du Par*, was built and

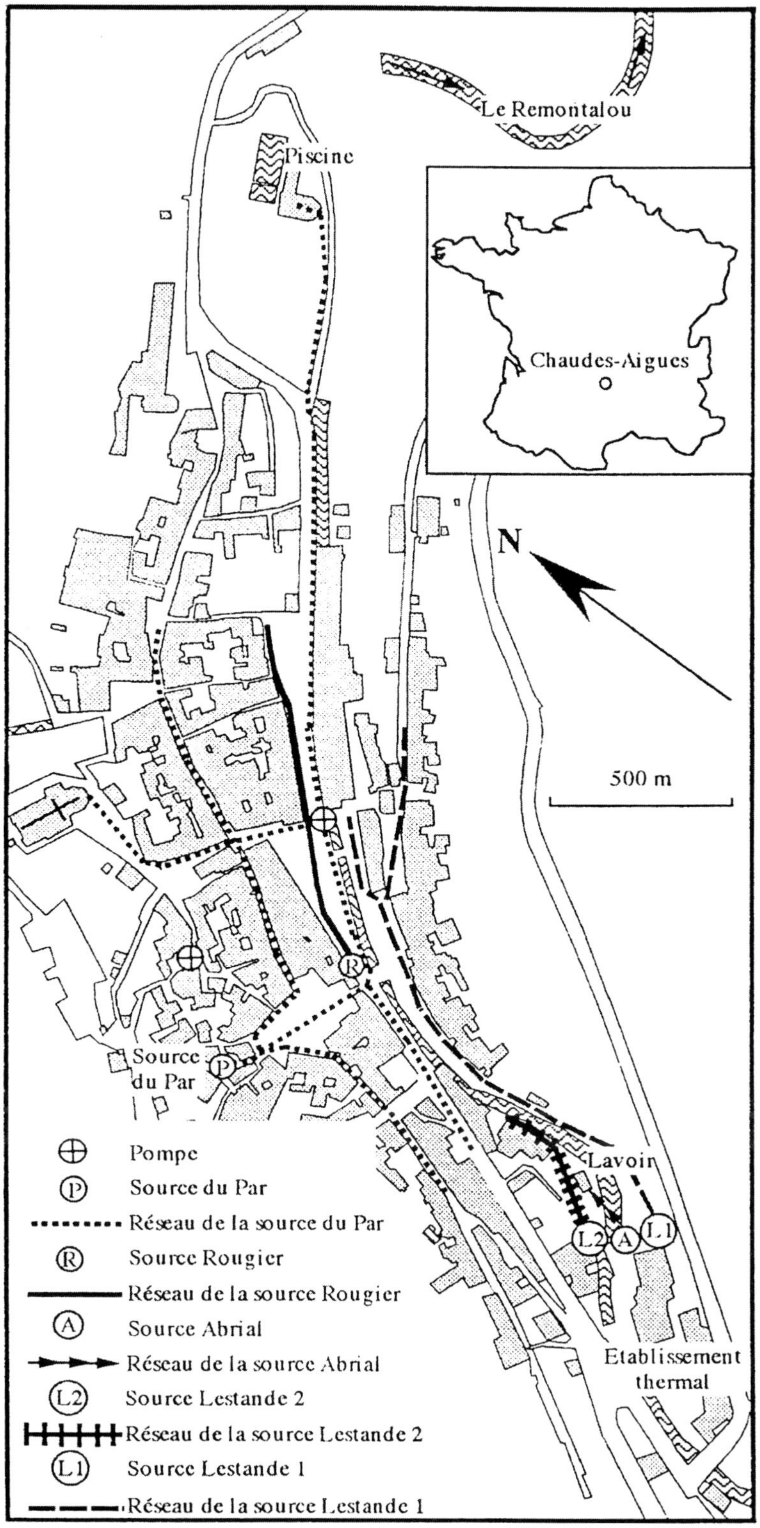

Diagram of the recent heat distribution network at Chaudes-Aigues.

immediately found a very receptive market.

- In 1964, a second therapeutic establishment replaced the first one.
- In 1964, a heat exchanger was installed and is still being used to heat a swimming pool to a temperature of 24° C.
- In 1972, a well drilled by the BRGM found 74° C water with a flow rate of about 100 liters per minute (~6 m^3/h). The discovery came just in time to meet the fast growing need for hot water, for the popularity of Chaudes-Aigues spas had spread throughout Western Europe.
- The first heat pumps appeared in 1976.
- In 1978, a 20 kilowatt micropowered electrical station, using Freon as a secondary fluid, was tested successfully.

Public wash house supplied by thermal waters.

Conclusion

Several new uses of Chaudes-Aigues thermal waters have been developed in the past two centuries. Among the most successful projects were a greenhouse for vegetables built in the 18th century and a dry-steam room in the Felgères' establishment used to incubate eggs. The inventor of this incubation system, Mr. Darcet, was awarded a medal by the *Société d'Encouragement.*

A long history on the use of natural heat can be traced back to this little town in Auvergne. Its inhabitants gradually have learned how to distribute geothermal waters and many ways to use their physical and chemical characteristics:

- Human comfort: home heating, cooking, and laundry
- Health care: thermal balneotherapy for various maladies and hydropathic cures
- Industry and crafts: scouring wool

For all these reasons, Chaudes-Aigues was considered to be the ideal site for the Museum of Geothermics and Water Cures, inaugurated in the spring of 1993.

Cited and Selected References

Azemard, A. 1863. *Proceedings of the Congrès Archéologique de France,* 30th session; Rodez, 2 June 1963. Rodez Imprimerie de Carrère Ainé.

Ballainvilliers. 1765. Etat de l'Auvergne en 1765 présenté à M. de l'Averdy. In *Tablettes historiques de l'Auvergne.* Bouillet, J. B. Clermont-Ferrand.

Banc, J. 1605. *La mémoire renouvelée des merveilles des eaux naturelles en faveur des nymphes françaises.* Pierre Sevestre, Paris.

Berthier, J. 1810. *Journal des mines.*

Bonnet de la Brageresse. 1782. Mémoire manuscrit sur les eaux de Chaudes-Aigues. *Mémoire de l'Académie de Médecine.*

Bosc d'Antic. 1771. Examen des eaux thermales de Chaudes-Aigues. In *Oeuvre de M. Bosc d'Antic.*

Bouillet, J. B. 1834. *Description historique et scientifique de la Haute Auvergne (Département du Cantal).* Baillère, J.B., Paris.

Chevallier, A. 1828. *Essai sur les eaux de Chaudes-Aigues, Département du Cantal, et analyse chimique des eaux thermales de cette ville.* Rapport sur un mémoire, fait au Ministère de l'Intérieur, par l'Académie Royale de Médecine. Imprimerie Royale.

Chevallier, A. 1868. *Notice sur Chaudes-Aigues, ses eaux minérales et leurs applications.* Mémoire manuscrit.

Clos (du). 1675. *Observations sur les eaux minérales de plusieurs provinces de France.* Académie Royale des Sciences.

Felgères, C. 1890. *Histoire de la baronnie de Chaudes-Aigues depuis ses origines (XIe siècle) jusqu'en 1789.* Aurillac.

Lecoq, H. 1836. *Chaudes-Aigues et ses eaux thermales.*

Legrand d'Aussy. 1794. *Voyage fait en 1787 et 1788, dans la ci-devant Haute et Basse Auvergne.* Letter 31.

Philander. 1545. Commentaire sur vitruve. In *Vitrusium.* Book 5, Chapt. 10.

The Authors:

Jean Pierre Gibert
2 Voie Du Wissous
21300 Massy France
Telephone: 33.1.69.82.72.46
Fax: 33.1.69.81.72.46

Florence Jaudin
BRGM
BP 6009
45060 Orléans Cedex 2
France
Telephone: 33.2.38.64.31.61
Fax: 33.2.38.64.33.33
E-mail: f.jaudin@brgm.fr

"In the presidential residence church at Bessastadir on the Alftanes Peninsula near Reykjavik there is a stained glass window, showing the incident during the voyage of St. Brendan when he and his followers witnessed a volcanic island emerging out of the ocean, most likely during a submarine eruption on the Reykjanes ridge. 'Believers in God, stand firm in faith unfeigned. We are now at hell's gates,' St. Brendan said, making the sign of the cross and praying, 'Lord Jesus Christ deliver us from this island.' The original painting on which this stained glass window is patterned is by an Icelandic painter, Finnur Jonsson (b. 1892). The boat shown here is not a leather boat *(curragh)*, but a wooden hull boat, riveted together with nails. Some people believe that the Irish were also able to build that kind of wooden boat just as the Vikings did."

These tales were written down about three centuries after St. Brendan died, and as a saint he was associated with several marvelous incidents and miracles, strengthening people's faith in his powers. The account of the voyages, probably handed down by oral tradition and written down in the sagas, was in fact a composite of historical fragments from voyages by Irish monks over a period of two centuries or more. The stories are recorded in *Navigatio Sancti Brendani*, composed in Latin by an unknown Irish writer, most likely in the 9th century. To all appearances, the book contains a report on a single voyage during the years 565 to 573. *Photo and text from* Iceland - A Portrait of Its Land and People, *by Hjálmar R. Bárðarson, 1989. Printed with permission*

19. Historical Aspects of Geothermal Utilization in Iceland

by
Ingvar Birgir Fridleifsson

Abstract: Volcanoes and hot springs were probably the only natural phenomena completely new to the first settlers when they came to Iceland in the 9th century A.D., as these were unknown in Northern Europe. The distribution of hot springs did not affect the location of farmhouses during the Age of Settlement, and the hot springs were gradually used for washing and bathing. The oldest known geothermal pipeline, a covered duct made of stone slabs, dates back to the 13th century. Space heating with geothermal water, however, did not begin until 1908. For several centuries, sulfur mined in geothermal fields was one of the main export products of Iceland. Hot springs were used at one locality for extracting salt from seawater between 1773-1793.

Reprinted by permission of the International Geothermal Association, from the *Proceedings of the World Geothermal Congress,* 1995, pp. 427-431.

Location of Farms and Hot Springs

Warm springs and steaming ground probably were among the natural phenomena that were most surprising to the first settlers when they came to Iceland in the 9th century. It is unlikely that any of the settlers had seen hot springs or active volcanoes before. The novelty of the hot springs is reflected in numerous place names, including *reyk* ("smoke"), *laug* ("hot spring"), and *hver* ("boiling spring"), becoming Reykir, Reykjavik, Reykjadalur, Laugaland, Laugarvatn, Laugardalur, Hveravellir, and Hveradalir. It is likely that the settlers were suspicious of the hot springs and steam to begin with. But many of them were Vikings who claimed to fear nothing, and one can assume that they soon started using the hot springs for bathing and washing.

Iceland has more than 250 separate thermal areas with over 600 main hot springs. The majority of these are in lowland areas in the southern, western, and northern parts of the island.

The distribution of the hot springs does not appear to have had any significant influence on the location of farmhouses during the Age of Settlements (874-930). The *Landnáma* (*The Book of Settlements*), which was written in the 12th century and covers the entire island, includes stories of some 430 principal settlers, about 100 of whom obtained land within the lands of earlier settlers (commonly kinsmen of the first settlers). Stories are told of the land claims of each settler, his farmstead, his ancestors in the old country, and his descendants in the new. Researchers have tried to locate the exact positions of individual farmhouses from the text of *Landnáma*. In many cases, there is little doubt about these locations. A comparison of a map of the settlement farms in the edition of *Hið íslenska fornritafélag* (*The Book of Settlements, 1968)* and maps with the locations of the hot springs does not at all indicate that the first settlers wanted to build their farmhouses near the hot springs. On the contrary, Ingólfur Arnarson (considered to be the first Norse settler in Iceland) built his farmhouses about 3.5 km from the hot springs in Reykjavík. Apparently, he preferred to have his farm close to a good beach for landing boats rather than at hot springs suitable for washing clothes and bathing. Why not? The women would take care of the washing anyway!

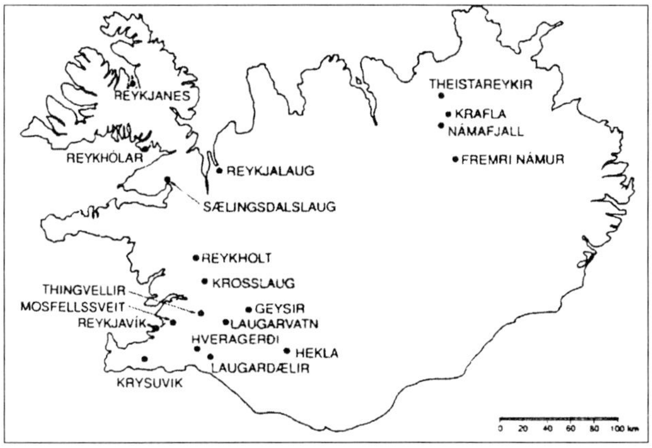

Map of Iceland with the main geothermal localities mentioned in the text.

Use of Hot Springs through the Ages

It is not known with any certainty when the hot springs were generally used for washing and bathing. Reference is frequently made in the sagas, written in the 12th-13th centuries, to bathing in hot springs. The sagas have several romantic descriptions of people bathing in the hot springs, in particular in *Laxdæla Saga* (*Íslendingasögur*, *Sagas of the Icelanders*, 1981), where the beautiful but strong minded Guðrún Ósvífursdóttir meets her hero Kjartan Ólafsson in the *Sælingsdalslaug* ("hot pool") in the twilight. In addition to washing and bathing, the hot springs were also used widely for bending wood and bones and for softening materials. Apart from the elaborate bathing pool of the saga writer Snorri Sturluson, which will be described later, the bathing pools seem to have been rather primitive. Commonly, people would take a bath in brooks where hot water from the often-boiling springs would be mixed with cold

Women washing clothes in the hot springs at Laugardalur, Iceland, around the beginning of the 20th century. *Courtesy of Ljosmyndasafn Reykjavikur/Hitaveita Reykjavikur*

water. Eggert Ólafsson and Bjarni Pálsson, in their travel book from 1752-1757, described the bathing pools in Laugarnes (now the site of the main swimming pool of Reykjavík), where the people of Reykjavík brought clothes for washing and sailors and workers came in large groups on Saturday and Sunday nights to bathe (Ólafsson and Pálsson, 1943).

Boiling food (meat, fish, eggs, and milk) and baking bread were also common in the hot springs. Ólafsson and Pálsson also mentioned that people used the hot clay from mud pools in Ölfus (probably in the geothermal field of the present-day Hveragerði hot spring village) to make wooden barrels for airtight food storage. Furthermore, they wrote about dry-baths or steam baths (saunas) in Thingeyjarsýsla in Northern Iceland and in Árnessýsla in Southern Iceland. Thorvaldur Thoroddsen, who was traveling in Iceland a century later (1882-1892), mentioned a dry-bath at Sturlureykir near Snorri's Reykholt in Western Iceland. He also mentioned that people diverted hot water into fields where they grew potatoes and cabbage, significantly increasing the vegetable yield. Here we have the first documented examples of soil heating in Iceland.

Sulfur mining in geothermal areas was of significant interest from the 13th to the late 18th century, and geothermal springs were used for salt extraction from seawater during 1776-1792. These activities are described later.

Hot Springs of National Interest

The first time that geothermal energy was of national significance in the history of Iceland was associated with the adoption of Christianity as the formal religion of Iceland around the year 1000. After a decree by an arbitrator, the Parliament decided that the people should give up the pagan gods of the Vikings and become Christians. This was accepted with some grunting. But when it came to the ceremony of baptism, which involved every individual being submerged in water, real trouble arose. The people at the Parliament absolutely refused to be baptized in cold water. According to *Kristni Saga* (*Íslendingasögur*, *Sagas of the Icelanders*, 1981),

The sketch at top shows the old church at Reykjahlið near Lake Myvatn in Northern Iceland. A lava flow in 1729 engulfed several farms but encircled the church, which stood on a small hill—"divine intervention?" *Photo and text by Bárðarson, 1989. Reprinted with permission*

the Saga of Christianity originally written in the 12th century, the people from Northern and Southern Iceland were baptized in a hot spring named *Reykjalaug* (later called *Vígðalaug*, "the consecrated spring") at Laugarvatn, some 20 km east of the Parliament site at Thingvellir. The people from the western parts were baptized in *Reykjalaug* in Lundarreykjadalur (later called *Krosslaug*, "the spring of the cross") some 30 km north of Thingvellir. Both springs were considered to have healing powers from then on.

Snorri's Bath and Conduits for Water and Steam

THE ONLY MAN-MADE BATH STILL REMAINING IN ICELAND FROM PAST CENTURIES IS *SNORRALAUG* ("Snorri's bath"), the bath of the famous historian and saga writer Snorri Sturluson (1178-1241) in Reykholt, Western Iceland. It is, in fact, one of the most remarkable archaeological monuments in Iceland. The pool is circular, slightly under 4 m in diameter, about 0.9 m deep, and made of hewn blocks of stone (silica sinter, which can be found about 2 km from Reykholt) that fit each other in the most exact manner. The floor is paved with the same stone, and a stone bench, capable of seating some 30 people, surrounds the inside of the bath. It is filled by hot water carried through an underground stone conduit from a boiling spring called Skrifla about 120 m away. In the olden days, it is said that the temperature of the water could be altered by the addition of cold water from a brook that now is dried up.

Snorralaug, in Reykholt. The hot water conduit is on the right. The entrance to a tunnel between the bath and the farmhouse is in the background. *Drawing by Helgi Torfason*

Archaeological excavations have revealed three conduits from the vicinity of Skrifla (Grímsson and Ólafsson, 1987). Two of these lead to Snorralaug and are considered to have been used for transporting water to the bath. The third goes from the morass of Skrifla up a slope to the farm site. This pipe is considered to have been used to transport steam from Skrifla to a place near the farm. It is not known whether the steam was used for a steam bath or for cooking and baking, or possibly for warming up a house. The three conduits are built in such a way that ditches were dug through the soil and humus layer (1-2 m deep), a narrow channel was made into the underlying clay at the bottom, and slabs of rock were used to cover the channel. In some cases, small slabs of rock were lined along the sides of the channel to support the larger slabs of rock used to cover the channel. The ditches were then filled with soil.

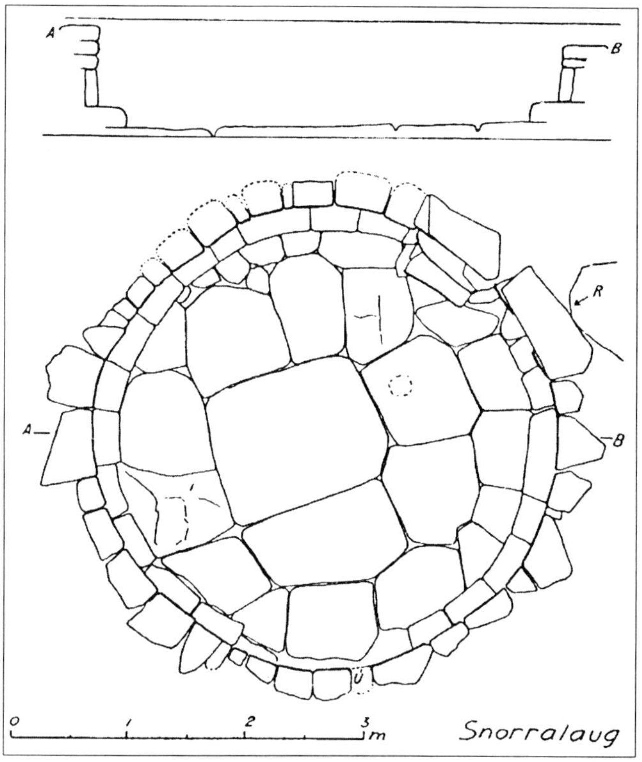

Plan and cross section of *Snorralaug* (Snorri's bath) before the restoration in 1959. R: Inflow of water from a 120 m underground conduit. Ú: Outlet for water. *From Grímsson, 1960. Reprinted with permission*

According to *Landnámabók* (*The Book of Settlements*) describing the situation late in the 10th century, there was no farm at Reykholt but only sheep sheds belonging to the settlement farm Breiðabólsstaður and a bath. As Tungu-Oddur took a bath there, this must have been before 960. Nothing is said about the position or nature of the bath, but this original bath in Reykholt must have been fed with hot water from either of the two boiling springs, Skrifla or Dynkur (Grímsson, 1960). Both springs are situated in a morass and exposed to winds on all sides. Both are far too hot for direct bathing. It is not known exactly when a ditch was first dug from Skrifla to bring the thermal water to the dry, solid ground sheltered from the northerly winds at the foot of the hill, which later became the Reykholt farm site. One can imagine that this happened in stages. The distance from the hot spring to Snorralaug is about 120 m.

In the latter part of the 12th century, Reykholt had become a rich church farm. Snorri Sturluson, a scholar and a powerful chieftain, moved there in 1206. According to *Sturlunga Saga*, he had a defense wall built around his dwellings. Reference is made to two gates on this wall. One was on the northern side toward the church; the other was the doorway of a corridor or a tunnel leading to the bath. This bath is mentioned several times in *Sturlunga Saga* (*Íslendingasögur*, *Sagas of the Icelanders*, 1981), including a narration of an event in 1228 picturing Snorri himself in the bath one evening, chatting with friends. But no information is given on its age, size, or structure. The bath in Reykholt is mentioned in several books from the 18th century, including the travel book of Eggert Ólafsson and Bjarni Pálsson (Ólafsson and Pálsson, 1943) who visited Reykholt in 1757. Snorralaug was rebuilt in 1858 (Thorsteinsson, 1971) and again in 1959 (Grímsson, 1960). The walls of the underground passage nearest to Snorri's bath were also restored.

Although there is no absolute proof, it is considered very likely that the pool and the pipelines were built by Snorri Sturluson himself. At least the pool existed during Snorri's time. Snorri was not only the most remarkable writer and historian known in Iceland's history; he was also an ambitious builder, as reflected by the defense wall around his farmhouse described in *Sturlunga Saga*. The model for the construction of the wall around the farm could easily have been sought in Norway, where Snorri was a frequent visitor. But where did he get the idea of piping the geothermal water? During the 11th and 12th centuries, pilgrims went frequently from Iceland to Rome. It is even known that a young couple, who recently had inherited Reykholt, died on a pilgrimage to Rome in the late 12th century. It is tempting to speculate that some scholar visiting Snorri Sturluson on return from a pilgrimage told him how people in Italy constructed subterranean conduits to transport hot and cold water. Perhaps this is the first transfer of geothermal technology from Italy to Iceland. The sad fact is that this technology of piping hot water and steam seems not to have spread from Reykholt to other parts of Iceland. For 700 years, people at farms with boiling springs continued to suffer in cold, unheated houses.

Hot Springs as Barometers

Eggert Ólafsson and Bjarni Pálsson, who traveled around Iceland in the mid-18th century, mentioned that hot springs have been considered good indicators of changes in the weather (Ólafsson and Pálsson, 1943). Reykjalaug in Miðfjörður in Northern Iceland can be taken as an example. When the temperature in the spring rose, people said that rain would be coming soon, even if the sky was blue and cloudless. They also reported that people in the vicinity of the Great Geyser in Haukadalur in Southern Iceland could foresee rain and storm from the height of the water column in the eruptions of the geyser. When the water column went as high as Laugafell in the background, then it was certain that a rainstorm was coming. With modern meteorology, people realize that the height is controlled by the air pressure. When a low depression goes over the country, the pressure on the surface of hot springs is reduced, and thus the hot aquifer pressure becomes relatively high. Additionally, depressions sweeping in from the ocean normally bring rain and storms. Our forefathers did not know this. Therefore it is understandable that people found hot springs mysterious and even magical.

The Nuisance of Hot Springs

The romantic aspect of the use of the hot springs reflected in the sagas appears to have dwindled with time, at least when it came to counting the perks of farms for the tax man. In *Jarðabók Árna Magnússonar and Páls Vídalíns* (*Description of the Farmsteads of Iceland*), written in 1703-1714, a description is given of every substantial farm in Iceland and the benefits and disadvantages associated with each farm listed (Magnússon and Vídalín, 1982). For all the farms with known geothermal activity that the present author looked up in the *Jarðabók*, the hot springs are considered a nuisance and a disadvantage. One must keep in mind, however, that the farmers describing the quality of their farmlands to the learned gentlemen at the beginning of the 18th century, probably had in mind that the descriptions might be used for assessing the taxation value of their farms.

It says of Laugardælir farm in Southern Iceland: "A part of the hay field is spoiled by a hot secretion which is caused by a nearby hot spring, and from this warm broth the grassroots are eaten up resulting in patches with no vegetation." This description was written on July 27, 1709. Presently this "hot secretion" area is the main production field of Hitaveita Selfoss (District Heating Service of Selfoss), with an installed capacity of 21 megawatts-thermal. The description of Hlaðgerðarkot in Mosfellssveit, which is now a part of the main production field of Hitaveita Reykjavíkur (Reykjavik District Heating Service), says: "The possibility of cutting turf (as building material) is limited. Peat resources for cooking fuel are both of bad quality and limited. The meadows are spoiled by scree. The land is small in size. Storms are fierce so that both houses and haystacks are in danger. The water is warm." There is a certain crescendo in this description, as the musicians would say. After describing all the bad qualities of the farmland, it is stated that even the water is not drinkable, as it is warm. At present, this area is to the Icelanders as oil fields are to the Arabs. In 1908, a farmhouse at Reykir near Hlaðgerðarkot was the first in Iceland to be heated with geothermal water.

Sulfur Mining

Although the Icelanders did not make much use of the geothermal energy, at least not when asked by the tax man, the King of Denmark certainly made a lot of money during the Middle Ages from sulfur mining in Iceland. Sulfur deposits are common in the high temperature fields of the country. The main sulfur mines were in Krýsuvík in Southwestern Iceland and in Námafjall, Krafla, Fremri-Námur, and Theistareykir in Northern Iceland. At an early stage, sulfur was exported from Iceland as an expensive commodity.

Iceland lost its independence to the King of Norway in 1262 and together with Norway to the Queen of Denmark in 1388. Apparently, the Archbishop of Niðarós (now Trondheim) in Norway acquired some kind of exclusive right to transport or buy sulfur from Iceland during the latter part of the 13th century. What the sulfur was used for is not known, as gunpowder was not invented until about 1400. It has been suggested that perhaps the Catholic Church was exporting sulfur from Iceland so that people in churches in Europe could get acquainted with the smells of Hell (Lýður Björnsson, 1994). Iceland's most famous volcano, Hekla, was well known to the Church in Europe during the Middle Ages as one of the gateways to Hell.

Later the exclusive right to export sulfur was taken by the Danish King. At the beginning of the 16th century, sulfur was exported from Iceland by merchants from Hamburg, Germany, and the King of Denmark had to buy sulfur for powder production at an extremely high price. In 1561, the king forbade the Icelanders to sell sulfur to foreigners or to anyone except merchants appointed by him. After this, the Danish government controlled the export of sulfur from Iceland through the 16th century (Sigurjónsson, 1967). The export of sulfur was very lucrative during this period, and it is reported that the king earned some 6000 *ríkisdalir* from a single shipload. This is equivalent to the price of 1,500-2,000 dairy cows, which cost some $2-3 million US in Iceland today. The sulfur export was at this time the main source of income to the Danish Crown in Iceland. At the beginning of the 17th century, the price of sulfur fell on the European market and the Crown stopped exporting it. Mining was much reduced during the 17th and 18th centuries, and the license for mining and export was sold to various individuals. The king finally gave the Icelanders the right to sulfur mining as a part of an Industrialization Plan (*Innréttingar*) in the 1760s.

Salt Production with Geothermal Heat

The first experiments in producing salt from seawater by vaporization with geothermal heat were conducted by Eggert Ólafsson and Bjarni Pálsson in 1753 at Reykhólar in Western Iceland (Ólafsson and Pálsson, 1943). In 1773, a team sent by the Danish Crown started an experiment in producing salt from seawater using boiling geothermal springs on the beach at Reykjanes in Northwestern Iceland. The experiment resulted in commercial production of salt that lasted until 1793 (Björnsson, 1978). In 1776, salt was produced using 32 pans, each about 126 x 126 x 32 cm, made mostly of lead. The salt was thought to be of reasonable quality and was used for salting fish and meat. The annual production was up to 298 *tunnur* (equivalent to nearly 90 tonnes). The total production during 1773-1793 was 3,711 *tunnur* (equivalent to 1,113 tonnes), or about 53 tonnes/year. Commercial production continued until 1793, when it was stopped due to economic reasons. The salt did not sell well, as it was contaminated with lead from the drying pans and the lead gave a dark color to the fish. Experiments with salt production in a geothermal field in Iceland started again in the 1960s.

Initial Geothermal Drilling

The first geothermal drilling occurred in Laugarnes in Reykjavík in 1755 and in Krýsuvík in Southwestern Iceland a year later. Eggert Ólafsson and Bjarni Pálsson brought to Iceland a hand-driven drilling tool owned by the Danish Science Society. They drilled some 4 m down to the bedrock in Laugarnes but to about 10 m depth in the Krýsuvík high-temperature field. In the last drill hole in Krýsuvík, they got only to a depth of some 3 m, but according to Eggert Ólafsson "at that stage the soil started moving and in spite of the narrow well around the drilling rod, a thin soup started squirting out of the well with great force. We were forced to stop and pulled the drill out. But at that stage the heat came out in full force and squirted a boiling mixture of water and clay 2-3 m into the air. In a short while this restlessness stopped and we expected that the heat had become quiet. But after a short while the heat force increased again and the well started erupting and boiling again. We saw then that we had created a new boiling spring with our operations." After this episode, geothermal drilling did not resume until in 1928 and at that time also in Laugarnes in Reykjavík.

Discussion

In a country such as Iceland, with no coal deposits and very limited deposits of firewood and peat, the hot springs held significant potential value for washing wool and clothing, bathing, and boiling food at individual farms. But it is sad to think of all the thousands of people shivering in their houses for 10 centuries on farms with ample free-flowing water most suitable to heat the houses. The mean annual temperature in Iceland is about 4° C and the average temperature of July (the warmest month) about 11° C. Some heating is needed in the houses almost around the year. At many farms, it would have been very easy to heat the houses by constructing under the farmhouses a simple pipeline, such as the one at Reykholt, thus providing floor heating.

By all indications, the technique of piping the hot water came to Reykholt in the early 13th century. But it was not until in 1908 that a farmer at Reykir in Mosfellssveit, Stefán B. Jónsson (1861-1928), who had spent 12 years in the USA, constructed a 2.3 km pipeline (1 inch steel pipe) from a hot spring to heat radiators in his house (Thórðarson, 1993). The spring was at a higher elevation than the house, and thus no pumping was required. The achievement of the farmer and self-educated carpenter and blacksmith Erlendur Gunnarsson (1853-1919) was even greater. He lived at Sturlureykir in Western Iceland, 4 km west of Reykholt, and had a boiling spring by his farm at an elevation of 6 m below the farmhouse. After much experimenting, he invented a simple mechanism to separate the steam from the water. He piped the steam to his house in 1911 and used it both for cooking and for heating the house. Initially, the radiators were heated by steam, but later the steam was used to heat water for the radiators in a closed system (Finnbogason, 1943). It is an interesting coincidence that the world's longest geothermal pipeline, 63 km and constructed in 1980, transports boiling water from the Deildartunga hot spring some 2 km west of Sturlureykir.

The historical aspects of geothermal development during the 20th century are outside the scope of this chapter. Hopefully, the history of geothermal resources development after 1900 will be dealt with properly in the near future, as many of the pioneers of this development in the various countries are still alive. A few items of historical interest as regards Iceland will be mentioned here.

The first municipal district heating service to use geothermal water, Hitaveita Reykjavíkur, was established in 1930. At present, about 86 percent of the Icelandic population lives in geothermally heated houses, and geothermal energy provides 46 percent of the total primary energy supply for the nation. Most school centers and villages established after 1930 are located near hot springs.

The first greenhouse was heated with geothermal water in 1924 at Reykir in Mosfellssveit. Most of the pioneering work for multipurpose use of geothermal energy, however, started in the Hveragerði hot spring village (Thorhallsson, 1988):

- 1930: the first dairy to pasteurize milk and produce cheese
- 1938: hay drying experiments
- 1939-1941: a plant to dry seaweed
- 1944: experimental electrical production
- 1946: a freezing plant for vegetables (using absorption refrigeration)
- 1947: commercial bread production, a wood-drying kiln
- 1955: a factory to cure concrete blocks and pipes
- 1975: a fish drying plant
- 1979: a candy factory

Acknowledgments

Sincere thanks for enlightening discussions are due to Lýdur Björnsson (sulfur mining and salt production), Gudmundur Ólafsson (archaeological excavations at Reykholt), and to Jón Bödvarsson who read the manuscript critically. Helgi Torfason is thanked for making a drawing of Reykholtslaug from a photograph.

Selected References

Bárðarson, H. R. 1989. *Iceland—A portrait of its land and people.* Hjálmar R. Bárðarson.

Björnsson, L. 1978. Salt production in the Westfjords (Saltvinnsla á Vestfjörðum). *Ársrit Sögufélags Ísfirðinga,* 21, 25-69. (In Icelandic).

Björnsson, L. 1994. Personal communication.

Finnbogason, G. 1943. *The history of industry in Iceland (Iðnsaga Íslendinga).* (In Icelandic).

Grímsson, Th. 1960. Restoration of Snorralaug (Gert við Snorralaug). *Árbók fornleifafélagsins,* 19-47. (In Icelandic with an English summary).

Grímsson, Th. and G. Ólafsson. 1987. Ancient pipelines in Reykholt (Fornar leiðslur í Reykholti í Borgarfirði). *Árbók fornleifafélagsins,* 99-121. (In Icelandic with an English summary).

Íslendingasögur. The sagas of the Icelanders. Reprint 1981. Reference is made to *Kristni Saga, Laxdœla Saga,* and *Sturlunga Saga.* Prentverk Odds Björnssonar hf, Iceland. (In Icelandic; all of these sagas are available in English translations in separate publications).

Landnámabók. The book of settlements. Reprint 1968. Hið íslenska fornritafélag, Reykjavík. (In Icelandic).

Magnússon, Á. and P. Vídalín. 1982. *Description of the farmsteads of Iceland 1703-1714 (Jarðabók Árna Magnússonar og Páls Vídalín).* Sögufélagið, Reykjavík. (In Icelandic).

Ólafsson, E. and B. Pálsson. 1943. *Book on travels in Iceland 1752-1757 (Ferðabók eggerts og bjarna).* Ísafoldarprentsmiðja, Reykjavík. (In Icelandic).

Sigurjónsson, A. 1967. *Ásverjasaga.* Helgafell, Reykjavík. (In Icelandic).

Thórðarson, S. 1993. Geothermal water provides comfort at all times (Hveravatn veitir hin mestu þægindi á öllum tímum). *Fréttabréf SÍH,* no. 91, 2-3, Reykjavík. (In Icelandic).

Thorhallsson, S. 1988. Experience in developing and utilizing geothermal resources in Iceland. *Geothermics,* 17, 205-223.

Thoroddsen, Th. 1958. *Report on research in Iceland 1882-1898 (Ferðabók).* Snæbjörn Jónsson & Co. hf. (In Icelandic).

Thorsteinsson, K. 1971. From the regions of Borgarfjörður (Úr byggðum Borgarfjarðar). *Prentsmiðjan Leiftur, 2.* Reykjavík. (In Icelandic).

The Author:

Ingvar Birgir Fridleifsson
United Nations University
Geothermal Training Programme
Orkustofnun, National Energy Authority
Grensasvegur 9, 108 Reykjavik, Iceland

Majestic Mt. Ararat, its volcanic peak shrouded in clouds, rises in Turkey just west of the Armenian border. At its base, nomad tents dot the plateau. ***Photo by Jeffrey Tayler, published with permission***

20. Transcaucasus: One of the Most Ancient Regions Using the Earth's Heat

by
Guram Buachidze
Sabir Aliev
Anatolii Ananyan

Abstract: Archaeological findings from several places in Georgia, Armenia, and Azerbaijan attest to the fact that thermal springs and some by-products of the Earth's energy found south of the Greater Caucasus Range were known and used by peoples since prehistoric times. These uses gradually intensified through the centuries, resulting in a tradition of thermal bathing in many localities with natural manifestations. Among these is Tbilisi, the capital of Georgia, whose etymology relates to the occurrence of thermal springs. Georgian historical events include long-lasting balneological practices. The region also has myths and legends that emerged due to the presence of volcanic features, thermal springs, and mud volcanoes, as well as the unusual manifestations known as "everburning lights."

How Geothermal Resources Affected Early Trancaucasian Peoples

The region south of the Greater Caucasus Range between the Black and Caspian Seas is one of the world's oldest centers of civilization. A human lower jaw discovered in a recent archaeological excavation in Georgia indicates that people were already settled in the Transcaucasian region about 1.9 million years ago. It is likely that prehistoric people in this region would have known and used thermal springs, such as those in the Vorotan Valley. They also might have noticed the imposing external manifestations of the Earth's heat: volcanic eruptions, lava flows, ash avalanches, tephra covers, and earthquakes that occurred in several areas of the Greater and Minor Caucasus Ranges during the Paleolithic and Neolithic.

Ancient legends, originating in Neolithic times and handed down from parent to child in early historic times, mentioned the demolition of prehistoric settlements by lava and their burial

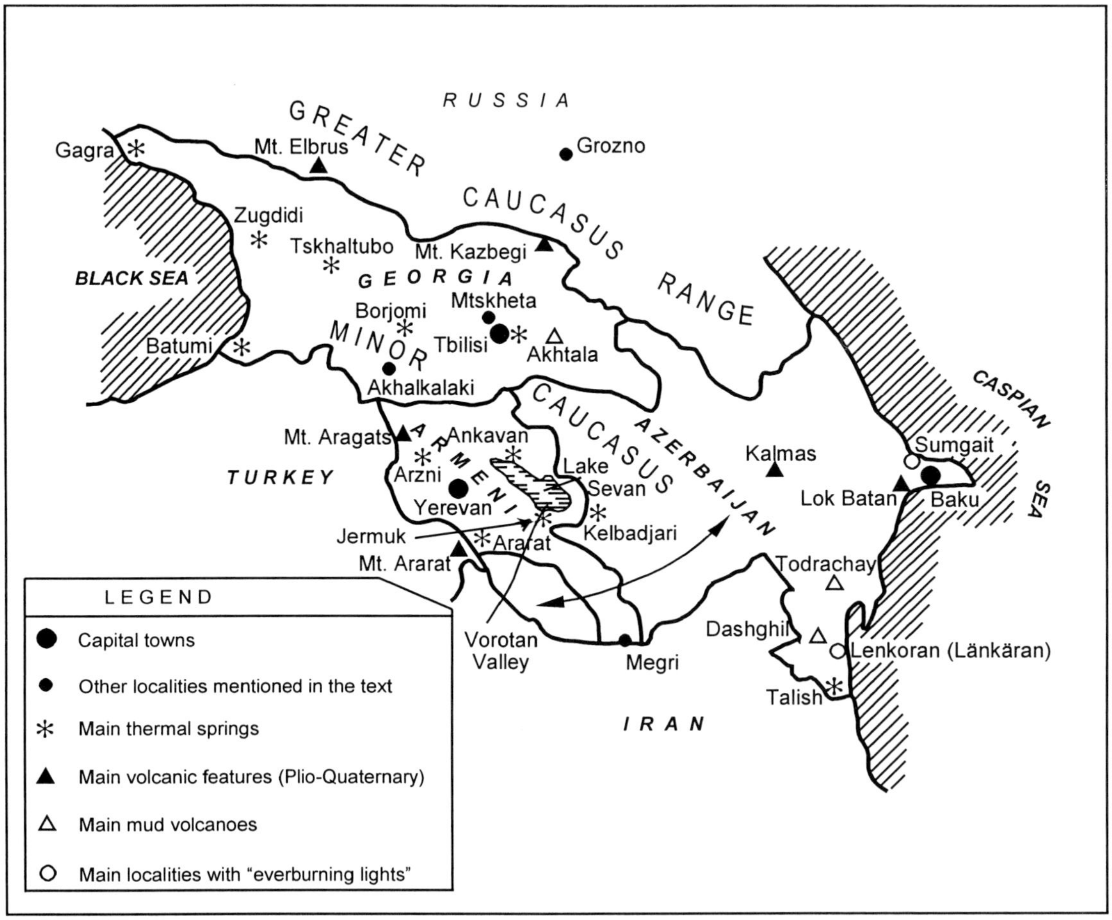

Main thermal manifestations and volcanic features of the Transcaucasus.

under a pyroclastic cover. These legends are not surprising: a Neolithic tomb has been recently discovered in Armenia, completely filled and covered by volcanic agglomerates dating from 6,000-7,000 years ago. As for the myths directly or indirectly related to the occurrence of volcanic features in the Caucasus region, worthy of mention is that of "Amirani," a legendary hero who, because of his conflict with the deities, was chained to the highest peak of the Greater Caucasus Range (Mt. Elbrus) and condemned to suffer the attacks of an eagle devouring his liver during the day, while the liver regenerated during the night. It is similar to the ancient Greek myth of Prometheus, who was condemned to the same punishment by Zeus after his "theft of fire" in disobedience of a divine order. Since the myth of Prometheus is a rather elaborate one, it probably evolved from the older Caucasian myth of "Amirani."

Some legends of the Transcaucasian region refer to Mt. Ararat, a prominent massif with a perpetual snowcap. The Ararat massif is the result of relatively recent (Plio-Quaternary) volcanic activity. Although there has been no eruption, strong earthquakes are known to have

The Great Caucasus Range is a segment of the Alpine-Himalayan chain that, because of its morphological prominence, is considered the southern geographical divide between Europe and Asia. A wide variety of geologic environments exists within and south of the Great Caucasus Range, between the Black and Caspian Seas, and inside the present-day territories of Georgia, Armenia, and Azerbaijan. This variety accounts for the notable differences in the regional heat flow field, whose values display sharp changes from zone to zone, and even within the same zone at relatively short distances.

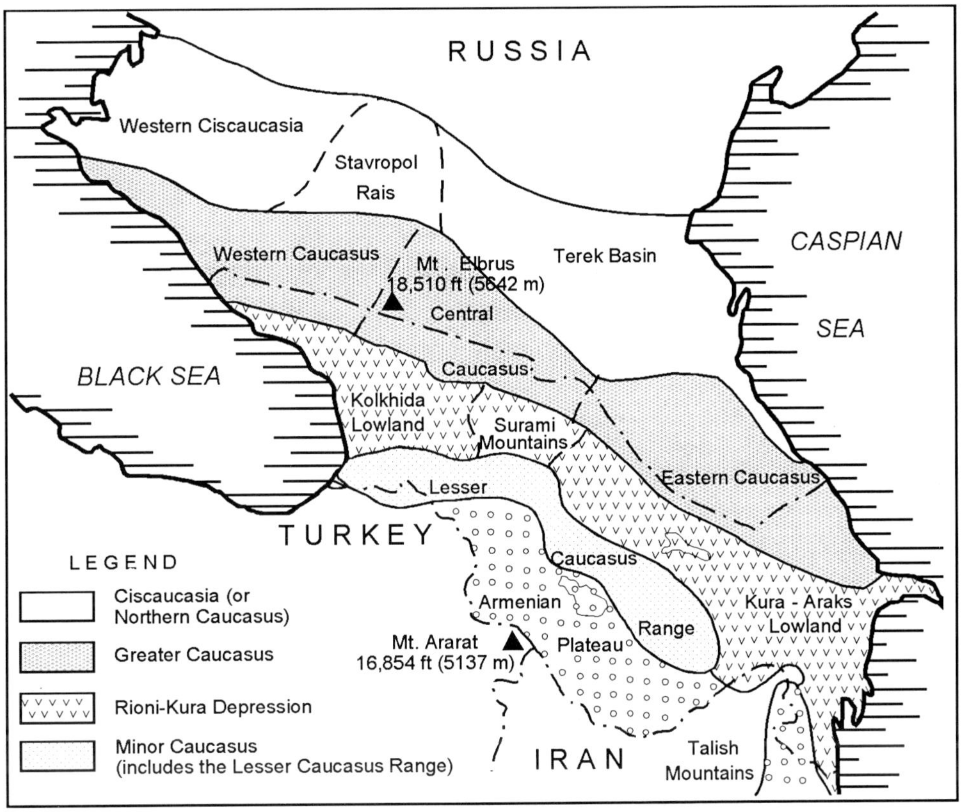

Mountains and lowland regions of the Caucasus.

Igneous processes that accompanied and followed tectonic activity in the Caucasus region starting from Pre-Cambrian times were rejuvenated during the Alpine orogenesis, resulting in the renewal of the regional heat field and the establishment of favorable conditions for the formation of local and relatively shallow heat flow anomalies in specific geologic sectors. These anomalies provided the basic conditions for active geothermal phenomena such as volcanic eruptions, earthquakes, mud volcanoes, and thermal springs in areas with particular hydrogeological and tectonic features.

The rejuvenation of the natural heat field was particularly strong during the Pliocene and the Quaternary. It involved the whole Transcaucasian region, especially the central sector of the Great Caucasus Range and the Lesser Caucasus, in the areas of Borjomi, Akhalkalaki, Sevan Lake, the Vorotan̈ River, and Megri (Paffengholts, 1948). These areas have many natural manifestations.

The most prominent geomorphological feature of the Great Caucasus Range is Mt. Elbrus, a volcano (5642 m) that started to form in the Pliocene and was completed in the Quaternary. Along its slopes, warm waters (15°-20° C) spring out perennially in permafrost areas above 4000 m, leading to the conclusion that many hydrogeological systems develop actively in these areas and take their temperatures at depth from a broad thermal anomaly beneath Mt. Elbrus. For this reason, some geologists think that Mt. Elbrus is a quiescent, but not yet extinct, volcano.

Comparable situations occur in the other prominent volcanic features of the region considered in this chapter, such as Mt. Ararat, 5137 m, located in Turkey near the convergence of the Turkish, Armenian, and Iranian frontiers.

---

occurred in the area in historical times. No nearby mountains detract from the solitary magnificence of this massif, which is formed by two adjoining peaks: Great Ararat and Little Ararat.

The "mountains of Ararat" are mentioned in the Bible (*Genesis* 8:4), but the Ararat massif proper is locally named *Masis* by the Armenians, *Agri-Dagh* by the Turks, and *Koh-i-Nuh* ("Noah's mountain") by the Persians. A Persian legend refers to Ararat as the "cradle of the human race." However, the Great Ararat summit is celebrated and known worldwide because it is believed to be the site upon which Noah's Ark came to rest, waiting for the waters of the universal flood to subside. Armenian legends maintain that the remains of the Ark were long visible on the mountain, not disappearing until 1840, when a tremendous earthquake followed by a huge avalanche affected the mountain's summit and totally destroyed the nearby village of Aghuri. According to tradition, Aghuri is the village where Noah built his altar and made ritual animal sacrifices to God soon after he and his companions could safely leave the Ark and regenerate the human race. The story of Noah's Ark is still a tradition among the Armenians, who believe themselves to represent the world's first human race after the Deluge. This is why Mt. Ararat was in antiquity, and still remains, a sacred mountain for the Armenians, venerated as the "mother of the world."

The Vorotan Valley, Armenia. The thermal springs in this area probably were used by prehistoric peoples.

Prehistoric people of the Transcaucasus region were also familiar with some of the by-products of the Earth's heat. This is evidenced by the widespread use of volcanic rocks as building materials and of obsidian for domestic tools (knives, scrapers) and hunting or warfare weaponry (axes, arrowheads). Moreover, due to its abundance in the region, it is likely that obsidian was used to barter for other kinds of materials or products from neighboring lands, starting from early Neolithic times.

Historical Use of Thermal Springs and Geothermal By-Products

Written accounts, archaeological discoveries, ruins, and ancient monuments prove that the systematic use of thermal waters for healing purposes dates back at least 2,500 years. The most important of these localities are described briefly.

Armenia

Archaeological excavations conducted a few decades ago in Arzni led to the discovery of open-air bathing pools dug in travertine deposits. Many Roman coins dating from 100 B.C. to 200 A.D. have been found inside and near the bathing pools. This indicates that commanders, rulers, soldiers, and common people using the thermal baths left coins as tokens of gratitude to divinities and water nymphs for successful healing (Demekhin, 1940). Construction of the pools dates back to the beginning of the Christian era, but thermal bathing for therapeutic purposes was probably practiced at Arzni since at least the 5th century B.C., well before Armenia came under Roman rule.

Ruins of old baths at Erebuni, Armenia.

Jermuk also has been frequented since early historic times. Stepanos Orbelyan, an ancient Armenian historian, reported that the *nakhararq* (barons and other nobles of the 3rd-5th centuries A.D.) used to take baths at Jermuk for healing, for rest and recovery after hunting parties, and for recreation (Yeremin, 1939). This spa has remained in more or less continuous operation since ancient times and is currently one of Armenia's most popular thermal resorts.

Ancient sanctuary cut in volcanic rock, Armenia. Christian symbols of a later period are seen at the top.

Ruins of very old baths, probably dating from pre-Christian times, have been discovered in Erebuni. The walls are made of juxtaposed, uncemented volcanic blocks. Bathing has continued here from ancient through recent times.

Thermal springs in Ankavan and Ararat have been in use since early historic times. Near one, a cultic structure cut from volcanic rock was built in pre-Christian times. The old structure was modified and later became a Christian temple. Near Ankavan and Ararat, imposing travertine formations up to 50 m thick outcrop over large areas. The travertine, made of almost pure $CaCO_3$, has a crystalline structure and is snow-white. The use of the travertine from these localities dates from prehistoric times, when early Armenians excavated this rock for carvings. In historic times, travertine quarries provided blocks and facing slabs for monuments and other buildings. This peculiar by-product of geothermal

Temple-palace of the sun's and Earth's heat at Garni, Armenia.

activity has been an important and long-lasting resource for Armenia: travertine from Ankavan and Ararat is still used by the Armenian building industry for facing material that is comparable to precious white marbles (Ananyan, 1959).

In Garni, the use of thermal water for bathing dates back to ancient times and has continued through the present. The importance that ancient Armenians attached to such natural geothermal resources is confirmed by their construction of votive temples.

Azerbaijan

The two most important thermal resorts of Azerbaijan are Kelbadjari and Talish. Each has a cluster of thermal springs, with temperatures of up to 72° C at Kelbadjari and up to 64° C at Talish. Written records on the use of these springs date back to the early Middle Ages. These thermal springs are described in ancient books, listed under the name of the main disease that was usually treated. For instance, *gotursu* (or *goturlu*) referred to thermal springs used for treating diseases of the skin, especially eczema; *elisu* was the spa for rheumatism treatment; and *khaltan* was the thermal spring used for treating liver diseases.

Georgia

Georgia has many thermal localities, each of which is formed by a cluster of springs (Buachidze, 1980). The most important of these localities are Tskhaltubo, Mtskheta, Borjomi, Gagra, Zugdidi, Batumi, and Tbilisi.

At Tskhaltubo, the hydrothermal water has a total flow rate of hundreds of liters per second. In this area of Karst topography, prehistoric peoples settled in caverns near the spring outlets, undoubtedly bathing in the thermal waters.

Around Mtskheta, the ancient capital of Georgia, are ruins of very old baths. The floors of some bathing rooms are paved with mosaics resembling the Roman style. Most likely these baths were already in use at the beginning of the Christian era, if not earlier.

Borjomi is a renowned thermal locality now, but its bathing tradition is very old. A few decades ago, during the excavation of the catch basin for the Yekaterina spring (located in the center of Borjomi), the ruins of six ancient baths were discovered at a depth of 7 m. The floor and walls of these ancient baths are slabs and blocks of volcanic rocks, respectively. Archaeologists estimate that the bath buildings date from the 1st century A.D., although thermal bathing by early Georgians at Borjomi is probably much older.

Thermal Balneology at Tbilisi

Tbilisi is the present capital of Georgia, but because of the importance given to thermal practices, it can be considered the historical capital of hot springs in the whole Transcaucasus region.

From its first days, the town was known as *Tphilisi* or *Thbilisi*. Some historians maintain that this name derives from *Thubali*, proto-historic Georgian tribes that are considered the fore-fathers of the Georgian nation (Djavakhischvili, 1983). However, a more reasonable and the most accepted etymology of Tbilisi is now that of *Thbili (*which means "warm" in Georgian), with a clear connection to the warm springs that exist in the suburbs and outskirts of the town. One of these springs still flows and can be seen in the center of the old town.

In historical times, these springs spawned legends related to the founding of Tbilisi. One of them tells of King Vakhtang Gorgasali (late 5th century), who lost his falcon while hunting in the woodlands south of Mtskheta, the capital of the Iberian Kingdom. He later found it in the pool of a nearby hot spring, together with a pheasant. Both birds were scalded, and the king interpreted this event as a divine sign and decided to establish the new capital of Iberia in Tbilisi.

Geothermal bath at the Dedotsikhe fortress at Tbilisi, Georgia.

Another legend, dating from early medieval times, is related to the healing properties of Tbilisi's thermal springs.

The legend recounts that a wounded bird fell into the pool of a thermal spring but was immediately healed by the warm water and could fly away at once.

Legends aside, recent archaeological excavations within the urban area indicate that people had already settled Tbilisi during Aeneolithic times, the end of the 4th millennium B.C. (Early Metal Age); therefore, it is hard to imagine that they did not know how to use the waters from the thermal springs nearby. Later archaeological evidence reveals the progressive development of the town, from the Bronze Age (3rd millennium B.C.) to the first centuries of the Christian era (Meskhi, 1958).

By the 4th century A.D., Tbilisi had already developed into a fortress-town and the stronghold of Mtskheta. A large thermal spa was built within the fortress. The spa was modernized in the late 5th century, when Tbilisi became the new capital of Iberia.

The Kingdom of Iberia began forming in the 2nd century B.C. Its territory grew to include modern Karthli and Kakheti and adjoining regions of Southwestern Georgia. Iberian statehood lasted until the Sasanidian King Khosrau I conquered the region and abolished the Iberian monarchy in the mid-6th century A.D.

During the whole period of the Iberian Kingdom, and for subsequent centuries until 1450, the practice of thermal bathing at Tbilisi was unaffected by historical events and political changes. The baths were frequented by an increasing number of persons, including people from Tbilisi and other Georgian localities, patients from neighboring Transcaucasian regions, and visitors or travelers from faraway lands. This is confirmed by reports and works of national and foreign writers, who stressed the effectiveness of balneotherapeutic treatments for a number of diseases, as well as the architectural beauty of some thermal establishments. Ibn-Haukali, a 9th century Arabic traveler, wrote: "there are some thermal baths in Tbilisi, which are not heated by any fires.... Some baths here are as nice as the Tiberian baths...." The author apparently was aware that some thermal baths in Europe and the Middle East were heated by wood fires. It is also likely that the author was making a comparison with the baths of Hammei Tiberias in Galilee, which were established in 20 A.D. by Herod Antipas and named after the Roman Emperor Tiberias (see Chapter 4, this volume).

Other references to the Tbilisi spas are found in accounts by Arabic merchants and traders living in Georgia during the 11th and 12th centuries. Al-Farik and Ibn al-Jauzi, for example, reported that "the entrance to the baths is free..."; however, it is not clear whether "free entrance" meant that the admission to the baths was free of charge, or if the baths were public.

Bath complex at Tbilisi, 17th century.

Balneotherapy continued at Tbilisi even after the Georgian territories fell under the rule of the Ottoman Turks in 1453 and expanded, with new bathing complexes added for the next five centuries. This custom included all classes of people, regardless of gender or religion. The Russian traveler Vassilis Gagara, for instance, wrote in 1640 that "both Christians and Muslims used to attend the baths...." Moreover, the famous French traveler Jean Chardìn (later Sir John), when passing through Tbilisi in 1672 during his last visit to the Middle East, noted the widespread use of the town baths. He also wrote in his travel journals about the existence of a special pipeline that conveyed hot water from a natural spring to the palace of the Viceroys of the House of Mukhran, who governed at Tbilisi under the aegis of the Iranian shahs from 1658 until 1723.

Official documents of the 17th century also describe the Tbilisi spas (Dumbadze, 1990). One of these documents is a certificate establishing rules regarding the use of thermal waters. At that time, the owners and operators of all baths were private entrepreneurs, and each spa took the name of its owner. Another official document of that period declared that the baths could be used by people of any social class and religion; however, men and women had to visit the spas on different days. In 1701, Turnephor classified the thermal springs according to temperatures at the natural outlets and systematically described the healing properties of each type of thermal water (Meskhi, 1958).

It is worth mentioning the case of an extraordinary spa whose water was considered suitable for treating urinary diseases and related disorders. Agha Mahmad Khan Kajar, commander of the Persian army and a eunuch, knew of this spa and ordered his soldiers not to sack the city's thermal complexes after conquering Tbilisi in 1795. He then tried to reverse his physiological condition by bathing at the spa. When a number of baths resulted in no change, he was overtaken by a fierce anger and ordered his hordes to raze and burn all bathing establishments in Tbilisi. At that time, Tbilisi had six thermal complexes, each of which covered a whole city block. Two similar complexes were located inside the fortress for use by dignitaries, elite socialites, and selected people close to the governor of Georgia.

In his delightful work of poems, "A Voyage to Arzrum," Russian poet Alexander Pushkin eulogized the Tbilisi baths in a panegyric written around 1836. In particular, he described the ritual of the massage, required of all spa visitors before bathing, as well as the "sacredness" of the ablution ceremony.

Bathing complex at Tbilisi, 19th century.

From 1806 on, after Georgia was incorporated into the Russian Empire, the spas were transformed by new, larger buildings and fully modernized. The lower levels included marble-faced basins and comfortable couches for bathing, whereas the halls, meeting rooms, and restaurants were located on the ground floor and upper levels. The spas thus became more than curing centers; they functioned as public clubs, business meeting areas, and social centers. After most business meetings, especially after the successful ones, a party was held with tasty meals and joyful toasts.

Until the end of the Russian Empire, the spas were multifunctional. Baths gradually became a social "must" for ladies, who succeeded in having one day per week, usually Tuesday, reserved for women only. Georgians visited the spas not only to treat diseases, but also for beauty treatments, exchanging the latest city gossip, chatting among themselves, and thoroughly evaluating the attractiveness and economic positions of marriageable young ladies.

In 1925, as a result of the new political situation arising from the formation of the Union of Soviet Socialist Republics, the spa buildings were collectivized and began deteriorating. Balneotherapy continued at a modest level. During the 1960s, however, with the founding of the Georgian Institute of Health Resorts and Balneotherapeutics, the Tbilisi spas lost their original health emphasis and assumed different public functions.

Use of Other Manifestations

In addition to volcanic features and thermal springs, other geothermal manifestations that have had an important impact on Trancaucasian populations are mud volcanoes and "everburning lights." These manifestations formed because of particular hydrogeologic conditions affecting underground structures bearing organic gas.

Mud Volcanoes

Although they are seen frequently in several inland places of the Transcaucasus tectonic depression, mud volcanoes are concentrated in several sites on the shores and the shallow submarine platform of the Caspian Sea in Azerbaijan. The Transcaucasus region contains 400 mud volcanoes; about 200 land and 100 submarine are located in Azerbaijan. Altogether, the 400 account for 40 percent of the world's mud volcanoes (Aliev and Mekhtiev, 1968). The most important among them are indicated on the map.

Prehistoric Transcaucasian peoples undoubtedly noticed the birth and growth of the mud volcanoes, particularly the tribes who settled on the western shores of the Caspian Sea. Surely these peoples were amazed to see spectacular eruptions, such as those occurring at Todrachay, Kyanizadag, and Kalmas—which rose more than 500 m above ground level—or at Lok Batan, which is considered the "world champion" in number of eruptions. It has been estimated that during peak activity, Lok Batan emits some 500 million cubic meters of gas and 20 million cubic meters of thermal water per year, in addition to large quantities of mud. The last eruption took place in 1990, and it was huge: one eruption produced 300 million cubic meters of gas (organic gas, CO_2), together with 150,000 cubic meters of muddy material and breccia.

The curative use of volcanic mud in Georgia and Azerbaijan can be traced back to medieval times, but it probably began much earlier. The most important locality in Eastern Georgia is Akhtala, which belongs to the Tyulki-Tapa group of mud volcanoes. Here, volcanic mud has been used for centuries to treat rheumatism, and warm water was used to treat women's diseases. Both these applications date to early times and still are very popular.

The Dashghil and Koturdag mud volcanoes are located in the southern sector of Azerbaijan. At Dashghil, ponds have been constructed for treating a number of skin diseases, since the mud eruptions are accompanied by outflows of a mixture of thermal water, organic gas, and oil films.

At Koturdag, a "paste" of viscous mud and very dense oil squeezes out slowly but continually from a 50 m long/15 m wide fracture across the crater's top. A huge amount of this paste has accumulated over the centuries, and it has long been used to prepare therapeutic muds to treat arthritis and skin diseases.

In addition to balneological applications of mud and thermal water, some of the mud volcanic extrusions of Georgia and Azerbaijan contain organic fluids and have been exploited locally to produce oil for lamps. In the 12th century at Lok Batan, shallow wells were hand dug to produce oil. Similar wells were excavated in the following centuries. They continued to be dug by hand until the early 20th century. With the passing of time, of course, the excavation technique improved, resulting in larger quantities of oil. One of the last wells of this type was excavated at the beginning of the 20th century and yielded some 20,000 tons of oil and gas per day.

Everburning Lights

"EVERBURNING LIGHTS" ARE MANIFESTATIONS THAT CONSIST OF PERENNIAL FLAMES SPRINGING to the surface from open fractures in certain areas near thermal springs related to mud volcanoes. The flames are supplied by an organic gas, such as methane, released from oil-bearing structures underlying confined thermal aquifers. Thus the everburning lights are related tangentially to geothermal phenomena and are particularly abundant in the southeastern sector of Azerbaijan.

The existence of the lights has been known since ancient times. Early and pre-Christian Transcaucasian peoples considered them to be a surface expression of some underground power, and they began a cult of fire worship. This cult may have started in early Neolithic or Upper Paleolithic times, based on 11,000 year old petroglyphs depicting fire features, which are found in Azerbaijan's Gobustan Natural Reserve. The cult spread across the Transcaucasus region to remote areas, but its center remained the Apsheron Peninsula near Baku, which bears the name, "Land of Everburning Lights."

The existence of a sect of fire worshipers has been documented in this area since the 6th century B.C. The sect had, and still has today, two special temples: Atashkyakh (near Sura Hani) and Yanardag (near Länkäran). These temples and other localities with everburning lights were visited by fire worshipers from many distant lands, such as Arabia, Persia, Afghanistan, India, and even China.

The ancient cult continues to flourish. Masudi, an Arabian geographer of the 10th century, beautifully depicted the temples and described in detail the nightly torchlight ceremony held at a certain time of the year on the shores of the Caspian Sea near Baku. In 1825 the Russian scientist Ekhwald, during a study visit to Baku, described the climax of the fire worshiping ceremony. He wrote, "In the temple there is a hole from which gas escapes.... At a certain point the mouth of the hole was covered with a carpet, resulting in the accumulation of gas inside.... Afterwards, a wisp of burning straw was thrown into the hole; a strong burst of fire occurred...and everybody was happy."

Selected References

Aliev, S. A. and S. F. Mekhtiev. 1968. Some considerations of the Caucasus geothermy. *Proceedings of the 23rd International Geological Congress,* 49-54.

Ananyan, A. L. 1959. Measurement of temperatures in Jermuk. *Transactions of the Academy of Sciences of Armenian SSR,* 1, 71-73.

Buachidze, G. J., N. A. Goderdzishvili, B. S. Mkheidze, M. P. Shaorshadze. 1980. *Thermal waters of Georgia.* Sabchotha Sakarthzelo, Tbilisi.

Demekhin, A. P. 1940. Arzni, Hydrological sketch. *Armenian Geosurvey,* 1, 1-113.

Djavakhischvili, I. 1983. *History of Kartveli nation.* State Lit., Tbilisi.

Dumbadze, M. 1990. *History of Tbilisi, vol 1.* Science, Tbilisi.

Meskhi, S. 1958. *History of Tbilisi.* State Lit., Tbilisi.

Paffengholts, K. N. 1948. *Geology of Armenia.* Ministry of Geology USSR, Leningrad.

Yeremin, S. T. 1939. *Trade ways of Armenia during the Sasanids.* Science, Leningrad.

The Authors:

Guram Buachidze
Institute of Hydrogeology
Academy of Sciences
31, Rustaveli Avenue, Tbilisi
380008 Georgia
Telephone: 995.32.223.289
Fax: 995.32.001.153

Sabir Aliev
Institute of Geology
Academy of Sciences
Baku, Azerbaijan
Telephone: 994.12.986.824

Anatolii Ananyan
University of Armenia
Yerevan, Armenia
Telephone: 7.8852.530.119

SV

Interpretation of "Alyonushka," the oil painting by Victor Vasnetsov, sketched by Valentina Svalova. The 1881 painting, illustrating the folk tale "Sister Alyonushka and Brother Ivanushka," shows a very sad Alyonushka after Ivanushka has been turned into a baby goat for drinking water from the magic well.

21. Geothermal Legends through History in Russia and the Former USSR: A Bridge to the Past

by
Valentina B. Svalova

Abstract: The use of geothermal resources in Russia and the former USSR dates back to antiquity and is always associated with legends. This chapter traces the history of hot and mineral springs from Crimean antique towns through Kiev Russia, Moscow State, and the Russian Empire to contemporary times.

This history can be seen as waves. One arose in the Mediterranean near volcanoes and hot springs, rolled through the territory of Russia, and reached Kamchatka, where it met another culture in a magic land of geysers and volcanoes. Another originated in ancient Greece, rolled through Western Europe, crossed the Atlantic Ocean with the first settlers, and met the American Indian culture. Finally, the two waves reached the Pacific Ocean from different regions of high thermal activity and met at the Bering Strait. Christopher Columbus discovered America from the east coast as Vitus Bering reached America from the west. Consequently, the peoples of Kamchatka have legends as fantastic as those of the American Indians and ancient Greeks.

Introduction

MANIFESTATIONS OF THE EARTH'S POWER—VOLCANOES, geysers, earthquakes—have always excited people's feelings of beauty and fear. Interest in the Earth's geothermal resources arose very long ago. For ages, people have used these natural resources. It is as difficult to answer when, where, and which people first discovered the medicinal properties of hydrothermal waters as it is impossible to ascertain who invented the plow and wheel. Evidently, the discovery of geothermal resources came to people naturally; very likely wild animals led ancient people to hot springs. For a long time, warriors treated their wounds in mineral hot springs.

A Bridge from Ancient Greece to the Antique Towns of Crimea

The curative properties of underground springs were known to Stone Age people. The mineral springs of Epidaurus in Greece could be considered the most ancient known spa in the world (6th century B.C.). Even now, the site still has inscribed plates with diagnoses and descriptions of treatments. A sanctuary of the god-healer Asclepius is placed 9 km from the port of Epidaurus.

Although it is impossible to trace the transformation of Asclepius from a hero at the beginning of the 5th century B.C. to a god by the end of the same century, the Greeks honored him as the creator of medicine; Homer glorified him as a wonderful physician. According to legend, Asclepius not only treated sick people but also resurrected the dead. He acquired his healing ability from Apollo, the centaur Cheiron, and Athena. The goddess presented him with an ampule containing Medusa's blood, taken from the left side of her body. This blood had power to resurrect the dead. Athena kept another ampule of Medusa's blood, taken from the right side, that gave her power to destroy life and start wars. This legend has influenced Russian folk tales about "dead" water.

The significance of Asclepius and his sanctuary spread, and pilgrims arriving at Epidaurus had to clean themselves with sacred spring water and offer a sacrifice. Later they spent a night near the sanctuary, where the god healed them during sleep, a miraculous curing. In the beginning of the 4th century B.C., a theater was built on a slope of rock near the sanctuary, and solemn celebrations devoted to Asclepius were held there every four years. Historians presume that the founder of medical science, Hippocrates (460-370 B.C.), also lived and worked there.

For ages, waterways linked the Mediterranean with the Black Sea coast. In the 6th-5th centuries B.C., the Greeks founded towns on the Black Sea: Kerkinida, Chersonese, Feodosiya, and Panticapaeum (now Yevpatoriya, Sevastopol, Feodosiya, and Kerch).

The southern Crimean coast is famous for spas and health resorts with mineral springs and medical muds. Chersonese was founded in 421 B.C. Archaeological excavations begun in 1827 revealed *thermae*, theater ruins, and a mint. In the opinion of some investigators, the founders of this town-state were representatives of a democratic party forced to leave their native town, Pont Gerakleya, by a conquering oligarchic party. A famous Chersonesean oath tells about the patriotism of its citizens. The text of the oath was cut on a marble plate in the 3rd century B.C., discovered during excavations.

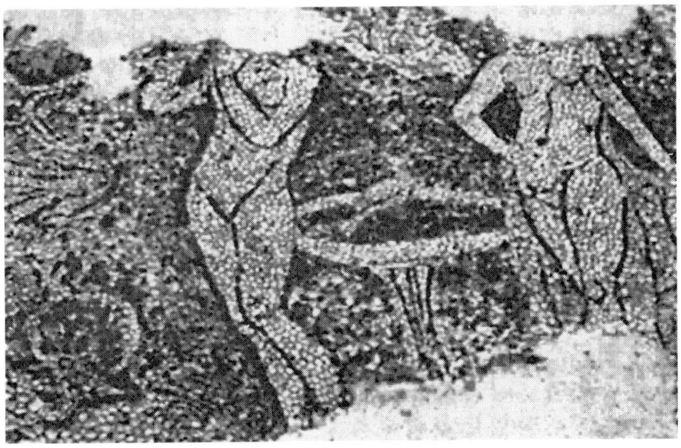
Chersonesean mosaic floor from a bathroom, 2nd century B.C.

The town had a good layout: streets edged with stone gutters crossed at right angles, and there were beautiful temples at the squares. Public buildings and the houses of rich people were decorated with columns and mosaic floors. An outstanding piece of Chersonesean artistry and skill was the mosaic floor from the bathroom of a rich citizen. In the center of the composition, two women are bathing with a washbasin between them. Two birds are also represented in the mosaic.

The craftsman's sector was located near town walls. There archaeologists have found the remains of ceramic production, in addition to metalworking, jewelry, and weaving. Metal medical instruments reveal information about the development of medicine in Chersonese, as do stone inscriptions. The antique world was full of such inscriptions, which are a unique, reliable way of recording and preserving information. During 150 years of excavations, archaeologists have found more than 500 Greek and Latin inscribed stones. These stone archives are like soundtracks of the most interesting pages of history. Here is the epitaph on Chersonesean stele dating to the 4th-3rd centuries B.C.:

"Father, physician Evkles from Tenedos,
Set up this tomb to his son Leschanorid."

Crimea came under the influence of the powerful Roman Empire in 100 A.D. The Romans, moving in different directions and conquering new lands, built fortresses, harbors, and roads everywhere. Such fortresses and roads were built in Kharaks, near the present famous health resort of Miskhor. As elsewhere, the Romans built *thermae*, the remains of which have been excavated by archaeologists. The Romans also set up a sanctuary and laid ceramic water pipes. As at other sites, the *thermae* were public places for the Romans. They came there to rest, relax, talk about life, and even read. Later, the fortress at Kharaks was destroyed. But in 1912, architect A. V. Shervud built the famous "Swallow's Nest," which has become the symbol of Crimea's southern coast.

The earliest data about the natural waters' peculiarities of the Black Sea region can be found in written documents of ancient Greek and Roman scientists. The Greek historian Gerodot (490/480-425 B.C.) recorded the first systematic description of the Scythian mode of life and pointed out the existence of salt springs on the left side of Ipanid (South Bug). Later, Aristotle (384-322 B.C.) wrote about the same spring. Vitruvius, the Roman architect and engineer of the 1st century B.C., also mentioned it. The Roman scientist Pliny the Elder (23/24-79 A.D.) wrote about a project to link the Black and Caspian Seas and mentioned the Crimean springs in his book *Natural History*. Historical documents show that ancient Russia was acquainted with Greek culture and the scientific heritage of Aristotle.

The "Swallow's Nest" on Crimea's southern coast. Many famous hot springs are in this area.

Legends and Reality

In Georgia and Armenia, in the region of Caucasian mineral springs, the remains of ancient baths were found which date from the Bronze Age. In Armenia, in the Agmagan range, stone symbolic dragons (*vishaps*) remain near underground springs dating from 2000 B.C. These dragons were set up as deities, guardians of underground water.

According to legend, Georgian Duke Vakhtang Gorgasali (4th century A.D.) was hunting in the forests of the Kura River Valley and wounded a partridge. When he found it, the partridge was boiled in hot water. Later, *thermae* were built near the Georgian hot springs, and their fame spread. It was here that Duke Gorgasali founded the town of Tbilisi (*Tbili* means "warm" in Georgian). The ancient water collection galleries remain in the hot spring region of Old Tbilisi, and it is still possible to visit the hot spring baths.

Almost all hot springs are surrounded by legends and traditions. According to a Turkmen legend, the herdsman Archman was turned out of his *aul* as a bearer of all illnesses. Looking for shelter, he found a warm spring. He drank water, bathed, and left the area in perfect health. Since that time, the efficacy of the Archman spring's medicinal properties has been elaborated on by new legends, real facts, and events. The area now includes a spa 130 km from Ashkhabad.

The old bathhouses of Tbilisi, Georgia.

According to another legend, the 15th century Turk, Uzun-pasha, ruled over Akkerman (now Belgorod-Dnestrovsky). He put the Ukrainian beauty Paraskoveya into a damp dungeon. Once when Uzun-pasha wanted to come into dungeon, the doors opened, the shackles fell away, and the girl ran to Dnestr liman. But *janizaries* overtook her, and when they raised their curved *yataghans* above her head, a miracle happened. Suddenly, the girl began rising and disappeared into the air. The *janizaries* turned into stones, and crystal-clear spring water erupted in her footprints. Centuries have passed, but the spring lives.

For ages, people of different countries and nationalities have named places and villages after hot springs. These names reflect thermal and mineral water usage through history. For example, the name of the famous spa *Khodga-Obi-Garm* means "holy hot water" in Tajik. *Yangan tau* means "hot mountain" in Bashkir. The spa name *Kuldur-Hot* came from a misrepresented Tungus word *Khul-Dgi-Uri*. And *Psekups* means "hot spring" in Kabardian.

Medicinal properties of hot mineral waters were reflected in old Russian folk tales about living and dead water. These tales show people's dreams about magic water healing wounds and reviving the deceased.

The Russian folk tale "Ivan-Tsarevich and Grey Wolf" tells about the waters' curative powers: "Ivan-Tsarevich lies dead and already ravens fly above him. Suddenly Grey Wolf ran, caught a raven with his nestling, and said: 'Fly, raven, for living and dead water. If you bring it, I'll set free your nestling.' Nothing else to do. Raven flew. He flew long and fast, and brought living and dead water. Grey Wolf sprinkled Ivan-Tsarevich's wounds with dead water, and they healed.

Wolf sprinkled him with living water and Ivan-Tsarevich came to life...." The Tretyakov Gallery in Moscow has a painting by the Russian painter Victor Vasnetsov (1848-1926), "Ivan-Tsarevich and Grey Wolf."

The Russian poet, Alexander Pushkin (1799-1837), used folk legends about living and dead water in his famous poem "Ruslan and Ludmilla." Ruslan was killed and then came back to life due to living and dead water. The Russian composer Mikhail Glinka (1804-1857) composed an opera, "Ruslan and Ludmilla," in 1842 describing the event. Pushkin's poem is set in Kiev Russia at the time of Grand Duke Vladimir-Red Sun (?-1015). Here the lines are about living and dead water, translated by the author:

In a silent solitary steppe,
Behind the wild mountains' far range,
The winds and terrible tempests dwelling,
Where even witch's daring sight
Fears late to deeply penetrate,
A fairy valley is concealed,
There are two springs in this fairy valley:
One gaily flows with waves alive,
Murmuring joyfully on stones,
Another streams with water dead;
It's calm around, winds sleep,
Even a freshness does not blow,
Centennial pines make no noise,
The birds don't fly and the doe dares not
To drink from secret waters there;
Since the world arose, a few spirits
Live, keeping silent in the valley
And guarding the dense and wild shore...
A hermit with two empty pitchers
Once appeared in front of the spirits;
They interrupted their long sleep
And went away, brimming with fear.
And bending forward he submerges
The vessels into virgin waves;
He filled them, in air disappeared
And came in two short moments
To the valley where lay Ruslan
In blood, immovable, lifeless, silent;

The old man stood still above the knight,
He sprinkled Ruslan with water dead,
And suddenly his wounds shone,
At once the corpse of marvelous beauty
Blossomed; after that the old man
Poured on the hero living water.
Now cheerful, new forces full,
Palpitating with young life,
Ruslan stands up, he looks around
With greedy eyes at clear daylight,
Past time gleams quick in front of him
As an awful shadow, a terrible dream.

Many Russian folk tales recount dreams about magic water. Ivan-fool, for example, turns into Ivan-Tsarevich after bathing in magic boiling water in the poem-tale "Konyok-Gorbunok" by Russian writer Petr Ershov (1815-1869). Anyone who visits the Tretyakov Gallery remembers the famous picture of Victor Vasnetsov, "Alyonushka," illustrating the folk tale "Sister Alyonushka and Brother Ivanushka," where Ivanushka is turned into a baby goat after drinking water from a magic well (see first illustration). At the end of the story, he is turned back into a boy. Very often, Russian parents say to their children, "If you don't listen to us, you will turn into a kid like Ivanushka, who did not listen to his sister Alyonushka and drank fairy water."

Reality and invention are mixed in famous legends about the Bakhchisarai Fountain. The powerful Tatar dynasty of Gyreys ruled in Crimea for more than 300 years (1427-1783). There is the mausoleum of Khan Kerim-Gyrey's beautiful Georgian wife in Bakhchisarai and the famous "Fountain of Tears," dedicated to her by the khan and constructed as marble cups into which spring water incessantly flows from a marble wall. Local people, however, say that his wife was not Georgian, but Polish—Maria from the Pototsky family.

The "Fountain of Tears."

The fountain has been memorialized, as well, in the fine arts. Pushkin wrote his famous poem "Bakhchisarai Fountain" in 1824. Soviet composer Boris Asafiev (1884-1949) created a ballet with the same name in 1934, and Russian painter Karl

Interpretation of "Bakhchisarai Fountain," an oil painting by Karl Brullov, 1849, sketched by Valentina Svalova. This famous painting of the legendary "Fountain of Tears" is one of the many works of art the fountain has inspired. The great poet Pushkin once wrote, "Fountain of love, fountain alive...."

Brullov (1799-1852) painted "Bakhchisarai Fountain" in 1849, which depicts a basin and fountain of spring water.

According to Pushkin's version of the legend, Khan Gyrey had a favorite wife, Zarema, in his harem. Then a new young wife, Maria, appeared. She was very sad but so beautiful that the khan forgot about Zarema. One night Maria died, and Zarema was killed by eunuchs-guards. From that time, Khan Gyrey left his harem and tried to forget those events by participating in cruel wars.

On returning from the wars, he built the "Fountain of Tears," which Pushkin described as

> Fountain of love, fountain alive,
> I brought as a gift for you two roses.
> I like unceasing murmur yours
> And so poetic tears those.

Now, after more than one and a half centuries, living roses don't wither at the fountain, for someone brings fresh ones every morning.

Coats-of-arms of Russian towns with symbols of salt boiling: (1) Starya Russa; (2) Ekaterinaburg; (3) Solikamsk.

Saline Springs and Salt Boiling

Mineral and hot springs were used not only for medical purposes but also for salt extraction. The oldest data about salt extraction date to Gerodot's time (5th century B.C.). It was one of the Greeks' trading products with Scythia. Later, salt was boiled from estuary waters in the Black Sea region and from saline springs north and east of Russia.

Salt boiling has been known in Novgorod Russia since the 11th century A.D. The people of Novgorod had searched for fresh and salt water and were familiar with concentrations of natural solutions, where salt separation depended on the temperature. Use of salt solutions spread, due to the preparation processes of paint for icons and astringents for building materials. The importance of salt water output was reflected in the names of Russian villages and coats-of-arms for Russian towns.

Since the 11th-12th centuries, the technique of drilling for brines with wooden tubes was used in Novgorod. These tubes reached a depth of 170 m. Underground water

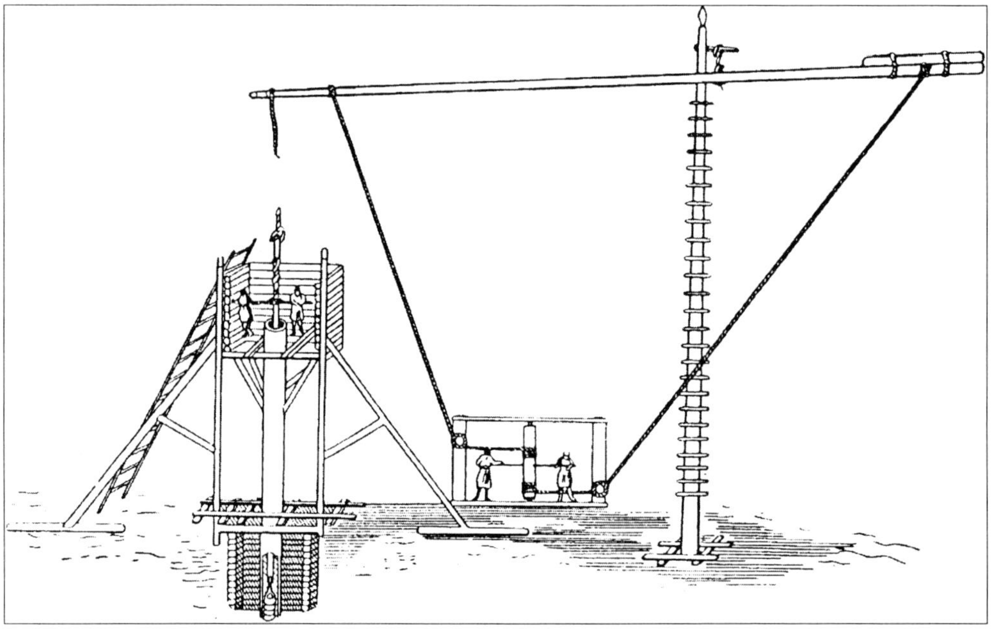
Salt well drilling with wooden tubes.

output and salt boiling were at a high technical level in Russia, as confirmed by the interest of foreigners. An old Swedish book was titled, "Some data about Russia, its roads, ways, fortresses with approaches to them, about boundaries, combined by the last Royal Embassy of the Tsars's court in Moscow and written by Erik Palmquist in 1674." In particular, the author described the saltern at the Mshaga River, near the town of Narva, and included a drawing.

Legends of the Urals

Underground waters are connected very closely with the formation of mineral and ore deposits. In the mid-18th century, Russian scientist M. V. Lomonosov elaborated on the roles of natural waters, solutions, water-rock interaction, elements migration, and water in geological processes and ore deposit formations. But ancient people already knew about metal ore deposits. The remains of precious and nonferrous metals output and smelting have been found in the Caucasus, Transcaucasia, Ural, Cisurals, Kazakhstan, Altai, Siberia, and Middle Asia. They are of different ages, the oldest dating back to 3000 B.C. In the Urals and Siberia, the signs of prehistoric ore-working, known as "chud mines," date from the middle of the 3rd millennium B.C. The mines prospered in the 12th century B.C. and closed in the 5th century A.D.

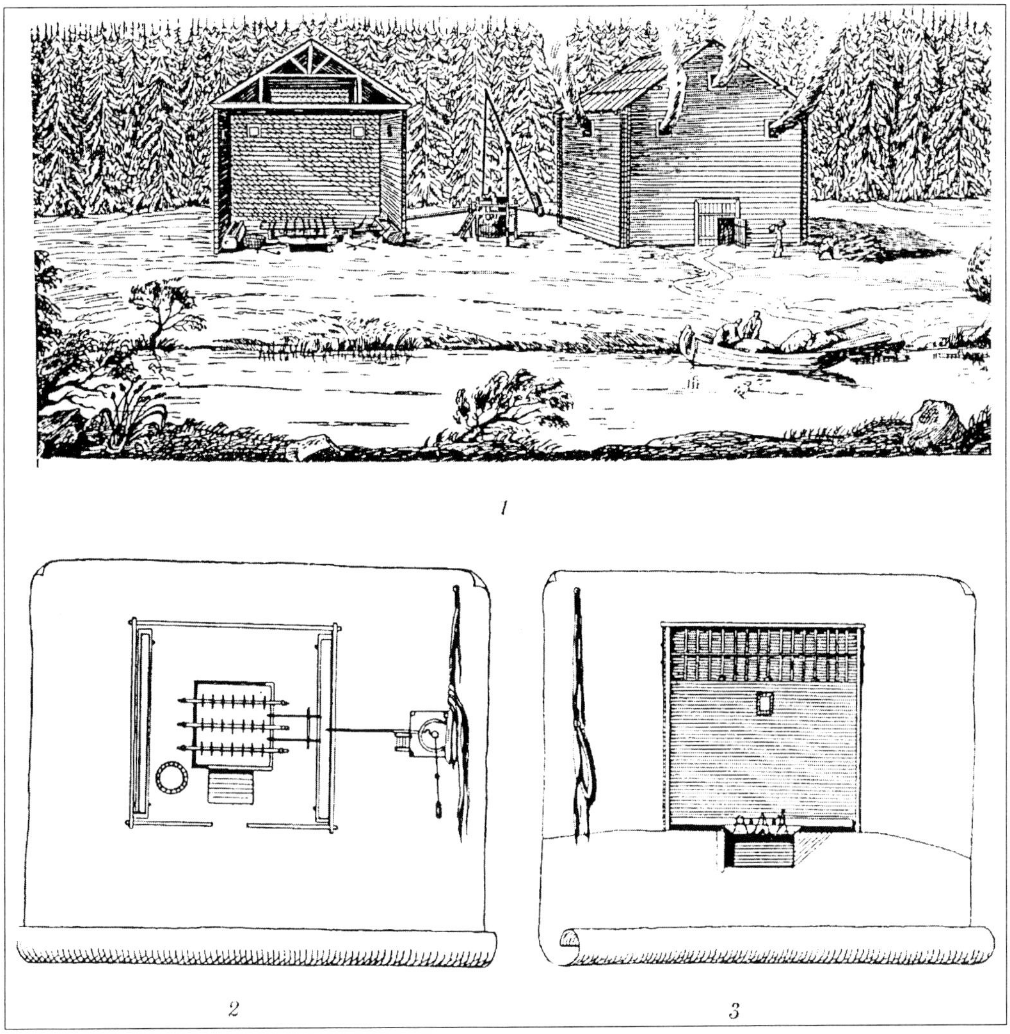

Drawing of a saltern near the Mshaga River made by Swede E. Palmquist, in 1674: (1) common view; (2) plan of the saltern; (3) profile of the building.

Thus the Ural region is very rich in mineral waters, ore deposits, and precious stones. The heaviest Russian golden nugget, "Big Triangle" weighing 36 kg, was found in the Urals in 1842. The area is a natural geological museum where different rocks and minerals lie in shallow beds—sometimes even at the surface—as in the Ilmen preserve. More than 250 minerals have been found there. Some of them are very rare and found only in that area. "Stone fever" began at the end of the 18th century when Cossack Prutov found the first Ilmen blue topaz. Later, amazonites were discovered in Ilmen. And the Urals were famous for malachite.

The immense natural resources of the Urals gave rise to a great number of fairy tales and legends. Russian writer Pavel Bazhov (1879-1950) collected 43 of these ancient stories in his book *Malachite Box*. According to legends, the riches of the Urals belong to the Mistress of Copper Mountain, who could turn into a lizard. She could reward and ruin. She enticed

Danila-master, who dreamed of making a stone flower, into the mountain by promising that he would learn a secret skill. This story was the basis for a ballet by Russian composer Sergei Prokofiev (1891-1953), "Legend about Stone Flower."

The Fairyland of Kamchatka

Geothermal waters attract people's attention mostly in volcanic regions, which have many hot springs and geysers and where the water temperature is higher than in other areas. Kamchatka is the "hottest" region of Russia. There are 127 volcanoes, with 22 active. The Kuril Islands have 100 volcanoes, with 21 active. The first data about the Kurils and Kamchatka were recorded by Siberian Cossack Vladimir Atlasov in 1697-1699. Here investigators discovered a unique fairyland of hot stones and lakes, boiling geysers, and volcanoes.

Atlasov's Cossacks walked on the hot land, bathed in hot lakes, and cooked food in boiling springs surrounded by glaciers and snows. They lowered meat and vegetables in wicker baskets into boiling water. It was not even necessary to salt the food as thermal waters are themselves salty.

In the 17th century, Moscow Russia, rapidly growing both in geographical size and trading areas, sent special expeditions to Siberia and the Far East. In 1648, the Russian explorer Semen Dezhnev discovered a strait between Asia and America, and on one trip, Atlasov reached Kamchatka. Tsar Peter I ordered several large expeditions, although he died before they became a reality. The most important were the Caspian Sea expedition (1715-1720), Messershmidt's expedition to Siberia (1719-1727), and Bering's expedition or the first Kamchatka expedition (1725-1730).

The Academy of Sciences, founded in 1725, also organized expeditions, and the resulting scientific reports included data about temperature measurements in water springs, permafrost zones, caves, and mines of the Urals, Siberia, Altai, and Volga River regions. One such expedition was the "Second Kamchatka" or the "First Academic" expedition (1733-1743), which included renowned scientists. As a result of this expedition, Stepan P. Krasheninnikov's *Description of the Land of Kamchatka* was published in 1756. It recorded the first scientific data about the rivers, lakes, hot springs, volcanoes, and geysers of Kamchatka, as well as descriptions of active volcanoes and geysers and temperature measurements for six groups of hot springs. Of particular interest is Krasheninnikov's description of the South Kamchatka region, around the Paugetka River near the Kambal and Koshelev volcanoes: "Springs gush out in many places as fountains mostly with great noise. Some springs are as lakes in big gaps. Small streams flow from them, combining with each other, dividing all this area into small islands, and falling into the Paudgy [Paugetka] River." In 1739, Krasheninnikov observed the eruption of the Tolbachik volcano.

S. P. Krasheninnikov and *Description of the Land of Kamchatka*

Science is obliged to Krasheninnikov for the first detailed investigations and reliable descriptions of Kamchatka. These provided a unique source of knowledge about this remote part of Russia. The son of a soldier in the army of Peter I, Krasheninnikov experienced his great success due to his own talent and ardor for science. In 1732, at the age of 21, he was sent to study at the Academy of Sciences as one of the best students of the Slav-Greek-Latin Academy. In August of 1733, he was among six students included in the Kamchatka expedition under the command of Vitus Bering.

Quickly, Krasheninnikov's expedition leaders, Gmelin and Miller, recognized his talent and decided to send him to Kamchatka in their place, as a most talented and capable student. Arriving at Kamchatka in October of 1737, Krasheninnikov immediately began to prepare for his first trip inside the peninsula. During his numerous expeditions, he collected rocks and minerals, plants and animals. The field for his investigations was very broad and included scientific observations and investigations, historic documents and archives research, descriptions of languages, modes of life, customs, and legends of the Kamchatka peoples.

Stepan P. Krasheninnikov, important early explorer of Kamchatka. He first arrived on the peninsula in 1737.

Krasheninnikov fulfilled his task perfectly. His book, *Description of the Land of Kamchatka*, is known as a brilliant work in world geographic literature, and it was translated into English, French, German, and Dutch. Unfortunately, many details of the book were omitted in translations. The book was illustrated with engravings by Grimmel, based on drawings by the painter Berkman, who took part in expedition. The book consists of four parts:

A volcano on Kamchatka, engraved by Grimmel from a painting by Berkman.

I. Kamchatka and neighboring countries

II. About advantages and disadvantages of Kamchatka

III. About peoples of Kamchatka

IV. About the subjugation of Kamchatka, about riots of different time and treasons, about the present state of Russian settlements

The Kamchadal, from I. F. Krusenshtern's atlas.

The Kamchadals were a nation of people living in Kamchatka. Krasheninnikov collected many stories and legends from them. Although a critic suggested transferring the Kamchadals' stories to the ethnographic section (Part III) of the book, Krasheninnikov did not agree, and the legends and tales of ancient people of Kamchatka permeate the entire narrative.

Legends and Stories of the People of Kamchatka

ARCHAEOLOGICAL EXCAVATIONS HAVE determined that people settled in

Kamchatka about 15,000 years ago. There were three main nationalities: the Kamchadals, Kurils, and Koryaks. The Kamchadals were as masterful at telling legends and tales as the ancient Greeks. They ascribed miracles to all high mountains, volcanoes, hot springs, and lakes. Malicious spirits dwelled in hot springs and souls of the deceased in volcanoes. Some legends recount moving mountains, which perhaps reflect earthquakes.

The Shiveluch volcano. *From Krasheninnikov, 1786*

Shiveluch Volcano. This mountain was once at the site of Kronotzk Lake, but it was forced to move to another place because the *susliks* disturbed it too much. Two big lakes behind Tiim Mountain were formed by the footprints of Shiveluch as it moved to the new place.

Kronotzk Lake. High mountains surround Kronotzk Lake, two of them higher than others. One of them is Kronotzkaya, and another one is anonymous (now Krasheninnikov volcano). As this

Krasheninnikov volcano in the foreground and Kronotzkaya volcano on the horizon. *From Aprodov*

volcano is flat on top, the Kamchadals tell that Shiveluch volcano, once located at the site of Kronotzk Lake, leaned upon the volcano when it got up, breaking off its top.

Silver Fir Forest. The Shemech River has many hot springs around it and a silver fir forest at its south bank, unique in Kamchatka. This forest is forbidden to the Kamchadals; none dare to even touch it. According to a legend confirmed by many examples, anyone who touches it dies a violent death. They tell also that this forest grew from the bodies of Kamchadal warriors. Starving, they were forced to eat the bark of trees and finally perished of starvation in the forest.

Quarrel of the Kamchadals with Sea Whales. Three cliffs resembling columns are in the sea, near the convergence of the Unagkig River. The cliffs were torn off the coast during an earthquake. At the same time, this part of the coast, together with a Kamchadal settlement, fell into the sea. According to legend, this settlement was destroyed by sea whales after the Kamchadals quarreled with the whales over a knife the whales demanded from them.

The Malicious Spirit Unkalyak. Unkalyak ("Stone Spirit") is a big gorge. The Koryaks tell that a malicious spirit, Unkalyak, lives there. Anyone who passes near this place must sacrifice a stone. If they don't, the spirit sends misfortune. Since people throw stones at the same place, there is now a big pile of stones.

God Kut and His Wife. The Koryaks consider Kut as the God and first inhabitant of Kamchatka. The nearest river to his *yurta* was called *Ualkal-va-em*, "Cheek River," as he marked his dwelling with a whale's jaw. Koryaks now put trees in front of their *yurtas* instead of whale's jaws. The Koryaks also tell the story of Kut's presence. Not far away is an unnamed lake with an island divided in two parts by a hollow. Kut collected birds' eggs on this island. The hollow arose as a result of Kut's fight with his wife over eggs. Kut's wife was lucky and found big birds eggs, but Kut found only small eggs. He was upset and considered her happiness as the reason for his failure. He wanted to take away her eggs, but she resisted. A fight ensued: Kut dragged her by the hair, wishing for revenge. As a result of that fight, the hollow arose in the island.

Stone Boats. Nine versts (a verst is 1067 m) from the upper reaches of the Ozernaya River (Lake River) there is a white rocky mountain with peaks that look like vertical boats. Local inhabitants tell that God Kut, the creator of Kamchatka, lived in this place and used stone boats for fishing in the sea and lake. When he left, he put his stone boats on the mountain. They were surrounded by such great reverence that people feared even to approach them.

Gamul Spirits. The Opala River flows from under Opala hill, the highest near the Penzhinskoye Sea (now Shelekov Gulf). It serves as a beacon for sailors of both seas, although it is 85 versts inland from the Penzhinskoye Sea. The Kamchadals respect it very much and tell terrible things about it. They fear not only to climb it but even to approach its foot, as many Gamul spirits live there. This is why a great number of sables and foxes live there. The Kamchadals say that there is a large lake on the top with many whale bones, as Gamuls eat whales.

Heart of Alaid. Near the south end of Kamchatka is a deserted island, Alaid, consisting of one high mountain with smoke above the peak. The Kurils living near Kuril Lake tell a legend about Alaid. The mountain was originally located in the middle of the lake. Because of its height, it blocked the light to other mountains and constant quarreling occurred. Alaid was forced to leave and reside in the sea. But Alaid left its heart in the lake, in memory of its former location. The stone is situated in the middle of Kuril Lake and is called "Heart-Stone." The path of Alaid is traced by

View of Kuril Lake with the island "Heart-Stone," also called the Heart of Alaid. The view is from the Iljin volcano. The Kambalnaja and Koshelev volcanoes are in the background. *Sketch by V. Svalova*

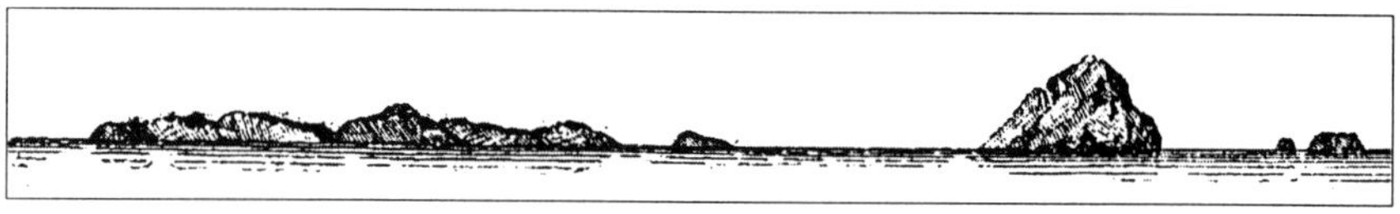

Alaid Island. *From Krasheninnikov, 1786*

Lake River, which arose when the mountain began to move to the sea. Water from the lake rushed after Alaid, carving a path to the sea.

Kamchatka Volcano. The Kamchadals suppose that Kamchatka hill (now Klyuchevskaya hill) is a dwelling of the deceased. It erupts when the dead heat their *yurtas*. The dead catch whales in an underground sea and eat them. They burn the whales' fat for lighting and their bones for firewood. Gamul spirits also live in this mountain, where they boil whales, catching them by night and bringing home 5 or 10 whales skewered on their fingers. This is confirmed by the numerous bones found at all fire-spitting mountains.

Bilyukay. Special spirits live in high mountains where snow always lies, and the main spirit is Bilyukay. The Kamchadals fear to pass near high mountains and volcanoes. Bilyukay drives before him partridges or black foxes. Those who see him will be lucky in hunting and fishing all their lives.

The Kamchadals and Cossacks think that eruptions are the omens of bloodshed. They prove it by many examples. They state that the longer and stronger the eruption, the greater the bloodshed.

Real Story. The Kamchadals suppose that all hot springs and volcanoes are the dwellings of spirits. So they feared to approach them and to show them to the Russian explorers. But Stepan Krasheninnikov knew about the hot springs, and inhabitants of the Shemyachinsky settlement were forced to explain the true reason why they hid their knowledge. Finally, they showed the hot springs to Krasheninnikov but did not come to them. When they saw how the Russians bathed in the springs, ate boiled meat, and drank the water, the Kamchadals thought that they would soon perish. After returning, the Kamchadals told about the audacity of the Russians and were surprised that even spirits could not harm them.

In Conclusion

These stories have covered vast expanses of time and territory, stretching from Russia's western to eastern frontiers. Some stories, perhaps, continued on their journey east to North America, and others returned from the east to Russia. The movement, part of their natures, is annealed in time.

References

Aprodov, V. A. 1982. *Volcanoes.* Misl, Moscow.

Bazhov, P. P. 1948. *Malachite box.* OGIZ Khud.lit., Moscow.

Berngard, M. L. 1986. *Corinth and Argolida.* Arkads, Warsaw.

Goleva, G. A. 1993. *Geysers and hot lakes of Kronotsk preserve.* Progress-Academy, Moscow.

Gordeev, D. I. 1953. *M.V. Lomonosov-founder of geological science.* MGU, Moscow.

Gordeev, D. I. 1954. The main stages of native hydrogeology history. In *Works of Hydrogeological Problems Laboratory, vol. 7.* AN SSSR, Moscow.

Dvorov, I. M. and V. I. Dvorov. 1976. *Thermal waters and their usage.* Prosvesheniye, Moscow.

Frolov, N. M. 1976. *Hydrogeothermics.* Nedra, Moscow.

Krasheninnikov, S. P. 1786. *Description of the land of Kamchatka, vol. 2.* 2nd ed. Empire Academy of Sciences, Saint Petersburg.

Krasheninnikov, S. P. 1786. *Description of the land of Kamchatka.* Reprint 1948. OGIZ Geographic Literature, Moscow.

Larionov, A. K. 1979. *Entertaining hydrogeology.* Nedra, Moscow.

Sherbakova, V. S. et al. 1979. *Tavrida Chersonese.* Tavriya, Simferopol.

Svyatlovsky, A. 1990. *Town at the Earth's end.* World of Art, Moscow.

Ustinova, T. I. 1955. *Kamchatka's geysers.* Geographic Literature, Moscow.

Zavaritsky, A. N. 1955. *Volcanoes of Kamchatka.* AN SSSR, Moscow.

The Author:

Valentina B. Svalova
Scientific Secretary of Geothermal Council of Russia
Head of International Projects Department
Russian Academy of Sciences
Institute of Environmental Geoscience (IEG RAS)
Bakulev Street, 8-104
11 7513 Moscow, Russia
Telephone: 207.80.83
Fax: 7.095.923.1886
E-mail: svalova@geoenv.msk.su

Bath pavilion for women, built in Fatepur by the Emperor Akbar in the 16th century. *D. Chandrasekharam*

22. A Prehistoric View of the Thermal Springs of India

by
D. Chandrasekharam

Abstract: Indian geothermal springs are the centers of myths and religious beliefs for ancient as well as modern Indian civilization. The ancient culture and mythology of India are still alive, and Indian geothermal spring sites bear testimony to this belief. Shiva, the god who tamed Ganga ("the water") during her descent from Heaven, is the ruling deity at the spring sites. Ancient Indian civilization's valuable contribution to modern India is recognition of the medicinal powers of thermal spring waters.

Reprinted by permission of the International Geothermal Association, from the *Proceedings of the World Geothermal Congress,* 1995, pp. 385-387.

Introduction

The Indian subcontinent hosts nearly 320 thermal springs, which are found along major faults and subduction zones. Indian history does not refer to the ages of the springs. However, since these springs are associated with basaltic volcanism and rifts, we may assume that they started flowing before the arrival of human civilization.

Ancient Indian civilization, one of the oldest in the world, is beset with myths and religious beliefs. Its mythology is entirely different from that of the western world, centering on gods, demigods, and spirits that remain part of the living culture in both scientific and nonscientific society. This becomes clear on visiting any of the 320 thermal springs in India. These thermal springs are considered to have divine power, to be gifts from God.

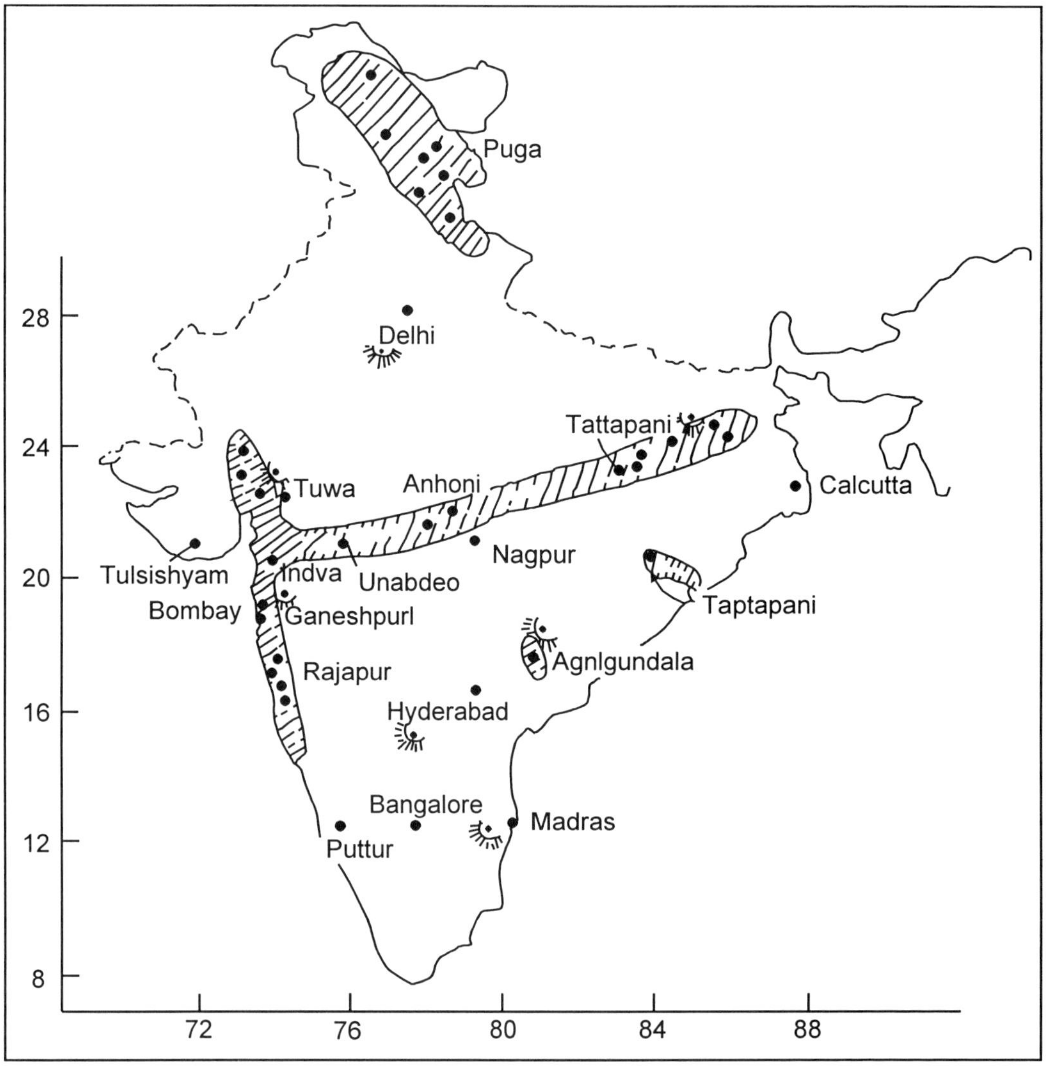

Map showing the location of thermal springs (dots) and certain important archaeological sites (half circles) in India. Hatched areas represent rift and collision zones.

Indian historical literature refers to the magical powers of "rarified waters." Megasthenes, a Greek ambassador who visited India in the 3rd century B.C., used this term to refer to Indian waters with extraordinary powers (Kalota, 1978). The medicinal value of thermal springs and the beliefs and myths associated with these holy waters also are recorded in Indian history, and available historical information is included in this chapter.

Archaeological Excavations

Many Indian archaeological sites are near thermal springs, as is seen on the map. The excavation in Devnimori, a village in Gujarat State, is one such site, located northeast of the

Tuwa thermal springs *(Lat: 22° 45-22° 50 N; Long: 73° 22-73° 29 E).* Archaeological excavations in the area have unearthed several ancient Shiva temples (Shivalinga temples, to be precise) and Buddhist *stupas* ("tombs").

From information collected by the author during interviews with pilgrims visiting the springs, a legendary belief among the villagers and pilgrims was that when exiled *Pandavas* (characters from the epic *Mahabharata*, the storehouse of myths) were passing through the area, Draupadhi requested one of her husbands to fetch some water to quench her thirst. In response to her request, Bhima, not finding any source of water in the area, punctured the earth with his arrow, making a hole through which water started to flow. The discovery of a Hanuman temple during the excavations lends support to this belief. Hindu settlers thus considered the place a sacred place of worship, and they constructed Shiva and Hanuman temples here (Hanuman and Bhima were *Vayu Putras*, sons of Vayu).

Ganga, consort of celestial gods, brought to earth by the saint Bhagiratha (12th century sculpture). *D. Chandrasekharam*

The descent of Ganga from Heaven. The snake represents Shiva's head (7th century sculpture). *D. Chandrasekharam*

The reason for making Shiva the deity of worship at this site (and at almost all thermal spring sites in India) is that the *ganga* ("the water") flows through the Goddess Ganga sitting over Shiva's head. Hindu mythology states that when Ganga, the consort of celestial gods and the daughter of *Himavan* ("the mighty Himalayas"), descended from the great Himalayas, Shiva stood below the mountains to break her fall and allowed the water to flow through his matted locks, dividing it into seven rivers. Thus, in Hindu mythology, Shiva and Ganga are inseparable.

Mask of Shiva covering Shivalinga, with Ganga on top of Shiva's head. *D. Chandrasekharam*

Four Shivalinga temples were excavated in Devnimori (Mehta and Chowdhary, 1966). Subsequent to Hindu settlers, the Buddhists who migrated to various parts of the country also considered Devnimori sacred, and they established camps there. Thus, Bimbisara, an ancient Indian ruler who held his first counsel after

Buddhist temple near Rajgir thermal spring, Bihar. The spring is at the lower right side. *D. Chandrasekharam*

Buddha's death, constructed a Buddhist temple near the site of Rajgir thermal springs.

The sacred beliefs held by ancient Indians maintain strong roots in modern Indian civilization: all the thermal spring sites have become sites of worship and centers of pilgrimage today, and at each spring site, a Shiva temple is built. Both ancient and modern Indian civilizations worship Shiva as *Linga,* an erect phallus symbolizing power and creation. The Shivalinga temples lend support to these beliefs at Vajreswari, Ganeshpuri, and Rajapur in Maharashtra; at Manikaran in Himachal Pradesh; and at Taptapani, Orissa.

Shiva worshiped in the form of *Linga* (1st-4th century sculpture). *D. Chandrasekharam*

Therapeutic Value of Thermal Springs

Besides giving holy status to thermal springs, ancient Indians realized their therapeutic value and considered the waters cures for certain diseases. Megasthenes mentioned the medicinal value of Indian waters, calling them "rarified waters" that gave Indians intellectual powers in the fields of arts and crafts and long lives in comparison with the Greeks. Though he did not define "rarified waters," we may interpret them as waters rich with minerals and gases. Thermal springs can be considered such waters, and this could be the reason thermal spring sites were used extensively for bathing. Ancient kings, including the Mughals who invaded India, constructed tanks for bathing at the springs. Thus the belief that the waters cure certain diseases is not new to modern India, and the therapeutic powers of some Indian springs are listed in the following table.

The ancient Indian medical science of *Ayurveda* describes water as the main component for curing many diseases. Similarly, the Hindu *Vedas,* four ancient sacred books of Hindu knowledge and belief written as psalms, chants, and sacred formulae, consider water as one of the *Panchabhutas,* or five divine elements: *ap* ("water"), *prithvi* ("earth"), *tejas* ("fire"), *vayu* ("air"), and *akasa* ("sky"). One of the four *Veda* texts was called the *Rigveda*, and in this volume water was recognized as the first element of the *Panchabhutas*. Eventually the theory of

Pilgrims having a holy bath in Rajgir thermal spring. *D. Chandrasekharam*

Indian thermal springs and their therapeutic value

State	Location	Therapeutic value
Maharashtra	Vajreswari	Cures leprosy, gout, goiter, paralysis
Maharashtra	Ganeshpuri	Cures leprosy, gout, goiter, paralysis
Maharashtra	Rajapur	Cures rheumatism
Orissa	Taptapani	Cures skin ailments
Madhya Pradesh	Anhoni	Cures skin diseases
Bihar	Rajgir	Cures rheumatism, paralysis, dyspepsia, leucoderma

the *Panchabhutas*, which is Upanisadic in origin, became the doctrine of the integrated concept of the Hindu orthodox system, one also followed by Buddhists and Jains, as is seen in the paintings from Bali, Indonesia (Subbarayappa, 1970).

Ancient Chinese civilization adopted a concept similar to *Panchabhutas* that was known as the "Five Elements": soil, water, fire, wood, and gold (Wang Ji-Yang, 1995). Thus the idea of water as the primary component in curing diseases and in keeping the body healthy was recognized since Vedic times and is still followed by modern Indians.

The Mughals, who invaded India in 1398, paid less attention to the medicinal and religious values of the thermal springs, preferring to develop beautiful gardens with all available sources of water (Rawlinson, 1954; Jellicoe and Jellicoe, 1975). After the fall of the Mughal dynasty, the Hindu faith in thermal springs revived and it continues today.

Post-independent India, too, realized the medicinal value of thermal springs and performed extensive chemical analyses of the waters (Gosh, 1954; Chaterjee, 1958; Banerjee, 1967). The curative powers of the waters lie in their high radon (about 9 millimicrocuries per liter) and sulfur concentrations (Banerjee, 1967).

The idea of exploiting Indian thermal springs for commercial purposes emerged in 1940 with the Geological Survey of India, which realized that the quality of water from Indian thermal springs was comparable to spring water imported from European countries (Chaterjee, 1958). Springs without religious significance were selected for commercial development and the waters from them classified as to medicinal or table use. Waters from Surajkund and Kawa Gandhawan in Bihar were medicinal waters; and waters from Phillips Kund, Rishi Kund, Rameswar Kund, and Lachmiswar Kund in Bihar were table waters. The last table shows Indian thermal spring waters with chemical and medicinal properties similar to those of European waters imported before World War II. M/s G. F. Kellner and Co. started bottling Indian thermal spring waters prior to World War II but later abandoned the project (Chaterjee, 1958).

Indian waters equivalent therapeutically to imported European waters

Name of thermal spring in India	**Equivalent therapeutic value to European water**
Rajgir, Bihar	Cure metabolic disorders
Bridkhal, Uttar Pradesh	Table water
Kawa Gandhawan, Bihar; Anhoni, Madhya Pradesh	Hyperacidity
Suraj Khund, Bihar; Sahasra Dhara, Uttar Pradesh	Dry eczema, neuralgia, catarrh of the lungs
Vajreswari, Bombay; Unhavre, Maharashtra	Blood pressure, arthritis, neuritis, urinary infections

Conclusion

Though modern science can explain the origin of these thermal springs and their mystical powers, modern Indian civilization still considers them as divine manifestations of God. This religious belief, together with the therapeutic powers of the thermal springs, made ancient Indians choose the sites as pilgrimage centers. The belief remains deeply rooted in modern Indian civilization, as confirmed by the large numbers of pilgrims visiting the sites today.

One of the remarkable features of Indian mythology is its homogeneity over the entire continent, a fact that is evident from the presence of Shiva temples at almost all of the thermal spring sites. Because India retains its ancient beliefs and myths, these sites continue to be centers of worship and pilgrimage. The depth to which these beliefs have penetrated all Indians can be seen from the number of people visiting the springs. The therapeutic powers of the springs further strengthen these beliefs among Indians, and the beliefs have spread to other Indian religions.

Selected References

Banerjee, S. 1967. Supplement to the mineral springs of India by Gosh. *Indian Miner,* 21, 288-327.

Chaterjee, P. K. 1958. Mineral springs of India. *Indian Miner,* 12, 116-125.

Gosh, P. K. 1954. Mineral springs of India. *Geological Survey of India Records,* 80, 541-558.

Jellicoe, G. and S. Jellicoe. 1975. *The landscape of man.* Thames and Hudson Ltd., London.

Kalota. 1978. *India as described by Megasthenes.* Concept Publishers, Delhi.

Mehta, R. N. and S. N. Chowdhary. 1966. *Excavations at Devnimori.* Department of Archaeology and Ancient History. M.S. University of Baroda.

Rawlinson, H. G. 1954. *India—A short cultural history.* The Cresset Press, London.

Subbarayappa, B. V. 1970. India's contributions to the history of science. In *India's Contribution to World Thought and Culture.* Vivekananda Rock Memorial Committee, Madras.

Wang Ji-Yang. 1995. Historical aspects of geothermal energy in China. *Proceedings of the World Geothermal Congress,* 389-394.

The Author:

D. Chandrasekharam
Department of Earth Sciences
Indian Institute of Technology
Bombay, India
Telephone: 91.22.5776568
Fax: 91.22.5783480
E-mail: dchandra@geos.iitb.ernet.in

Liubeiting ("floating cup pavilion"), where hot spring waters move through gracefully undulating stone channels. A cup of liquor or wine, placed in the channel on one side of the pool, would be warmed by the time it had floated to the other side. *Drawing by V. Svalova*

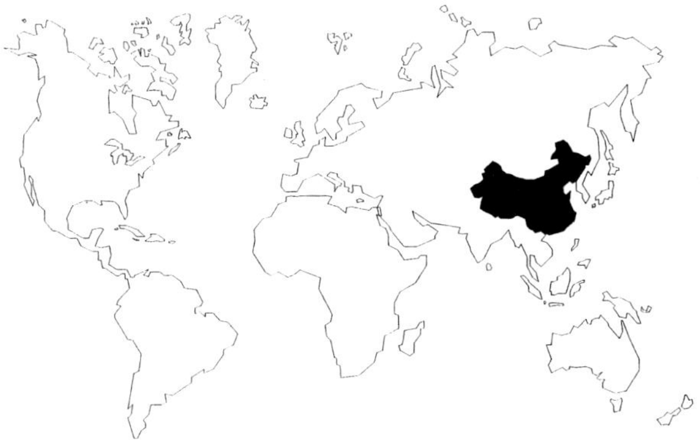

23. Historical Aspects of Geothermal Waters in China

by
Wang Ji-yang

Abstract: China has a long history of using geothermal waters. Early peoples used hot springs for irrigation and domestic purposes. During the Han Dynasty, salt was extracted from thermal waters in the Zigong area of Sichuan Province. In the Ming Dynasty, Li Shi-zhen, a famous medical doctor, used hot spring waters for treating diseases. This chapter discusses historical aspects of geothermal waters used in China, including legends and myths about hot springs, hot springs used for recreation and luxury, tea and wine culture at hot springs, and hot spring use for medical treatments.

Reprinted by permission of the International Geothermal Association, from the *Proceedings of the World Geothermal Congress,* 1995, pp. 389-394.

INTRODUCTION

CHINA HAS OVER A 3,000 YEAR TRADITION OF USING GEOTHERMAL waters. Although early peoples used hot springs for irrigation (both for heat and water) and domestic purposes, such as washing clothes and cooking food, most hot spring use has focused on disease treatments and recreation. Li Shi-zhen, a famous medical doctor from the Ming Dynasty (1368-1644), advised people, "If you get ill, go to a hot spring area and take a bath." In Xiaotangshan, a hot spring area near Beijing, two thermal water pools were sunk in 1666, the fifth year of the Emperor Kangxi of the Ching Dynasty, for the Empress Dowager Cixi, especially for her bathing. In Huaqingchi, a hot spring area not far from Xi'an City, the ancient capital of the Tang Dynasty, quite a fancy bathing house was built for the private use of the Imperial Concubine, Yang Gui-fei.

While using hot spring waters, early peoples tried to understand the origin, the genesis of the springs. Li Bai, the great poet in the

Tang Dynasty, wrote, "Where underground fire exists, there must be hot springs." Another great poet from the Song Dynasty (1127-1279), Su Shi, said, "Where fire-mountain exists, there would be boiling springs."

In addition to hot springs, thermal waters were used for salt extraction in some inland areas where sea water salt was not available. During the Han Dynasty (206 B.C.-220 A.D.), salt was extracted from thermal waters in Zigong City, Sichuan Province. Afterward, the city became the major salt-supply area for southwestern China, gaining the name of "Salt Capital."

Drilling for hot brine with cable tools in Zigong City during the Yuan Dynasty about 1200 A.D. This may be one of the world's first depictions of drilling. *From* Energy in China, *published by the Ministry of Energy, 1989*

Legends and Myths about Hot Springs

A number of legends and myths about hot springs exist from the early days of Chinese civilization. Although there were no divinity cults, several mythical stories exist in Chinese history. This is the most famous:

There appeared a magic hot spring, called the "Fairy Spring," northwest of Lishan Mountain near Xi'an City. The hot spring was magical and could treat almost every kind of disease.

However, before you got into it, you had to pray to the hot spring by sacrificing animals, such as oxen, pigs, and sheep; otherwise, your disease could not be treated but became even worse. One day, Qin Shi-huang, the first Emperor in China of the Qin Dynasty (221-206 B.C.), traveled to that spring and jumped into the hot spring water quickly, without praying. Quickly, Qin Shi-huang felt uncomfortable and found that his skin was swollen. At that instant, a fairy girl emerged from the hot spring waters and asked Qin Shi-huang to apologize. Qin Shi-huang apologized for not following the rule of the hot spring and he prayed. Immediately, he felt well again and enjoyed the hot waters.

Another magical hot spring was found in Zhejiang Province. During the Tang Dynasty, the daughter of the governor of Zhejiang Province became seriously ill and no treatment could be found. After she went to the hot spring and took a bath, the disease was gone and she recovered rapidly.

A third magical hot spring was on Lushan Mountain in Jiangxi Province. This spring was very good for skin and sexual diseases. It is believed to be the only hot spring in China recorded as treating sexual diseases in ancient times.

This famous poem describes an era of Chinese history when both mankind and nature bled, a time when the hills were red with war and the volcanoes with erupting fire.

The Red Hills

by Pao Chao (died 466 A.D.)

Red hills lie athwart us as a menace in the west,
And fiery mountains glare terrible in the south.
The body burns, the head aches and throbs:
If a bird light here, its soul forthwith departs.
Warm springs
Pour from cloudy pools
And hot smoke issues between the rocks.
The sun and moon are perpetually obscured:
The rain and dew never stay dry.
There are red serpents a hundred feet long,
And black snakes ten girths round.
The sand-spitters shoot their poison at the sunbeams:
The flying insects are ill with the shifting glare.
The hungry monkeys dare not come down to eat:
The morning birds dare not set out to fly.
At the Ching River many die of poison:
Crossing the Lu one is lucky if one is only ill.
Our living feet walk on dead ground:
Our high wills surmount the snares of Fate.
The Spear-boat General[1] got but little honour:
The Wave-subduer[2] met with scant reward.
If our Prince still grudges the things that are easy to give,[3]
Can he hope that his soldiers will give what is hardest to give?[4]

[1]Hou Yen (1st century B.C.)
[2]Ma Yüan (1st century A.D.)
[3]Rewards and titles
[4]Life

Hot Springs for Recreation and Luxury

The most famous hot spring used for luxurious recreation was Huaqingchi hot spring from the Tang Dynasty. Development started at the hot spring with bathing and treatment facilities built in the Qin Dynasty. In 644 A.D. of the Tang Dynasty, a palace called *Tangquan*, which means "thermal spring," was built around the spring, which is at the foot of Lishan Mountain. In 747 the name of the palace was changed to *Huaqing*, "glorious purity," meaning that the water looked crystal clear and was sweet tasting, and everybody would be kept young forever by drinking the spring water everyday and bathing in it all the time.

Huaqing ("glorious purity") hot spring in the foothills of Lishan Mountain. The most luxurious imperial palace during the Tang Dynasty was Huaqing Palace, built in 747. *Drawing by V. Svalova*

This palace was the most luxurious imperial palace in the Tang Dynasty. The Emperor Tang Xuan-zong and his beloved Concubine Yang Gui-fei visited it often, spending most of their leisure time there and taking a daily bath. Lines from the poem "Changhenge," by Bai Ju-yi, a famous poet of that time, vividly describe the luxurious life of Emperor Tang and Concubine Yang at Huaqing Palace: "It feels so good bathing in Huaqing hot spring in the early days of the still-cold spring season, and the spring water can clean everything out of the body, even the 'make-up' of Concubine Yang."

Another famous place for luxurious hot spring recreation is the Xiaotangshan hot spring area, 25 kilometers north of downtown Beijing. This hot spring area was first recorded in *The Annals of the Yuan Dynasty,* about 800 years ago. Since then, the royal family, the noble lords, and high-ranking officials in the Ming (1368-1644) and Qing (1644-1911) Dynasties have visited this area often and even built mansions here. Two hot spring pools were sunk in 1666, and the remains of the Empress Dowager Cixi's bathing pool still can be seen.

Close-up of two hot spring pools in the Xiaotangshan hot spring area near Beijing. The pools were built in 1666, the fifth year of Emperor Kangxi of the Ching Dynasty, for the Empress Dowager Cixi. *Drawing by V. Svalova*

Tea, Wine, and Liquor Culture at Hot Springs

It is well-known that the so-called "tea, wine, and liquor culture" in China has a long history and that high-quality waters were—and still are—of significant importance for good tea, good wine, and good liquor. The most famous Chinese liquor, *Maotai*; the high-quality Chinese beer, *Qingtao*; and *Zhangyu*, a good wine, are all made from the spring water. In ancient times, people also used hot spring water for tea, such as hydrogen sulfide hot spring water from Huangshan Mountain in Anhui Province. It was believed that tea made from this water was good for the health.

In Hangzhou City, famous for the beautiful West Lake and an excellent green tea, *Longjing*, which means "dragon's well," there exists a famous spring called *Hupao*, "running tiger." The spring water is so clear, pure, and sweet that the *Longjing* tea made here is marvelous. It is said in Hangzhou that the "two uniques" (*Longjing* tea and *Hupao* spring) make the West Lake more beautiful. The Japanese tea ceremony may have derived from the ancient Chinese tea culture.

Hupao ("running tiger") spring in Hangzhou. *Drawing by V. Svalova*

In Zunhua County, Hebei Province, a famous hot spring called *Tangquan*, which means "hot-water spring," is located east of the Eastern Royal Tombs of the Qing Dynasty. Legend says that the Emperor Tang Taizong of the Tang Dynasty stayed here while on an expedition to the east. He bestowed on the spring the name *Fuquan*, "lucky spring," after watching his soldiers take a bath here and recover quickly from illness and fatigue. To celebrate, he gave a banquet near the hot spring and used the flowing hot spring water of about 60° C to warm the liquor. Soon after, the Fuquan Temple was built, and the Fuquan mansion plus the spring pool followed. In addition, a pavilion called *Liubeiting* ("floating cup pavilion") was built next to the spring pool in memory of the event. From the photo, it can be seen

Liubeiting ("floating cup pavilion"), next to the hot-water pool of *Tangquan* ("hot water spring") spring in Zunhua County, Hebei Province.

From Dragon to Frog: The First Earthquake Recorder

The first instrument to record earthquakes known in any detail was a scientific device invented in about 132 A.D. by a great Chinese scientist of the Han Dynasty, Zhang Hen. His invention was called a seismoscope, which, unlike a seismograph, did not record the complete history of the earthquake, only the direction of the strongest impulse.

The instrument was beautiful. Golden metal dragons were affixed upside down at intervals around a large golden vessel and golden frogs with open mouths stretching upwards were placed on the floor beneath each dragon. Balls were held in the dragons' mouths, which were probably connected by linkages to a vertical pendulum, as shown in the drawing. Shaking from earthquakes would release a ball.

For example, an earthquake moving from right to left would move the container to the left (assuming it is attached to the floor). However, the pendulum would remain stationary (Newton's 1st law of motion) and strike against the right side of the container, hitting the linkage and opening one dragon's lower jaw. This would allow the ball to fall into the frog's mouth, recording the direction of the quake. The linkage in the cross section is unknown but assumed, based on records.

It was reputed that when the device was first brought to the Chinese court, a ball fell into a frog's mouth one day but no earthquake was felt. Zhang Hen was in disgrace. Days later, a horseman came racing up from a distant place in the direction of the fallen ball, telling of a great earthquake: the seismoscope had worked. Zhang Hen was elevated again to a position of respect, but the use of his instrument seems to have died out in a short while.

Text by John W. Lund based on a reference from Earthquakes *by Bruce A. Bolt, published by W.H. Freeman and Company, New York, 1993. Drawing by Toni Boyd and John W. Lund*

A seismoscope displayed at an exhibit in San Francisco, California.

that a cup holding liquor or wine warms up while floating from one side of the pool to the other. This is an unique aspect of Chinese culture from ancient times. One can imagine how fantastic such a banquet would be!

Another interesting story about Chinese wine and liquor culture and hot springs comes from Western China, along the "Silk Road" to Europe. A famous spring there, called *Jinquan* ("golden spring"), is in Gansu Province on the edge of Jiuquan County. During the West Han Dynasty (140-117 B.C.), the great general Huo Qu-bing commanded his soldiers to move quickly to the west and win the war. Wudi, an emperor of the Han Dynasty, gave liquor for the victory celebration. Clearly, it was impossible to share the liquor with so many soldiers. Instead of drinking alone, General Huo decided to pour the liquor into the spring water and share the "spring-liquor" with the whole army. Since then, the spring has been called *Jiuquan* ("liquor spring") and has become more and more famous as the story has spread.

Hot Springs for Treatments

It is clear, then, that hot springs often cure diseases and in early days people used hot springs to treat various diseases. With time, people studied the effects of hot spring water on human beings and tried to interpret the mechanism of "water-body" interaction using the theory of Chinese traditional medicine. For instance, Ban Gu of the East Han Dynasty thought that hot springs belonged to "fire," of the so-called "Five Behaviors of the Universe" (gold, wood, water, fire, and soil). Therefore, the "cold" diseases, such as arthritis, rheumatism, and digestive diseases, must be treated with hot spring water. On the other hand, the "fever" diseases, such as typhoid and yellow fever, should never be treated with hot springs. Sulfur hot springs are very good for skin diseases, because skin diseases are always treated with sulfur alone in traditional Chinese medicine.

In Inner Mongolia, a "holy" hot water spring, the Arsan, has a temperature of about 50° C. The spring water is effective for treating fibrous tissue syndrome, arthritis, and psoriasis. As a result, the local herders believe it is "holy water."

Tanggangzi, which means "hot water hill," is the hot spring most famous for medical treatment in China. It is in the southern suburb of Anshan City, the largest steel-producing city, in Liaoning Province. The spring was found in 644 of the Tang Dynasty, and a county called *Tangchi*, "hot

Arsan ("hot holy water") hot spring in Inner Mongolia. *Drawing by V. Svalova*

water pond," was set up around the hot spring area in the Liao and Jin Dynasties. During the Qing Dynasty, more and more people came here for bathing and hot mud treatments. In 1950, a sanitarium was built here, and today the Tanggangzi Sanitarium is a well-appointed physiotherapeutic hospital, one of the largest physiotherapeutic centers in China.

Conclusion

THUS CHINA HAS USED GEOTHERMAL WATERS FOR OVER 3,000 YEARS. HOT SPRINGS WERE GENERally regarded as valuable natural gifts and many cities, towns, and villages have grown up around hot spring areas. The most famous ones are the cities of Fuzhou and Zhangzhou, and the former is now the capital of Fujian Province.

Fuzhou has a "hot spring" street in the central downtown area of the city, and local residents have used the hot spring water for bathing since the early days of settlement. Zhangzhou City is located in the so-called "Golden Triangle" of southern Fujian Province. Many people living in Taiwan have emigrated from here, and Mrs. Akino, the former president of the Philippines, originally came from the Zhangzhou area. Several years ago, she made a short visit to her hometown and took a hot spring bath. Today, Zhangzhou is the biggest Chinese center for nonelectrical uses of geothermal resources. Besides spas and bathing houses, fish farming and vegetable drying are the main activities.

The only Chinese hot springs inspiring terror and fear are found in Taiwan and in the Tengchong volcanic area of southwestern Yunnan Province. Located about 12 kilometers west of Taibei (Taipei) City, Beitou hot spring is famous for its Geothermal Valley (Hell Valley) because of its mystical, terrible nature. Geologically, Beitou hot spring is related to the Datun active volcano and has a number of high-temperature geothermal manifestations, which gave the valley its name. However, the Beitou hot spring water is very good for treating beriberi, favus of the scalp, and scabies because the water contains a lot of sulfur. Historically, sulfur mining in Taiwan was concentrated in the Datun volcanic zone, including the Beitou hot spring area. The earliest record of sulfur mining in Taiwan occurred in the Yuan Dynasty, about 800 years ago.

A second hot spring with a slightly mystical nature is situated in Tainan County of southern Taiwan. Celebrated for its "Water-Fire-in-One" spectacle, hot water with a temperature of about 75° C and natural gas gush out of a fracture at the same time. If the gas is ignited, flames will burn and flicker on the water surface, forming a "Water-Fire-in-One" picture.

Hot springs and other geothermal manifestations, such as "sulfur pond" (or "boiling pot"), "hot sea," and "frog mouth" fumarole, from the Tengchong volcanic area in southwestern Yunnan Province on the border with Burma and close to Thailand, are most spectacular and scenic but never cause any fear or terror. The same thing is true for the many geothermal manifestations in Tibet, such as geysers, steaming ground, and hydrothermal explosions.

Acknowledgments

The author is indebted to Drs. Raffaele Cataldi and John Lund. My colleagues, Profs. Chen Mo-xiang, Wang Ji'an, and Liao Zi-jie, are thanked for providing material and valuable comments on the manuscript.

Selected References

Chen Yan-bin. 1993. *Hot springs in China.* Zhoughua Publishing House.

Hu Su-min, Lu Jin-kai, and Zhu Guang-jian. 1992. *Famous springs in China.* Geological Publishing House, Beijing.

Liao Zi-jie. 1975. The history of geothermal development in Taiwan. *Geological Review,* 31, no. 3, 285-288.

Li Zhong-zun. 1987. The history of discovery, exploitation and utilization of geothermal energy in China. In *Selection of Hydrogeology and Engineering Geology.* Geological Publishing House, Beijing.

The Author:

Wang Ji-yang
Institute of Geology
Academia Sinica
Beijing 100029, China
Telephone: 86.10.620.27766
Fax: 86.10.64919140
E-mail: JYWLPX@public3.bta.net.cn

Commemorative Shinjeong Stele at Onyang hot spring.
The stele was erected in the reign of King Songjong (1469-1494).

24. Historical Review of Hot Spring Waters in the Republic of Korea

by
Byoung Woo Yum

Abstract: This chapter focuses on ancient uses of hot spring waters in Korea. Because the Southern Korean Peninsula has very poor geothermal resources due to its stable tectonic setting, geothermal springs, commonly known as hot springs, are restricted to a few locations. Koreans love to bathe in hot springs, using them through millennia for medical therapy, recuperation, and relaxation. Citations about historical uses of hot springs and mineral waters include 173 articles in 20 historical books and annals, such as the *Samguk sagi, Samguk yusa, Koryo sa, Choson wangjo sillok,* and *Tongguk yoji sungnam.*

Introduction

For many centuries, low-temperature, nonvolcanic geothermal waters, commonly called hot springs, have been used for medical therapy and bathing in 14 locations in the southern half of the Korean peninsula, in the "old geothermal areas" (Yum, 1994). Historical records of Korean hot springs were first reviewed by Cheong (1983). The histories and legends of hot spring uses before the Industrial Revolution, starting with the Japanese invasion in 1910, are found in several very old historical books and annals, cited separately at the end of the chapter. The legends often include tales of wounded animals, such as deer, cranes, and other birds that are cured by bathing in hot springs.

Much hot spring water is warm and slippery, and people are convinced that the water can cure eye and skin diseases; a hot spring called "Heaven's Blessing" has been used, with limited medical benefits, to treat such illnesses. People who live near such hot springs believe that the springs have the power to cure every disease, and tales of miracles pass quickly from ear to ear. Thus the ponds are crowded, filled by patients with many diseases, including leprosy, who dig to expand the ponds, making small villages. The original dwellers may not like this, fearing the spread of contagious infections brought by the diseased people. They sometimes get rid of the ponds, soaking a dead dog in them during the night. This effort to protect the villages reflects an aspect of the Korean national character, for the village is the basic unit of Korean society. Korean national traits are reflected in other legends, as well.

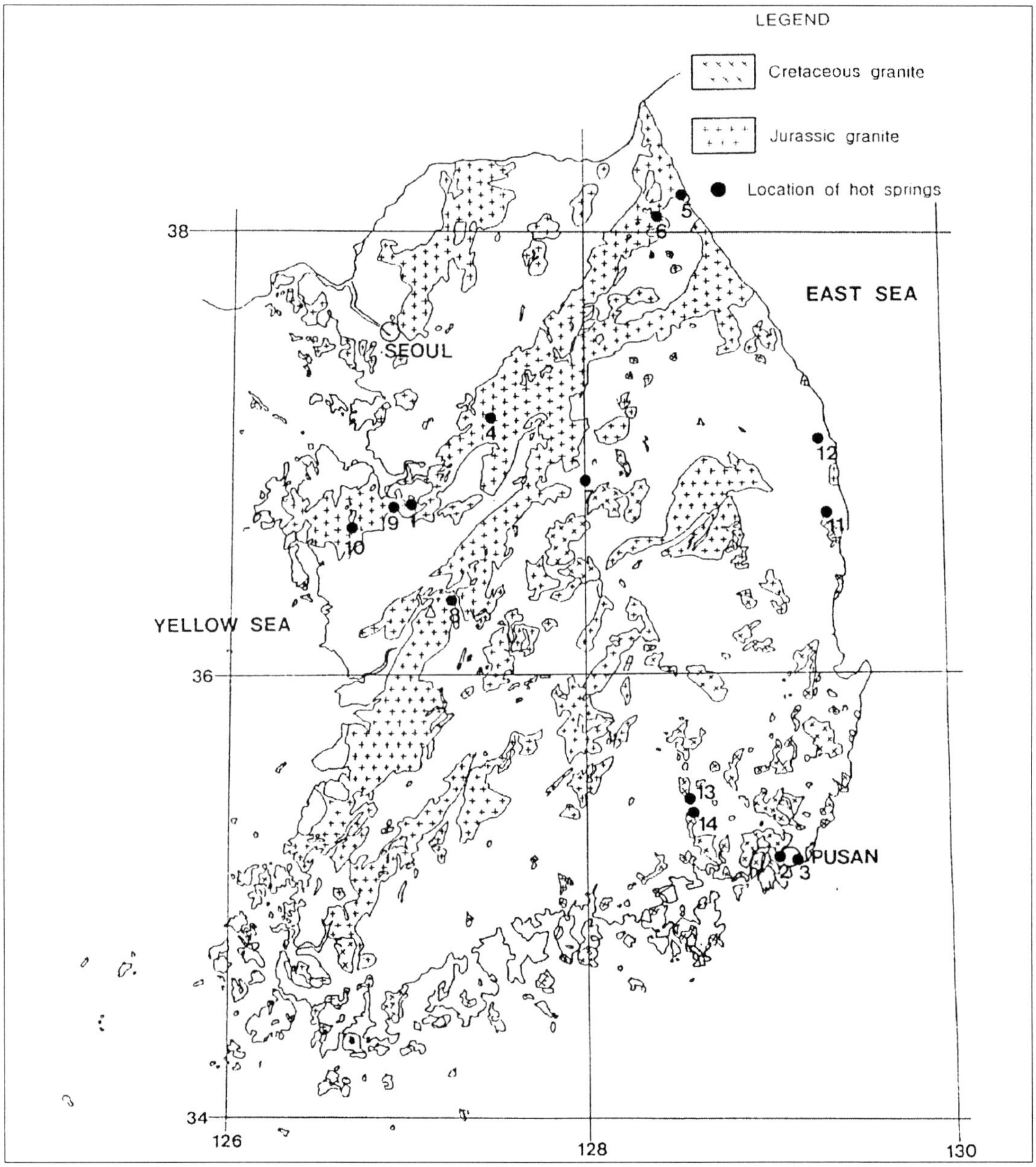

Map of ancient Korean hot springs, shown in relation to Korean granites, modified from Chwae, 1995, printed with permission: 1. Onyang, 2. Donglae, 3. Haeundae, 4. Icheon, 5. Cheoksan, 6. Osaek, 7. Suanbo, 8. Yusong, 9. Dogo, 10. Deoksan, 11. Baekam, 12. Deokgoo, 13. Bukok, and 14. Mageumsan hot springs. Each hot spring is discussed in the text.

Geologic and Tectonic Settings of the Korean Peninsula

The Korean Peninsula, which extends southward from Northeastern Asia, has a close affinity with the Asian continent in geology and tectonic setting. It is regarded as a stable, cratonic land mass. The Korean Peninsula has no geological resemblance to the islands of Japan, which are characterized by active volcanism (Kim, 1987).

Most Korean hot springs are located in granitic areas, Jurassic and Cretaceous in age, and are low-temperature geothermal resources, nonvolcanic in origin. The chemistry of geothermal waters verifies that the hot spring waters are alkaline (pH = 8.0-9.5), fluoride-rich (5-15 mg/l), mainly low in total dissolved solids (130-800 mg/l), and of the Na-HCO_3 type (Lim, 1995). The map shows a close relationship between the granites and the hot spring locations.

Historical Records for Korean Hot Springs

Historical records show that Korean hot spring and mineral water resources were used in 43 locations—the number is larger than expected and includes the northern Korean Peninsula territories—according to a total of 173 articles in 20 books and annals, such as *Samguk sagi*, *Samguk yusa*, *Koryo sa*, *Choson wangjo sillok*, and *Tongguk yoji sungnam*. The oldest record, found in *Tongsa kangmok*, says that the brother of Sochon Wang (reign 270-292) bathed in a hot spring in 286 A.D. during the Koguryo period (37 B.C.-668). (Note: The word "Wang" and the suffixes "-jo" and "-jong" indicate the name of a king.) These records tell us that the royal family enjoyed bathing in hot springs from the early Three Kingdom period (18 B.C.-668).

In the Paekche period (18 B.C.-660), the histories of Onjeong-gun ("Warm Spring Province") and Onsoo-gun ("Warm Water Province") were included in several books, such as *Samguk sagi*, *Koryo sa*, and *Tongguk yoji sungnam*. Another book, *Samguk yusa*, recorded that a prime minister during the reign of Sinmun Wang (681-692) bathed in Donglae hot spring.

Korean Dynasties (Lee, 1984, printed with permission)

Ancient Korea, 2333 B.C.-108 B.C.
Three Confederated Kingdoms of Samhan (the Three Han States):
- Shilla, 57 B.C.-935 A.D.
- Koguryo, 37 B.C.-668 A.D.
- Paekjae, 18 B.C.-660 A.D.

Kaya, 42-562 A.D.
Parhae Kingdom, 669-928
Unified Shilla, 618-935
Koryo, 918-1392
Choson, 1392-1910
Japanese annexation, 1910
Republic of Korea, 1948

Through these records, we know that the royal family and people in the governing groups enjoyed bathing in hot springs from the time of the Three Kingdom period, and that ordinary people were permitted to do so from the reigns of Sejong (1418-1450) and Munjong (1450-1452) in the Choson period (1392-1910).

Reaching the hot springs required a 7- to 10-day journey from the capital, and rulers used the trip to view the lives of ordinary people. The rulers enjoyed the visits to the hot baths, and on their return journeys were inclined to be merciful. For example, an earlier king named Munjong (reign 1046-1083) in the Koryo period lowered several provincial taxes on the way back to the capital, and Sonjong (reign 1083-1094) gave vacations to officers who wanted to bring their old sick parents to the hot springs (*Koryo sa*). King Sonjong ordered a search for new hot springs around the capital city, decreeing that the finder would receive several benefits in social status, money, and an exemption from labor and duty.

The fact that every governor had to manage and repair bathing facilities at hot springs located in his territory has provided important records regarding the management and laws of hot spring use (*Kyongguk daejon*, *Daejon hoetong*). Records in *Koryo sa*, *Sejongsillok chiri chi*, and *Tongguk yoji sungnam* comment on the history and name-origin of the hot springs in 43 locations, offer a hot spring commentary (*Oju yonmun changjon sango*), and give locations and characteristics of hot springs and mineral springs (*Yijesokgo*).

The flowing hot spring waters of Korea were enjoyed mostly by the king, his family, and the nobility, a pattern that continued up to the Japanese occupation in 1910. Then, the Japanese developed and modified hot spring use to resemble modern customs in Japan.

Onyang Hot Spring

In Paekche (18 B.C.-660), the Onyang hot spring area was called *Onjeong* ("warm well"). The name was changed to *Onsoo* ("warm water") in the Koryo period (918-1392) and finally to *Onyang* ("warm village") (*Tongguk yoji sungnam*, Book 19). A legend says that a curious farmer found the hot spring while trying to learn why grasses in the area grew well, the surface water dried up easily, and the land did not freeze in the winter. Taejo (reign 1392-1398), founder of the Choson Dynasty, and his sons visited Onyang and built the special royal rooms for bathing. Hungson Taewongun later rebuilt the facilities.

Onyang has two historical monuments. One, the Shinjeong Stele, (the initial chapter photo), is Chungchongnam-do Cultural Property and Monument No. 229. The stele, which is an upright stone tablet, records the miraculous gushing-up of two wells during visits to Onyang by Sejo (reign 1455-1468). In 1464, well into his reign, Sejo arrived at Onyang on March 1 after visiting the Venerable Hyegak, a famous Buddhist priest in Pokchon Temple on Songni Mountain. On the fourth day of his stay in Onyang, the king noticed a gushing flow of water in the garden of the royal lodging. He had a well dug on the spot and the water in the well was ice cold, clear, sweet, and soft. As news of this miraculous well spread, all eight provinces of the kingdom sent the king letters of congratulations on the auspicious well. Six years later, in 1468, Sejo stayed in Onyang again with his mother, and he found another outflowing spring. A second well was dug named *Shinjeong* ("divine well"), and he ordered a second stele set up to record the miraculous phenomenon. Shinjeong Stele was erected in 1475, the seventh year of the rule of Songjong, who was a son of Sejo. Im Won-jun wrote the text on the stele, and Yi Suk-ham was the calligrapher. The stele tells us that the Onyang was famous for old mineral and hot springs.

Younggwoe-dae Stele at Onyang hot spring. The stele is 100 cm tall and 16 cm wide.

Another monument is Younggwoe-dae Stele, which is a Chungchongnam-do Cultural Property and Monument No. 228. Younggwoe-dae was an archery ground where Crown Prince Sado practiced archery when he went with his father, Yongjo (reign 1724-1776), to the hot spring in Onyang. Chongjo, the son of Sado, ordered the Onyang county chief to rebuild the archery ground and plant three trees in memory of his father. He also ordered the Younggwoe-dae Stele set up, commemorating his father in the 19th year of his reign, in 1795. The calligraphy of the title was executed by the king himself, while the inscription was written by Yun Haeng-im, an official of the palace library.

Donglae Hot Spring

According to historic records, Donglae hot spring has been in use since the Shilla period (57 B.C.-668). Because this spring has been famous for its recuperative powers since ancient times, it has attracted many people over the years. Legend says that a white crane always enjoyed bathing around the spring, and people dug a small pond for bathing at the hot spring. The prime minister, Chungwongong, during the reign of Sinmun Wang (681-692), enjoyed bathing here (*Samguk yusa*). *Tongguk yoji sungnam* mentions that a king of the Shilla period bathed in this spring and ordered copper pillars placed at its four corners to indicate the hot spring boundaries. The temperature of the spring was too hot for bathing and high enough to cook a chicken. *Yeonsangok hanglok* records that this spring was separated by stones into sections for men and women and that each room was built for five to six people to use, with hot water supplied from raised conduits.

Stele in Donglae, erected in 1766 upon the reconstruction of the public bath. Height: 144 cm.

A stele erected in October 1766, the 42nd year of the reign of Youngjo (1724-1776) in the Choson period, commemorates the renovation of the hot spring facilities by P. L. Kang, a magistrate of Donglae Province. The stele reads that a bathhouse was built in 1691, in the 17th year of the reign of Sukjong (1674-1720), that Kang rebuilt the house containing two public baths, one for men and one for women, and installed a caretaker's house, a gate, and the stele. An annual memorial service called "The Great Dragon Rite," begun in the Shilla period, is still held here. The 303rd service was on September 9, 1995, of the lunar calendar.

Ancient granitic bathtub in Donglae (inner space: 135 x 70 x 35 cm), hand-carved from massive granite in the 1760s. Before the granitic tubs were carved, wooden tubs were used for bathing. A 19 x 26 cm guidebook for Donglae hot spring, written in Korean, is used for scale.

Haeundae Hot Spring

HAEUNDAE HOT SPRING IS LOCATED ON THE BEACH, ABOUT 10 KM FROM DONGLAE HOT SPRING. The water quality is blackish and the total dissolved solids equal 4,500-5,500 mg/l. The spring was developed in the Shilla period (57 B.C.-668), and the nobility enjoyed bathing there very much. Queen Chinsong Yowang (reign 887-897), her noblemen, and their court ladies enjoyed pleasure-seeking parties every season; a faithful retainer closed the spring after a flood. Legend says that the spring was famous for curing leprosy and that patients came stealthily to the spring, bathing there at night.

Icheon Hot Spring

Icheon hot spring has been called *Oncheon bamee* ("hot spring and snake village") for 500 years. In the 1870s, a farmer found the warm spring in his rice field and cured his eye trouble by washing his eyes in the water. Since then, the hot spring has been called *Yaktang* ("medicine spring") and used for therapy.

Cheoksan Hot Spring

Legend says that a few cranes came to Cheoksan hot spring to let a female crane wash her wounded legs. Later, when she was cured completely, they flew north and people called this place *Haksapyeong* ("crane and sand field"). Another legend says that two old red, wounded snakes recovered completely after wintering at this spring.

Osaek Hot Spring

Osaek hot spring and mineral water (a carbonate water) were famous and used by patients with skin diseases from the late Choson Dynasty (1392-1910). This hot spring is famous for its high altitude (800 m); the other historical springs are at about sea level.

Suanbo Hot Spring

Two hundred years ago, a vagabond with skin diseases found a small hot spring surrounded by piled straw in a rice field. He drank the warm water and bathed in it, recovering easily in a short while. When he left the spring, curious farmers dug to expand the spring. Historical books give the hot spring different names (*Koryo sa*, *Sejongsillok chiri chi*, and *Tongguk yoji sungnam*).

Yusong Hot Spring

Yusong hot spring was found in the late Choson period (1392-1910). When King Taejo (reign 1392-1398) was searching for the capital city of the Choson Dynasty, he visited Mount Kyeryong, near Taejon, and stayed a few days to bathe there (*Tongguk yoji sungnam*). Taejong (reign 1400-1418) also visited the hot spring and enjoyed bathing in its waters.

Dogo Hot Spring

Two stories explain the discovery of Dogo hot spring. One says that the spring water was used to cure wounded soldiers in the Three Kingdom period and that the village was called *Shilla-ri* ("village of Shilla"). Another legend says that a lady of Shilla prayed to Buddha for a son and that she gave birth to a son who had the upper body of a cow and the legs of a human being. As she mourned for her son, she searched for a way to cure him, finally returning her son's body to normal by washing him in the spring water. These fantastic stories tell us that Dogo hot spring has been famous since the early Shilla period (57 B.C.-935).

Deoksan Hot Spring

The story of Deoksan hot spring is very similar to that of Cheoksan hot spring. There was a white crane with wounded wings and legs who came to Deoksan. He could fly after washing himself in the hot spring for three days. Farmers dug and expanded the spring, which has been used for bathing ever since. They called this place *Onchon-gol* ("hot spring village") (*Chungbo*). They said that these warm and slippery waters were very helpful, especially for treating skin diseases and neuralgic pains.

Baekam Hot Spring

In ancient times, a hunter pursuing a wounded deer with a spear accidentally found this hot spring flowing warm mineral water. Monks living in nearby Baekam Temple built a facility with tubs so patients seeking cures for various diseases could bathe in the waters. A governor in the area during the reign of King Myongjong (1170-1197) in the Koryo Dynasty (918-1392) built a bathhouse, which is now gone, with gigantic granite bathtubs. The governor hired a manager to control the facilities and to make them convenient for public use. The facilities were rebuilt several times, but no historical records exist for the reconstructions.

The documents *Koryo sa*, *Sejongsillok chiri chi*, and *Tongguk yoji sungnam* all say that a drop of water from Pyeonghae hot spring (another name for Baekam hot spring) would cure every patient and that a man could turn into an angel by washing his armpit in this water.

Deokgoo Hot Spring

A legend says that a Mr. Jeon, while leading a hunting party in the late Koryo period (about 600 years ago), discovered Deokgoo hot spring and that everyone in the party enjoyed bathing there after the hunt.

Bukok Hot Spring

The word *bukok* means "iron pot," and Bukok hot spring was named after the topography of its site, which looks like a traditional iron pot. There were three small springs at Bukok, which were cold in the summer season and warm in the winter. The water was famous for curing skin diseases. So many patients and lepers came to the village that the villagers finally buried the spring, closing it for good. The spring was uncovered in 1960s, and its temperature (80° C) is the highest of all the hot springs on the Southern Korean Peninsula.

Mageumsan Hot Spring

A legend says that mineral and hot spring waters had flowed in this area since the Choson period (1392-1910). The water at Mageumsan hot spring was very famous for curing all diseases by drinking and bathing, but a severe public nuisance caused by developing the springs resulted in its closure. The hot spring was developed further by a Japanese investor in 1910.

Present Status and Development

Most of these hot springs were flowing naturally until 1910, when further development began. Today, dozens of boreholes have been drilled at these ancient sites, and resort cities have been constructed, with many facilities using the waters, such as hotels, condominiums, public baths, and entertainment areas. All 14 of the famous, historical hot spring sites of Korea discussed here now suffer shortages of hot water, and the groundwater tables are declining rapidly 50-250 m beneath the surface. Reevaluation and reassessment are seriously needed to consider how to best use the hot springs. The gross pumping rate at every hot spring site should be controlled by precise hydrogeologic investigations.

As the demand for geothermal waters has increased tremendously, so has exploitation and development. By 1996, new geothermal investigations were conducted in boreholes drilled at more than 150 locations. The boreholes are tested by the Korean Institute of Geology, Mining, and Materials (KIGAM), which has published technical reports that discuss geology, geography, borehole characteristics, geophysical and temperature logs, long-term and step-drawdown tests, recovery tests, and chemical analyses of pumped water. The new locations are characterized by low temperatures (25°-35° C) and depths of 400-1500 m (Yum, 1994). Unfortunately, with extensive exploitation and development of new boreholes, the deep groundwater environment has deteriorated.

Discussion and Conclusion

Granitic rocks are the main heat sources for hot springs in the Republic of Korea. Korean hot springs can be classified into two groups: the 14 old, historic hot springs and the new hot springs at over 150 locations, developed by drilling boreholes. The old hot springs, showing high temperatures and relatively high total dissolved solids, are fed with waters from much shallower depths than the new.

Since the Shilla period, as shown in 173 articles in ancient books and annals—such as the *Samguk sagi*, *Samguk yusa*, *Koryo sa*, *Choson wangjo sillok*, and *Tongguk yoji sungnam*—the Korean people have been very familiar with hot springs, regarding them as a way to cure diseases. The ordinary Koreans of long ago used hot springs mainly for medical therapy. The ancient Korean governing classes also regarded hot springs baths as sources of recuperation and relaxation.

Before 1910, the use of hot springs in Korea was limited to bathing in the naturally outflowing hot spring waters. Today, the situation is almost the same: 99 percent of hot spring waters are used for bathing; the other 1 percent is used for local district heating projects, including greenhouses, and for aquiculture. In Korea, hot spring bathing is the most economical way to use hot spring waters.

Cited Historical Books and Annals

Choson wangjo sillok (Annals of the dynasty of Choson) were drafted by historians in the Office for Annals Compilation (Chunchuwan). To ensure their safe transmission to posterity, copies were placed in special repositories *(sago)* in widely separated parts of the country.

Chungbo (Loyal treasure).

Koryo sa (History of Koryo) is structured as annals, treatises, and biographies. A greatly revised final version was completed in 1451.

Oju yonmun changjon sango (Random expatiations), by Yi Kyu-gyong (1788-1856), covers the whole spectrum of scholarly endeavor: astronomy, geography, government, economics, society, history, and other disciplines.

Samguk sagi (History of the three kingdoms), the oldest extant history, was compiled by Kim Pu-sik at the command of Injong in 1145.

Samguk yusa (Memorabilia of the Three Kingdoms) was written by the monk Iryon (1206-1289).

Sejongsillok chiri chi (Geographical description and veritable records gazetteer) appeared during the reign of Sejong (1418-1450).

Tongguk yoji sungnam (Survey of eastern country [Korean] geography) was compiled in 1481, under headings such as pavilions, temples, shrines, historical remains, famed officials who had governed a particular district, other historical figures associated with the area, poetical creations, and other compositions.

Tongsa kangmok (Annotated account of Korean history) was written from an orthodox Confucian point of view by An Chong-bok.

Yeonsangok hanglok (Tales of mountains and valleys).

Yijesokgo.

Yongjae chonghwa (Assorted writings of Yongjae) is the literature of tales and anecdotes by Song Hyon.

MODERN SELECTED REFERENCES

Cheong, Tae Soo. 1983. *The present status of geothermal spring.* In *Oncheonji,* Ministry of the Interior. The Dong-Ee Publishing Company, Seoul. (In Korean).

Chwae, U. C., K. B. Kim, S. H. Hong, B. J. Lee, J. H. Hwang, K. H. Park, S. K. Hwang, P. Y. Choi, K. Y. Song, and M. S. Jin. 1995. *Geological map of Korea* (1:1,000,000). KIGAM.

Kim, O. J. 1987. *Tectonic settings of the Korean peninsula.* In *Geology of Korea.* D. S. Lee, ed. The Kyohak-Sa Publishing Company, Seoul.

Lee, Ki-baik. 1984. *A new history of Korea.* E. W. Wagner with E. J. Shultz., trans. Iljogak, Seoul.

Lim, J. U. 1995. *Geothermal potential in the Republic of Korea.* In *Terrestrial heat flow and geothermal energy in Asia.* M. L. Gupta and M. Yamano, eds. Oxford & IBH Publishing Company, New Delhi.

Yum, B. W. 1994, April-June. South Korea/geothermal summary. *IGA NEWS, Newsletter of the International Geothermal Association,* 7.

The Author:

Byoung Woo Yum
Senior Researcher
KIGAM (Korean Institute of Geology, Mining, and Materials)
30, Kajung-dong, Yusong-gu
Taejon, 305-350, Republic of Korea
Telephone: 82.42.868.3375
Fax: 82.42.861.9719
E-mail: ybw@kigam.re.kr

"Sightseeing at Hakone Spa," by Utagawa Kuniyashu in the Bunsei period, 1818-1829; woodblock print. A party of women relaxes with food, tea, and conversation by the stone-enclosed hot spring in the background. The tiny vertical lines rising from the hot spring water indicate heat. The woman seated in the center foreground is holding a long-stemmed pipe, filling its bowl with tobacco leaves from her silken pouch. *Courtesy of the Shogangi Temple. Photo by Mr. Suzuki, Hakone Board of Education*

25. Japanese Geothermal Waters throughout History

by
Mitsuru Sekioka

Abstract: From prehistoric times to just before the industrial age, the history of geothermal waters in Japan is the history of hot spring bathing. The oldest vestiges, dating from the pre-pottery period before 11,000 B.C., can be found near Oyu hot spring, Akita Prefecture. Archaeological evidence and bathing myths and legends appear in several stories from ancient times. Of particular importance in early records are the visits made by Emperors to hot spring spas. From the beginning of the Heian era in 794 to the late Edo era, ending in 1868, some texts describe the primary characteristics of bathing in hot springs, including ways of making artificial hot spring water.

Reprinted by permission of the International Geothermal Association, from the *Proceedings of the World Geothermal Congress*, 1995, pp. 395-399.

Introduction

As a volcanic country, Japan is blessed with many hot springs. Vestiges from hot spring bathing in prehistoric ages have been excavated in every part of the country. These imply that ancient peoples had a habit of bathing in hot springs.

In the preindustrial ages, the only recognition of geothermal heat was the water in hot springs used for balneotherapy. Because baths in hot springs were shown to hold various medical benefits, the discoverers of hot springs in myths and legends were frequently gods, Buddhas, high priests, and marvelous animals.

This chapter describes prehistoric vestiges of hot spring baths and recounts typical myths and legends of volcanoes and hot springs. There are details of bathing as it is described in old, historical books, and retellings of imperial visits to spas, for they symbolize the Japanese enjoyment of hot spring bathing.

The history of hot spring bathing is described from the years 794 to 1868, and some conclusions and remarks are included, along with some hot spring locations.

Prehistoric Periods

Evidence indicates that ancient peoples lived in Japan, cultures that used neither pottery nor polished stone implements. Old stone implements used during the third ice age, about 200,000 years ago, were excavated at a site in Niisato, Gumma Prefecture.

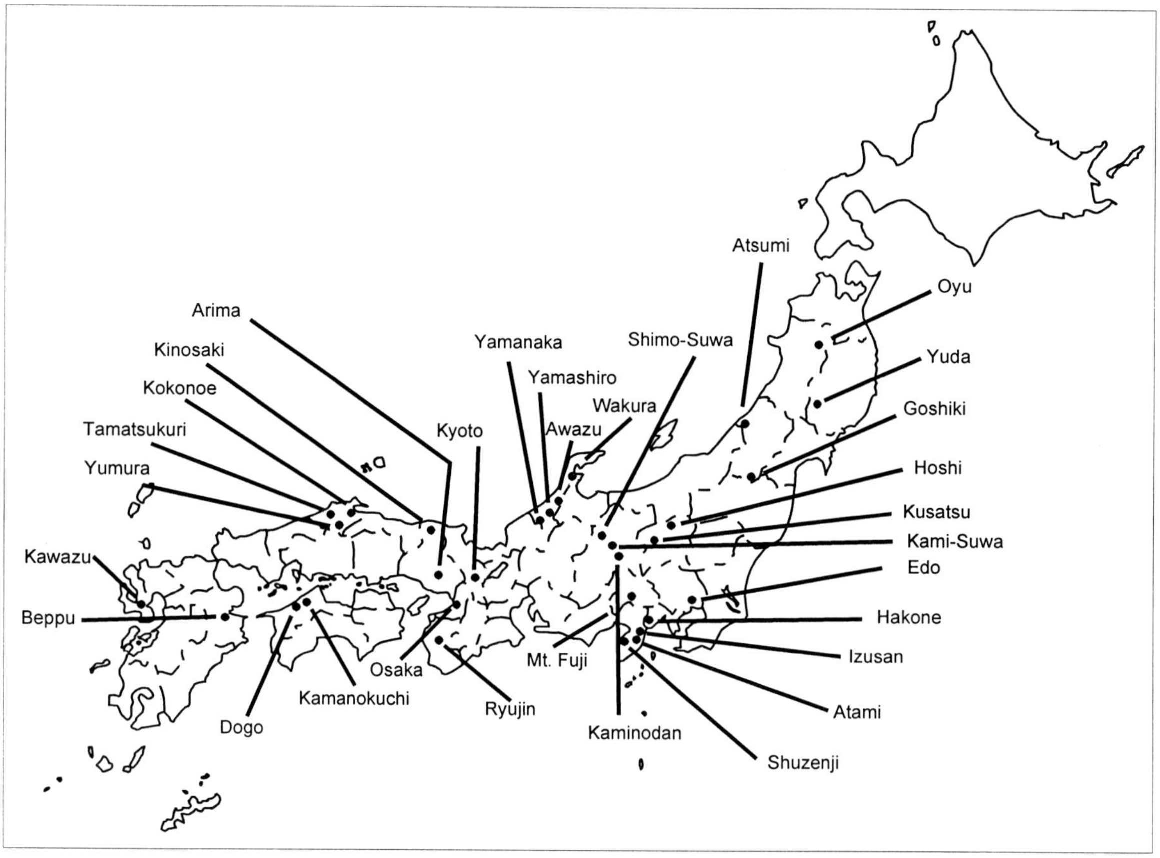

Location of the hot springs and the three big cities cited in the text, Edo (Tokyo), Kyoto, and Osaka.

The ancient Japanese peoples lived in a warm and humid climate, particularly in the summer. They must have felt especially refreshed in mind and body after taking hot and cold baths. It is easy to imagine that ancient peoples brought up in a country blessed with pure springs and clear rivers took dips in them to forget the summer heat and to sweat.

Moreover, since many hot springs are distributed along volcanic zones all over Japan, the people probably enjoyed bathing in them year round. Thus, bathing practices came into existence. Once people learned that bathing in hot springs had medicinal value, bathing may have become connected with natural reverence. This reverence for hot springs, in turn, promoted bathing in hot springs, and the process gradually gave the Japanese people a passion for hot spring baths.

Large stones smelling of hydrogen sulfide and arranged circularly were found at a depth of 5.5 m in Suwa, Nagano Prefecture. Experts believe that the stones were used about 6,000 years ago as the frame for hot spring bathing, for broken pieces of pottery and polished stone tools excavated together here were coated with encrustations.

Ruins of an ancient colony from about 6,000 years ago were excavated on the eastern side of Lake Suwa, Nagano Prefecture, and large stones with encrustations were found buried 6 m deep. The site is believed to be the ruins of a public hot spring bath for ancient peoples. Moreover, even today hot spring waters well up from the lake bottom near the eastern shore of Lake Suwa. About 10,000 years ago, the water level of the lake was around 2 m lower than at present, and the ancient peoples may have used the hot water to boil fish seized from nearby rivers and to warm themselves by bathing.

Many other vestiges of ancient Japanese culture have been unearthed near present-day hot springs sites. Prominent sites include Yuda hot spring, Iwate Prefecture (during the pre-pottery period before 11,000 B.C.); Oyu hot spring, Akita Prefecture, Kaminodan hot spring, Nagano Prefecture, and Kawazu hot spring, Nagasaki Prefecture (during the Jomon period from 11,000 B.C. to 300 B.C.); and Kokonoe hot spring, Shimane Prefecture, and Kamanokuchi hot spring, Ehime Prefecture (during the Yayoi period from 300 B.C. to 300 A.D.).

Myths and Traditions

Volcanoes

Since Japan has many volcanoes, many myths and traditions are associated with them. The two imperial histories that include myths, *Kojiki* published in 712 and *Nihonshoki* published in 720, tell of Prince Izanagi and Princess Izanami giving birth to the gods of house, tree, wind, sea, river, mountain, and field, followed by the Japanese countryside, consisting of many islands. Their youngest child was the god of fire, who burned the genitals of Izanami and she died. The god may have created the volcanic islands, which appeared last after the other islands, and it is well known in geoscience that the formation of the volcanic islands is very recent in geologic time, relative to the other islands.

These two histories also describe Prince Susano. He was exiled from heaven due to his sin, descended to a village in Izumo (presently in Shimane Prefecture), and saw a crying old couple hovering around their daughter. They replied to his question that a big red-eyed snake with eight hands and tails separated by its tree- and shrub-coated body lying across eight hills had descended from a mountain to attack the village and that it ate a young girl every year. Their daughter's turn came this year. Susano took pity on them. He changed her into a small comb to keep within his hair after he married her. Susano asked the old couple to make a strong liquor by fermenting the brew eight times and then pouring it into eight large tanks placed in front of eight gates in the fence enclosing the "snake living mountain." After a while, the snake came to drink the eight tanks of liquor, emptying them. When the snake was dead drunk, Susano attacked and scattered the snake with his sword until a nearby river was dyed red with blood. Last of all, cutting off the tail, he discovered a big sword. The scene is similar to that of a volcanic eruption. If the snake is replaced with an erupting volcano, red scorching lava from the crater can be envisioned flowing along eight ravines.

A Volcanic Folding Screen

Another volcanic legend, unique in its charm, is recounted by Takumi Tsukamoto in *Folk Legends of Aso*.

"On the Southern Japanese island of Kyushu, behind the blood-red hot spring called Chinoike-jinko, there rises Mt.Yomineyama. Here is the story of its birth.

"Long, long ago, there was a young empress of great grace and beauty named Asotsuhimeno-mikoto. When she became pregnant, her husband, Lord Takeiwatatsuno-mikoto (alias Aso Daimyojin, or the Great Gracious Lord of Aso) began to take better care of her.

"Now that her time was near, she naturally had become very large and fat around her middle. And so she became very worried that others might see her with this altered figure. The lord was all the more pleased with his wife, who looked very abashed like a blushing bride.

"'All right, I'll do something for her before the night is out,' he swore to himself. And overnight he built a mountain to keep his pregnant wife from being seen by others. As the mountain arose in a single night, it was named *Yomineyama* ('Night Peak Mountain').

"Since the mountain stands behind the spa, it is like a *byobu* ('folding screen') that hides from view the young girls bathing in the spa's open-air bath." *(A photograph of the hot spring is in this chapter.)*

Mount Aso, Calming the God

Mount Aso, an active volcano 5223 meters high on the Southern Japanese island of Kyushu, has an immense caldera, among the largest in the world. The volcano has influenced not only the physical lives of people but their imaginations as well. The following account is by Takumi Tsukamoto, from his book *Folk Legends of Aso.*

"Abundant folk legends were born and cultivated in the long history of the Aso people. Archaeological evidence reveals that a great many settlements existed in and around the Aso caldera in the Iron Age, about 300 B.C. to 200 A.D. By the 3rd century in Ichinomiya, Aso Valley, there was a clan state called Kuni.

"In those days, Mt. Aso was worshiped because it was believed that a god lived within it. Its eruption was dreaded as an expression of the god's anger. Prayers were offered morning and evening in hopes of keeping the god calm. This worship eventually led to the establishment of the Aso Shrine, one the oldest in Japan, in the 4th century A.D. Enshrined there is Takeiwatatsuno-mikoto, called the Great Gracious God of Aso. The Kuni Clan gradually expanded and later became know as the Aso Clan. The Yamato Imperial Court, headed by the ancestors of the present Imperial Family, having brought most of the Japanese islands under its control by the middle of the 4th century, appointed the head of the Aso Family as the governor of the Aso Area.

"In the 5th century, leaders of this clan were not only the officials of Shinto rites at the Aso Shrine, they were also the rulers of the entire Aso Valley. The Aso Family's divine authority as the custodians of a 'volcanic god' helped to advance the family's position.

"By the 12th century, the Aso Clan dominated not just the Aso and Nango Valleys, but all of Aso County and beyond. An organization of samurai headed by the Aso Clan helped them extend their influence even further.

"The peasants of the area lived very hard lives. They suffered from the cold of the mountains and also from the volcanic ash, *yona*, which often severely damaged rice, wheat, and vegetables. They had to work hard to reclaim and cultivate the poor land. Aso was under the reign of the Aso Family until the late 16th century when the family's power was severely curtailed by an angry General Hideyoshi Toyotomi."

The Aso Shrine, a Shinto shrine of the highest class established in the 4th century A.D., is still cared for by members of the Aso Family. Over 400 branch shrines fall under the influence of this shrine. *S. Hodgson*

Since ancient times, the Japanese have had faith in the gods. Because they believed that a volcanic eruption was due to the anger of a god living within the volcano, they would build a shrine on the volcano and decorate it with the order of merit from the Emperor.

Mount Takachiho in the Kirishima volcanic range has been handed down through tradition as the site at which the ancestor of the Emperor descended from the Plain of High Heaven, or the Japanese Olympus.

These episodes imply that the ancient Japanese might have combined ideas of gods and mankind through the mysterious presence and activities of volcanoes. This is also suggested in a story that occurred about 1150, when a Buddhist priest saw a Buddhist saint, Sengen, disguised as a woman in self-exile at the top of Mt. Fuji, the most famous volcano in Japan and the highest (3776 m). The priest realized that the Buddhist saint, Sengen, was identical to Prince Asama, the god of Mt. Fuji. Thus began the mixing in Japan of Buddhist and Shinto beliefs, which have persisted into the modern era—beliefs that were born from volcanoes.

Hot Springs

Old hot springs in Japan frequently have their own myths and traditions. The following are well known.

Kami-Suwa and Shimo-Suwa hot springs, Nagano Prefecture. When Prince Takeminakata and Princess Yasakatome were divorced, they were residing at the main palace of the Suwa Shrine on the southeastern side of Lake Suwa. As a sort of consolation, the princess received a cotton ball wet from soaking in a hot spring near the main palace. Holding the ball, she passed over Lake Suwa and put the ball in the autumn palace of the shrine. Here a hot spring began to well up, and this hot spring was named "Cotton" hot spring (Shimo-Suwa hot spring, at present). At the same time, waters in the hot spring near the main palace decreased in temperature and flow.

In addition, a spring gushing great quantities of hot water appeared at a site where many drops of hot water dripped from the cotton ball the princess held. Today, this is Kami-Suwa hot spring. Since drops dripped over the lake as well, some hot springs gushed from the lake bottom. Finally, so many drops dripped out that the cotton ball gradually dried up and Cotton hot spring became a low-temperature spring with weak water flow.

EARTHQUAKE
by
Kokan Shiren
(1278-1346)

Still things moving,
firm become unfirm,
land like ocean waves,
house like a boat—
a time to be fearful,
but to delight as well:
no wind, yet the wind-bells
keep on ringing.

ATAMI
by
Chugan Engetsu
(1300-1375)

[A rare view of the famous hot spring resort near Kamakura as it was in the 14th century, with a glimpse of the island of Hatsushima in the distance.]

Midnight dreams broken by the hissing roar—
hot water boiling from the roots of the cliff;
pipes this way and that lead the water, houses wreathed in steam,
every inn fitted with a bath, rooms let out to travelers.
By the sea's border land is warm—winter it never snows,
though cold days on mountain paths, on treads through frost at dawn.
A far-off island in fine rain, black with clouds and fog;
over red tides I watch the moon sink dimly out of sight.

TWENTY-FOUR HOKKU
(An excerpt)
by
Takai Kito
(1741-1789)

As I went by the foot of Mount Asama, there were no trees and grasses growing though the springs and autumns of three years had passed since the volcanic fires, and even boulders were buried under the ashes. After going over Usui Pass, my palanquin bearers told me that the red soil showing about ten feet below us was just about where the old road used to be. In the fields and paddies burned sand was raked into piles, and things looked quite desolate:

Even the soil is withered and saddening in this winter field.

Atami hot spring, Shizuoka Prefecture. Before the descent to the Earth by the descendants of the Sun-Goddess, the ancestor of the Japanese Emperor, Prince Omunachi, felt pity for the early deaths of Japanese people in that time. He taught Prince Sukunahikona about preparing medicine and bathing treatments. He was sent as the god of the hot spring presently called Atami hot spring in Izu Province, Shizuoka Prefecture. The hot spring boiled and gushed hot water twice a day, and all kinds of diseases were treated there by soaking the patient in a tub filled with hot water.

Dogo hot spring, Ehime Prefecture. When Prince Omunachi dipped Prince Sukunahikona in a state of apparent death into a hot spring, Prince Sukunahikona was revived. At that time, the hot spring—now called Dogo hot spring—was believed to flow through an underground waterway from Beppu hot spring in Oita Prefecture, across the sea.

Arima hot spring, Hyogo Prefecture. Princes Omunachi and Sukunahikona discovered Arima hot spring from fumes in a ravine.

Discovery Legends

OLD SPAS ARE ALMOST ALWAYS ASSOCIATED WITH THE LEGENDS OF A DISCOVERER. THE RESULTS of the classification of these legends by discoverer are as follows (the prefecture names are in parentheses):

- Gods and Buddhas: Arima (Hyogo), Dogo (Ehime), and Tamatsukuri (Shimane)
- High priests: Ryujin (Wakayama): Goshiki and Atsumi (Yamagata); Atami, Izusan, and Shuzenji (Shizuoka); Kusatsu and Hoshi (Gumma); Awazu, Yamanaka, and Yamashiro (Ishikawa); and Yumura (Shimane)
- Marvelous animals: Yuda (Iwate) and Kinosaki (Hyogo)

Fudoki

IN 713, THE IMPERIAL COURT OF JAPAN ORDERED ALL JAPANESE PROVINCES TO COMPILE A *Fudoki,* a description of the natural features of the province. The origin and description of hot springs are found in several *Fudokis*, as follows.

Fudoki of Izumo Province, Shimane Prefecture (733). The *Fudoki* says that when the governor of Izumo went up to the Imperial Court in Yamato to present New Year's greetings to the Emperor every new year, he always performed his ablutions at the site of a hot spring in Izumo. Usually, men and women gathered at the hot spring to enjoy baths, drink sake, and celebrate. The hot spring, today known as Tamatsukuri, was popularly believed to cure many diseases. The *Fudoki* also described Yumura, Yunokawa, and Ushio hot springs.

Fudoki of Bungo Province, Oita Prefecture. The *Fudoki* said that the water of Red hot spring is red and that it provides red mud used to paint pillars of houses. (In ancient times, the color red was held in esteem to avert evil.) The muddy red water turns clear as it runs eastward from the spring, depositing the red mud on the bottom of the river. Today, Red hot spring is called *Chinoike-jigoku* ("blood pond in hell") hot spring in Beppu. Here is where the hot spring gushed hot water and red clay mud, colored from ferric sulfide.

Chinoike-jigoku, "Blood Pond in Hell." The 78° C thermal waters are in a pond of 1,300 square yards. About 1800 kiloliters of hot water flow each day. *S. Hodgson*

The *Fudoki* said that when a man cries in a loud voice near the erupting hole, very hot water with steam will boil out and wither the surrounding vegetation. This geyser survives in Beppu today.

Fudokis of Settsu Province, Hyogo Prefecture. This *Fudoki* describes Arima hot spring in simple form.

Imperial Visits to Hot Springs

The first imperial visit to hot springs was made by Emperor Jomei in 631. His Majesty the Emperor went to Arima in September and returned to the Imperial Court in December. Hot springs were so appealing to him that he returned to Arima in October of 637 and stayed there until the next year. Two years later, in 639, he visited Dogo hot spring in Ehime Prefecture.

Legends say that the Emperor Keiko (71-130) visited Dogo hot spring, Ehime Prefecture, and others. The Prince Shotoku also visited Dogo in 596.

Hot Springs in History

The Heian Era (794-1192)

During the Heian era, one of the largest changes in Japan occurred when rice paddy lands were developed in warm and mild climates. Many new villages appeared and there were frequent population movements to them. Increased agricultural production from newly reclaimed paddy fields fostered the growth of commerce and, in his reign, the Emperor extended his control.

Noblemen frequently went to spas for bathing and for medicinal purposes. A poem composed by Lord Kanesuke Fujiwara, a famous poet, on his way to Kinosaki spa, Hyogo, was found in an old anthology, titled *Kokin,* published in 913. History books record that in 998, Lord Yuken Fujiwara took a vacation to improve his health in a hot spring in Nagano Prefecture. Lord Michinaga Fujiwara went to Arima spa for the baths in 1024, and Lord Sadatsugu Hamuro went to Arima spa in 1086.

The Kamakura Era (1192-1393)

In the Kamakura era, when samurai or Japanese warriors governed the country, noblemen still went to spas for baths. It is on record that Lord Karumitsu Fujiwara went to a spa in 1231. In September of 1287, Emperor Gofukakusa had hot spring water brought to him from Arima spa, over a distance of about 100 km, so he could bathe in the Imperial Palace in Kyoto. There are many descriptions of samurai who bathed at hot springs to heal injuries incurred in wars.

An episode from a famous history book, *Heike Monogatari*, describes a Buddhist priest exiled to a lonely southern island. He climbed a nearby volcano and dug for sulfur, exchanging it for food with nearby merchants. During this period, various parts of the country offered presents of sulfur to the Imperial Court as raw material for medicines and gunpowder.

The Muromachi Era (1393-1574)

In the Muromachi era, temples and shrines had large territories all over the country and custodians came and went everywhere to administer them. Conditions at nearby hot springs were included in their reports. At the same time, there were many tours, including pilgrimages to several spas.

The Azuchi-Momoyama Era (1574-1603)

In the Azuchi-Momoyama era, or the late Japanese warlike age, many warlords and subordinate personnel were in the habit of salving wounds by bathing in hot springs in their territories. The warlords were known to have several of their own secret spas for healing wounds.

The Edo Era (1603-1868)

From early in the Edo era, gifts of hot spring waters were made from Atami spa to the sovereign entrusted with full power by the Emperor. This was done several times a year by transporting 300 kiloliters of water in barrels for about 100 km. The gifts included water from Hakone spa, about 100 km west of Edo (modern-day Tokyo). Transporting barrels of hot spring water began in 1644. Under such circumstances, the *Daimyos* ("feudal lords"), warriors, and the general populace were attracted to bathing in Hakone and Atami spas.

Since about 1654, the Kaga local government (Ishikawa Prefecture) exported hot spring water barreled from the Wakura spa to Kyoto, Osaka, and Edo, where it was sold at a relatively expensive rate. This greatly enriched the local government.

During the Edo era, Japan had one of the most advanced publishing industries in the world, well before the invention of modern printing machines. Two medical books, one called *Yojokun* and published in 1710, are very famous in Japan. They include descriptions of diseases for which hot spring bathing, by spa, is efficacious. Guidebooks for many spas were printed from 1710 to 1730, describing medical benefits, ways of bathing, lines of travel, nearby sightseeing resorts, and souvenirs.

In the mid-Edo era, medical and pharmaceutical practices began including experimental techniques. A famous medical doctor, Konzan Goto, selected Kinosaki spa, Hyogo, as the experimental site for balneotherapy. He tested the medical benefits of hot springs for edema, lumbago, gout, palsy, paralysis, beriberi, gonorrhea, anal fistula, and menopause. The four volumes he and his follower, Shuan Kagawa, published in 1734 said that balneotherapy had medical benefits for chronic diseases, an opinion still held today.

Because the doctors said that Kinosaki hot spring offered such excellent medical benefits, many people came from all around to take the hot spring cure and the spa flourished. Another follower of Dr. Goto, Tsuan Yamamura, compared various hot springs in many places. He concluded that the medical benefits of Kusatsu hot spring equaled those of Kinosaki, and, in fact, a beautiful woodblock print of the time shows a bathing scene from Kusatsu spa in the Edo era. Tsuan Yamamura realized that travel to any spa was difficult for a sick person. Thus for home bathing, he developed a formula to make artificial hot spring water by adding sulfur to rice bran concentrated in salt water.

Other artificial hot spring waters were created by medical doctors who had studied western sciences in the Dutch language. Kondai Utsugi made artificial waters by heating pulverized sulfur, alunite, and arsenious sulfide in salt water. He claimed the water held the same medical benefits as Arima hot spring water. Finding his artificial water very expensive for most people, he lowered the price by altering the formula, dissolving natural geyserite into salt water.

In the late Edo era, between 1828 to 1843, Yoan Utagawa chemically analyzed water samples from 36 spas, classifying them as to acidity, salinity, sulfur, and chalybeate (iron salts) content, and identifying the medical benefits derived from each one.

Conclusions and Remarks

As the author is not an historian and English is not his mother language, the author has gone through all sorts of hardship. But it is a pleasure to compare Japan's geothermal history with the histories of other countries in this book. Since each section of the chapter was composed mainly by citing Yatsuiwa (1993), the other references listed were not noted in the text.

Acknowledgments

THE AUTHOR WISHES TO GIVE HEARTY THANKS TO MS. M. YATSUIWA AND MR. K. ONO, SEIKYU Publishing Company, for their permission to translate some parts of *Thermal Springs and the Japanese* into English to compose this chapter. The author also expresses his thanks to Prof. John W. Lund of the Oregon Institute of Technology for editorial suggestions. Dr. Y. Kanroji of the Hot Spring Research Center kindly provided materials concerning old hot springs in Japan.

Selected References

Aoki, K. 1965. *A history of Japan, vol. 1.* Chuokoronsha, Tokyo. (In Japanese).

Inagaki, M. 1990. Kamisuwa hot spring. *Geothermal Energy,* 15, 109-123. (In Japanese).

Inoue, M. 1965. *A history of Japan, vol. 3.* Chuokoronsha, Tokyo. (In Japanese).

Kinosaki-cho. 1988. *A history of Kinosaki-cho.* (In Japanese).

Kusatsu-cho. 1976. *On Kusatsu hot spring.* (In Japanese).

Tanaka, Y. 1986. *Mechanism of a volcano.* Hakuryukan, Tokyo. (In Japanese).

Tsakamoto, T. 1985. *Folk legends of Aso.* Japan.

Yatsuiwa, M. 1993. *Thermal springs and the Japanese.* Seikyusha, Tokyo. (In Japanese).

Yuhara, K. 1993. Atami hot spring. *Geothermal Energy,* 18, 253-269. (In Japanese).

The Author:

Mitsuru Sekioka
15-14 Kotsubo 7
Zushi Kanagawa 249, Japan
Telephone/Fax: 81.467.23.1098
E-mail: sekioka@red.an.egg.or.jp

A rare and stunning view of the steaming summit of Mount Mayon, looking north. *B. J. Barker*

26. Geothermal Legacy from the Early Filipinos and Later Migrants

by
Leonardo M. Ote
Arturo P. Alcaraz
Agnes C. de Jesus

Abstract: The early Filipinos did not make much use of the terrestrial heat so abundantly manifested in the region. However, they may have used a mixture of oil and bits of sulfur-rich rocks for medicinal and mystical purposes. References to fire or heat were associated with volcanoes, which early Filipinos revered as homes for their gods and as symbols of fair maidens in myths and legends. Later migrants to the islands used the Earth's heat for dressing slaughtered poultry and swine and steam-cooking rice and root crops. With the arrival of occidental cultures, spas were built for health and recreational purposes. Modern uses of geothermal heat are directed towards agricultural-industrial applications and electrical power generation.

To improve the socioeconomic well-being of indigenous cultural communities, the Philippine government has taken steps to resolve concerns about geothermal development in their areas. Consultation with indigenous peoples and the payment of geothermal royalties in recognition of their rights over their ancestral lands will ensure fulfillment of the energy development objective.

Prologue

The Philippine archipelago consists of about 7,100 islands and islets on the western margin of the Pacific Ocean. The two largest islands of Luzon and Mindanao, with land areas of 105,000 and 95,000 square kilometers, respectively, represent two-thirds of the total land area of the Philippines, which has about 67 million people. The combined land and water area is about 1,800,000 square kilometers, with five times as much water as land. For comparison, the Philippines is about the size of Italy, slightly larger than the United Kingdom, and slightly smaller than Japan.

Since its entry two decades ago into the geothermal energy community, the Philippines has managed to develop its geothermal resources from almost nothing to its position today as the second largest producer of geothermal electrical power in the world. The present installed generating capacity of 1,948 megawatts supplies 16 percent of the nation's electrical power needs. On the large islands of Leyte and Negros, geothermal energy is almost the only source of electrical power generation. With such large scale use of geothermal energy in the Philippines, the authors sought to learn if the early Filipinos also used the Earth's heat, which is so abundant in their lands.

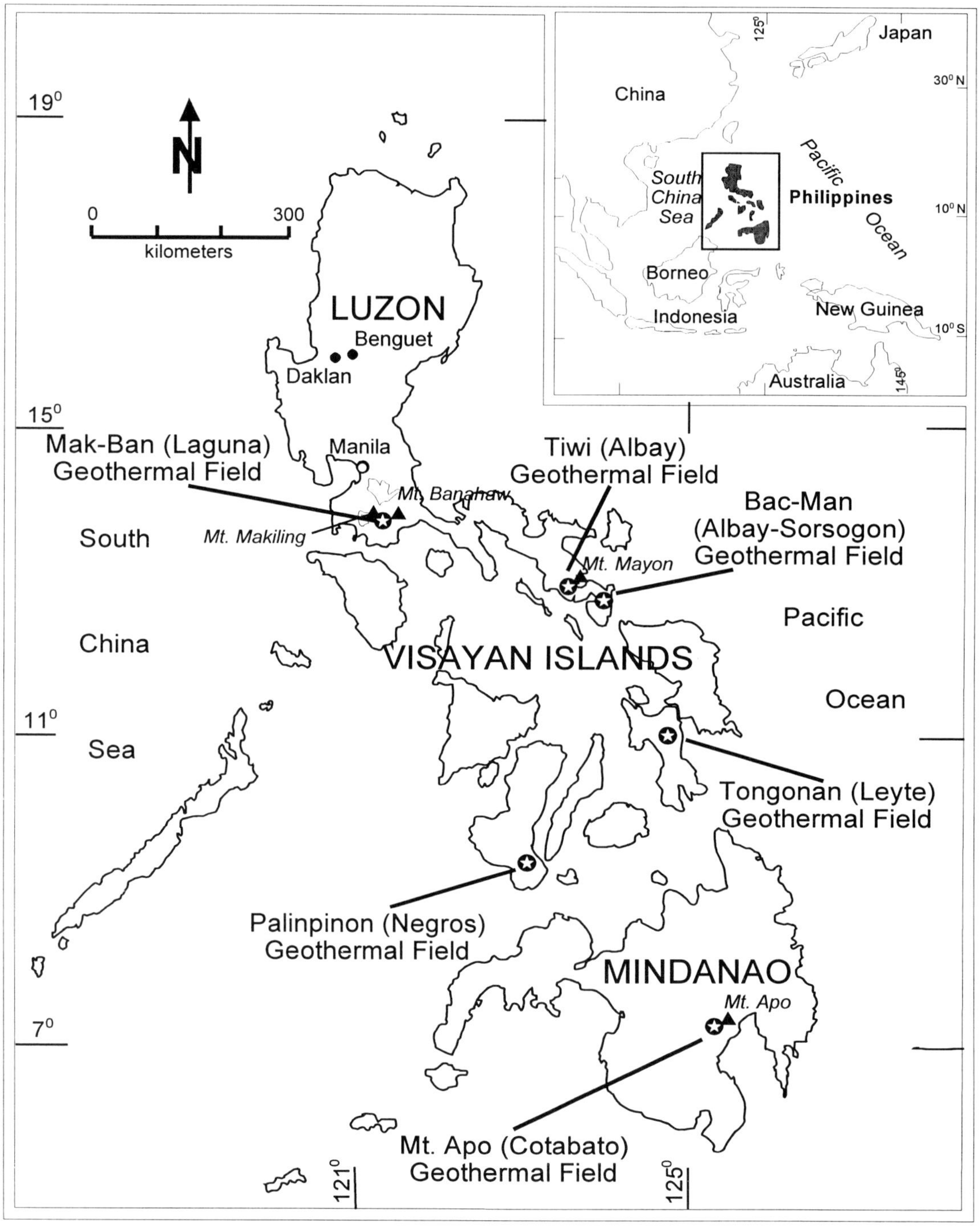

Map for sites mentioned in the text, including geothermal fields.

Earth's Heat Awareness by Early Filipinos

Descendants of the early Filipinos, the people who inhabited the Philippines before the migration waves from nearby lands began, are found along the length and breadth of the archipelago. Collectively, their settlements are now referred to as indigenous cultural communities. According to the Indigenous People Desk of the Department of Environment and Natural Resources, 110 ethno-linguistic groups exist in the Philippines, comprising about 12 million people. These communities have their own languages and cultures, which set them apart from the present-day mainstream Filipinos of Malayan, Oriental, and Western descent. Since they live mostly in upland areas of the country, they are known as the highlanders, as opposed to the lowlanders, those living in cities, towns, and other urban centers.

In trying to determine whether or not the early Filipinos were aware of terrestrial heat and how they used it, it was necessary to delve into the rich cultures and languages of the indigenous cultural communities. Our studies indicate a paucity of written information referring directly to the Earth's heat. Perhaps this is because written information is unavailable, due to the unfortunate circumstances of illiteracy, displacement from ancestral lands, the nomadic lives of indigenous peoples, and the rapid social changes that erode folk traditions (Manuel, 1985; Togon, 1995). Any reference to fire or heat is usually associated with volcanoes, which early Filipinos revered as homes for their gods and symbols of fair maidens in myths and legends. Many of these stories have been passed orally from generation to generation and now are found in the few books of Philippine mythology.

On the Origin of the Philippine Archipelago

Two schools of thought exist on how the Philippine archipelago came into being. One group subscribes to the idea that the Philippines was once a part of mainland Asia through a land bridge, since submerged. The other group maintains that the archipelago was formed largely through magmatic activity from the ocean floor that produced volcanic islands, some of which eventually coalesced into larger land masses. Each idea has geologic evidence behind it, and perhaps both are correct and can be unified into a plausible whole.

It is interesting to note, however, that most legends of why the Philippines has so many islands do not mention volcanism as the cause, but rather gigantic battles between the sea and sky, with the latter raining rocks on the former to become the islands (Jocano, 1969). Today, the Philippine archipelago has three major geographical units—Luzon, Visayas, and Mindanao—and the myth that follows offers a second idea of how the country was broken into many islands.

Long ago, when the Philippines was not yet divided into small islands but was one wide, solid island extending to the tip of northern Borneo, there lived in the Ilocos region of Luzon two huge beings, Angalo and Angarab. The husband and wife were so big that when they walked, the entire island shook. One day Angalo and Angarab went to gather clams at the southern end of the island, which was now in the Sulu Sea. When they opened the clams to eat the meat, they found sparkling stones and pearls inside. The luster of the pebbles caught the fancy of the couple, and they gathered more clams and removed the brilliant stones. After they had collected a considerable number, they started home.

Upon arriving at the middle of the island, a question arose: who should have the most glittering stones and pearls? The question started a quarrel that ended in a fight, with the two stamping their big feet and shouting at each other. The pressure from the heavy steps and the vibrations from the angry shouts caused mountains and hills to crumble, and the land was leveled and cracked. Once the fight got serious, big pieces of land went flying in all directions, becoming the islands of Luzon and Mindanao. The scene of the fight itself was horribly smashed, and the pieces became the Visayan Islands.

On Gods and Deities of Ancient Filipinos

Explanations of how things began were exceedingly diverse among the early Filipinos, according to Professor F. L. Jocano of the Centro Escolar University Research and Development Center, for these people lived as independent groups and developed distinct customs and traditions. However, all early Filipinos shared common responses of awe and fear to unexplainable natural occurrences. They believed that mysterious natural phenomena were caused by the struggle of powerful, intelligent beings with human traits and qualities, a concept that molded their thoughts and the creation of their gods.

Bathala is the highest deity of the indigenous cultural communities of Luzon. It is said that his coming was heralded by flood, fire, and violent earthquakes. Surprisingly, these are the same natural calamities that afflict the Philippines due to its particular geographical position in the

Pacific: typhoons and hurricanes bring floods, and geological circumstances cause volcanic eruptions and earthquakes.

Bathala exacted obedience and reverence from those who believed in him, but he was also just and merciful. The ancient Filipinos learned to rely on his benevolence, and perhaps from this outlook came the *bahala na*, or "leave it to God" philosophy of life of the Filipinos.

The indigenous cultural communities of Mindanao regard as their supreme deity a spirit called Apo Sandawa. Mindanao tribesmen are animists and thus believe in the power of spirits. According to them, Apo Sandawa resides on Mount Apo, the highest volcano on the island of Mindanao, and from there oversees the welfare and lives of the indigenous tribes living around the mountain. M. Bernard writes that people who live near the slope call it *Apo*, meaning lord ("superior and powerful") and ancient ancestor, or "grandfather of the mountains" (Bernard, 1985). To the other tribes, Apo was a dreaded mountain, the home of Mandarangan, god of the fire, who dwelt upon a throne of fire, always thirsting for human blood. When calamity threatened or occurred, human offerings were made to Mandarangan. The human offerings

Summit of Mount Apo, 2,954 feet (909 m) high. This tallest of Philippine volcanoes holds court over Southern Mindanao and is considered sacred by a dozen indigenous cultural communities.

communicate with other gods, and the ceremony was undertaken to secure a good harvest, victory over the enemy, or cure of an illness (Gloria, 1987). This practice has disappeared, but offerings of food, money, and tobacco still are made (Togon, 1995; Bernard, 1985).

By 1983, geological exploratory efforts uncovered a potential geothermal resource in the area of Mount Apo. Before it could be developed, the worship of the mountain god by the indigenous cultural groups of Mindanao had to be addressed. The fears and concerns of the groups were given due consideration and response before the country began developing the geothermal resources in 1992.

On Volcanoes and Fair Maidens

No mountain god, however, had to be appeased before developing the Makiling-Banahaw Geothermal Field south of Manila, the capital of the Philippines. Legend says that Mount Makiling, the volcano on whose flanks the energy resource is located, is the home—not of a god—but of a beautiful enchantress named Maria Makiling.

Maria Makiling was not a creature of flesh and blood but a nymph with a heart of gold. She helped the needy, showering them with love and kindness. But Maria, spirit that she was, fell in love with a mortal, a handsome young farmer named Ban-aw who lived on a nearby volcano. Although Maria helped Ban-aw to become prosperous, he decided to marry a village lass.

On the eve of his wedding, Maria went to Ban-aw and told him of her love. Ban-aw promised her that if she would give him a basketful of gold for his bride-to-be, he would break his engagement and join Maria Makiling, instead. Maria gave him the gold.

Maria waited and waited for Ban-aw to return, but he never did. One day the villagers saw Maria no more, but for many years, they would hear her calling the name of Ban-aw, and each time after her wails and cries were heard, a storm would come. The volcano where Ban-aw, the faithless mortal, went to live now is called Banahaw. Thus the geothermal area developed between the two volcanoes is named the Makiling-Banahaw, or Mak-Ban Geothermal Field.

A similar story concerns Mayon volcano in Albay province of Luzon, regarded as a premier tourist attraction because of its perfect cone and grandeur. The volcano also provides a beautiful

Mount Makiling, taken from the west and overlooking sugar cane fields. *B. J. Barker*

Mount Banahaw, home of the faithless mortal, rises in the distance. The Mak-Ban Geothermal Field is in the valley, close to Mt. Bulalo where the photo was taken. *B. J. Barker*

background for Tiwi and Bacon-Manito geothermal power plants. A popular legend of Mount Mayon concerns a chieftain's daughter who was so pretty that people called her *Magayon*, meaning "beautiful." As fate would have it, she fell in love with one of her father's soldiers who was not of noble birth. This displeased her father; he preferred that she marry Maraut, the son of another chieftain.

Majestic Mayon volcano of Albay, Philippines. In the foreground are the Bacon-Manito Geothermal Field and the cooling towers of the geothermal power plant.

Magayon and her soldier-lover decided to elope. However, Maraut overheard their plans and informed Magayon's father. Enraged, he sent his henchmen to chase the eloping pair. The pursuers killed the young man with an arrow, while Magayon ran into a nest of poisonous snakes and died from their bites.

The grieving father buried his daughter on the spot where she died and a strange thing happened that night. A storm arose suddenly with thunder and lightning. When morning came, the mound on top of Magayon's grave had grown higher, into a perfect cone, which the people called Mayon, in memory of Magayon, the unfortunate maiden in love.

Two more mentions were found of how ancient peoples used terrestrial heat. One concerns a mixture of oil and bits of sulfur-rich rocks, known locally as *dawa*. *Dawa* is believed to be medicinal, and its use was surrounded with mysticism (Gloria, 1987; Payne, 1985). In addition, the Earth's heat helped early Filipinos to make pottery, as some rocks were altered hydrothermally into potting clay. But most geothermal legacies left by early Filipinos were intangible, auras of romanticism surrounding volcanic areas and tales of volcanoes as homes for the gods.

Later Migrants

Information from the annals of later migrants in the geopolitical history of the islands makes mention of some other uses of the Earth's heat. Among these were (and still are) using hot springs for dressing slaughtered poultry and swine and for steam-cooking rice and root crops. When migrants from occidental cultures arrived, spas were built for health and recreational purposes.

There are examples of modern uses of the Earth's heat for nonelectrical purposes. A one-megawatt thermal plant for agricultural and industrial applications, commissioned in 1994, was built as an adjunct to the geothermal development in Palinpinon Geothermal Field on Negros island. A 0.75 megawatt geothermal plant began operating in 1998 in Manito, Albay, also for agricultural and industrial purposes.

Aids for Locating Geothermal Prospects

Unknowingly, past generations of Filipinos provided clues for locating geothermal prospects. These were the geothermal names given to localities or geographic features connected with terrestrial heat, such as *Mainit* ("hot"), *Los Baños* ("the baths"), *Magaso* ("steaming"), *Putinglupa* ("altered ground or kaolin"), *Cagua* ("kettle"), *Aslum* ("bitter"), and *Azufre* ("sulfur").

Exploratory help from indigenous cultural groups came in another form. Much of the interior of the Philippines has not been mapped in detail, and hot springs and other thermal manifestations are often in difficult, rugged terrain. Thus the information and physical guidance furnished by indigenous cultural groups have helped locate geothermal prospects.

Contact with tribal members in indigenous cultural communities and use of some as guides facilitated the entry of field exploration parties into places of tribal domain. Contact also served to introduce these groups to uses of the Earth's heat, and consultation with tribal elders of the immediate area usually preceded any exploratory activity.

Winning over Indigenous Cultural Communities

The development of Philippine geothermal resources has encountered stiff opposition in areas dominated by indigenous cultural communities, who have focused on legal, environmental, and cultural issues. The Mindanao I Geothermal Project on Mount Apo was one of the more controversial developments in this regard. Ote and de Jesus (1995) discuss the efforts undertaken by the governmental developer to resolve these concerns through a series of initiatives, transforming the project into one that is legally, environmentally, and socially acceptable.

Not only were the legal and environmental issues solved but so were the cultural issues, which were paramount if the project was to proceed. The tribal groups around Mount Apo regarded the volcano as the abode of their mountain god, Apo Sandawa. They held that any development on the volcano's slopes would be a desecration of holy ground. However, consultations with legitimate tribal leaders within and outside the project site and visits by the leaders to existing geothermal projects in 1990 convinced them of beneficial effects the undertaking would bring their communities.

In recognition of tribal rights and at the request of the tribes, the project developer endorsed the enactment of an ancestral domain law and contracted to pay the tribes one centavo (US $.00038) per kilowatt-hour of electrical generation, as royalty for using ancestral lands. Upon recommendation by the tribes within the project site, propitiatory rites and ceremonies, called *pamaas* in the local Manobo dialect, were held to seal the agreement. The *pamaas* were led by a tribal leader to seek approval for the project from the mountain god, Apo Sandawa. Amidst prayer chants and tribal music, sacrificial food in the form of meat from seven white animals was partaken of by all those involved in the project.

The rituals were reminiscent of the ceremony performed by the Igorots, an indigenous cultural group of the Cordilleras in Luzon, when the tribe granted permission to drill exploratory geothermal wells in Daklan, Benguet, in the 1980s. The rite, called *cañao*, was held to appease the

Propitiatory rites and ceremonies, called *pamaas*, are performed by the Manobos, an indigenous cultural community, together with PNOC-EDC officials, to seek approval from the mountain god, Apo Sandawa, for a geothermal project to be built on the slopes of his dwelling place, Mount Apo.

spirits dwelling in that area and those whose burial grounds are near the project site. The participating parties were dressed in tribal clothes, feasted on sacrificial food especially prepared for the occasion, and danced to native chants of prayers.

Epilogue

The Earth's heat appears to have played only a small role in the lives of the early inhabitants of the Philippines. Today that role is changing. Electricity generated from geothermal resources will reach the indigenous cultural communities, which have long awaited the benefits of modern civilization. The royalties that the tribes receive from developers who generate electricity on their ancestral lands will be ample to improve their socioeconomic well-being, including expanding their arts and crafts. Then they can raise praises and thanks to their mountain gods for having lavishly endowed their lands with the Earth's heat, whose development has indeed proven a boon, both to themselves and to the energy self-reliance program of the Philippine nation.

Acknowledgments

THE AUTHORS WISH TO EXPRESS THEIR GRATITUDE AND APPRECIATION TO THE MANAGEMENT OF THE PNOC-Energy Development Corporation for its encouragement and support of the study, and to Messrs. H. P. Ferrer and F. G. Delfin for their critical review of the chapter. In addition, the authors and editors wish to thank Benjamin J. Barker for his three photographs.

SELECTED REFERENCES

Bernard, M. A. 1985. Mt. Apo, The mysterious mountain. *GIMBA* 2, no. 2, 8-11.

Demetrio, F. R. 1978. *Myths and symbols Philippines.* National Book Store Pub., Manila, Philippines.

Department of Energy. 1995. *Philippine energy plan 1996-2025.*

Gloria, H. K. 1987. *The Bogobo: Their ethnohistory and acculturation.* New Day Publishers, Quezon City, Philippines.

Jocano, F. L. 1969. *Outline of the Philippine mythology.* Centro Escolar University Research and Development Center, Manila, Philippines.

Manuel, A. 1985. *Guide for the study of Philippine folklore.* E. L. Abueg, Quezon City, Philippines.

Ote, L. M. and A. C. de Jesus. 1995. *Mt. Apo geothermal project, A learning experience in sustainable development.* Paper presented at the 1995 World Geothermal Congress, Florence.

Payne, K. W. 1985. *The sulphur eaters: Illness, its ritual and the social order among the Tagabawa of South Central Mindanao.* University of California, Berkeley. (Ph.D. dissertation).

Togon, A. S. 1995. Folklore awareness and knowledge among the Mt. Apo ethnic tribes. In *Cultural study of tribes around Mt. Apo.* Makati, Philippines. (Study commissioned by PNOC-EDC).

The Authors:

Leonardo M. Ote
Arturo P. Alcaraz
Agnes C. de Jesus
PNOC-Energy Development Corporation
Merritt Road, Fort Bonifacio, Makati City
Philippines
Telephone: 63.2.893.1320
Fax: 63.2.815.2747

Head of a Dhyani Buddha, volcanic stone, Shailendra Dynasty, c. 800 A.D., Java, Indonesia. This dynasty built the Borobudur and the statue may be from the Borobudur itself. The texture of the volcanic stone contributes to the soft blurring of the facial features and the tightly-wound curls contrast strongly against the smooth, soft planes of the face. Height, 12 ½ inches; width, 10 inches; depth, 9 inches. ***Asian Art Museum of San Francisco, The Avery Brundage Collection, B60S379***

27. Indonesia: Land of Volcanoes, Heart of Fire

by
Wimpy S. Tjetiep
Asnawir Nasution
Susan F. Hodgson

"To obtain heavenly elixir, *amerta*, the gods sent Anantabhoga to uproot mighty Mandara mountain. The tortoise Akupa went under the milky sea, to be the fulcrum on which the mountain would rest. 'Be kind, great ocean, and let your waters be parted by this island. Great will be the rejoicing in the three worlds when *amerta* has been distilled.' Divine serpent Basuki was the rope, wound about Mandara mountain. As the rope was pulled the mountain would turn to churn the milky sea.

"King of the gods Indra sat, steadying the mountain, on its peak. Ready to begin work, the gods held the tail end of the serpent, the demons its head. As they pulled in turns, poisoned fire appeared in the snake's hot breath. But their striving did not diminish, as both sides struggled mightily, joyously, to obtain the elixir. The sea rumbled as thunder, and all nature on Mandara mountain was in turmoil for the churning...."

—From a Balinese version of the *Adiparwa*, the creation sequence and first book of the great Indian epic, *Mahabarata. Translation by Mary Zurbuchen and I Wayan Wija (Lueras, 1987)*

Origins and Influences

The story of Indonesia is woven on a geothermal loom. Many countries possess areas with geothermal resources, but few have almost all national territory on volcanic terrain. In Indonesia almost every island is volcanic, and volcanoes and other geothermal phenomena reverberate from the core of ancestral memory. The chapter starts with geothermal influences from over a millennium and a half ago when Indian religious traditions first came to Indonesia, reaches into mists of prehistory for ancient legends and beliefs, and returns at last to modern-day practices.

It was gradually, early in the first millennium, that Indian culture came to Indonesia, introduced by early Indian traders who described Indian Hinduism and Buddhism to the islanders and entertained their own countrymen back at home with tales of the rich, volcanic islands. From this mutual exchange of cultural and material goods on a constantly increasing scale, which on Indonesia's part extended to trade with China and additional Buddhist influences from that quarter, a peaceful amalgamation began of Indonesian, Hindu, and Buddhist beliefs. According to Claire Holt (1967), such blending continues today about 1,600 years later, and the present religion of Bali is an offshoot of the union.

Many intricate structures built throughout Indonesia in basalt and other volcanic rocks express the evolving religious consensus, and these structures reached their creative peak in the 7th through the 10th centuries. Much designing and building was supervised by Indian Hindu or Buddhist priest-architects, commissioned by Indonesian political rulers to base the undertakings on Indian treatises of the arts, called *shastras*. The earliest reference to priest-architects is from Indonesia's earliest known stone inscription, found in East Borneo and dated about 400 A.D. The Indonesian builders and the Indian-trained carvers must have given their own countrymen new knowledge of the great Indian epics and Buddhist legends and fables, creating inspiration

Pura Besakih, the greatest Balinese temple, is on the southwestern slope of the volcano Gunung Agung. The huge complex was begun in the 8th century and building has continued ever since. Note the split gate, the *candi bentar*, in front of the Bale Pegat, a shrine with three seats enthroning the three aspects of God.

for the chants of the storytellers and designs sculpted in relief on the narrative stones decorating Java's ancient sanctuaries (Holt, 1967).

Close up of the *candi bentar*, the split gate, whose carefully carved sides blend with the multi-roofed temple pagodas.

A Cosmic Mountain is important in many early Hindu creation epics, and perhaps Indian cosmology matched naturally with the Indonesian concept of a sacred mountain, such as the volcano Mount Semeru on the island of Java and the volcano Gunung Agung on the island of Bali, considered the seat of the gods. Certainly the Cosmic Mountain theme has influenced many religious and secular forms of expression on Bali, such as the *candi bentar* ("split gates") found on the island. These stylized temple entryways of carved volcanic rock resemble triangular, erect stone slabs, shaped like volcanoes with the mountain slopes carved from the base to the top into ever-smaller upturned curves. Split vertically down the center and pushed apart slightly for passage, the halves are said to represent the two parts of the legendary Cosmic Mountain separated by the Hindu god Shiva, who placed both halves in Bali, one as the volcano Gunung Agung and the other the volcano Gunung Batur, the island's second highest peak (Lueras, 1987).

Another mountain symbol found in Bali is the tapered temple pagoda itself, or *meru*, which rises from a basalt base in a series of stylized roofs, three to eleven in number, each smaller than the one below, the count depending on the Hindu deity or deities honored (Lueras, 1987). Such pagodas are part of Bali's pavilion courtyard architecture, which was influenced by the Javanese in the 15th century, a time when many Hindus fled the growing influences of Islam on that island.

IT IS NOTEWORTHY THAT THE VOLCANIC THEMES FEATURED IN INDONESIAN BELIEFS AND THE volcanic influences—sometimes cataclysmic—woven throughout Indonesia's everyday life have not led to an emotive cosmology filled with eruptive violence and impassioned chaos. From the earliest times, the Balinese have believed in an ordered universe stretching from the heavens and the volcanic peaks, home of divine spirits who bring prosperity and good fortune, "down to the plunging depths of the sea, home for threatening, harmful forces. Everything within nature has direction, rank, and place. All that is holy is associated with height, the mountains, and the direction upstream toward the majestic volcano Gunung Agung. All that is threatening or harmful belongs to the forces of the underworld, the fathomless ocean and the direction downstream toward the sea" (Black et al., 1993). The fertile plains between the two extremes are home for

Map of Indonesia, showing the locations of the major islands in the archipelago.

the Balinese, who strive through ritual to maintain harmony between the poles (Black et al., 1993). Balance, order, and control without exaggerated contrast prevail in uniquely Indonesian Buddhist and Hindu expression, without the drama of Indian evocation. The great Indian poet Rabindranath Tagore once exclaimed on a 1927 visit to Indonesia, "I see India everywhere, but I do not recognize it."

Perhaps the most famous and inspired Buddhist monument in Indonesia is the Borobudur, found on Java, near the City of Jogyakarta. The 7-acre shrine, built on a hilltop about 800 A.D., rises

Volcanoes in Indonesia, the Geology

"Beginning possibly around the end of the Cretaceous period, perhaps 63 million years ago, the deep-sea area and parts of the Sunda Shelf were compressed and folded, thrown up into two parallel ridges, which here and there appear above water as islands. This folding extends northwest beyond Indonesia. Its northward course is marked by the Nicobar and the Andaman Islands, the Arakan Range of Western Burma, and finally by the Himalayas, which swing westward. Each of the two Indonesian ridges has the shape of a great arc with a northwardly turning hook at its eastern end. The more westerly of these two ridges is marked by small islands that parallel the west coast of Sumatra, as well as by Sumba, Timor, and the Tanimbar (Timor Laut) group, with Ceram and Buru forming the hook. The eastern ridge is marked principally by Sumatra, Java, and a single line of islands from Bali to Wetar, with the Bandas at the tip of this arc's hook. The two ridges are different in certain geologic features, notably in the degree of volcanic activity. On the islands of the westerly arc, such activity has nearly or quite ceased, and ominous volcanic peaks are not a prominent feature of the landscape. In contrast, the islands of the easterly arc are dotted with volcanoes. Indeed, from almost any point in this arc at least one volcanic cone is likely to be visible, perhaps slumbering blue-gray on the distant horizon, or perhaps actively trailing a plume of smoke by day, and reddening the sky by night. A majority of Indonesians spend their lives in sight of a volcanic crater.

"Indonesian volcanoes are not confined to the easterly ridge, for they are also present on Sulawesi and in the Northern Moluccas. In the stretch of islands from Sumatra and Java eastward through the Bali-Wetar chain, and thence northward through Sulawesi and the Moluccas, more than 400 volcanoes exist, and more than 100 of these have been active in historic times. About 77 of them are regarded as still active, but the figure is tentative, for a crater will occasionally come to life after decades or even centuries of quiescence. For example, in the year 1646 the little volcanic island of Makian, west of Halmahera, was rent by a violent eruption, which opened a chasm into the heart of the crater. In 1860, when the English naturalist-explorer Alfred Russel Wallace visited Makian, he found it clothed to the summit in vegetation, and with a dozen populous villages on its flanks. But in 1862, after 215 quiet years, Makian exploded again, killing most of the inhabitants, showering the islands with ash, and darkening the air with dust at Ternate 35 miles away.

"As another example, in 1883 the volcanic island of Krakatau (more accurately Rakata), west of Java, blew up with inconceivable violence, a blast in the 10,000-megaton range. Six cubic miles of Earth, rock, forest, farms, people, livestock, and wildlife were blown into the upper atmosphere to drift as dust for years, providing the world with unusually colorful sunsets.... Gunung Agung ('prominent mountain') on Bali, quiet for more than a century, erupted three times in 1963, taking 1,700 lives and leaving 87,000 people homeless; and then nearby Batur, quiet since the 1920s, exploded and forced 1,200 people to flee their villages. Papandajang in Java is now about 8,600 feet high, but it was once much higher; in the year 1772 it blew its summit away, destroying 40 villages.

"[Except for a nonvolcanic peak in New Guinea called Irian Barat], the highest elevations of Indonesia are provided by volcanoes, active or otherwise. On Sumatra, Kerintji is 12,484 feet high. Almost as high, at 12,224 feet, is Rindjani on Lombok, just east of Bali. Rindjani is often called simply Lombok Peak, for its towering presence dominates the comparatively small island.

"The Indonesian population is strongly concentrated in regions of recent or continuing volcanic activity.... This situation would at first appear surprising, but it is easily explained. Young volcanic soils are by far the most productive ones, indeed the only ones that could support a large population of farmers. The rural inhabitants of Indonesia feel that any given volcano is likely to erupt only at long intervals. In the meantime, why starve on poor land? So the farmers...move to fields by a volcano for the promise of richer soils" (Neill, 1973).

on four rectangular and three rounded terraces of andesite—a volcanic rock—forming themselves a mountain in the volcanic skyline, a home for the gods. A supreme creative achievement, the Borobudur represents Mahayanist Buddhist belief in the form of a *mandala*, an aid to meditation, and climbing the monument replicates the journey of mankind from Earthly travail to eternity. Walking in contemplation around the terraces up to the top of this shrine, one passes on the lower four levels sacred texts illustrated by 1,460 sculptures on over 1½ miles of bas-relief panels depicting the pleasures of Earthly life. Along the three upper circular terraces, the panels end and the atmosphere changes to the serene, unadorned simplicity of cosmic bliss. One passes Dhyani Buddhas meditating, protected by open *stupas* (see photo at the beginning of the chapter). "On the highest terrace of all a single enormous *stupa* rises, enclosing— invisible to the human eye—that which stands for the Buddha himself—empty space—for the Buddha is the beginning, the center, and the end of all life" (Vlekke, 1960).

The Borobudur, near the City of Jogyakarta on Java. The shrine was built about 800 A.D. *Information Division, Embassy of Indonesia*

Hindu art in Indonesia is equally famous and equally influenced by volcanic surroundings and materials. The supreme emanating force in Hinduism is the god Shiva, the destroyer. At a Javanese Shivite site in a collapsed caldera on the heights of the Dieng plateau, remains of temple groups dating perhaps from the early 800s stand on a volcanic terrain of hot springs and sulfurous fumes. The temple city in this sacred place once held not only the priests and their attendants but many pilgrims, including royalty. It was a "proper abode for divine ancestors from whom Javanese kings eventually claimed their descent" (Holt, 1967).

One of the most sacred places in every temple on the island of Bali is a shrine for offerings dedicated to the volcano Gunung Agung. These shrines have conical roofs peaked with a spire

Looking across the plain from a high terrace of the Borobudur. *Information Division, Embassy of Indonesia*

of twisted fiber. High on the southwestern slope of Gunung Agung is Bali's most sacred "mother temple," the Hindu complex of Pura Besakih, first built in the 8th century and added to ever since. The rite of Eka Dasa Rudra, the largest and most sacred ceremony in Bali, is held to purify the universe once every 100 years at the Pura Besakih temple. During the ceremony, animal sacrifices are thrown directly into Gunung Agung's crater (Lueras, 1987, with reference to Peter Jennings, 1969). Colors associated with the shrine are red, white, and black: red symbolizing the Earth as lava, white symbolizing light, and black symbolizing water and outer space (Black et al., 1993).

In 1963, preparations for the Eka Dasa Rudra ceremony had begun when fire and ash spewed from Gunung Agung, dormant for over a century, a sign that a priestess interpreted as a sacred portent sent to purify the temple. By the time of the great sacrifice, the volcano was issuing thick, dark columns of smoke and shortly afterward, Gunung Agung erupted, destroying hundreds of homes, taking about 1,700 lives, and leaving about 87,000 people homeless. The priests then decided the calculations had been wrong and reheld the ceremony in 1979 without incident at the ancient site, most likely a terraced sanctuary from prehistoric times where the god of Gunung Agung was worshiped and received offerings (Black et al., 1993).

Rocks ejected during volcanic eruptions are called volcanic bombs and volcanic blocks. These spinning mixtures of lava and rock, shaped by flight in air, usually are sized somewhere between a fist and a football, with some vastly larger and weighing tons. After an eruption, the people of Bali select from the larger rocks sacred shrine objects, new sites for homage and regular offerings (Lueras, 1987).

Incredible view of the steep peak and the crater of Merapi volcano, Java. Propitiatory offerings of precious silk textiles are thrown into the crater. The very active volcano is kept under volcanic watch. *Information Division, Embassy of Indonesia*

From the beginning in Indonesia, when people at the end of the Stone Age first lived together in permanent settlements, belief in animism probably existed, the conviction that natural phenomena and objects, like wind, rocks, and trees, have souls and were formed by good and evil supernatural forces. At pyramidal *maraes*, stone platforms of volcanic rock, the early people of Bali gathered to invoke the blessings of natural phenomena, including the mountains. They called on spirits of their ancestors in the high places, at the tops of volcanoes (Lueras, 1987). They also used these high places for city planning, for ancient peoples "placed linear mountain villages on axes aligned radially from the centers of Bali's three great volcanoes" (Black, et al., 1993).

Still today, "high volcanoes are regarded to have a life and spirit of their own, and are venerated. The ruling princes of Central Java send elaborate propitiatory offerings, including precious silk textiles, to be thrown into the craters of the towering Lawu, Merapi, and Merbabu volcanoes. In East Java, the Tenggerese hold annual ritual processions to the top of Mt. Bromo, making offerings at the edge of its crater" (Holt, 1967).

Geothermal Legends

Some legends tell how the volcanoes came to be. One says that Mount Mahameru, the Indian World Mountain and abode of the gods, was brought to Indonesia. The highest mountain of East Java is called Semeru, and "a description of the legendary transportation of the gods of the Semeru to Java is given in a 16th century East Javanese work, *Tantu Panggelaran.* An attempt to plant the mountain in the west caused the island to tip, so that for the sake of balance it was moved eastward. On the way pieces kept chipping off its lower rim, and thus the mountains Lawu, Wilis, Kelud, Kawi, Ardjuna, and Welirang were formed. The mountain's damaged foot made it difficult to steady it on the spot; it shook, and the top cracked off and fell away, becoming Mt. Penanggungan" (Holt, 1967).

A West Javanese legend from the famous Sundanese Sang Kuriang myth explains how the truncated volcano Takubang Prahu was created. It was made by an unwelcomed suitor who hurriedly tried to build a mountain overnight on demand of a woman whom he desired. Unbeknownst to him she was his mother, and she frustrated his labors by causing a cock to crow prematurely, falsely signaling the arrival of morning (Holt, 1967).

Not only were early religious rites and legends centered around volcanoes and volcanic rocks, but practical uses were made of volcanic resources, such as obsidian. Knife blades, points, and scraper-like instruments fashioned of obsidian, probably dating from the last Paleolithic age, were discovered in two caves in Central Sumatra (Heine-Geldern, 1935). Such paleolithic uses may continue today in New Guinea, where a sago pick-head of obsidian was seen and where leaf-shaped points of obsidian may tip the shafts of hunting arrows (Neill, 1973).

Hot Springs

Indonesian hot springs, sites for activity in the earliest times, still are used for religious ritual, medical cures, and pure enjoyment. In ancient times, the Balinese believed in omens. Hot springs and fumaroles were considered sacred places and most were in remote areas, isolated tropical jungles on the volcanic slopes, usually reached after hours of walking. Here dying animals were sometimes found, along with strong smells of sulfuric acid. The villagers considered the sites mysterious places that could kill innocent animals, a sign, they believed, from God to do the right thing. To inform visitors to be careful and to be on their best behavior, the people built *puras*, special places for praying, close to hot springs and fumaroles. On the slopes of Mount Batukau, one of the three highest volcanoes on Bali, the holy hot spring of Air Panas issues from a river bank. Spirits are believed to frequent unusual natural areas such as this, and at a small temple people pray and leave offerings (Black et al., 1993).

In Lombok, four natural hot spring areas are in isolated areas of tropical jungle, on the flanks of the inactive Nanggi volcanic complex. Most natives of Lombok are of the Sasak Tribe, who believe in God and believe that any disease may be cured at hot springs. In a special month, close to Ramadan, a leader takes hundreds of old and young people from the tribe to the hot springs. They spend one or two days camping by the springs, submerging themselves in small hot spring pools, hoping that this, plus faith and prayer, will cure their ills.

On the islands of Sumatra, Nusa Tenggara, and North Sulawesi farmers usually bathe in the hot springs on returning home from the fields. They believe that the hot mineral waters will cure many diseases. Some hot springs have carbonate deposits useful for tooth ailments, especially those of the elderly.

The great fumarole (in Dikit Geothermal Field) in the Humluyang caldera, Central Barisan volcano, Jambi, island of Sumatera. Hot springs and fumaroles cross the caldera along a fault, and farmers use the hot water to treat the fibers they weave into handicrafts. *N. Akbar, 1978*

On the island of Sumatera, farmers in the Kerinci area of Jambi submerge rattan and wild pandanus leaves into hot springs for several days to increase the strength, durability, and pliability of the fibers. The rattan and leaves are woven into handicrafts, such as mats and baskets.

Some hydrothermally altered rocks turn to white clay. In Indonesia, this clay usually is found around fumaroles in remote areas. Villagers use the clay to paint walls instead of purchasing paint, for the clay is free and readily available to dig out and carry away. Such clay is used in small villages like Sembalun on the eastern part of Lombok and Marga Bayur on the southern part of Sumatera.

In the fumarole areas of Tompaso and Bora, Northern and Central Sulawesi farmers use the heat from fumaroles to cook corn, sweet potatoes, and eggs.

From the beginning, geothermal resources in Indonesia have nourished both body and spirit, offering pleasure and danger, beauty and horror to a populace who understands well that nobody knows what the next moment will bring in a terrain of volcanoes with a heart of fire.

Acknowledgments

All extracts reproduced from *BALI, The Ultimate Island* by Leonard Lueras and R. Ian Lloyd are copyright © 1987 by Times Editions Pte Ltd. and reprinted by permission of St. Martin's Press, Inc.

Material from *Art in Indonesia* by Claire Holt is published with permission of Cornell University Press.

All extracts from *Insight Guides Bali*, by Star Black et al., are used by permission of the publishers, Apa Publications GmbH & Co. Verlag.

The geologic essay is from *TWENTIETH-CENTURY INDONESIA* by Wilfred T. Neill. Copyright © 1973 by Columbia University Press. Reprinted with permission of the publisher.

We would like to thank the Information Division of the Embassy of Indonesia for helping secure some of the photographs.

Selected References

Black, S., W.A. Hanna, R. Ballinger, J. Darling, F. Eiseman, D. Harnish, M. Wijaya, and M. Zurbuchen. 1993. *Insight guides Bali.* Apa Publications (H.K.) Ltd., Singapore.

Heine-Geldern, R. 1935. *Sumatra, Its history and people.* E. M. Loeb, ed. Oxford University Press, London.

Holt, C. 1967. *Art in Indonesia.* Cornell University Press, Ithaca, New York.

Lueras, L. 1987. *Bali, The ultimate island.* St. Martin's Press, Inc., New York.

Neill, W. T. 1973. *Twentieth-century Indonesia.* Columbia University Press, New York and London.

Vlekke, B. H. M. 1960. *Nusantara, A history of Indonesia.* Quadrangle Books, Inc. and W. van Hoeve Ltd., Chicago, The Hague, and Bandung.

The Authors:

Wimpy S. Tjetiep
and
Asnawir Nasution
Volcanological Survey of Indonesia
Jalan Diponegoro No. 57
Bandung, Indonesia
Telephone: 62.22.771402
Fax: 62.22.70276

Susan F. Hodgson
California Department of Conservation
Division of Oil, Gas, and Geothermal Resources
801 K Street, MS 20-20
Sacramento, CA 95614-3530 USA
Telephone: 916.445.9686
Fax: 916.323.0424
E-mail: shodgson@consrv.ca.gov

"Bella, a Maori guide, sitting at the edge of Rotorua geysers."
By William Henry Jackson, 1895. The Library of Congress

28. Traditional Use of Geothermal Resources by New Zealand Maori

by
C. M. Severne

Abstract: Geothermal resources were revered in traditional times throughout *Aotearoa* ("New Zealand"). In the Taupo Volcanic Zone, the Maori people used geothermal resources for many traditional purposes, including cooking, heating, bathing, and agriculture. The Maori people regarded these resources as natural treasures that were preserved for future generations by guardians (*kaitiaki*) living in the area. The understanding between Maori and the *taha wairua* ("spiritual dimension") did not allow improper use of or disrespect for physical resources to go unpunished.

Development of geothermal resources by people other than Maori has raised a number of issues regarding *kaitiakitanga* ("guardianship") and *rangatiratanga* ("full authority") over the resources. The information in this chapter includes research conducted by the New Zealand Waitangi Tribunal and information from early explorers.

Reprinted by permission of the International Geothermal Association, from the *Proceedings of the World Geothermal Congress,* 1995, pp. 401-404.

Introduction

'Ko Tongariro te maunga	*Tongariro is the mountain*
Ko Tongariro te awa	*Tongariro is the river*
Ko Taupo te Moana	*Taupo is the sea*
Ko Ngati Tuwharetoa te iwi	*Tuwharetoa is the tribe*
Ko Te Heuheu te Tangata	*Te Heuheu is the man*
Tihei Mauriora'	

ALL NATURAL RESOURCES IN THE MAORI TRADITION WERE *taonga*, valuable natural treasures derived from gods. In a very special way, Maori were aware that their possession of them was on behalf of someone else in the future. Myths and legends support a holistic view encompassing spiritual and physical dimensions.

In 1840, many of the Maori leaders and representatives of the Queen of England signed the Treaty of Waitangi. The treaty extinguished Maori sovereignty and established that of the Crown. Natural resources are covered in the treaty by the term *taonga katoa*, which implies Maori treasures, namely knowledge, language, heritage resources, and more, all inherited from ancestors. This important term was translated in the English version of the treaty as "other properties." The treaty was believed to signify a partnership, but this was never attained due to similar examples of misunderstandings and conditions not honored.

As the treaty was not honored by the government, Maori have watched the development of laws enacted by the government deprive them of privileges secured in the Treaty of Waitangi. In 1975, the Waitangi Tribunal was established to make recommendations to the Crown on claims relating to the treaty. To prepare claims for the hearings, *Whakapapa* are recited; this identifies the *iwi* or *hapu* who have *mana whenua* and so *kaitiakitanga* and *rangatiratanga* over the resources. The ownership of the geothermal resource is still not resolved, although claims for Ngawha and the Rotorua areas have been heard.

MAORI GLOSSARY (additional Maori words are explained in the text.)

hapu - aggregation of *whanau'* (family), where *whanau* is the association of close relatives
iwi - Maori tribe sum total of its *hapu*
kaitiaki - steward or caretaker
ngawha - geysering boiling pools
puia - boiling pools
rangatiratanga or *tino rangatiratanga* - full authority
tangatawhenua - people of the land
taniwha - supernatural being, can dwell above or underground, has many forms
taonga - treasure, prized thing of cultural and/or spiritual significance
tapu - sacred
tohunga - high priest
waiariki - warm pools
whakapapa - genealogy
whanau - family
whaunaunga - extended family

This traditional view of resources has always differed fundamentally from that of the European settlers (*Pakeha*) who came to Aotearoa (New Zealand) in the 19th century. Maori and *Pakeha* each had very different world views, for the Maori traditionally did not compartmentalize resources into separate components.

The preservation of physical resources for future generations is a part of the concept of *kaitiakitanga*. Exploitation of the geothermal resources cannot be confined to traditional uses, as Maori have a commitment to provide for their families in a changing economic environment. The development of geothermal resources by other parties has resulted in complex situations between the developer and *tangatawhenua*, due to the lack of understanding and recognition

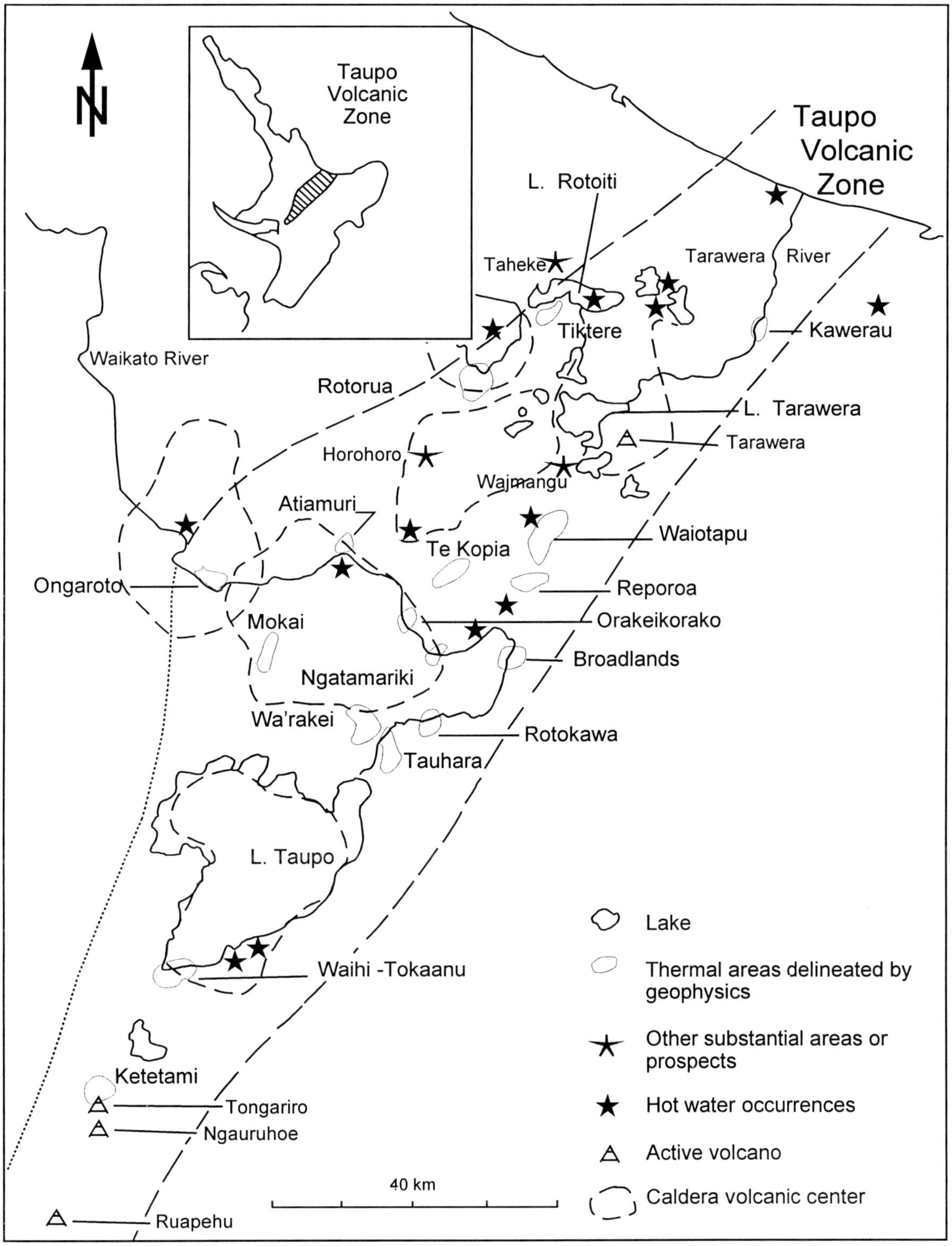

Taupo Volcanic Zone, North Island, New Zealand.

of *rangatiratanga* and *kaitiakitanga* by the developers. Development requiring massive geothermal fluid withdrawal has destroyed many of the manifestations traditionally used by Maori. Some measures can be enforced to preserve thermal features. The deep geothermal resources themselves are not in danger of overexploitation; at present only the top 1-2 km of an enormous heat resource is mined (M. P. Hochstein, pers. comm.).

In Maoridom, it was believed that human existence was subordinate to the physical and spiritual components that combined to form the total environment. Geothermal resources are a case where resources were not isolated. Traditional evidence indicates that thermal springs were the face or eye of the geothermal *taonga*. Maori understood that thermal springs and the underlying geothermal activity were one and the same thing and that particular pools were connected to other surface manifestations. Maori people have long connected the *ngawha* and *puia* with the centers of volcanic activity that are still active.

The geothermal resource in the Taupo Volcanic Zone is a highly esteemed *taonga* of the Ngati Awa, Te Arawa, and Ngati Tuwharetoa tribes, who have centuries of association with the resource. The tradition attached to the tribal areas of *wahi tapu* ("sacredness") is explained later.

Legends

The geothermal resources encountered in the Taupo Volcanic Zone are referred to as the legacy of Ngatoroirangi. This account is one that is associated with the descendants from the Te Arawa canoe, *Mai Maketu ki Tongariro*, "from the prow of the canoe at Maketu to the stern at Tongariro" (Stokes, 1991).

Mountains, rivers, and lakes are physical features recorded in myths and legends since time immemorial. Each feature of the landscape had special meaning and significance. Maori legends and songs speak of the spiritual connection of the Maori with this resource. Maori extended their deep sense of spirituality to the whole of creation. In their myths and legends, they acknowledged gods and other beings who bequeath all of nature's resources to them (Wai. 22, 1988).

Ngatoroirangi, *Tohunga* ("high priest") of Te Arawa canoe, journeyed inland from Maketu in the Bay of Plenty to Tongariro. On the slopes of Tongariro, nearly freezing to death, he called to his sister for help. The following account of the legacy of Ngatoroirangi was recorded by Dr. Ferdinand von Hochstetter in 1959:

> "I reproduce the peculiar legend as I heard it from the mouth of the Chieftain Te Heuheu at Lake Taupo. Among the first voyagers who came from Hawaiiki to New Zealand, was the chief Ngatoroirangi ('heaven runner,' 'the traveler in the heavens'). He landed at Maketu on the east coast of the North Island. Thence, he made his way with the slave Ngauruhoe to explore the new country. He travels through the district, stamps springs of water from the ground to succor the mountains and glimpses to the south a great mountain, Tongariro (literally, 'towards the south'). This mountain he determines to climb in order to survey the whole country from its summit. He comes to Lake Taupo in the inland plains. Here the bushes tear up his great cloth of kiekie leaves. The shreds strike root and become Kowhai trees. Then he climbs the snow clad Tongariro; higher up, however, it is so cold that chieftain and slave are in danger of freezing. Ngatoroirangi, therefore, calls his sisters who remain at Wakari [Whakaari] to dispatch him fire. The sisters hear the call and send sacred unquenchable fire that they brought with them from Hawaiiki. They send it by means of two *taniwha* (spirits of the mountain and water, living underground), Pupu and Te Haeata [Te Pupu and Te Hoata], by a subterranean passage to the peak of Tongariro. The fire comes just in time to save the chieftain. When, however, the latter wishes to offer it to his slave, so that he can warm himself, Ngauruhoe was dead."
>
> "To this present day, the hole through which the fire rises in the mountain (the active crater of Tongariro) is known by the name of the slave, Ngauruhoe. Since, however, the fire was the sacred fire from Hawaiiki, it still burns strongly today, and burns along the whole stretch between Wakari [Whakaari] and Tongariro at Motou-Hora, Okakaru, Roto-ehu, Rotoiti, Rotorua, Rotomahana, Paeroa, Orakeikorako, and Taupo, at every point where it shot up when the two *taniwha* brought it underground. Hence the innumerable hot springs" (Hochstetter, 1959). [Corrected spelling in brackets.]

This account is one of many that exist. Similar versions and variations involve the sisters traveling with the heat, sending gods or demons, or the sisters being called to in Hawaiiki, not Whakaari. Always, the route the two *taniwha* took is consistent with the account of the Chieftain Te Heuheu.

The following version also is often heard:

> "The sisters, who were back in Hawaiiki, heard his call and set out at once. They rested and lit a fire at Whakaari (White Island, which is an active volcano in the Bay of Plenty), and then they made their way underground to Tongariro. Sparks fell from their fire at Waiotapu, Ohaaki, Rotokawa, Tapuaeharuru, and Tokaanu, creating hot springs, geysers, and boiling mud pools in these places. At Tongariro, the sister's fire warmed Ngatoroirangi, but it came too late for Ngauruhoe; the crater of the mountain became his tomb, and its peak is known by his name. Then Ngatoroirangi took the fire and hurled it into the crater, where it still burns. His sisters returned to the Bay of Plenty, initiating, as they went, thermal activity at Waimahana, Whakarewarewa, Ohinemutu, and Tikitere" (Orbell, 1985).

Physical features were named to commemorate myths that surround them. For example, when Ngatoroirangi thought he was dying of cold, his words were "*Ka Riro Au Te Taonga*."

Ngatoroirangi was chanting these words on the summit of the mountain we call Tongariro (Maxwell, 1991).

Myths also associate the mountains with movement. One of the best known myths is the tale of the four warrior mountains' love for Pihanga. In the days when the world was young, the great mountains stood on the heart of Ika a Maui (North Island, New Zealand). They were gods and warriors. Taranaki stood southwest of Ngauruhoe, Tauhara, and Edgecumbe, which was to the northeast.

Pihanga was the sought after mountain; her fame and beauty had spread to the far corners of the land. Each of the four mountains wooed her and wished her to be his wife. The mountains battled for her hand and Tongariro emerged victorious. The defeated departed. Taranaki traveled toward the setting sun, carving out the mighty Whanganui River. Mount Edgecumbe journeyed northwards, and Tauhara, heavy in heart and lingering to gaze back at Pihanga, today stands on the northern shore of Lake Taupo (Potton, 1987).

The sighting of a phantom canoe on Lake Tarawera is deeply embedded in the mythology surrounding the 1886 eruption of Mt. Tarawera. Sophia, a well-respected guide at the Terraces, returned from guiding visitors one day and saw a war canoe full of warriors. The canoe was moving swiftly, as if pursued by the enemy. It was an omen and Sophia met with the landlord of the Hotel at Wairoa, saying, "I shall never go back to the Terraces; something is going to happen." A few days later, the forces of nature broke out and Tarawera erupted, along with Lake Rotomahana, destroying the Terraces and surrounding villages; 150 people were killed. Sophia reportedly believed the eruption was punishment from a demon called Tama-o-hoi, who long before had been imprisoned in the mountain by Ngatoroirangi.

Kaitiaki (Caretakers of the Resource)

In many cases, particular resources formed a central cultural component between *hapu* or *iwi* and their traditional territory. Particular springs in a region may have been used by all *hapu* from the time of the ancestors, who used the springs in early times for ceremonies. Events occurring through time accumulated into a history, enhancing the magical qualities of a pool.

Swimmers in a warm spring at Orakei Korako, with a geyser in the background, from *Te Ika A Maui: or, New Zealand and Its Inhabitants*, by the Rev. Richard Taylor. Maori villages were often built in thermal areas. The warm pools were popular bathing places and food was easily cooked in baskets lowered into hot pools. *Auckland Institute and Museum*

The *hapu* living in the immediate vicinity of the pools or springs acted as the *kaitiaki* ("steward or guardian") on behalf of all the *hapu*. One of their major functions was to nurture and preserve the resource for future generations. Use by other *hapu* may have been due to the geographic position of the springs, traditionally a crossroads for Maori traveling in the region. The seasonal uses by the larger *iwi* will be discussed further.

There was a system of *tapu* ("sacred rules") that, combined with the Maori belief in gods having overall responsibility for nature's resources, served effectively to protect these resources from improper exploitation. To disregard or disobey any of the rules of *tapu* was to court calamity and disaster (Wai. 304, 1993).

The spiritual properties of an area are important and are *wahi tapu* ("sacred") due to incidents involving ancestors. The *wairua* ("spirit") of a feature or area has qualities to be respected and feared, because if not properly acknowledged it can cause ill to those who defile it.

Direct Use

In traditional Maori terms, hot springs were not a single isolated phenomenon; hot springs were seen as the face or eye of the resource. The surface components of the resource have been given specific names, ranging from boiling mud pools (*ngawha paruparu*) to sulfur deposits (*kupapa*) (Maxwell, 1991). *Puia*, *ngawha*, and *waiariki* are Maori terms used to distinguish among geysering pools, boiling pools, and warm pools.

Icelanders distinguish among *hverjar*, *namur*, and *laugar* as Maori distinguish among *puia*, *ngawha*, and *waiariki*. The *puias* of New Zealand correspond to springs called *hverjar* in Iceland. *Hverjars* can be either permanently flowing boiling springs or intermittent hot-water discharges. Surface manifestations at Tokaanu, Orakeikorako, and Rotorua include many intermittent geyser-like springs, which Maori referred to as *puia*. *Puia* is also used to describe a crater or volcano, for example the crater lake at Ruapehu, an andesitic volcano on the southern end of the Taupo Volcanic Zone (Hochstetter, 1959).

The areas with geothermal activity provided warmth, heat for cooking, therapeutic healing qualities from mud and hot pools, *kokowai* ("red ochre"), and sites for *paruparu* ("black dye") preparation. In general, the favored sites for permanent settlement were within easy reach of forest, river or lake water, arable land, and geothermal areas (Stokes, 1991).

Medicinal benefits from geothermal areas have been known for centuries, and I outline a few. As these medicines are *taonga* to the various *hapu*, and should remain so, my discussion is brief. Bathing in pools was believed therapeutic for various skin diseases, such as eczema and rashes. Various pools were used to treat different ailments, and leprosy, arthritic, and rheumatic diseases could be treated in different pools from the same area.

Sulfur was mixed with pork fat and applied directly to any skin disease, *hakihaki*. It was also ingested whole to combat internal diseases. Maori had special medicines for people burnt by hot water or steam. At Rotorua, "shells from fresh-water shellfish (*kakahi*) were ground to the consistency of talcum powder, *rekereke* was applied next, then rendered fat was added to seal the burn" (Maxwell, 1991). The ingestion of some thermal waters (*e.g.*, Waitangi at Rotorua)

The Cold Basins, White Terraces, Rotomahana, from *Illustrated London News*, 1880. Pools like these were used for healing, relaxation, and a chance to *korero* ("talk"). *Auckland Institute and Museum*

was said to help cure stomach complaints. Often the ingested waters have been predominantly bicarbonate-chloride waters.

As mentioned before, some of the pools and springs are *tapu*. The *wairua* ("spirit") of the manifestation can be due to past events or to current activities. Examples of *tapu* manifestations are pools used for burial or bathing the dead, areas used for childbirth, and pools used by menstruating women.

Areas were often used seasonally by the whole *iwi* for digging the raw material needed to make *kokowai*, a red dye made from thermally altered clay, prepared by heating the clay and mixing it with oil and/or water, depending on use. The periodic use of thermal areas for digging roots and collecting clay for *kokowai* did not necessarily indicate continuous occupation. The treasures the Maori held dear (*taonga*) were not often mentioned to travelers for fear of exploitation.

Case Studies: Well-Known Areas Traditionally Occupied by Maori

Te Tarata and Otukapuarangi (Pink and White Terraces)

Te Tarata and Otukapuarangi were already the focus of an international tourist industry by the 1850s. The Ngati Tuhourangi people became an affluent tribe as the tourist industry in New Zealand developed when the world became aware of the unsurpassed geothermal spectacles, of which Hochstetter said, "They baffled description."

The following observations were made by Dieffenbach on his travels: "When we arrived on the crest of these hills, the view which opened was one of the grandest I had ever beheld. Let the reader imagine a deep lake of a blue color surrounded by verdant hills, in several islets, some showing the bare rocks others covered with shrubs, while on all of them steam issued from a hundred openings between the green foliage without impairing its freshness; on the opposite side a flight of broad steps of the color of white marble with a rosy tint, and a cascade of boiling water falling over them into the lake" (Dieffenbach, 1843).

At Rotomahana in 1882-83 prior to the eruption, it was observed that the local Maori had taken full advantage of the resource. On the edge of the lake, square shallow baths had been constructed, the sides formed of large stone slabs set upright, edge to edge, and the water led into them by ducts from various springs that bubbled up everywhere from the ground. A platform in this area was constructed by laying broad stone slabs over a spring for drying karaka and tawa berries. The stones were heated to a temperature necessary for drying (Wilson, 1884).

Rotomahana was a Ngati Tuhourangi stronghold right up to the Tarawera eruption in 1886. This meant that this tribe had a monopoly over the booming tourist industry before the Terraces were destroyed in the eruption. Ngati Rangitihi, another local tribe, and Tuhourangi disputed over lands at Rotomahana for many years, but reached a peace in 1855. The lives of Tuhourangi and Ngati Rangitihi were changed drastically after the eruption. When the surrounding arable land and the Terraces they so relied on for income were destroyed, the tribes relocated around Rotorua and in the Coromandal.

Waihi

On the southern end of Lake Taupo, Hochstetter visited Waihi where hot springs discharge on the shore of the lake. The local inhabitants, Ngati Turumakina, had fitted up for themselves several bathing places where they filled basins with water at temperatures of 34° C. The foreshore near the village was changed forever in 1846 and 1910 by large landslides, and the 1846 event destroyed the ancient village of Te Rapa.

The following account of Te Rapa was recorded by Dieffenbach (1839-1841):

> "Where the shore joins the delta of the Waikato [Dieffenbach refers to the Tongariro River as Waikato in his records], there is a narrow belt of flat land on which stands the village of Te Rapa. Behind it, the hills rise to about 100 feet above the lake. In ascending, the ground is found to be of high temperature; the surface is often bare, or scantily covered with mosses or lichens. It is formed of red and white clay of soft and alkaline nature, which natives use instead of soap, and sometimes eat. Gaseous effluvia seem to have converted the rock of the hill, which is basalt and sometimes amygdaloid, into this clay. When we approach the top of this amphitheater of hills, the scene which presents itself is very striking. Vapors issue from hundreds of crevices, and in most of these places, there are shallow springs, the bottom of which is a soft mud into which a stick can easily be driven ten feet. The temperature of the water is 200°-212° F. In some springs, it has an argillaceous and others a sulphurous taste. A subterranean noise is continually heard, resembling the working of a steam engine or the blast of an iron foundry. By placing some ferns over a crevice, and their food (potatoes, kumara, or pork) upon it, natives have a ready and convenient oven" (Dieffenbach, 1843).

Above the shores at Waihi is the thermal area Hipaua, named after the largest thermal feature, the Hipaua fumarole. The area is characterized by bare steaming ground, fumarolic activity, and occasional pools. Here at the foot of steaming hills, the ancient village of Te Rapa was overwhelmed by a landslide and the great Maori Chief Te Heuheu Tukino II (*Mananui*) was killed, along with 60 of his *whanau*.

During the month of May 1846, Taupo experienced an unusually heavy rainfall that flooded rivers throughout the district. The heavy rain caused small slips to occur, with the result that in the course of three days the little valley (Waimatai) formed a lake. As the lake rose, there were further slips. Finally the barrier burst and the water, red clay, mud, and rocks went thundering down the hill toward the lake and Te Rapa. The avalanche swept everything before it and overwhelmed Te Rapa. It happened during the early morning hours when people were asleep and all were buried under the muddy thermal clay except for a few who fled to the nearby hills. It is said that during the night before the landslide, there was a thunderstorm. Chief Te Heuheu rose at midnight, went outside, and recited incantations to his gods to pacify the elements. His ancestral spirits did not heed him and instead invoked Ruaimoko, the subterranean god, to destroy him.

The remains of the chief and his wife were taken to Pukawa, north of Waihi, until their final resting place at the top of Tongariro was prepared. The choice of a resting place on Tongariro was opposed by a neighboring chief, who considered that such a burial would give force to a claim by the descendants of Te Heuheu to a mountain to which the Te Heuheu line did not the have sole right. However, the bodies were secretly deposited on the mountain.

A resolution of the tension created within Tuwharetoa by the mountain burial was to come 60 years later. In 1910, in order to fulfill the terms to create the Tongariro National Park, the remains were brought down from the mountain to be buried in a vault provided by the government. When the bones were moved to Waihi, one of the carriers broke the law of *tapu* by jumping over a cooking fire. When he was returning with a party the following day, a landslide again came down from the thermal area, Hipaua, and he was killed.

Wairakei

THE GEOTHERMAL RESOURCES AT WAIRAKEI WERE USED BY SEVERAL TRIBES IN TUWHARETOA, and Ngati Rauhoto maintain rights of *kaitiakitanga* and *rangatiratanga* over the Wairakei area. It has been reported that the encampments of temporary dwellers processed *kokowai* and fished for trout, among other activities (Stokes, 1991).

Kokowai was a valuable commodity used in ceremonial exchanges of goods that cemented tribal relations. However, the use of *kokowai* declined when European dyes were introduced. The resources of Wairakei that were highly valued included *kokowai* and the healing powers of the hot pools, *Matarakutia*. Ferns were also collected at different times of the year. Areas of *wahi tapu* exist all around the Wairakei thermal area, with the exact location known only to the members of the *hapu*. Karapiti ("Craters of the Moon") is an area of special concern to the *hapu*. There are legends surrounding the large fumarole, destroyed under roof collapse in 1981, for which the area was named. The story involves a maiden with her child, jumping into the fumarole.

Kiriohinekai steam was so named by local people for its smoothing effect on skin after bathing, and Tuwharetoa used *Matarakutia* to care for leprosy or *ngerengere*. *Piroriro*, "new skin," was the source spring that had curative properties for cleansing and invigorating (Stokes, 1991). The tragedy of development in the Wairakei field has been the loss of 22 geysers. By May 1964, six years after commissioning the geothermal plant in 1958, only one geyser was still playing. With continued extraction of steam and ground subsidence in the area, the geysers, hot springs, and mud pools have been replaced by steaming ground and a few fumaroles. The *taonga*, or treasures, in this area were numerous.

Discussion and Conclusion

We have discussed how Maori communities used geothermal resources as an integral part of their lives, for bathing, cooking, cleaning, healing, mediating, burial preparation, and sometimes for burial itself.

Maori have watched the implementation of the laws enacted by the government from 1840 to the present; all have tended to deprive them of privileges secured in the Treaty of Waitangi. In 1975, the Waitangi Tribunal Act was passed, and this has enabled the Maori to prepare claims to the government for the land and resources lost to them.The ownership of the geothermal resource is still not qualified, although claims for Ngawha and the Rotorua area have been heard and recommendations made.

The detail of Maori description and understanding of the geothermal resources is hardly surprising; their lives and customs were integrated with their physical environment. Knowing that the resource extended below the surface, Maori took a characteristically holistic approach to the *taonga*.

In many cases, the geothermal *taonga* formed a central component of the cultural relationship between *hapu* and *iwi* and their traditional territory. Springs may have been used by all *hapu* of a region from the time of the ancestors. Such history of tribal usage would reflect the magical qualities of the waters of the springs. It might also reflect the fact that the springs were traditionally a crossroads for Maori traveling. The *hapu* who traditionally lived in the immediate vicinity of particular springs may have tended the springs on behalf of the whole *iwi*.

Maori believed there to be an interconnection between the springs and other surface manifestations. The geothermal resource is seen by them as one entity, with the surface and subsurface components inextricably linked. The legacy of Ngatoroirangi has been carried down through the centuries orally and more recently as written transcript. Whaakari (White Island) was the first place that the resource appeared, the beginning of the trail of heat through the Taupo Volcanic Zone.

Acknowledgments

I would like to thank the following people: ECNZ Geothermal for sponsoring the presentation of the paper on which this chapter is based in Florence in 1995; Tuaropaki Trust, who initiated the paper; Professors D. H. Freeston, M. P. Hochstein, P. R. L. Browne, and the other staff members of the University of Auckland Geothermal Institute for their support and guidance; *Ngati Hine whanau* for their support.

Selected References

Dieffenbach, E. 1843. *Travels in New Zealand.* Reprint 1974. Capper Press, Christchurch.

Electricity Corporation. 1990. *Water right application and impact assessment.* Wairakei Geothermal Power Station.

Grace, J. T. 1959. *Tuwharetoa.* Reprint 1992. Reed Books.

Hochstetter, F. von. 1959. *Geology of New Zealand.* C. A. Fleming, trans./ed. Government Printer, Wellington.

Lynsar, F. B. 1928. *New Zealand, The dear old Maori land.* Brett Printing, Auckland.

Maxwell, P. 1991. The Maori use of the geothermal resource. In *Report to Waitangi Tribunal* (Wai 153).

Meade, R. H. 1870. *Maoriland.* Reprint 1984. Capper Press, Christchurch.

Orbell, M. 1985. *The natural world of the Maori.* William Colins Publishers, Auckland.

Potton, G. 1987. *Tongariro, A sacred gift.* Lansdowne Press, Auckland.

Sharp, A. 1971. *Duperrey's visit to New Zealand in 1924.* Alexander Turnbull Library.

Stokes, E. M. 1991. *Wairakei geothermal area: Some historical perspectives.* University of Waikato, Hamilton.

Waitangi Tribunal Report. 1988. *Muriwhenua fishing report* (Wai 22). Department of Justice, Wellington.

Waitangi Tribunal Report. 1993. *Ngawha geothermal resource* (Wai 304). Department of Justice, Wellington.

Wilson, A. 1884. *Maoriland.* Union Steam Ship Company of New Zealand.

The Author:

C. M. Severne
Geothermal Institute
The University of Auckland
Private Bag 92019
Auckland, New Zealand

Madam Pele, by Herb Kawainui Kane, 1987. *Reprinted with permission*

29. Historical Impacts of Geothermal Resources on the People of North America

by
John W. Lund

Abstract: The Indians of North America considered hot springs as sacred places, and they were great believers in the miraculous healing powers of the heat and mineral waters. Hot spring areas were also known as neutral ground, places where warriors could go and rest unmolested by other tribes. Even though archaeological finds date the Native American presence at hot springs over 10,000 years ago, there is no recorded history of this prior to the arrival of the Europeans in the 1500s.

Many legends concerning geothermal activities are part of Native American oral history, such as those about Madam Pele, the Hawaiian goddess of volcanic fire, and the story of the battle between Skell and Llao describing the eruptions of Mount Mazama (Crater Lake) and Mount Shasta.

Introduction

The history of geothermal use prior to the Industrial Revolution centers itself mainly in the volcanic region of Western North America. A few isolated locations are also found in the middle and eastern part of the continent. The use of geothermal resources during this period was by Paleo-Indians and Indians, also referred to as Native Americans. Even though archaeological finds date Native American presence at hot springs for over 10,000 years, there is no recorded history prior to the arrival of Europeans in the 1500s. Rock paintings (pictographs) and rock carvings (petroglyphs) are the only sources of writing available for study; and, as far as the author knows, none depict geothermal phenomena. With the Europeans' discovery of geothermal phenomena, such as volcanoes and hot springs, the ownership and use patterns changed considerably and many became commercial operations. Thus, the use of geothermal springs and related phenomena has gone through three stages of development in North America: use by

Indians as sacred places, development by early European settlers to emulate the spas of Europe, and use in modern times as places of relaxation and fitness and as energy sources for heating and cooling.

Early Settlements

The time of the first human incursions into North American is subject to debate; however, it is generally associated with Beringia, the land bridge between Siberia and Alaska

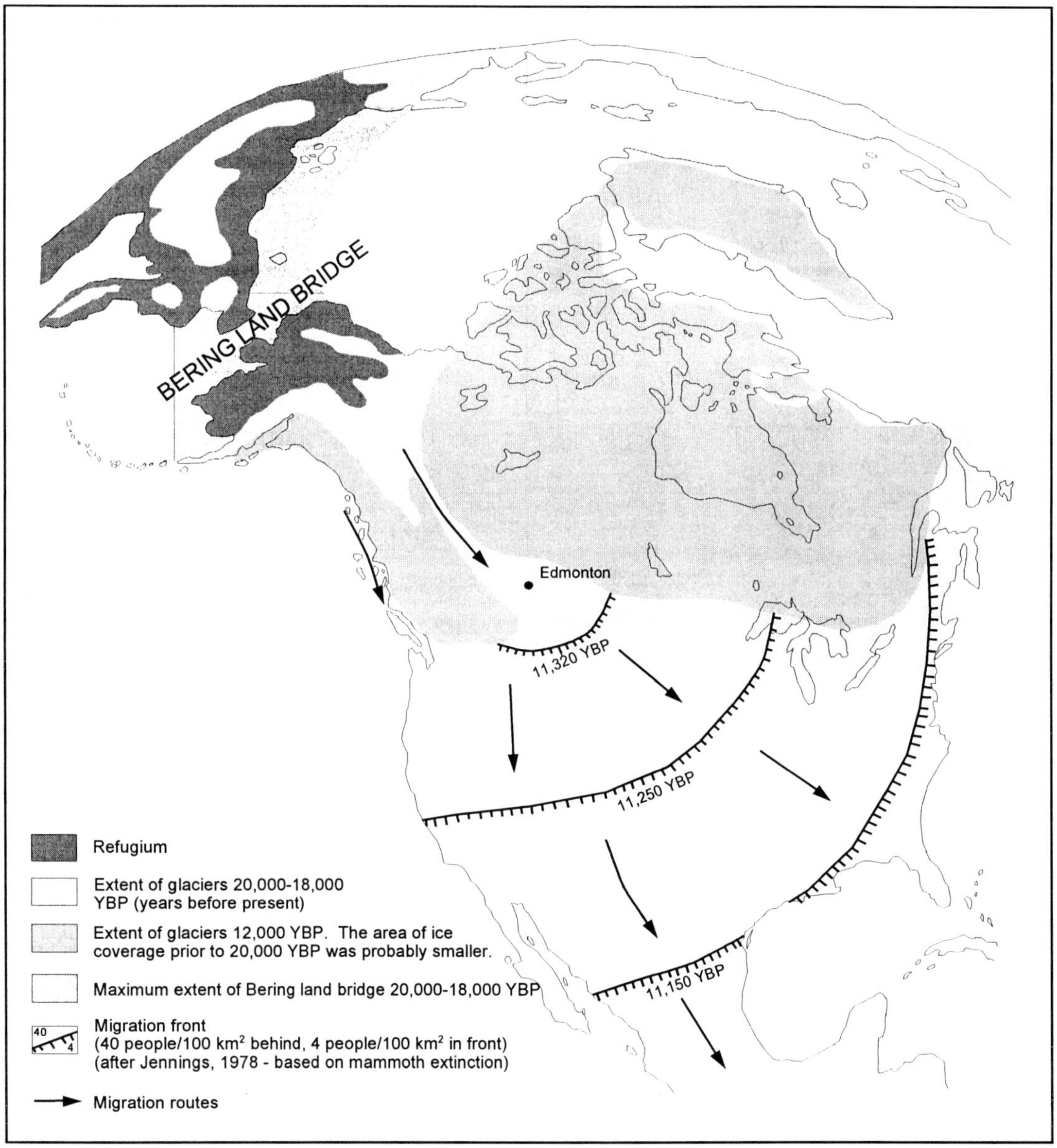

Bering land bridge, *Beringia*, after Jennings, 1978. *Reprinted with permission*

that was formed during the Pleistocene epoch. In the last stage of this glacial period, the Wisconsin or Weichsel glacial advance, so much of the ocean's water was tied up in glaciers that the oceans were about 100 m lower. As a result, a dry land bridge up to 2000 km wide formed between the Bering and Chukchi Seas. The land bridge probably existed from 50,000 to 40,000 and from 25,000 to 14,000 years ago. It vanished altogether about 8000 B.C. (Jennings, 1978).

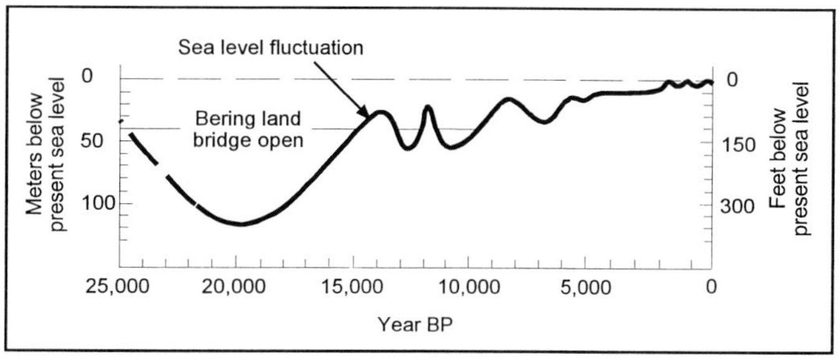

Beringia chronology, after Jennings, 1978. *Reprinted with permission*

Small bands of hunters from Eastern Asia probably followed game migrating eastward across this wide area characterized by long winters, short summers, and continual winds. However, the summer thaw exposed many marshy areas and lush meadows, conditions somewhat attractive for migrations. The ice-free summer bridge led into the ice-free interior of Alaska called *Refugium*, The Refuge (Jennings, 1978). Here the first Paleo-Indian site was discovered, the oldest dated site in North America: Old Crow Flats near the Alaskan-Canadian border is dated between 40,000 and 25,000 years old, depending upon the reference (Dumond, 1978; Fagan, 1985).

Later, ice-free corridors opened into the central portion of North America between the Cordilleran and Laurentide ice sheets. This corridor is estimated to have existed before 18,000 and after 12,000 years ago. It passed into the central plains around present-day Edmonton, Canada. More recent evidence points to a second ice-free route along Canada's western coast (Recer, 1997). A human skull dating from about 7800 B.C. was found recently on Prince of Wales Island. The route along the coast was opened when about 60 meters of the continental shelf was uplifted by the weight of 2500 meters of ice atop the coastal range, deforming the land like someone sitting in the middle of a waterbed. The route was open for about 2,000 years and closed about 9,500 years ago. Some advanced parties of Paleo-Indian hunters, pursuing game such as mammoth and bison, may have moved into the central plains as early as 20,000 to 25,000 years ago, as evidenced by dated sites in the Ohio Valley (Meadowcroft Cave) and by hunted remains found as far south as Mexico. The dates of these sites are questionable, with the first well-attested settlement dated at 13,000 B.C. (Fagan, 1985). However, significant groups of Paleo-Indian hunters of the Llano and Folson-Plano groups had only moved as far as Edmonton by 9500 B.C.

The "Old North Trail," as this ancient route into the United States is now called, started in Canada near Edmonton and extended south along the eastern slopes of the Rocky Mountains, near present-day Calgary, Alberta, and Helena, Montana, before splitting in two. One branch went into Central Utah and the other into Central Colorado near Denver and Pueblo (Stark, 1997). This route is known to be over 3000 km long and may have been used by humans for 10,000 years, according to archaeologist Brian Reeves (Stark, 1997). The route, included in the oral history of the Blackfeet Indian Nation, was probably used until the 1930s; portions of it can still be identified by travois ruts, tipis rings, and stone cairns. It was a trade route for carrying obsidian and chert from quarries in Wyoming and Montana to Canada. The trail is now under consideration as an entry in the National Register of Historic Places by the Montana Historic Preservation Office.

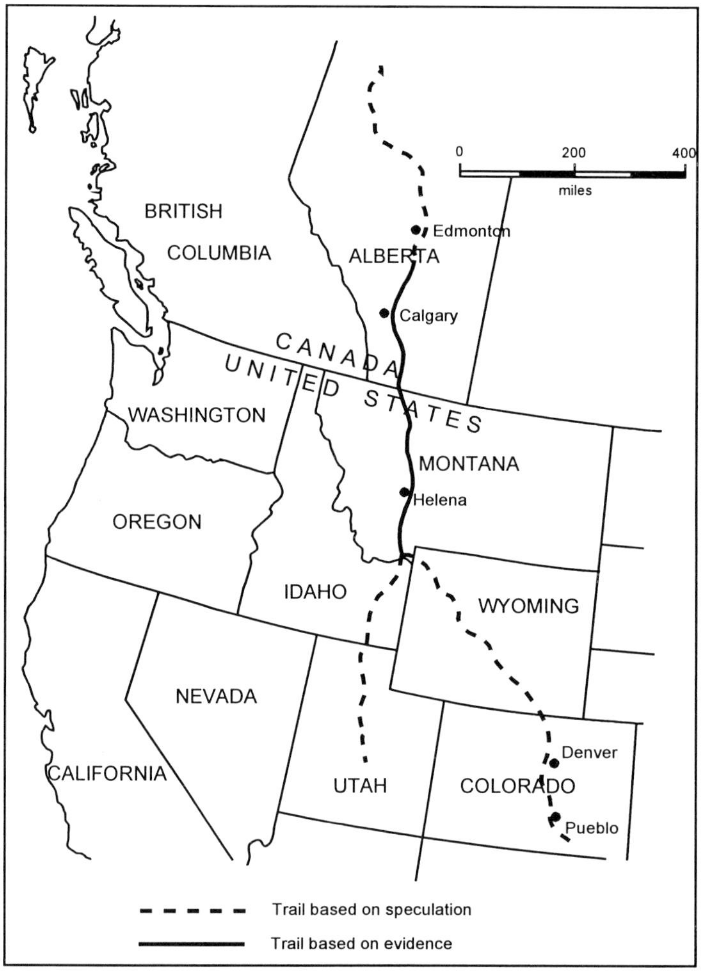

The Old North Trail, after Peter Stark, *Smithsonian*, July 1997

The earliest documented sites were Paleo-Indian sites, such as Clovis and Folsom, New Mexico; Lindenmeier, Colorado; Borax Lake, north of San Francisco; Tule Springs, near Las Vegas, Nevada; and Fort Rock Cave and Alvord Lake, in Eastern Oregon. They date from 11,000 to 9000 B.C. The Paleo-Indians are best known to have existed in the Great Plains and the Great Basin provinces. About 9,000 to 10,000 years ago, the Paleo-Indians were replaced by specialized hunter-gatherer cultures, or Indians. Most of Canada, due to the Cordilleran ice sheet, was not occupied until this time. Numerous sites are dated after this period, but plains and basin sites are difficult to find due to the migratory nature of the Indians searching for game.

The Indians dominated North America until they were essentially replaced by the Europeans in the east in the early 1700s and in the west around the mid-1800s.

The debate is not over. Recent language and DNA research suggests several migratory waves have existed, some as early as 30,000 to 40,000 years ago (Cole, 1998). This is supported by the discovery of a 12,500-year-old settlement in Monte Verde, Chile, which would have taken Asian settlers at least 6,000 years to reach. These southern settlers may have spread north after the last ice age (14,000 to 22,000 years ago) as a new wave of migrants was arriving across the Bering Strait, interacting with them. A final wave, the ancestors of modern-day Eskimos, arrived from Asia about 5,000 years ago, staying in what are now Alaska, Canada, and Greenland. Some speculate that human arrival in the New World occurred as early as 250,000 years ago (Perlman, 1998).

Indian Legends

Many legends were transmitted orally from generation to generation by the Native Americans. Some are still recited by these peoples, but others survive because they were modified and recorded by early European settlers. Some legends depict quite accurately geologic events that took place hundreds and even thousands of years ago, with the addition of various gods and people to explain why the events occurred. The legends interpreted geologic processes as part of a unified world view that blended physical and spiritual components. Mountains were not simply inert piles of rock; they were living presences with distinctive personalities. In native traditions, heaven, Earth, and human consciousness were bound together in a rhythm of life, a sense of wholeness. The new discipline of geomythology studies ancient myths that reflect natural events such as earthquakes and volcanic eruptions (Harris, 1990). Several noted legends related to volcanic phenomena and to hot springs are retold here.

Hawaii—The Goddess Pele

Pele rules the volcanoes of Hawaii, and people have no power to resist her. She is *Pele-honua-mea* of the sacred land, *Pele-'ai-honua*, eater of land, and *Ka-'ula-o-ke-ahi*, the spirit of the redness of the fire. In folklore she may appear as a tall, beautiful young woman or as an old woman, wrinkled and bent with age, sometimes accompanied by a white dog. When enraged, she may appear as a woman aflame or as pure flame. Her personality is volcanic:

unpredictable, impulsive, given to sudden rages and violence. Hers is both the power to destroy and the power to create new land (Kane, 1987).

When Pele came to the island of Hawaii, seeking a permanent home, she found another god of fire already in possession of the land. Ai-laau was known and feared by all the people. *Ai* means the "one who eats or devours"; *laau* means a "tree or forest." Therefore, Ai-laau was the fire god devouring forests who, by reputation, often laid waste to the districts of Southern Hawaii by lava flows and fire. Even though he was destructive, he was also part of the creative forces, for he built islands for future life through volcanic activity. Ai-laau was the fire god who formed the Hawaiian Island chain that Pele was later to visit. When Pele came to Hawaii, she desired to see Ai-laau, to find a resting place at the end of her journey. But he had left in fear of her, as he had seen her toiling by the sea. Pele thus made her home in his place, the crater of Kilauea (Westervelt, 1963).

Numerous legends surround Pele's wanderlust, her reasons for traveling to the Hawaiian Islands from the South Sea islands called *Ka-la-kee-nui-a-Kane*. Some reasons were peaceful and others were based on quarrels with her family. She usually lands on Kauai and works her way down the chain to the big island of Hawaii. Several of the legends are reported by Westervelt (1963). The one that follows was reported by Kane in 1987.

Tradition states that Pele came, as did the Polynesian discoverers, in a sailing canoe from the ancient homeland in the islands of the Tahiti group. She was said to have been driven from this ancestral home by her elder sister, goddess of the sea and water, for seducing her husband. Her first landfall was in the northern islands of the Hawaiian archipelago. Pele needed a deep pit for her home wherein the sacred fires could be protected. However, her sister always followed and put out the fire, and Pele moved southeastward along the island chain, digging new craters. Each effort was flooded out by her sister. Today on the old island of Kauai, the craters have become wet swamps. Volcanic events become progressively younger as we move like Pele, down the island chain toward the big island of Hawaii. This myth coincides with the modern geological theory of plate tectonics, which suggests the Hawaiian Islands are built in assembly-line fashion as the ocean floor slides northwestward over a "hot spot" in the oceanic crust.

Since Pele's sister Na-maka-o-Kaha'i was older, she was also more powerful, for water was believed to be more powerful than fire. The two finally locked in battle on the island of Maui and Pele was torn apart. A hill named *Ka-iwi-o-Pele* ("the bones of Pele"), standing at the site

Madam Pele and her sister Na-maka-o-Kaha'i. By Herb Kawainui Kane, 1987. *Reprinted with permission*

of the battle, is believed to be her mortal remains. With Pele's death, her spirit was freed and elevated to godly status, taking flight to the island of Hawaii. Her permanent home is on Mauna Loa, the Earth's largest mountain, with an elevation of over 9000 meters—about 4000 m above sea level and 5000 m below.

This legend was brought home to the author during a lecture tour over several of the islands where the use of geothermal energy was discussed. Our lecture group was on Maui, the island just northwest of Hawaii, and we were proposing as part of a geothermal project to drill on the slopes of the volcano Hale-a-kala, which last erupted in 1790 but was most active 400,000 to 800,000 years ago. After our presentation, a native Hawaiian elder stood up and lectured us for over an hour on volcanic geology from Pele's point of view. He warned that drilling on Maui would bring Pele back from the island of Hawaii and cause destruction to the residents of Maui.

Pele still lives in the hearts and minds of the native Hawaiians. Her personification is in the natural phenomena of volcanic activity. Hawaiians refer to her as Tutu Pele (an affectionate term for grandparents) and look upon her with respect. In a recent eruption where lava destroyed part of the village of Kalapana, a Hawaiian resident who had lost his land said, "I love my home, live here all my life, and my family for generations. But if Tutu like take it, it's her land" (Kane,

1987). As part of the respect for Pele, older Hawaiians still offer her the first 'ohelo berries picked, before eating any themselves. Some still bring gifts of flowers, food, and drink and set them on the rim of Halema'uma'u crater within the Kilauea caldera. They also refrain from any acts that might be disrespectful to Pele. Tourist who have taken souvenirs of rock home with them from Hawaii Volcanoes National Park have often returned the rocks to the National Park Service with notes describing misfortunes that have occurred since their removal. They hope by this action to be released from Pele's terrible spell.

Oregon—The Battle of Llao and Skell

American Indians commonly attributed natural catastrophies to the actions of invisible spirits who controlled the physical world. The Klamath Indians of Southern Oregon devised a remarkable myth to explain the origin of Crater Lake (Harris, 1990). This legend was told to a young soldier in 1865 by Lalek, an aged Klamath chief. He emphasized that the story was ancient, from a time when his people lived in rock houses and white men ran wild in the woods. The story explains the cataclysmic eruption of Mount Mazama, its collapse, and the formation of the Crater Lake caldera about 7,700 years ago.

It is a story about Llao, Chief of the Below World, Skell, Chief of the Above World, and Loha, a beautiful young woman. Llao saw this woman one day when standing atop Mount Mazama and begged her to return with him to his lodge in the underworld. Even though he promised her eternal life, the young maid refused to accompany him. Because her people would not force Loha to accept Llao's offer, he thundered angrily and tried to destroy her people with fire. Skell came to the rescue by descending from heaven to the summit of Mount Shasta, which is 200 km south of Mount Mazama. In the battle that followed, the sky glowed and turned dark, the Earth shook, and fire flowed from Llao's mountain, burning the forests. Loha's tribe, driven from their home, was forced to flee south into the waters of Klamath Lake.

Two brave medicine men tried to save their people by hurling themselves into the fiery mouth of the underworld. Noting their sacrifice, Skell again shook the Earth, causing Llao's mountain to crash down upon him. When finally the dark clouds of ash cleared and light had returned, the lofty peak was gone, replaced by a giant hole. The curse of fire was lifted and the caldera gradually filled with rainwater, forming Crater Lake. This myth is similar to the Greek story of Hades, god of the underworld, who pursued the maiden Persephone and carried her away to his underworld kingdom.

The Llao-Skell myth closely describes the geologic events that occurred in the formation of the Crater Lake caldera. It states correctly that Mount Mazama did not lose its summit by blowing itself apart but by emptying much of its underground magma reservoir, allowing the volcanic cone to collapse. The eruption in the Crater Lake caldera formed the cinder cone called Wizard Island, which represents the maid Loha. The myth also describes frequent eruptions of Mount Shasta over the last 10,000 years, as Skell visited Shasta often, starting fires upon each visit.

Washington—Mount St. Helens

An old Klickitat Indian myth talks about the volcanic struggle between Mount Adams (Phato) and Mount Hood (Wyeast), sons of the Great Spirit, who fought each other because of their love for a beautiful Indian maiden Loo-wit. Chief Phato ruled the land north of the Columbia River and Chief Wyeast the land south in Willamette Valley. The Great Spirit built a bridge over the Columbia (Bridge of the Gods) so that the families of the two brothers could visit each other. With time, the two peoples quarreled, and the Great Spirit punished them by keeping the sun from shining. After a time the Great Spirit felt sorry for the people and gave them fire through an old woman who had kept herself from the wrongdoing of her people. As a reward, she was turned into the beautiful maiden, Loo-wit.

Wyeast and Phato grew jealous of each other over Loo-wit, becoming so angry that they fought. Their people also fought each other and many were killed. The Great Spirit was so angered by the fighting that he broke the Bridge of the Gods, which collapsed into the Columbia River. He changed the two chiefs into mountains—Mount Hood and Mount Adams—and they continued to quarrel. Their fighting caused sheets of flame to burst forth from the peaks, and they hurled hot rocks at each other. Many of the rocks fell into the river and formed the narrows at The Dalles. Lo-wit was changed into a snow-capped peak—Mount St. Helens—which still has the youth and beauty promised by the Great Spirit [ed. note: this was written before the 1980 eruption]. The rocks and white water where the Bridge of the Gods fell are known as the Cascades of the Columbia (Clark, 1953).

Considerable evidence suggests that such a massive earthen dam did exist, temporarily blocking the Columbia's flow. Radiocarbon dating reveals that the slide occurred about 1100 A.D., triggered by an earthquake. About two cubic kilometers of rock buried nearly 36 square kilometers of the Columbia Gorge, creating a natural dam more than 60 meters high. It may have taken years for the impounded water to overtop the dam and wash most of it away.

Meanwhile, it served as a land bridge for the Klickitats, Willamettes, and other tribes who enjoyed easy access to each other's property, as reflected in the legend (Harris, 1990).

Oregon—Mount Hood

According to legend, Mount Hood once towered so high that when the sun shone on its south side a shadow stretched north for a day's journey. Inside this majestic mountain was a lodge of evil spirits who sent forth streams of liquid fire to destroy the Indians' homes. In those days the Indians were taller than they are now, as tall as the pine and fir trees, and their chief was such a giant that his warriors could walk under his outstretched arms.

In a vision, the chief was told he must conquer the evil forces that lived in the mountain to prevent his people from being annihilated. The brave chief climbed the mountain and found a large hole at the summit from which fire and smoke poured out. He hurled boulders into the opening in the hopes of killing the evil spirits, but they threw the boulders back, now heated red hot, along with smoke and fire. The heated rocks rose toward the sky and fell a long distance away. The exchange continued for many days, and when the chief looked down in the valley he saw that it was desolate, choked with lava and ash. Grief stricken, he sank to the Earth and was buried by a stream of lava. Some of the tribe survived, but their starving children and grandchildren did not grow as tall, and their descendants will only be restored to their former height when a great chief comes who can conquer the demons of fire in the mountain. Today the chief's sorrowful profile, with its distinctive scalp lock, can be seen halfway down the northern side of the mountain, appropriately called the "Chief's Face" (Clark, 1953; Harris, 1988).

Montana—Sleeping Child Hot Springs

Sleeping Child Hot Springs is in the Rocky Mountains south of Missoula, Montana. The beauty of this hot springs was discovered by the Nez Perce Indians in the 1870s. Chief Joseph of the Nez Perce led his tribe out of the reservation and through the Lolo Pass into the Bitter Root Valley, fleeing General Howard and a group of soldiers. Trying to avoid a confrontation, Chief Joseph split his tribe into smaller groups. One group traveled through Sleeping Child Creek. Facing a possible battle, they left their infants by the hot springs. When they returned, the infants were safe and peacefully sleeping, protected by the natural hot springs. And so Sleeping Child Hot Springs got its name.

Montana—Lolo Hot Springs

In the same area as Sleeping Child Hot Springs, west of Missoula, Montana, is Lolo Hot Springs, which were known and used extensively by local Indians. The hot springs were noted by Lewis and Clark in 1806 on their return east, after traversing the rugged area near Lolo Pass (Bergon, 1989). Lewis described the hot springs in his journal entry for June 29, 1806 [grammar, spelling, and capitalization of the original text are retained]:

> "these warm springs are situated at the base of a hill of no considerable hight on the N. side and near the bank of travellers rest creek which at that place is about 10 yards wide. these springs issue from the bottoms and through the interstices of a grey freestone rock, the rock rises in iregular mas[s]y clifts in a circular range arround the springs on their lower side. immediately above the springs on the creek there is a handsome little quamas plain of about 10 acres. the prinsipal spring is about the temperature of the warmest baths used at the hot springs in Virginia. In this bath which had been prepared by the Indians by stoping the run with stone and gravel, I bathed and remained in 19 minutes, it was with dificulty I could remain thus long and it caused a profuse sweat two other bold springs adjacent to this are much warmer, their heat being so great as to make the hand of a person smart extreemly when immerced. I think the temperature of these springs about the same as the hotest of hot springs in Virginia. both the men and indians amused themselves with the use of a bath this evening. I observed that the indians after remaining in the hot bath as long as they could bear it ran and plunged themselves into the creek the water of which is now as cold as ice can make it; after remaining here a few minutes they returned again to the warm bath, repeating this transision several times but always ending with the warm bath."

Poster from Sleeping Child Hot Springs. *Reprinted with permission*

The Virginia springs Lewis referred to were in Hot Springs, Virginia, where spring temperatures range between 36° C and 41° C (Lund, 1996). Lolo Hot Springs, today with a well developed pool and resort on the edge of a wilderness area, uses 43° C and 47° C water from two springs.

West Virginia—White Sulfur Springs

A large, historic, health-oriented mineral springs resort, The Greenbrier, occupies 2600 hectares in an upland valley of the Allegheny Mountains near the West Virginia-Virginia border. The natural mineral waters have a temperature of 17° C. The resort and spa have a 200-year history (Lund, 1996). Originally, the forests near the springs were inhabited by Shawnee Indians, who hunted around the springs in a small marsh with salt deposits, which attracted animals. Early European settlers reported the Indians valued the curative powers of the hot spring waters.

In 1778 a Mrs. Anderson, on hearing the benefits of these mineral waters, was carried 25 km on a litter to the springs in an attempt to cure her chronic rheumatism. In the Indian custom, a tree was felled, hollowed out as an improvised tub, and filled with spring water heated with hot stones. Mrs. Anderson bathed in the water and drank from the spring. In a few weeks the pain from the rheumatism receded, and the news of her recovery spread rapidly to other settlers in the region. As a result, numerous log cabins were built around the spring in the next few years to house crowds of visitors. Later this developed into a rustic spa at Bowyer's Sulfur Springs, which by the early 1800s was a "grand" resort (Conte, 1989; Lund, 1996).

An Indian legend about these springs speaks of two young lovers who came to the valley to escape the notice of their elders. Catching them, an enraged chief shot two arrows, one killing the boy and the other barely missing the girl. Where the second arrow hit the ground, a sulfur spring appeared and legend says that when the last drop of water is drunk from the spring, the lovers will be restored to life (Conte, 1989).

The Influence of Hot Springs

The Indians of North America considered hot springs and fumaroles as sacred places. This was one of their mysteries of the universe or *Wakan Tanka* in the Lakota/Dakota/Nakota language, meaning "Great Mystery," "Ultimate Mystery," "The Unknowable," or "Great Sacredness" (Mona, Dakota Indian Nation, 1996). They were great believers in the miraculous healing powers of heat and mineral waters. Every major hot spring in the United States and Canada has a record of Indian usage.

Hot springs were regarded as neutral grounds, places where warriors from any tribe could rest unmolested and recuperate from battle. Often the spring was guarded jealously, kept secret from the Europeans as long as possible. Sometimes battles were fought between Indians and settlers to preserve these rights. The "Fountain of Youth" sought by early Spanish explorers, such as Ponce de León and Hernando DeSoto, perhaps evolved from accounts of healing properties of hot spring waters (Lund, 1993).

An example of a battle fought over a hot spring is recorded on a sign near Evan's Plunge in Hot Springs, South Dakota (Lund, 1997):

> "Long before the white man discovered the valley of healing waters, the Sioux and Cheyenne Indian tribes fought for possession of the natural warm water springs. Legend tells us that the battle raged on the high peak above the springs and the Sioux emerged victorious.
>
> "The mammoth spring at the end of the interior of the plunge is known as the 'Old Original Indian Springs.' Here the Indians drank and bathed in its warm healing water."

The early European settlers in the 1700s and 1800s used the natural hot springs. Later, they realized their commercial value and developed many into spas in the European tradition. Many spas were successful, such as those at Saratoga Springs, New York; White Sulfur Springs, West Virginia; Hot Springs, Virginia; Warm Springs, Georgia; Hot Springs, Arkansas; Calistoga, California; and Harrison Hot Springs, British Columbia. However, the United States had neither governmental, trade union, or social security programs nor a national health insurance program that supported such spas. Thus, in spite of the proven benefits of spa therapy, the United States lagged behind many other countries in developing mineral springs, even ones owned by states and the Federal government. By the 1940s, the interest in spas languished, and most of the majestic resorts declined and closed. Recent interest in hot springs, in soaking and in physical fitness, has renewed spa development.

Examples of some of the early uses of hot springs by Native Americans and early European explorers follow.

Georgia—Warm Springs

Warm Springs is the most famous of Georgia's seven known hot springs. It has the largest flow rate—up to 58 liters per second—and an average temperature of 31° C. The main mineral constituent is bicarbonate. For the local Creek Indians, the springs were probably a place for healing where the Indians of all tribes were allowed to bring their sick and wounded to drink the waters and bathe in the mud. The Iroquois Indians from New York (1500 km north) visited the Creeks in Georgia, calling the area "the land where the waters are warm." A number of trails maintained by the Indians ran through the area. The trails were used by all travelers, many of whom stopped at the springs. Later, the trails became military and post roads, and a number of taverns with rustic accommodations were built along them, including one at Warm Springs in the early 1800s.

Settlers obtained the land in 1825 in a treaty signed with the Lower Creek Indians. The original springs were called Meriweather Warm Springs, after General David Meriweather of Revolutionary War fame. After several stages of commercial development, the springs became the site of a polio treatment center, made famous by President Franklin Delano Roosevelt (Lund, 1993 and 1996).

New York—Saratoga Springs

Saratoga Springs is about 40 km north of Albany, New York, just south of Adirondack Park. About 18 springs discharge 13° C carbonated mineral water along the Saratoga fault between Whitehall and Albany, a distance of 105 km.

The springs have been used in spas for drinking and bathing, considered a cure for everything from skin disorders to digestive problems. The water and carbon dioxide are bottled and sold commercially. The Mohawk and Iroquois Indian tribes used this area for hunting and frequented the springs, especially High Rock Spring. The Mohawks, a fierce warrior tribe, fought to defend their hunting grounds around the springs on many occasions. They called the area Kayaderosseras. Other tribes probably used the springs, which were on the path between Quebec and the Mohawk Valley.

The first written record of the springs was in 1642, when the Mohawks, returning from a raid on Quebec, brought with them a Jesuit priest whom they had captured. They stopped by a spring

Mohawk Indians at High Rock Springs, New York, from Swanner, 1988. *Reprinted with permission*

and were refreshed by the bubbling waters. There is an unsubstantiated story of Sir William Johnson being carried to the springs in 1767 by Mohawk braves for treatment of a wound he had received in the Battle of Lake George in 1755. Probably the first settlers to visit the spring were surveyors in the 1760s. Perhaps people were attracted to the area by the great number of animals who came to lick the salt deposits. The spring was included in the Kayaderosseras Patent in 1771 and purchased by European settlers. After a number of different ownerships and overexploitation, a majority of the land was taken over by the State of New York and is now administered as a state park (Swanner, 1988).

Arkansas—Hot Springs

Hot Springs, Arkansas, was one of the most popular commercial spa areas in the United States, created to imitate the great spas in Europe—so popular that in 1832 it became a Federal reservation and finally, in 1921, a national park. It is the only national park in the United States created just to protect hot springs for spa use. This natural geothermal resource consists of 47 springs producing a total of about four million liters per day of 61° C water.

The United States National Park Service estimates that humans have used the Arkansas hot springs for at least 10,000 years. The "Valley of the Vapors" was an honored and sacred place to the Indians, who believed it the home of the "Great Spirit" who brought forth the healing warmth of "Mother Earth," warming the waters by his breath. Like many North American hot springs, this was neutral ground where warriors of all tribes could rest and bathe peacefully. Stone artifacts found near the springs show that the Indians used the waters extensively. When early European explorers and East Coast settlers pushed the Indians westward, the Quapaws and part of the Cherokee Nation wandered into the area. Legend reports that DeSoto or one of his scouts from the 1541 expedition was the first European to visit the site. Evidently, DeSoto was in the area for he claimed the territory for Spain, and in the late 1700s, French trappers, hunters, and traders were familiar with the area (Bedinger, 1988).

The French actually ruled the area until 1763 when the territory west of the Mississippi was ceded back to Spain. It was retaken by France in 1800 under Napoleon Bonaparte and finally purchased by the United States as part of the Louisiana Territory in 1803. During this period no written records are known of travels to the hot springs, but the springs were probably visited many times, for crude log cabins and huts were found at the site in 1804. The spring water was first analyzed in 1804 and permanent settlements began in 1807. The first settler, a plantation owner named Jean Emanuel Prudhomme, was introduced to the springs for his health by the Natchitoches Indians, and remained in the area for a year. The springs became famous for curing ailments and in 1830 the first bathhouse was opened for visitors. The Federal reservation followed and the development peaked in the early to middle 1900s. The National Park Service still controls the geothermal resource and commercial development in the park of about 2000 hectares (Lund, 1993 and 1996).

Wyoming—Thermopolis

Thermopolis, a Greek word for "Hot City," is located in North-central Wyoming at the mouth of the Wind River Canyon, about 160 km south of Yellowstone National Park. At least eight hot springs have created large travertine terraces along the river in the area. The Big Horn Spring, called the largest mineral hot spring in the world, flows at a rate of 120 liters per second at a temperature of 56° C.

Based on evidence associated with Legend Rock petroglyphs in the sandstone cliffs about 40 km northwest of town, Indians have lived in the area for at least 2,000 years. The hot springs were known as "healing waters" and *Bah-gue-wana* or "smoking waters" by the Shoshone. Chief Washakie of the Shoshone tribe (who later signed the treaty with the Federal government) reportedly had a bathhouse erected over Black Sulfur Springs at Thermopolis. The Big Horn Spring was included in the Shoshone Indian Reservation Treaty of 1868 and became known for "health giving properties."

The United States Congress was asked to set the area aside as a national park or reservation, and in 1896, the Federal government signed a treaty with the Shoshone and Arapahoe Indians to allow public hot spring use. One-fourth of the water from the Big Horn Spring was set aside for free public use, and spring management was given to the State of Wyoming in 1897. The springs became Hot Springs State Park, and the State Bathhouse was built to fulfill treaty conditions. Today, Hot Springs State Park provides geothermal water to commercial establishments such as the Pioneer Center (a retirement home for state residents) and The Gottsche Rehabilitation Center (a center for therapy) (Lund, 1993 and 1996).

The Great Geyser of the Fire-Hole Basin. *Thomas Moran, reprinted with permission of Scribner, a Division of Simon & Schuster from* SCRIBNER'S MAGAZINE, *vol. ii, no. 2, New York: Charles Scribner's Sons, 1871*

Wyoming—Yellowstone National Park

SINCE 1808, WHEN JOHN COLTER DESCRIBED THE HOT springs of "Colter's Hell," attention has focused on Yellowstone. Explorers "discovered" many of the springs early in the 1800s following Colter's visit, which was reported in a series of articles by N. P. Langford in *Scribner's Monthly* (1871). Nathaniel Langford was one of nine Montanans who undertook an exploration of Yellowstone in 1870. They spread word of their "discovery" through newspaper articles, a Congressional report, and Langford's lecture tour to several cities, for he had an agreement with Jay Cooke to publicize Yellowstone. Cooke was the financial agent for the Northern Pacific Railroad, which wanted to encourage tourism in the American West. Eventual public interest in the preservation of the many spectacular thermal features in Wyoming's northwestern corner found by Colter led to the 1872 establishment of Yellowstone National Park, the nation's first.

Surprisingly, the local Indians knew little of Yellowstone's geothermal features and seldom visited the area. The local Shoshone, called *Tukuarika* or, more commonly,

"Sheepeaters," were alone in permanently occupying the area. Even they knew nothing about the spectacular Firehole Geyser Basin and exhibited more astonishment and wonder than any of the early Europeans guided into the area (Chittenden, 1895).

The country probably was *terra incognita* for the vast body of Indians who dwelt around it, and Europeans generally thought that "superstitious fears" kept the Indians away. However, the reasons reported by Chittenden in 1895 appear more practical:

> "There was nothing to induce the Indians to visit the Park country. For three-fourths of the year that country is inaccessible on account of snow. It is covered with dense forests, which in most places are so filled with fallen timber and tangled underbrush as to be practically impassable. As a game country in those early days it could not compare with the lower surrounding valleys. As a highway of communication between the valleys of the Missouri, Snake, Yellowstone, and Bighorn Rivers, it was no thoroughfare. The great routes, except the Bannock Trail already described, lay on the outside. All the conditions, therefore, which might attract the Indians to this region were wanting. Even those sentimental influences, such as love of sublime scenery and a curiosity to see the strange freaks of nature, evidently had less weight with them than with their paleface brethren."

Early trappers and mountain men had visited the Geyser Basin in the early 1800s and some wrote reports appearing as early as 1842. Most of their stories were discounted as exaggerations and not believed. James Bridger, who had visited the area often as early as 1824, described many of the thermal features in the 1840s. Although determined to have his narratives published, no one would accept his stories. In 1856, the editor of the *Kansas City Journal* had written an article using Bridger's description, but the editor's friends ridiculed the whole thing as incredible and the story was suppressed. The editor, Colonel R. T. Van Horn, said Bridger told him that the Geyser Basin "was a place where hell bubbled up." The Colonel was very interested in the matter at the time and took notes of the conversation. Unfortunately, the notes never saw print after a man who claimed to know Bridger told the editor that he would be laughed out of town if he printed "any of old Jim Bridger's lies." Years later, Van Horn apologized to Bridger (Chittenden, 1895).

Wyoming—Guernsey

Even though Guernsey is the location of only a few small springs near the southeastern corner of Wyoming, these are probably the most famous thermal features in the state. Although early immigrants may have heard vague tales of "Colter's Hell" and of Yellowstone country, thousands had actually seen Guernsey Springs. To the immigrants of the famous

Oregon Trail, this spring was a welcome landmark, a place to soak sore feet and wash dusty clothes—and it was appropriately named "Immigrants' Washtub." In 1858, soldiers at nearby Ft. Laramie set up a lime kiln and used the limestone from which the springs rise. The springs then became known as "Lime Kiln Springs." Today, the 21° C springs serve as watering holes for cattle (Breckenridge and Hinckley, 1978).

California—Calistoga

Located about 120 km north of San Francisco, California, this resource is at the northern end of the famous Napa Valley and at the southern end of The Geysers Geothermal Field. The area is noted for numerous fumaroles and hot springs and the surface temperatures are near 100° C. The area was settled first by the Southern, Central, and Western Pomo, the Western and Northern Wappo, and the Lake Miwok Indians. All spoke different languages and had distinct cultures. These early peoples came from a wide area to use the natural hot springs, fumaroles, and heated muds to soothe aches and pains. To Native Americans, this was *Tu-la-ha-lu-si*, "the beautiful land," and the hot spongy turf was *Coo-lay-no-maock*, "the oven place," according to a local historian. As in many western geothermal areas, the Indians were first to use and appreciate geothermal resources. They took cinnabar from local deposits to use for red war paint, and later mercury ore was mined by settlers (Archuleta, 1977).

The Indian Hot Springs Ground at Calistoga, from Archuleta, 1977. *Courtesy of the Sharpsteen Museum, Calistoga, California*

In the early 1800s Spanish explorers looked for a mission site in this area, which they naturally called *agua caliente,* "hot water," a name given to many other geothermal locales in the Southwestern United States. The missionaries planted valley fields with grape vines—the fruits destined for sacramental wines—and sowed golden mustard, still blooming in orchards and vineyards each spring. In 1862, Samuel Brannan established a resort and spa in Calistoga similar to the one at Saratoga Hot Springs in New York. The name "Calistoga" came from a misspoken phrase in a public address by Brannan about the locale. Meaning to say "the Saratoga of California," he instead intoned "the Calistoga of Sarafornia," and the name stuck. Today, the well known spa community is in the heart of the Napa Valley wine country.

British Columbia—Harrison Hot Springs

GEOTHERMAL RESOURCES IN CANADA ARE MAINLY IN BRITISH COLUMBIA, AND ONE OF THE most famous locales is Harrison Hot Springs, about 80 km east of Vancouver. The springs and resort are referred to as "The Spa of Canada." Two hot springs, at 50° and 65° C, are located near the southern end of Lake Harrison. The Indians knew of the springs and of Keekwully Tybee, who sent up the medicine waters all hot from below. Believing the springs of boiling water (*Warum Chuck*) were of supernatural origin, they regarded the hot water in the lake with reverence and awe. The Indians believed that whoever drank the water received mystical powers of endurance over others. Some of the stories say the waters will boil as long as there is sickness in the land. The natives referred to the lake as Lake Qualts, which means "hot water" (Rendall, 1981).

The springs were probably discovered by Europeans in 1859, when one of a nearly frozen group of miners sailing on the lake toppled into the water from weakness. He was so happy in the warm water that the rest joined him before they continued their journey. In the late 1800s, a hotel and bathhouse were built adjacent to the springs. Springs literature boasted that the hot sulfur springs provided "a sure cure for paralysis, rheumatism, syphilis, diabetes, neuralgia, skin diseases, mercurial poisoning, dipsomania, and all diseases of the womb, liver, and kidneys, besides many other maladies to which human flesh is heir." Today, The Harrison Hot Springs Hotel, without the promise of cures, operates as a luxury resort at this location.

Alaska

INDIANS AND ESKIMOS THROUGHOUT ALASKA, FROM THE SEWARD PENINSULA TO THE SOUTHEASTern islands, were aware of geothermal springs. Kruzgamepa Hot Springs, 80 km north of Nome, now called Pilgrim, was used for bathing years before settlers arrived. People visiting Chief Shakes Hot Springs, near Wrangell in Southeastern Alaska, still use the wooden cribs placed there by the Tlingit Indians before the Russians arrived. Some springs held mystical powers, such as Kilo Hot Springs in North-central Alaska. The pools at Hooniah or Tanakee Hot Springs, one of the best known resorts in the early part of the century in Southeastern Alaska, are used by Native Americans today, although the resort is still quite rustic (Markle, 1979).

When the Russians began colonizing Alaska, they followed the Indian example of leaving most hot springs as they found them. Most Russian and early American settlements were not near a geothermal resource. This was due in part to the nomadic fur traders who did not tarry long at any one place. They traveled mainly along the coast and most of the springs were inland. In the late 1800s, with the discovery of gold, many of the hot springs were developed by miners who had moved inland. Chena, Manley, and Circle Hot Springs in the Fairbanks area are examples of miner-developed spas still operating today. Many others have declined and are little used today.

The Influence of Volcanoes

Arizona—Sunset Crater

About 1075 A.D., the eruption of Sunset Crater east of Flagstaff, Arizona, spread a blanket of volcanic cinders and ash over a large part of the lands occupied by the Sinagua Indians, whose living patterns changed markedly because of it. After the initial disruption, the Sunset Crater cinder fall is thought to have provided moisture-conserving mulch that made the area more attractive for dry farmers to grow maize. The larger number of sites in the immediate posteruptive period and the appearance of artifacts from other Indian tribes (Anasazi, Mogollon, and Hohokam) suggest that there was large-scale migration from neighboring areas at this time (Lipe, 1978).

Oregon—Mount Mazama

People have inhabited Eastern Oregon since about 9000 B.C. However, evidence of this occupation all but disappeared after the eruption of Mount Mazama around 7,700 years ago, when Crater Lake was formed and a layer of volcanic ash and pumice over eight meters deep in places blanketed the landscape. No significant reoccupation of the area occurred until about 3000 B.C., and at that time tools, baskets, and plant food resources changed markedly (Aikens, 1978). The eruption of Mount Mazama was carbon-14 dated based on charred sandals found buried beneath pumice and ash in Fort Rock Cave, about 100 km northeast of the mountain.

Washington—Mount St. Helens

Mount St. Helens had three recent major stages of volcanic activity, from about 10,000 to 8710 B.C., about 1900 to 450 B.C., and intermittently between 800 and 1850 A.D. (Davis, 1995). Others report the intermediate episode lasted from 4000 B.C. to 1700 A.D., with three major periods each lasting 500 to 700 years (Lewarch and Benson, 1991). Long-term land use by Native Americans began around 5000 B.C. and continued for about 3,000 years. Then a break in occupational density appeared, perhaps a function of large-scale vulcanism on Mount St. Helens. There is a lack of radiometrically dated artifacts between 1640 B.C. and 160 A.D., even with an increasing population level in the surrounding plateau region of Washington and Oregon. Tephra deposits have been measured nearly a meter deep in some places due to the volcanic eruption, which would substantially limit occupation of the area. Dated components occur again after the first century A.D.

Hawaii—Kilauea

In 1790, the rule of Hawaii was contested by several chiefs. The two main contenders were Keoua, considered a dangerous upstart by the other chiefs, and Kamehameha, the future

Keoua's army destroyed by the eruption of Kilauea. By Herb Kawainui Kane, 1987. *Reprinted with permission*

ruler. After several inconclusive battles on the big island of Hawaii, Keoua was returning to his home district with his army, accompanied by wives and children of the soldiers. They were walking along the flank of Kilauea volcano when part of Keoua's army was completely destroyed by a rain of hot ashes, rocks, and poisonous gases from the mountain. Badly shaken by this disaster, no doubt believing the Pele had turned against him, Keoua lost the will to continue the war and Kamehameha became the ruler (Kane, 1987).

Use of Volcanic and Hot Spring Material

Indians have fashioned obsidian and basalt into tools and weapons since the first migrations into Western North America. The implements and extensive middens of obsidian flakes can be found throughout the west, protected from "pot hunters" by Federal law. Muds colored by sulfur and mercury deposits were used to decorate pottery, skins, and rock surfaces with pictures or writing (pictographs and petroglyphs). Using obsidian hydration rates measured on freshly fractured pieces, prehistoric sites can be dated as far back as 800,000 years ago (Fagan, 1985).

Arkansas—Hot Springs

The Tunicas Indians of the Caddo Nation lived around the present-day Hot Springs National Park in Arkansas. For centuries, they mined a very hard, evenly textured flint called novaculite (a form of silica similar to chert) in quarries adjacent to the hot springs. The Indians used it for arrowheads, spearheads, and tools. Europeans mined novaculite to make whetstones, and this "Arkansas Stone" is still mined outside the park, highly prized for its uniform sharpening characteristics (Bedinger, 1988).

California—Coso

Obsidian was also highly prized by the Indians of California for tools and weapons due to the sharp cutting edges that could be fashioned. Obsidian deposits were available on the volcanic Great Basin side of the Sierra Nevada and obsidian was traded across the mountains, west into California's vast Central Valley. Such materials have been dated as early as 3000 B.C.

One source, located near Coso Hot Springs in Southern California on the grounds of the China Lake Naval Weapons Center (also the site of a geothermal power plant), had inclusions of white cristobalite, an amorphous silica. Samples from this source have been found throughout the Central Valley. The author was at this site in the 1960s, amazed at the large obsidian deposit and midden. The author later discovered obsidian pieces from the site at nearby Little Lake, attributed to the Pinto culture estimated to be between 2,500 and 7,000 years old. Other obsidian trading, moving east to west across the Cascades, occurred from a source in Newberry Crater, Central Oregon.

Rock Writings

Petroglyphs (rock carvings) and pictographs (rock paintings), together called rock writings, are abundant throughout Western North America and Hawaii. Their ages and meanings are generally a mystery, as present-day North American Indians and native Hawaiians have lost the oral history associated with these figures. Most are carved or painted on volcanic rocks, mainly basalts such as those found on the Columbia River plateau and in Hawaii, but some welded tuffs (fused volcanic ash) have been used at Lava Beds National Monument in California. The basalts most commonly used were cooled pahoehoe lava (ropy in form) or columnar basalts, both fairly fine-grained, providing a smooth surface on which to incise or paint.

No figures show hot springs, geysers, or volcanoes, which seems unusual as these phenomena were noticeable and certainly influenced the lives of early inhabitants. Why is this so? Discussions with personnel at the Bishop Museum in Hawaii and others indicate that the reasons appear to be cultural. The common people made the rock writings, and they mostly depicted items associated with everyday living: animals, people, hunts, battles, trails, birth, water sites, and even a surf boarder (Cox and Stasack, 1970; Martineau, 1973). Such rock writings could represent ancient "chat rooms," bulletin boards, or places to record and boast of achievements. We know these early peoples didn't understand the causes of eruptions and other mysterious volcanic phenomena, and the priests and shamans may have prohibited the common people from depicting them, feeling that rock writings were beneath their social status. Thus, through cultural forces, we are left with no known rock illustrations of geothermal manifestations.

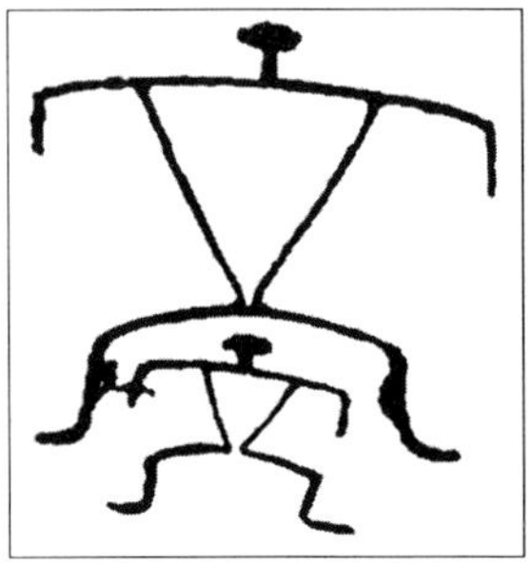

Family Hawaii (Kaeo trail)

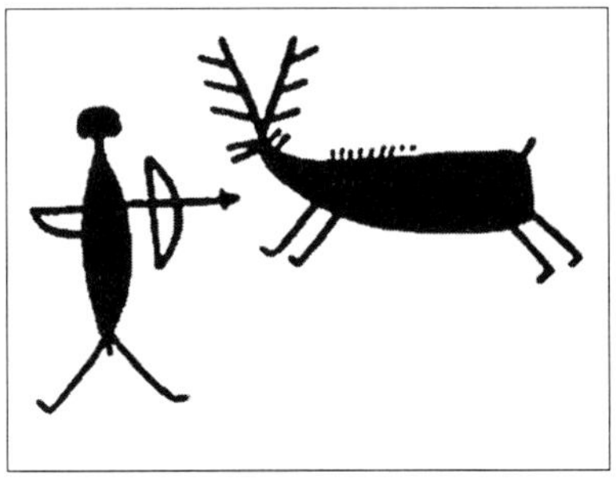

Rock Island, Washington

Ohio—Hopewell

Over 136 kg of worked obsidian fragments were found at the Hopewell burial site in Southern Ohio. These included between 250 and 500 magnificent spears, probably made about 2,000 years ago. No obsidian deposits are found in the Midwest, and the source of this obsidian was traced to what is now Yellowstone National Park, about 2500 km to the northwest. Many other unusual items such as copper, mica, galena, and chlorite also have been found in these burial sites, reflecting the wide geographical knowledge of much of the Eastern United States that some early inhabitants possessed. Trade helps to explain the apparent speed with which new ideas and techniques moved across long distances (Griffin, 1978).

California—Calistoga

The Indians at the northern end of the Napa Valley, where the hot springs and fumaroles are found, used obsidian from nearby Glass Mountain to trade for fish and other sea products from the coastal Indians. Indians came from all over to the curative waters of *Coo-la-no-maock*, "the oven place," and built sweat houses over hot pools. The Indians also traded cinnabar, used for vermilion war paint, and settlers later mined the ample mercury deposits in the area.

Conclusions

North American Indians have a long history of understanding geothermal phenomena and using its resources, going back at least 10,000 years. Many hot springs, geysers, and fumaroles were sacred places to Native Americans, who had a special respect and understanding of the natural environment. Unfortunately, much oral history and many legends about these features have been lost, leaving us with archaeological evidence, speculation, and a few legends told to and recorded by early European settlers.

Many geothermal sites are used today by Native Americans, and these uses must be considered in environmental impact studies, such as ones undertaken for California geothermal power projects at Coso in Southern California and Medicine Lake at Glass Mountain in Northern California. At Coso, during the environmental study phase, the developers considered Indian hot springs use because, in the words of Earl Lent of the Owens Valley Paiute-Shoshone, "Coso is not merely a word or a geographic location. It is a place of healing and, perhaps most important, it is a crucial link between our people and the way and culture of our ancestors." Coso is known for the spirituality of the springs, Indian burial locations, and its status as a prayer site. Native Americans were concerned that any geothermal development might affect the spirituality of the prayer site and that activity might cause unwanted intrusion, such as noise, during ceremonies that occur four to twelve times a year. The developers were able to work out a solution with the Indians, and the geothermal power project is operating.

In conclusion, the present trend of hot spring users to "get back to nature" and preserve and protect natural geothermal resources is not unlike the philosophy of our earliest settlers.

References

Aikens, C. M. 1978. The Far West. In *Ancient Native Americans,* J. D. Jennings, ed. W. H. Freeman and Co., San Francisco.

Archuleta, K. 1977. *The Brannan saga—Early-day Calistoga.* Smith McKay Printing Co., San Jose, California.

Bedinger, M. S. 1988. *Valley of the Vapors—Hot Springs National Park.* Eastern National Park and Monument Association, Philadelphia, Pennsylvania.

Bergon, F., ed. 1989. *The journals of Lewis and Clark.* Penguin Nature Classics, Penguin Books, New York.

Breckenridge, R. M. and B. S. Hinckley. 1978. *Thermal springs of Wyoming.* The Geological Survey of Wyoming, Bulletin 60, Laramie, Wyoming.

Chittenden, H. M. 1895. *The Yellowstone National Park.* Reprint 1964, R. A. Bartlett, ed. University of Oklahoma Press, Norman, Oklahoma.

Clark, E. E. 1953. *Indian legends of the Pacific Northwest.* University of California Press, Berkeley, California.

Cole, R. 1998, March 17. New theories change beliefs about how people got here. *Herald and News,* Klamath Falls, Oregon.

Conte, R. S. 1989. *The history of the Greenbrier—America's resort.* Pictorial Histories Publishing, Charleston, West Virginia.

Cox, J. H. and E. Stasack. 1970. *Hawaiian petroglyphs.* Bishop Museum Press, Honolulu, Hawaii.

Davis, L. G. 1995. *Volcanism, climate change, and prehistoric cultural succession in Southern Washington and North-Central Idaho.* Corvallis, Oregon. (M.A. thesis, Oregon State University).

Dumond, D. E. 1978. Alaska and the Northwest coast. In *Ancient Native Americans,* J. D. Jennings, ed. W. H. Freeman and Co., San Francisco.

Fagan, B. M. 1985. *People of the Earth.* Little, Brown and Co., Boston.

Griffin, J. R. 1978. The Midlands and Northeastern United States. In *Ancient Native Americans,* J. D. Jennings, ed. W. H. Freeman and Co., San Francisco.

Harris, S. L. 1988. *Fire mountains of the West: The Cascade and Mono Lake volcanoes.* Mountain Press Publishing Co., Missoula, Montana.

Harris, S. L. 1990. *Agents of chaos.* Mountain Press Publishing Co., Missoula, Montana.

Jennings, J. D. 1978. Origins. In *Ancient Native Americans,* J. D. Jennings, ed. W. H. Freeman and Co., San Francisco.

Kane, H. K. 1987. *Pele—Goddess of Hawaii's volcanoes.* The Kawainui Press, Captain Cook, Hawaii.

Lewarch, D. E. and J. R. Benson. 1991. Long-term land use patterns in the Southern Washington Cascade range. *Archaeology in Washington*, 3, 32- 34.

Lipe, W. D. 1978. The Southwest. In *Ancient Native Americans*, J. D. Jennings, ed. W. H. Freeman and Co., San Francisco.

Lund, J. W. 1993. Spas and balneology in the United States. *Geo-Heat Center Quarterly Bulletin*, 14, no. 4, 1-3.

Lund, J. W. 1996. Balneological use of thermal and mineral waters in the United States. *Geothermics*, 25, no. 1, 103-147.

Lund, J. W. 1997. Hot Springs, South Dakota. *Geo-Heat Center Quarterly Bulletin*, 18, no. 4, 6-11.

Markle, D. 1979. *Geothermal energy in Alaska: Site data base and development status.* Geo-Heat Center, Klamath Falls, Oregon.

Martineau, L. 1973. *The rocks begin to speak.* KC Publications, Las Vegas, Nevada.

Mona, Dakota Indian Nation. 1996, July 27. Personal communication by e-mail.

Perlman, D. 1998, February 17. Scientists push back date man arrived in new world. *Chronicle*, San Francisco.

Recer, P. 1997, July 4. Canadian coast yields evidence of migration corridor from Asia. *The Oregonian*, Portland, Oregon.

Rendall, B. 1981. *History of Harrison Hot Springs and Port Douglas area—Healing waters.* Harrison Lake Historical Society, Canada.

Stark, P. 1997. The Old North Trail. *Smithsonian*, 28, no. 4, 54-69.

Swanner, G. M. 1988. *Saratoga—Queen of spas.* North Country Books, Inc., Utica, New York.

Westervelt, W. D. 1963. *Hawaiian legends of volcanoes.* Charles E. Tuttle Company, Rutland, Vermont.

The Author:

John W. Lund, Director
Geo-Heat Center
Oregon Institute of Technology
Klamath Falls, Oregon 97601 USA
Telephone: 541.885.1750
Fax: 541.885.1754
E-mail: lundj@oit.edu

The Witches' Cauldron near The Geysers Resort Hotel. Thomas Moran, 1873. *after one half of a stereopticon view by Eadweard Muybridge, 1870*

30. Hot Springs and Fumaroles, Early Days at The Geysers

by
Susan F. Hodgson

Pomo basketry language for
ka teá katip´, obsidian chips.

LIKE STRANDS OF GRASSES WOVEN TOGETHER INTO A POMO BASKET, EVENTS AT THE GEYSERS Geothermal Field overlap, shaping each other through time. In the first era of geothermal history at the Northern California site, The Geysers was untouched wilderness. About 12,000 years ago, this era ended and the second began when Native Americans came at the start of what is called the Paleo-Indian period. Now the area was inhabited by humans for the first time. By the mid-1800s when so many settlers came, six Indian tribes lived at The Geysers: the Southern, Central, and Eastern Pomo; the Western and Northern Wappo; and the Lake Miwok. All spoke different languages, had distinct cultures, and lived in well-defined areas of the field (Fredrickson, 1974).

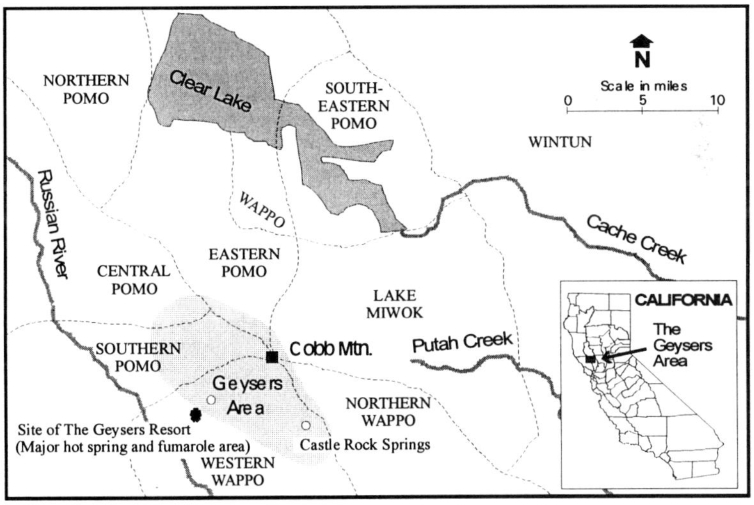

Indian tribes in The Geysers area. *Fredrickson, 1974*

Two views of "The Old Indian Steam Bath" near The Geysers Resort Hotel, 1871, one half of stereopticon views by world-famous photographer Eadweard Muybridge. An animal hide could be thrown over the stick structure, trapping the steam. These are the only known photographs of an authentic Indian structure at The Geysers.

Most likely the steam bath in the photo is the one referred to by Benjamin Avery in 1873 in the following passage. "A ravine [near Big Sulphur Creek] contains a clear hot spring, which was formerly built over with stones and sticks by the Indians, and the steam used as a sanitary agent. It is still known as the Indian Spring."

Robert Thompson told in 1877 how the Indians built scaffolds or gratings over steaming pools, on which they placed the sick, especially those suffering from rheumatism.

Ernest Finley wrote in 1937 that a pool still is called the "Indian's sweatbath," and another steam bath used by indigenous peoples is located about two miles below the first.

Drawing of the Old Indian Steam Bath, published in 1890 by Dr. Winslow Anderson.

Trails open to all the tribes crisscrossed The Geysers. Some trails led to the hot springs and fumaroles, which were used for healing purposes and may have held ceremonial significance (Fredrickson, et al., Sept. 1978). Early writers noted that archaeological sites at The Geysers often were found near hot springs (Meneffee, 1873).

Castle Rock Springs, in the eastern part of The Geysers, was one of the main hot spring areas. It was used for healing, primarily by the Lake Miwok living near Middletown. This use can be traced to at least the beginning of the historic period. An elderly Lake Miwok knew that the hot springs were used before his time and told of a nearby burial ground that he could no longer find (Fredrickson et al., August 1978).

Although Wappo territory included The Big Geysers—the hot spring and fumarole area along Big Sulphur Creek where the Geysers Resort Hotel would be located—the area was open to other tribes (Fredrickson et al., August 1978). Wappo invalids, wrote Stephen Powers in 1871 or 1872, "were accustomed to wallow in the hot, steaming mud and pools, receiving benefit therefrom into their bodies."

An ethnographic field survey indicated that the Wappo collected sulphur salt at The Big Geysers. Sulphur salt is what the Wappo called the mineral-rich residue of evaporated hot spring waters. The Wappo mixed sulphur salt with the ashes of burnt stalks of cow parsnip and ate this with acorn bread, "presumably for medical purposes" (Fredrickson et al., August 1978). The Wappo word, *te'ke*, means "mineral left as a deposit after the evaporation of the water from the springs at The Geysers in Sonoma County" (Barrett, 1908). *Te'ke* may refer to sulphur salt.

A Wappo village called *tekena'ntsonoma* was just north of The Geysers "near the head of the main branch of Sulphur Creek and at a point about 12 miles a little south of east of Cloverdale." Again, the word *te'ke* is defined as above, *nan* means a well or other deep hole containing water, *tso* means ground, and *no'ma* means village (Barrett, 1908).

An ethnographic field survey found that the area around the former Geysers Resort Hotel was held especially sacred, that "its waters were used medicinally to treat a multitude of illnesses. An Indian woman said that her grandmother described the different types of water available near the former resort, saying 'Every kind of water came out of the mountain, ice cold, lukewarm, and boiling hot.'"

The woman said that a Cloverdale man, sick for a year, was cured by the following treatment at this site. Basins were excavated near sources of cold, lukewarm, and hot water, and small channels were constructed leading to each. Proper water temperatures were obtained by opening and closing the channels. On a litter made of willow branches, the man was brought to the area from the Cloverdale Pomo village of Makamo, near the confluence of Big Sulphur Creek and the Russian River. He stayed at The Geysers for four days and four nights, attended by an Indian doctor. At the conclusion, the man walked out without assistance, completely cured (Fredrickson et al., Sept. 1978).

William Bell Elliott, who followed a grizzly bear to Geyser Canyon in 1847, told his grandson, "Pop" McCray of Cloverdale, that the Indians he met at The Geysers had cut shallow niches in the bank of Big Sulphur Creek, which flows on the southwestern boundary. The sick were laid inside the niches on ladder-like stretchers, according to an interview from the late 1920s.

Stephen Powers was a writer and anthropologist who visited The Geysers in his travels among California's Indians in the summers of 1871 and 1872. His account of their lives includes the following legend that he reprinted from the *San Francisco Bulletin.* The truth of the legend is not known, but the seriousness of Powers' project suggests that he thought the legend was important.

A Legend of The Geysers

"In passing up the gorge in which are situated the Pluton Geysers [Geyser Canyon by the former Geysers Resort Hotel], you will notice a human head carved in stone. It bears so striking a resemblance to a half-finished piece of statuary that the most casual observer asks its history. This is the legend as told by the Indians who inhabit the Coast Range:

'The discovery of the Geysers is a comparatively modern event. From the time when the memory of man runneth not to the contrary, peaceful tribes of Indians inhabited the rich, luxuriant valley of Russian River and its tributaries. With hunting and fishing, with clover, wild oats, and acorns, with the various roots, berries, and fruits provided by Nature, they lived a happy, contented life. The dense chaparral which covers the mountains and lines the cañons of the region surrounding the Geysers effectually concealed these wonderful springs. It was since the

Spaniards and Mexicans began to settle the country and fatten their immense herds upon the rank herbage that the Indians were compelled to put forth greater exertions for food. Two of their young men were hunting on the south side of the river, below where Cloverdale now stands, when they caught sight of an unusually large grizzly bear. Simultaneously they fired their sharp-barbed arrows into the monster's side. He dropped as if dead, but well knowing it to be a habit of the grizzly to fall to the ground upon receiving the slightest wound, they again let fly their flint-headed shafts, and again struck the bear. Sorely wounded, the animal instinctively staggered toward the thick underbrush, leaving a trail of blood behind. Sure of their game, the hunters followed the blood stains into the chaparral and up the cañon....

'The sun had moved far down the heavens, and the lofty western mountain shut out his beams from the gorge. At sight of their dying game, the Indians gave a loud, exultant shout.... In their eager haste they had not noticed the thousand minute jets of steam issuing from the hillside, nor did they hear the hoarse, rushing sound that filled the cañon with a continuous roar, until just as they reached the body.

'Halting, amazed, they found themselves standing on the brink of the Witches' Cauldron, in the midst of the hissing, seething Geysers. One horrified, ghastly look at the smoky, steaming hillside; one breath of the puffing, sulphurous vapor; one terrified glance at the trembling, springy earth, and the frightened hunters darted back down the cañon. With stoical skepticism the aged chief and council listened to the tale the hunters told as the tribe gathered around the camp-fire. Earth that smoked! Water that boiled and bubbled without fire! Steam that issued from holes in the ground with a noise like the rushing of the storm-wind! Impossible! But the two young braves were noted for courage and truthfulness, and at last they prevailed on a score of their fellows to return with them. It was all true. There lay the dead bear by the black, seething waters that were hotter than fire could make them. After a thorough examination, the medicine-men concluded that the strange mineral waters must have rare healing properties. Booths of willows were erected over the jets of steam, and the sick laid thereon. The cañon became a favorite resort, and all the Coast Range tribes came hither with their invalids. Many wonderful cures were effected, and yet, occasionally, things happened that convinced the superstitious medicine-men that the place was under the control of an evil spirit.

'Finally, one cloudy night, a strange, rumbling sound rose through the darkness, and earth trembled violently. After that no one approached the spot for many days.

'It is a common belief among the Coast Indians that evil spirits frequently dwell within the bodies of grizzlies. It was now universally believed that the spirit of the slaughtered bear had charge of the Geysers. There were many sick and dying with a strange plague, or pestilence that had suddenly appeared among the tribe. Something must be done. Many urged a return, at all hazards, to the medicinal springs; others held that the angry demon of the gorge had sent the pestilence upon them. At last a gray-haired seer whose hand was skilled in all cunning craft was persuaded to try to appease the spirit by making a graven image near the Witches' Cauldron. Enough of the idolatrous traditions of their ancestors were remembered to enable them to have faith in this strange attempt at propitiation. Day after day the good old sculptor went all alone to the cañon, and chiseled away the rock until the semblance of a human face appeared. As the work neared completion, he often lingered later, in his anxiety to finish the statue. It was believed that when the task was entirely ended the demon would retire, and let the people be healed. A few more days and the finishing strokes would be made on the figure. Everyone was full of hope. The old man was working at the dawn, and when the evening came and the twilight shadows stole down the mountain and up the ravine he had not returned. Suddenly a weird, hollow moan seemed to tremble on the shuddering air, and at the next instant the earth shook so violently that the cliffs toppled from their base. The terrible shocks were felt several times during the night, and when the sun arose the old seer was gone from earth. The cold, stony face of the image alone remained. Not the slightest trace was ever discovered of the faithful sculptor; yet during the night new springs had burst forth three-quarters of a mile down the river. Here the sick were brought, and from that day to the present time the Indians used only the lower springs. Scaffolds are raised above the steam-jets three or four feet, and willows and brush are laid across. On these the sick are placed, and the mineral vapors encircle and heal them.'

"Years after, the white men came to the great valley of the Russian River, and in due time were guided to the springs. The Indian guides would not go farther than the lower springs, but the pale-faces found the image still guarding the ravine. Enterprise and love of gain have built a beautiful hotel across from the Geysers, and hundreds of tourists annually flock thither.

"The Indians, however, firmly believe that the wrathful demon still holds sway, and they can never be induced to approach the gorge of the main Geysers."

It was this "beautiful hotel" that opened the next era of history at The Geysers, the age of organized tourism.

Mountains, in Pomo basketry language.

The Resort

Let us imagine we are sitting on the oak-shaded piazza of The Geysers Resort Hotel, pictured in the corner below in 1870 with the proud proprietor, J.C. Susenbeth. As leaves rustle in the afternoon breezes, we remember the history of The Geysers, the branches of the past weaving through those of the present. We gaze across Big Sulphur Creek at the southern edge of the geothermal field, the same steep canyons crossed by William Bell Elliott in 1847 when, as a member of John C. Fremont's survey party, he tracked and killed a marauding grizzly bear, Old Slewfoot.

Earliest known depiction of The Geysers Resort Hotel. *Hutchings' California Magazine, 1860*

Sources differ for exact dates, but between 1848 and 1854 tourists started coming to The Geysers and lodging was built. Robert A. Thompson wrote that the first was a house called the Old Homestead, built by M. Levy on a beautiful flat just west of Geyser Canyon. This was the first house in the area. Levy's flat, which he later promoted as a "Leafy Dell," was the same spot where Elliott killed the grizzly.

J. C. Susenbeth, proprietor, stands in front of The Geysers Resort Hotel in 1870. His ad read, "A good table is kept at the Hotel, and the best of liquors and cigars will be found at the bar." *One half of a stereopticon view, by Eadweard Muybridge*

Thompson said that in 1854 Major Ewing erected a cloth house across from Geyser Canyon where the hotel would be built. Levy, seeing this as a better location than his own, consolidated his interest with Major Ewing's. Soon a sawmill was brought in and construction was begun on the first hotel wing, which was built by James Alexander Mead.

Geyser Canyon, site of the fabled hot springs and fumaroles, is in the background. Better roads proved the key to commercial success at The Geysers. In 1869, a toll road was built from Knight's Valley (east of Healdsburg) to The Geysers Resort Hotel and a stage line was put on the route. In 1874, a toll road was built from Cloverdale up Sulphur Creek, opening for business the following season, according to Robert Thompson in 1877. In 1871, a stagecoach ride to The Geysers with Clark Foss cost $6. *One half of a stereopticon view, The Library of Congress*

The hotel did not prosper in its first decade. William Brewer, a geologist visiting in 1861, wrote in his journal, "A company preempted a claim of 160 acres, embracing the principal springs and the surrounding grounds, built quite a fine hotel on a most picturesque spot, and at an enormous expense made a wagon road to them, leading over mountains over three thousand feet high."

Brewer adds, "But the road was such a hard one, the charges at the hotel so extortionate, and the stories of the wonderful geysers so much magnified, that in this land of 'sights' they fell into bad repute and the whole affair proved a great pecuniary loss. The hotel is kept up during the summer, but the wagon road is no longer practicable for wagons and is merely used as a trail for riding on horseback or mules."

However, these setbacks proved temporary. "The first hotel register, kept in 1854, had 20 names. From then on the number increased every year, and in 1875 there were 3,500 names enrolled," wrote Robert A. Thompson in 1877.

What was The Geysers Resort Hotel like in its prime, from the late 1860s to the early 1880s? Major B. C. Truman wrote in 1883, "The hotel was so full during June, July, and August 1882, that people were compelled to sleep in bath-houses and on billiard tables. The hotel building is not at all modern in its structure, but is roomy and airy. There are broad piazzas on which the room doors all open; and Mrs. Sherwood, the jewel of a housekeeper, sees that everything is as neat and clean and sweet as a daisy.

"The table is as good as the market can make it; and, in their season, there is fish, and bear meat, and venison. We could recommend no better place for reasonable recreation, absolute *re*-creation, and perfect rest, than the Geysers. Instead of fashion, there is freedom; instead of expense, economy; instead of watering-place luxury, country comfort."

Some found the hotel too rustic. One guest, the beautiful Mrs. Frank Leslie, complained that the hotel was "constructed upon a novel and decidedly breezy plan," with thin board partitions that allowed guests to hear themselves discussed with "frankness and candor."

Perhaps neither Major Truman nor Mrs. Leslie knew that rattlesnakes were hung outside the kitchen door, valued by the Chinese cooks for medicine and meat.

When the resort owner's children wished to go on a picnic, a Chinese cook would mix Boston brown bread batter in a coffee tin and give them raw eggs to carry. The children would take these to a boiling spring by the side of the creek and put in the bread to bake. In a few hours when it was done, they would boil the eggs and enjoy a delicious lunch.

After 1885, the popularity of The Geysers Resort Hotel declined rapidly. During the next decade, unsuccessful attempts were made to maintain patronage with appeals to lower-income vacationers. One ad called the resort "the grandest, most beneficial for health, and cheapest pleasure trip in the world." Although the hotel's great days were never revived, a steady stream of visitors continued for well over 75 years.

Dr. Winslow Anderson noted in his book published in 1890 that "these famous springs [at The Geysers Resort Hotel] have been the objects of wonder and admiration to all the many thousands who visit them yearly."

Indeed, the guest list shone. Lured by advice that a trip to Northern California was incomplete without a visit to The Geysers, the rich and famous streamed to the resort—among them J. Pierpont Morgan, Ulysses S. Grant (his daughter-in-law worked in the resort as a chambermaid and a cabin was built and named in his honor), William Howard Taft, Horace Greeley, Theodore Roosevelt, William Jennings Bryan, Lotta Crabtree (the celebrated actress), Madame Melba (the operatic star), Mark Twain, Jack London, Luther Burbank, Giuseppe Garibaldi, and the Prince of Wales (afterwards King Edward VII), for whom a cement tub was constructed at one of the springs.

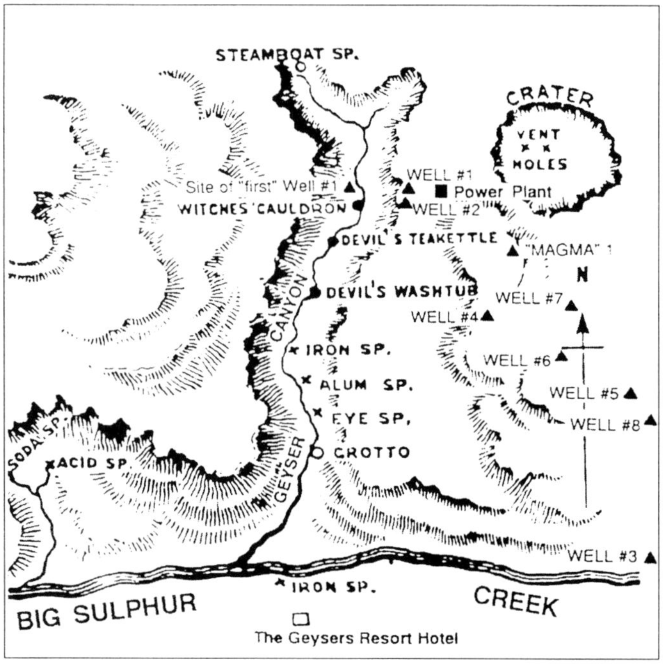

This map of Geyser Canyon, published in the late 1800s, includes fanciful names for the thermal features. The Geysers Resort Hotel, the eight geothermal wells drilled by John Grant in the 1920s, and "Magma" 1, the first modern well drilled in 1955, are added for reference.

View from the top of Geyser Canyon, looking toward The Geysers Resort Hotel, by now an extensive complex. Drawing from Dr. Winslow Anderson, 1890. Dr. Anderson advised hotel guests that "Each thermal and mineral spring differs materially in composition and temperature from any that may have been used before.... Place yourself, therefore, under the management of the competent keepers of the baths and the attendant physician, and do not use your own judgment about a matter which you but imperfectly understand."

"Do not," he added, "commence a course of treatment at the springs by bathing once or twice daily. The American fashion, as noted elsewhere, of hastening and rushing through everything may do well enough for business, but where the life of an individual or the treatment of an obstinate disease is at issue this plan is not only deleterious, but may prove fatal to the life of the patient."

Geyser Canyon, *Hutchings' California Magazine, 1860*

Benjamin Avery wrote, "The best time to visit Geyser Canyon is early morning, before the sun has risen above the mountain-tops.... Columns and clouds of steam may then be seen rising to a height of two hundred feet or more, obscuring the landscape like a fog just rolling in from the sea.

"One returns to the hotel after a morning tramp through Geyser Canyon and along Big Sulphur Creek with an enormous appetite, and is glad to rest for a few hours. Afterward, there are delightful strolls up and down the creek, and good trout-fishing for those who will go far enough." *Scribner's Monthly, 1873*

One half of a stereopticon view, by Andrew Price

These hotel guests, posing by the Witches' Cauldron, hiked up Geyser Canyon with a stout stick, called a 'Geyser pony' to marvel at the steam and hot springs. "Everyone takes a 'Geyser pony'... to help him or her over the rocks and springs, and then all start on the trail single file," wrote Benjamin Avery. "[On the way] the guide, who perhaps is the jolly landlord himself, points out a chalybeate spring of fine tonic properties, whose waters his guests imbibe, mixed with soda-water.

"The hot ground under the feet [in Geyser Canyon]; the subterranean rumblings; the throbs and thuds near some of the largest and most energetic steam-vents; the warmly moist atmosphere, filled with acidulous and sulphurous vapors, sometimes charged with strong odors of sulphuretted hydrogen; the screaming, roaring, hissing, gurgling, and bubbling of the various springs; all contribute to make the scene as repellent to some natures as it is grand and exciting to others....There are no spouting fountains in the canyon, but numerous bubbling springs, that sink and rise with spasmodic action. These number a hundred or two, and are of varying temperature and constituents." *Scribner's Monthly, 1873*

THIS PHOTOGRAPH OF WOMEN COOKING EGGS WAS TAKEN BY EADWEARD MUYBRIDGE AT THE Witches' Cauldron in 1870. "The Witches' Cauldron is a black cavernous opening in the solid rock, about seven feet across, and of unknown depth, filled with a thick inky liquid, boiling hot, that tumbles and roars under the pressure of escaping steam...that seems to proceed from some Plutonic reservoir. One irresistibly thinks of the hellbroth in *Macbeth*" (Avery, 1873).

"The rocks for several feet above this infernal fountain, over which its contents have splashed, were covered with innumerable crystals and stalagmites of pale sulphur. In the year 1861 this cauldron, from some unknown cause, was emptied of its contents and filled with steam. The proprietor of the hotel at the place, fearing that it would thus be deprived of one of its greatest attractions, caused a small stream of water to be led into the cauldron, curious himself to see what would be the result. The instant the cool water came in contact with the lower portion of the cavity a fearful commotion ensued. The ground, for several rods about, shook with violence, and in a few minutes after, the inflowing water was ejected with stunning reports, and thrown to the height of nearly one hundred feet. In about three hours after the water was shut off the viscid fluid reappeared, and has continued to boil and bubble ever since" (Cronise, 1868).

In the late 1880s, another author noted that an explosion at the Witches' Cauldron sounded like the report of a heavy cannon. When a ton or two of rocks blew from the uncharted depths, the noise was heard five miles away.

Eadweard Muybridge, soon to be world famous, was perhaps first to photograph The Geysers. Many of Muybridge's photographs of The Geysers are well known, yet not the fact that he made them.

After his work at The Geysers, Muybridge invented a way to take sequential photographs of animals and people in motion. Later, working with his locomotion studies on photographic strips, he invented motion pictures.

"Eye of newt and toe of frog,
Wool of bat and tongue of dog...
Double, double, toil and trouble;
Fire burn and cauldron bubble."
—William Shakespeare

Another especially famous Muybridge photograph of The Geysers was taken at the Witches' Cauldron. "A clever photographer, Mr. Muybridge, conceived the idea of grouping three lady visitors about this cauldron, with hands linked, and alpenstocks held like magic wands, in which position he photographed them," wrote Benjamin Avery in an article entitled *The Geysers of California*, published by *Scribner's Monthly* in 1873.

One half of a stereopticon view, by Eadweard Muybridge, 1870, titled "The Witches' Cauldron, *Macbeth*, Act IV, Scene I."

Scribner's hired Thomas Moran, not yet the world-famous artist he would be, to illustrate Avery's article by drawing some of the Muybridge photographs. The drawings Thomas Moran made are widely reprinted and never credited to the artist, perhaps because he signed them with only his initials at the time when his career was just beginning.

Drawing by Thomas Moran of the Muybridge photograph for *Scribner's Monthly*, 1873. Note Moran's overlapping initials (T, Y, M) in the lower right-hand corner. The *Y* stands for "Yellowstone."

Thomas Moran would gain international renown for his many paintings of Yellowstone National Park and the Grand Canyon, which he visited time and again. Moran Point, Mount Moran, and Moran Canyon—famed western sites—all are named after him.

"WE FELT LIKE BOILED ANGELS" A FIRST-HAND ACCOUNT

"During May 1880, Fred. Somers, a well-known writer; Charlie Stoddard, the poet; Julian Rix, the artist, and another *trio* of valetudinarians, one day shook the San Francisco dust from their sandals, and hied themselves to the Geysers for recreation and recuperation, which emancipation they must have abundantly enjoyed, according to Somers' account of their little 'racket,' which appeared in the *San Francisco Argonaut* on May 15, 1880, a portion of which we reproduce, as follows:

> 'Inside, the house is partitioned off into male and female departments, with similar accommodations for each. Under the bath-house is a jet of steam, that bursts out of the ground just reeking with sulphur and iron and all the sweet-smelling minerals in the bowels of the earth. Over this jet a half-hogshead has been placed, and, by means of a wooden spout, the steam is conveyed to what may be called the *perspiration parlor*. Shedding our clothes, we investigated this apartment. It was a sulphur-stained room, with a settee against the wall, and a few chairs with raw-hide bottoms. Forsyth showed us the steam spout, and the adjustable valve, and how, when the valve was pushed in the steam would crawl under the floor, and climb over the back of the innocent-looking settee, and flay you alive. Then, having explained things, and invited us to take seats without cushions or anything, he proceeded to give us a practical illustration. He pushed the valve, and while the conversation went on the steam began to transact business. It came up the spout, and crawled in under the floor, and came out under and over the back of the settee just as has been described, and it made things very tropical. In less than five minutes that little room looked, and felt, and smelt, like a miniature Hades, with Forsyth in the box office. Hardened and accustomed to this temperature, he smiled through the vapor, while the six of us smothered in the sulphur. The skin rolled off in flakes as we frantically caressed our ribs. Every pore was open and howling with the heat. Finally, when the perspiration was pouring beautifully, and we were yelling for forks to try ourselves and see if we were not done, Stoddard broke up the *sèance* by fainting. We dragged him under the tepid shower in the next room, and he recovered. Then, red as lobsters, we went through the rest of the performance—with a slight suspicion, however, that we were getting more than the regulation dose. The tepid showers were delightful. The water came direct from the warm springs in the bank, and, tempered with minerals, was as soft and soothing as oil. From the warm showers we filed down a steep flight of stairs and took a plunge into a large tank of sulphur water as clear as crystal, and also toned by the hot springs to a degree suited to the steaming body of the plunger. Here we lay off to cool, as it were, and then, passing under the cool showers, were ready for the divans, or lounges, arranged after the manner of the Turkish baths. Amid the curling smoke of cigarettes we pronounced this natural bath very good.
>
> 'Fresh from God's Hammam, we felt like boiled angels. The skin was as soft as silk; the liver fluttered with joy; there was a tingling sensation along the spine; and, finally, there succeeded a sleep that tangled up the whole splendid sensation'" (Truman, 1883).

THE LEGEND OF "HAPPY JACK"

THIS 1922 PHOTO OF "HAPPY JACK" EDWARDS WAS TAKEN AT HIS HOMESITE NORTH OF GEYSER Canyon, near The Geysers Resort. "Happy Jack," with a long white beard, is standing at photo right. Today, 12 modern wells in The Geysers Geothermal Field are named after him.

Geothermal Resources Council photograph

Ernest Finley wrote in 1937, "Among the passengers on a stage that rolled into The Geysers [around 1908] was a wealthy San Francisco lawyer. He is said to have gone to the place to obtain relief from arthritis, for one reason, but there was another: his affliction was not limited to physical suffering, and he also sought surcease of sorrow following a romance that had gone wrong. Near the top of the peak directly above The Geysers he built a rude cabin, with only the hard earth as a flooring, and in this primitive abode he lived without companionship for more than a score of years.

"From the day that he entered the region until the day of his death the man never left Pluton Canyon [Geyser Canyon]. Now and then he would descend the mountain to obtain food and necessary supplies. Nature must have soothed the pangs of unrequited love as well as the ills to which flesh is heir in the case of this latter-day hermit, for he appeared always to be in jovial mood on visiting the hotel or when curious tourists climbed the mountain to talk with him and inspect the strange conditions under which he lived. In time he came to be known far and wide merely as 'Happy Jack.'

"The hermit died in 1928 and was buried on the mountain he loved, which had brought him relief from both bodily torture and anguish of spirit, with the moaning sibilance of 'Steamboat Geyser' as a never-ceasing requiem. The site of his burial place, about one hundred feet from the little cabin he occupied, which is still standing, is marked by an expensive concrete sepulcher. The stone at the head of the vault, conspicuous by its strange surroundings, bears the simple inscription: John Edwards—'Happy Jack'—1928."

Until 1912, a stage ride began and ended most visits to The Geysers Resort Hotel. One early tourist wrote at the end of a vacation;

"And we will ride away from the Geysers, grateful that we have seen its marvels and terrors, and the more grateful that the Creator hides from us, by so much ever-renewing loveliness on the bosom of the world, the awful fact which the 'Professor' has so concisely stated, that we live on a globe which has a 'crust of fossils and a heart of fire.'"
Bancroft's Tourists' Guide, 1871.

The hogback road from The Geysers. (W. Anderson, 1890)

Permissions

Photographs by Eadweard J. Muybridge, courtesy, The Bancroft Library.

Photographs by Andrew Price, courtesy, The Library of Congress.

Etching by Thomas Moran and excerpts by Benjamin P. Avery are reprinted with the permission of Scribner, a Division of Simon & Schuster from SCRIBNER'S MAGAZINE, vol. vi, no. 6, New York: Charles Scribner's Sons, 1873.

Excerpts from *History of Sonoma County, California, Its People and Its Resources*, by Ernest Finley, are reprinted with permission of *The Press Democrat*, Santa Rosa, California.

"A Legend of The Geysers" is from *Tribes of California* by Stephen Powers, with introduction and annotations by Robert F. Heizer. Copyright 1976 by The Regents of the University of California. Reprinted by permission of the University of California Press.

Excerpts from William Brewer's journal are from *Up and Down California in 1860-1864*, edited by Francis P. Farquhar. Copyright 1949 by The Regents of the University of California. Reprinted by permission of the University of California Press.

Selected References

Anderson, W. 1890. *Mineral springs and health resorts of California, with a complete chemical analysis of every important mineral water in the world.* Bancroft Company, San Francisco. (The book first sold for $1.50, "With special rates to Spring owners").

Avery, B. P. 1873, October. "The Geysers of California." *Scribner's Magazine.*

Bancroft's tourists' guide around the bay, (north), The Geysers. 1871. A. L. Bancroft & Company, San Francisco.

Barrett, S. A. 1908. *The ethno-geography of the Pomo and neighboring Indians.* University of California Press, Berkeley, California.

Brewer, W. H. 1860-1864. *Up and down California in 1860-1864, The journal of William H. Brewer.* Reprint 1930, Francis P. Farquhar, ed. University of California Press, Berkeley, California.

Cronise, T. F. 1868. *The natural wealth of California.* H. H. Bancroft & Co., San Francisco.

Finley, E. L., ed. 1937. History of Sonoma County, California, its people and its resources. *The Press Democrat,* Santa Rosa, California.

Fredrickson, D. A. 1974, June 4. *An archaeological survey of ten small land parcels in The Geysers Geothermal Field, Lake and Sonoma Counties, California.* Prepared for Pacific Gas and Electric Company; manuscript filed at the Northwest Information Center, Sonoma State University. Access subject to approval.

Fredrickson, D. A., W. H. Wright III, M. S. Ziegenbein, M. M. Druckman, and H. Ritter. 1978, August 16. *Paleontological resource investigations in Units 18, 19, 20, and 21 areas of The Geysers region, Lake and Sonoma Counties, California.* Submitted to Pacific Gas and Electric Company; manuscript filed at the Northwest Information Center, Sonoma State University. Access subject to approval.

Fredrickson, D. A., D. W. Peri, S. M. Patterson, and S. L. McMurray. 1978, September. *Ethnographic and historical survey of the geothermal leasehold Unit 17, Lake and Sonoma Counties, California.* Prepared for Pacific Gas and Electric Company; manuscript filed at the Northwest Information Center, Sonoma State University. Access subject to approval.

Hodgson, S. F. 1988. 1927 and 1988: Views of The Geysers Geothermal Field. *The Geothermal Hot Line,* 18, no. 1, 1-9.

Hodgson, S. F. 1992. A Geysers album. In *Monograph on The Geysers Geothermal Field.* Geothermal Resources Council, Davis, California.

Hodgson, S. F. 1993. *A Geysers album.* Videotape. California Department of Conservation, Division of Oil and Gas, Sacramento, California.

Hodgson, S. F. 1996. Eadweard Muybridge, Early geothermal photographer. *The Geothermal Hot Line,* 22, no. 2.

Hodgson, S. F. 1996. Eadweard Muybridge to Thomas Moran. *The Geothermal Hot Line,* 22, no. 2.

Hodgson, S. F. 1997. *A Geysers album: Five eras of geothermal history.* California Department of Conservation, Division of Oil, Gas, and Geothermal Resources, Sacramento, California.

Langhart, E. *Personal archives of The Geysers Geothermal Field.*

Meneffee, C. A. 1873. *Historical and descriptive sketch book of Napa, Sonoma, Lake and Mendocino.* Reporter Publishing House, Napa City, California.

Muybridge, Eadweard. *Photographic collection.* The Bancroft Library, Berkeley.

Panoramics. 1860. A trip to the California Geysers. *Hutchings' California Magazine,* 4, no. 7.

Powers, S. 1877. Tribes of California. In *Contributions to North America Ethnology, vol. 3.* Department of the Interior, US Geographical and Geological Survey of the Rocky Mountain Region. Reprint 1976. University of California Press, Berkeley and Los Angeles.

Thompson, R. A. 1877. *Historical and descriptive sketch of Sonoma County, California.* L. H. Everts & Co., Philadelphia, Pennsylvania.

Truman, Major B. C. 1883. Tourists' illustrated guide to the celebrated summer and winter resorts of California adjacent to and upon the lines of the Central and Southern Railroads. H. S. Crocker & Co., San Francisco.

The Author:

Susan F. Hodgson
California Department of Conservation
Division of Oil, Gas, and Geothermal Resources
801 K Street, MS 20-20
Sacramento, CA 95814-3530 USA
Telephone: 916.445.9686
Fax: 916.323.0424
E-mail: shodgson@consrv.ca.gov

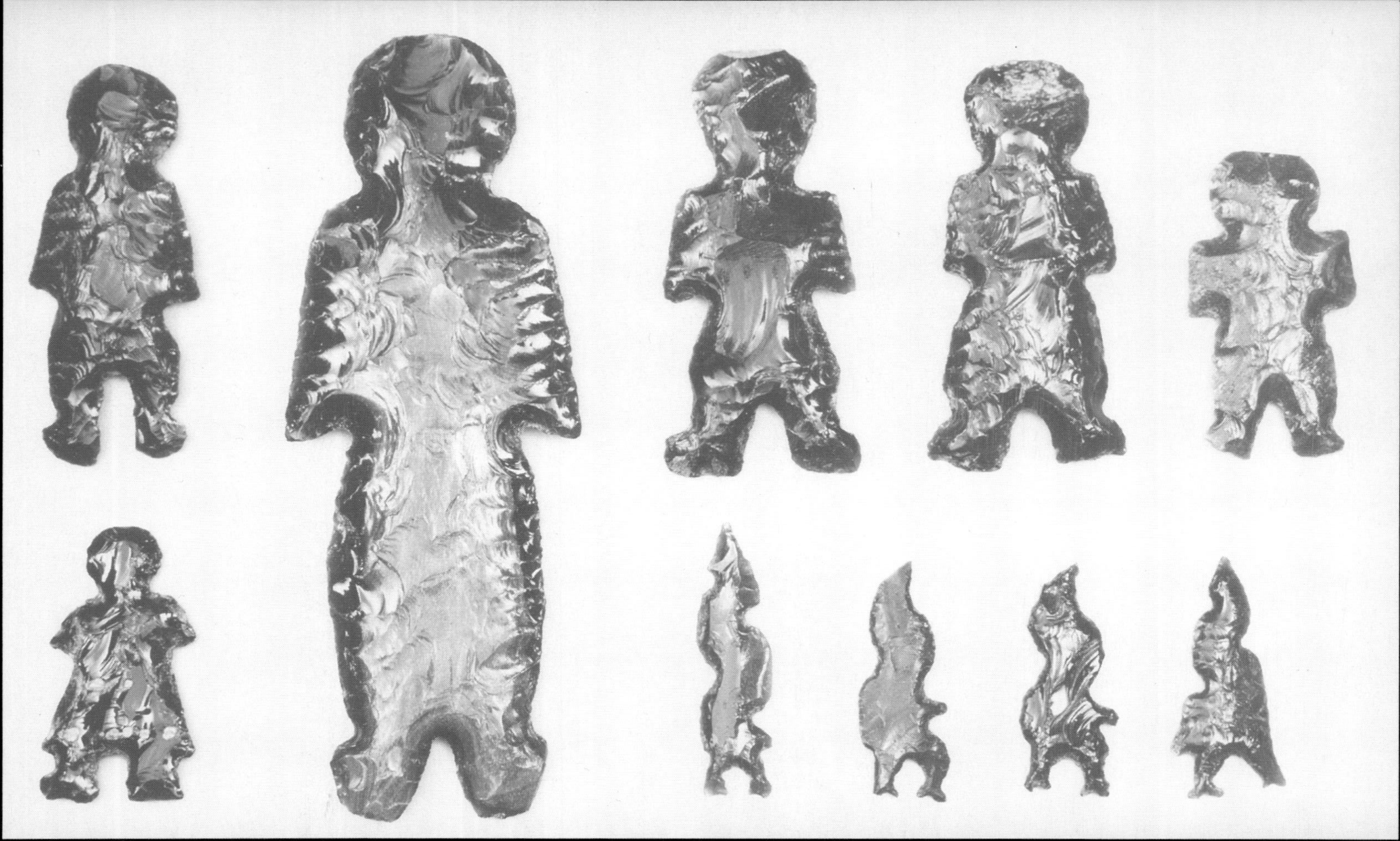

In Mesoamerican cosmogony, death marked the passage from the sensate realm into the abstract. As geothermal phenomena influenced peoples' daily beliefs and activities so strongly, it is fitting that these funerary offerings were flaked from greenish-black obsidian, a geothermal resource.

The offerings are ¾ to 6 inches high and part of a set of 58 objects, 3rd to 7th century A.D., from the Early Classic period. Said to be from Cuautitlán in the Basin of Mexico, six of the pieces represent the human form. The upper body design suggests that the arms are crossed in front. Similar objects have been excavated from the Pyramid of the Sun at Teotihuacán (see photo, this chapter). ***The Saint Louis Art Museum***

31. Cosmogony and Uses of Geothermal Resources in Mesoamerica

Abstract: Many Mesoamerican cultures developed in the Mexican Volcanic Belt, which runs east-west across Southern Mexico. For thousands of years, this was a place particularly rich in active geothermal manifestations, where intense contacts between mankind and geothermal forces deeply influenced the material lives of Mesoamericans and played an important role in basic decisions, such as selecting where to live. Vapor baths, cooking in steam at fumaroles or on naturally hot rocks, crop irrigation, and therapeutic applications were all ways that the Earth's natural heat was used in Mesoamerica.

Mythical and religious interpretations of geothermal phenomena occurred early on, reflecting a profound respect toward terrestrial heat. People's relationships with geothermal resources were both beneficial and harmful, constructive and destructive. Volcanic explosions drove away populations, who then returned, drawn by the fertile soils, volcanoes, lakes, and natural heat. Mesoamerican peoples understood this duality, which deeply influenced their cosmic view.

Reprinted by permission of the International Geothermal Association, from the *Proceedings of the World Geothermal Congress,* 1995, pp. 417-422.

by
Mario César Suárez Arriaga
Raffaele Cataldi
Susan F. Hodgson

"Mexico has as many volcanoes as England does blackberries."
—*10th International Geological Congress, Guadalajara, 1906*

Introduction

Cosmogony is the science of the creation and evolution of the universe. This chapter treats some aspects, recently discovered, of the close relationship that existed between the development of Mesoamerican cosmogony and geothermal resources.

When opening any book on the history of science, one rarely finds a chapter dealing with science in pre-Hispanic America. However, Mesoamerican cultures created not only original art but also hydraulics, architecture, and mathematics. They developed autonomously the concept of zero, an advanced numerical system, and they made astronomical calculations with astonishing precision. Today, we know that scientific knowledge was a daily reality for Mesoamerican peoples. Less well known is their general concept of the world, how they questioned origin, being, and destiny. Their beliefs and knowledge originated from ideas about the cosmos. We want to demonstrate that their cosmogony arose, at least partially, from intense experiences with geothermal phenomena. To comprehend this vision, one must look at the natural environment, religion, and Mesoamerican mysticism.

For at least 4,000 years, human communities have resided in zones with geothermal manifestations throughout the Americas. Archaeological investigations show that stable settlements grew at the feet of all Mesoamerican volcanoes (Burgassi et al., 1992). We can infer that since remote times, humans coexisted with volcanoes, with geothermal manifestations, and with other phenomena related to the Earth's heat.

Being predominantly volcanic, Mesoamerica was a region that included earthquakes, thermal waters, fumaroles, boiling muds, steaming grounds, and volcanic eruptions. Surface geothermal manifestations and the energy of the Earth were well known to ancient Americans. For millennia since prehistoric times, people had close contact with the energy of the Earth. Therefore, geothermal resources, in the widest sense of the word, deeply influenced the material, spiritual, and cosmic development of the Mesoamerican people.

A religious attitude toward volcanoes was widespread among pre-Columbian populations. Because of their magnificence and eruptive natures, volcanoes were considered sacred cult objects, often because so many volcanoes were nearby.

We have classified into two groups the available information about geothermal development in Mesoamerica, and a faithful approach to describing influences of geothermal phenomena incorporates both. Information from the first group is scarce, for it is based on direct facts described rigorously and is seriously limited by not going beyond documented evidence. One important original source was the Náhuatl texts compiled by the Spanish historian, Fray Bernardino de Sahagún, who began in 1534 to transcribe narratives of Náhuatl elders reciting what they had learned in Aztec schools.

The second, more abundant type of information is subjective and indirect, based on myths, legends, and oral testimonies that possess meanings simultaneously poetic, mystical, and religious. These references, which cannot be demonstrated, are based on intuitive faith in the authenticity of a story that liberates the imagination.

This chapter is a preliminary outline, not an exhaustive treatment of the subject. Our survey has a double objective: to explore the influence exercised by geothermal resources on the material and spiritual development of Mesoamerica and on the relationship between humans and territory, and to emphasize the historical antiquity of the influences by geothermal phenomena on cognitive processes.

The history of geothermal resources in the Americas still needs to be written; this brief document sketches that possibility for the first time. The theme is left open for new studies, in hopes that a better understanding may develop of the ancient history of Mesoamerica. "The culture of the old Mexicans, suddenly annihilated, was one which humanity could have been proud to create.... From time to time, in the infinite of the whole and amid the enormous indifference of the world, some men assembled in society and invented something that surpassed them, a civilization. They were the creators of cultures. And the Indians of Anáhuac, at the feet of their volcanoes, at the shores of their lakes, could be counted among those men" (Soustelle, 1955).

From Prehistory through the Postclassic Period

The origin of people in the Americas apparently does not introduce any epistemological problem, for researchers have not found human remains from predecessors of *homo sapiens* in the Americas, as they have in Africa, Asia, and Europe. Today, it is accepted that the proto-Americans came from Asia some 50,000 years ago, passing through the Bering Strait. The first immigrations could have been the "Amurians" or inhabitants of the Amur River shore in Siberia (Ruiz, 1987; Museo Nacional, 1996). A Mexican historian suggests these people could have come from southern zones in Central Russia (Sodi, 1992). The Cahuilla, Sioux, Tolteca, and the Teotihuacano have facial features resembling the Amurians. That first passage could have occurred during the Wisconsinan glaciation that lasted from 70,000 to 30,000 years B.C., a date that agrees with the oldest ages of human fossils found in America (Carmona, 1993). The Mongoloid immigration perhaps included the forerunners of the Olmecas and the Maya and probably occurred around the year 12,500 B.C., at the end of the Woodfordiense interglacial period.

The first contacts of proto-American *homo sapiens* with geothermal phenomena probably happened in Southwestern Alaska, with nomads passing through the northern portion of the

"Ring of Fire," an area between the North American and the Pacific plates. As new populations advanced southward into the Cascade Range, they found numerous active volcanoes and other geothermal manifestations. Eventually, their slow expansion reached the imposing manifestations of what is today Yellowstone National Park and The Geysers Geothermal Field in the Northwestern United States. Continuing southward into the enormous deserts of Arizona, New Mexico, and Texas, some ancestral nomads arrived at the great morphotectonic depression between the Imperial and the Mexicali Valleys at the Mexican border with the United States. By now, familiarity with geothermal resources should have been high.

Archaeological evidence indicates that North American Indians lived near hot springs for over 10,000 years and that all kinds of geothermal manifestations were sacred places for them, sites of refuge for rest and recuperation from warfare, havens for the elderly, and areas for cooking (Lund, 1995). The Indians believed that the Great Spirit lived at geothermal sites, a being who provided heat and "the rejuvenating warmth of mother earth" (Lund, 1995). The breath of the Great Spirit was manifested through steam discharges and fumaroles.

In the remote and distant past, prehistoric peoples used volcanic rock, such as basalt, silica, and obsidian, to manufacture weapons and tools. Some 3,000 to 3,500 years ago, with the beginning of stable agricultural settlements, the first uses of natural heat arose. Tales and legends from pre-Hispanic sources tell of the use of thermal waters and mud to clean and cure the body, to temper arrows and lances for hunting and war (Suárez and Cataldi, 1993). Food was toasted or steamed by very hot stones. Ears of corn, potatoes, and meats were cooked this way.

The first archaic groups in North America spread slowly into the central and southern parts of Mexico, seeking improved soils and climates. Yarza de la Torre points out that volcanic activity in Mexico generally began about 62 to 63 million years ago in the Tertiary era, decreasing first in the Western Sierra Madre and lasting longer in the region between 18° and 22° latitude (Burgassi et al., 1992).

This region, called the Mexican Volcanic Belt, forms an east-west band across Southern Mexico. It is characterized by an imposing series of 3,000 volcanic phenomena, 10 of which are active today. The sector is particularly rich in active geothermal manifestations such as thermal springs, fumaroles, gas exhalations, mud lakes, and volcanoes, and in hydrothermal deposits such as kaolin, sulfur, and iron oxide. This temperate region, with an altitude of 1500 to 3000 meters, became a choice area for settlements (Ruiz, 1987). Here, in the heart of Mesoamerica, a deep relationship grew between people and geothermal phenomena as civilizations evolved to high cultural levels through the centuries.

American civilizations developed apart from other influences, separated from Europe by the Atlantic Ocean and from Asia by the Pacific; they remained isolated for thousands of years from the discoveries and cultural mixtures of the Old World, and this separation made them unique.

Although many different cultures developed in Mesoamerica, one can speak of a Mesoamerican civilization because a uniformity developed, a shared cultural inheritance. All Mesoamerican peoples had a common mental horizon: similar religious beliefs, a theocratic-military social structure, and a shared general concept of the cosmos. Mesoamerican peoples revered the same gods but gave them different names. Their cult objects represented things captured by the senses: volcano, sun, moon, wind, fire, water, corn, and then death—the end of the sensate realm and the start of the abstract. Mesoamericans knew peak periods and decadence, ascent and destruction, with no break in global continuity of a civilization built in a universe deeply religious, recurrent, circular, where all returns to its origin and everything begins again, uninterrupted by the outside world (Paz, 1992).

The history of Mesoamerica begins with the Prehistoric period, 10,000 B.C. to 2000 B.C.

- Formative or Preclassic, from 2000 B.C. to 100 B.C.
- Classic, from 100 B.C. to 900 A.D.
- Postclassic or Historic, from 900 A.D. to August 13, 1521. On this day Tenochtitlán surrendered to the Spanish invaders and the great destruction began (Sodi, 1992).

Mesoamerican peoples settled into permanent communities once they began cultivating corn. They were outstanding painters, sculptors, artisans of gold and copper, architects, poets, and creators of feathered embroideries and codices.

Religion that included sacrifice formed a basis for all Mesoamerican cultures, and rigor and asceticism were underlying concepts. The Aztecs called their supreme god Ometeótl, the Lord of the Duality, with feminine and masculine aspects. The double divinity was the creative principle, the origin of all beings and of all things, and the infernal counterpart was the Lord and

Lady of Death. Such dualism is found at the base of Mesoamerican thought and is the root of Mesoamerican cosmology and ethics.

The relationship between humans and geothermal resources grew in America autonomously but more slowly than in other parts of the world (Cataldi et al., 1992a and 1992b). Investigations and archaeological discoveries prove that 3,000 years ago, communities existed in many areas with geothermal manifestations and at the feet of all volcanoes in Mesoamerica.

Thus we can infer that from very remote times, people established a fundamental—and sometimes dramatic—relationship with volcanoes, geothermal surface manifestations, and with geothermal phenomena in general. Investigators of two Costa Rican volcanoes, Arenal and Miravalles, showed that life in the first settlements was regulated by volcanic eruptions. The eruptions drove off populations, which then returned to rebuild, drawn by the rich soil and other attractions in geothermal areas (Burgassi et al., 1992).

Ceramics discovered at Miravalles indicate the presence of an agrarian population from the year 800 B.C. Sepulchral badges, shaped as very thin plaques, were extracted from laminar formations of volcanic rock and used by pre-Hispanic populations between 800 and 500 B.C. (Burgassi et al., 1992). The geothermal zone of La Primavera, Mexico, is in a volcanic caldera with multiple domes, hot springs, and fumaroles. One of the domes, named Colli, contained obsidian arrowheads, and the hill is consecrated to the god of fire (G. Tibon et al., 1967). A piece of obsidian was found that characterizes Iztlacoliuhqui, "the one twisted of obsidian," the god of punishment.

Various indirect references link Mesoamerican peoples to geothermal resources. Most consist of names of places, gods, and narratives about religious myths (Suárez and Cataldi, 1993; Hodgson, 1995; Hernández et al., 1995). Some traditions and legends were transmitted orally and some were transcribed by Spanish friars. Many others, like most of the codices, were hopelessly lost. As an example, the indigenous informants of Sahagún, a Spanish historian, told him in about 1534 how travelers had arrived from remote regions, through the coasts of the Gulf of Mexico, coming to settle in a mythical place named Tamoanchán, where the Náhuatl culture first flourished:

> "Here is the story that the old men used to tell:
> 'In a certain time that no one can any longer recount,
> and which now nobody can remember....

> Over the water, in their boats they came, in many groups,
> And arrived at the edge of the water, on the north coast,
> And where their boats were left, is called Panutla....
> Off they went immediately following the water's edge,
> Searching for the white and smoking mountains....'"

Why did these people seek to settle near volcanoes? The informants never said. The story suggests a deliberate decision, typical of the relationship between mankind and geothermal phenomena in America.

Cooking with pots in steam began with the development and diffusion of pottery, probably starting from the Late Formative period from 800 to 100 B.C. The pots were placed upon boiling water, on spouts of vapor, or on hot ground—a method still used in the geothermal zone of Ixtlán de los Hervores in Central Mexico.

Cooking food with steam is a traditional use of natural heat in Mexican geothermal areas. The Náhuatl term *Atotonilco*, which means "place of the boiling waters" (*atl*, "water"; *totonil*, "hot"; *co*, "place"), is a word used throughout Mexico. Twenty-five ancient villages possess this name referring to thermal waters, to their healing, divine, and magical properties. We can infer that, at least a millennium before the Spanish Conquest, many human settlements grew up here because geothermal manifestations were nearby. In fact, geothermal resources were used for cooking and cures throughout Central America (Suárez and Cataldi, 1993).

From time immemorial, American peoples took vapor baths by throwing cold water on hot rocks taken from the mouths of hydrothermal manifestations. They socialized, enjoying the vapors. Eventually, hot springs and thermal muds were used medicinally at some towns with the name Atotonilco and throughout the Mexican Volcanic Belt.

With time, some Mexican peoples began to use a structure called the *temazcal* for religious purposes (González, 1991). The *temazcal*, a dome constructed with volcanic stone and mortar, was warmed from the outside by showering cold water on piled rocks, heated by fire. The narrow corridor at the entrance was adorned with an image of Teteo Innan, the great goddess of the Earth and of childbirth, and the *temazcal* was used by pregnant women and babies.

Soon thermal baths were used widely in Mesoamerica. There were hundreds, all assigned with medicinal and divine properties with thermal waters between 20° C and 40° C.

Mexico had four famous gardens before the Conquest, and the most grand and beautiful was Oaxtepec. Located south of Mexico City, in the valley of Cuautla, it was founded by the Olmecas and preserved by all successive conquerors, finally becoming the summer residence of Aztec kings and the foremost botanical garden on the continent. One Aztec ruler, King Moctezuma I, sent messengers throughout the empire with orders to fill the garden with the most beautiful and rarest of plants, trees, and medicinal herbs. By the reign of King Moctezuma II, Oaxtepec had reached its greatest splendor and the emperor built temples, rooms, gardens, pools, and many *temazcal* baths for his noblemen and warriors.

The garden had abundant sulfurous thermal waters at temperatures of about 25° C that Moctezuma's warriors used for healing. The waters made a strong impression on the Spanish when they saw them (Bernal, 1532). The thermal waters were used, as well, to irrigate the garden. Oaxtepec was the first pre-Hispanic garden cultivated in Mesoamerica by a method known today as geothermal hydroponics.

Many Aztec structures were built with blocks of volcanic rock, often basalt. The upper classes of Aztec society made their homes with *tezontle*, a hard volcanic rock extracted from stony ground in what is today southern Mexico City. Mesoamerican minerals related to geothermal processes were used for weapons and tools and held mythical and religious meanings. For example cinnabar, the principal ore of mercury often associated with geothermal regions, was used to redden the bones of important personages, providing them with heat in the afterlife. Tezcatlipoca, one of the more important divinities, was characterized by a sculpture in obsidian and given the name of "smoky mirror." Obsidian was associated with death, magic, war, and fertility.

Mesoamerican Cosmogony and Calendar

For the Nahuas, the original supreme being was Cipáctli, a feminine monster that inhabited the sea. Two gods converted into serpents cut Cipáctli in two portions. From the first half they made the underground world and from the second, heaven. For the Maya, the cosmos was a structure formed by horizontal planes in which sacred powers lived who decided the destinies of all. For both cultures, the divine energies were land, corn, volcanoes, wind, rain, lightning, sun, and stars.

Two stable, visible bodies in the heavens are the sun and the Milky Way. For the Aztecs, Huitzilopochtli, the god of war born on Earth, represented the sun. His dismembered sister, the goddess Coyolxauhqui, characterized the Milky Way (Aguilera, 1979). This divine couple in the Aztec pantheon was considered the founder of the universe.

For the Nahuas, the Earth was a cylinder surrounded by water. The universe originated in the center of the Earth, in its navel where the god of terrestrial fire resided. From that navel, the four cardinal points were drawn. Above the Earth existed thirteen heavens and below were nine underground hells. Through the first heavens went the courses of the sun, moon, stars, and comets. The thirteenth heaven was the site of the divine duality, Ometéotl, the principle of generation and conservation for the cosmos, who controlled the universe with fire, earth, water, and air through the navel of the world.

Water was always related to the origin of Mesoamerican people, the basic element in one of the first epochs of their existence (León Portilla, 1992). At the horizon, the waters of the Earth joined the waters of heaven to envelop the world. That world was called *Cemanahuac*, "that which is surrounded by water." The cosmic meaning of water was so important that in the Náhuatl language, *atl* ("water") came to represent the phoneme "a." Tláloc, the god of rain, was worshiped throughout Mesoamerica, called "Chac" by the Maya, "Cocijo" by the Mixtecas, and "Tajín" by the Totonacas. In Náhuatl, "city" is translated as *atl in tepetl*, literally, "water-mountain" (León Portilla, 1992). Pilgrimages were made to volcanoes for several festivals marking dates on the solar calendar. The existence of lakes in craters of some old volcanoes reinforced the belief in the relationship volcano-water, which implicitly contains the idea that the mountains are large water reservoirs, kept by the gods of rain.

> "All the high mounts, where the clouds join to make rain, are gods.
> To each one of them an image is made....
> Like the image of the volcano Popocatépetl ('smoky mountain')
> Or of her whose name is Iztaccíhuatl ('white woman'),
> Or the image of the mount Poyauhtécatl ('that which is from the region of the fog')...."
>
> —*From the* Florentine Codex

Ancient Mesoamericans believed that man and woman were necessary collaborators in the living universe. Both depended on nature in the same way that nature depended on them. Heat was the force that sustained life. Heat was equal to life; cold was equal to death.

In many chronicles, in the oral tradition, in indigenous manuscripts, and on reliefs of pre-Hispanic monuments, the idea is found that our era was preceded by four prior eras—named suns by the Aztecs. Each era ended with the destruction of the world by different, terrible cataclysm generated from the energy of the Earth. The bygone eras were those of *atl* ("water"), *ocelotl* ("jaguar"), *quiahuitl* ("rain"), and *ehecatl* ("wind") (León Portilla, 1961). Each era had great mythical importance, and the meaning and the reason for the end of each follow.

The five suns, or eras, in the center of the Aztec calendar. *Reprinted with permission from* Geotermia, Revista Mexicana de Geoenergía, *vol. 9, p. 33, published by the Comisión Federal de Electricidad, Mexico*

Atl-Tonatiuh ("Water Sun"). The first people were made from ashes. In the first catastrophe, humanity was destroyed by water in the form of floods, and people were converted into fish.

Ocelotl-Tonatiuh ("Jaguar Sun"). In the second era, the sun stopped at noon, interrupting its path. Suddenly night appeared and heaven was oppressed. Now giants lived, who, in spite of great corpulence, were weak. When they fell by accident, they fell forever.

Quiahuitl-Tonatiuh ("Rain of Fire Sun"). In the third era, the rock boiled, burning the people in a rain of fire streaming down from volcanic eruptions (Tlequiahuitl) that destroyed the world. People in this era met a tragic end, changed into turkeys. The date 4-Quiahuitl, which records the end of this time, is placed under the protection of Tláloc, god of rain. He was also the god of fire, which fell from the sky as lightning rays and as volcanic eruptions (León Portilla, 1983).

Hieroglyphic of Fire

In the Aztec hieroglyphic for the third era, the "Rain of Fire Sun," the god of volcanic fire, elegantly arrayed and with lava pouring from his mouth, falls face downward between adjacent volcanic slopes outlined in escaping tongues of lava. He will envelop in flame a man and woman sitting beneath him in the protective circle of their home, still unaware of his descent. Turkeys fly around the couple, presaging the end of the third era when mankind is changed into turkeys.

Ehecatl-Tonatiuh ("Wind Sun"). During this era, everything was destroyed by the wind in the form of terrible hurricanes, snowfalls, and glaciers. Humans were converted into monkeys and scattered through the mountains. At the end of the fourth era, the gods met in Teotihuacán to recreate the world. With the sacrifice of one god, who flung himself into the fire of an active volcano, the Fifth Sun emerged.

Pyramid of the Sun, Teotihuacán, Mexico. *M. C. Suárez Arriaga*

Ollin-Tonatiuh ("Movement Sun"). This was the last of the eras, the era of the Fifth Sun in which we live and one that will be devastated by earthquakes. The divinity of the Fifth Sun is Xiuhtecutli, god of fire, who links the heat of the Earth with that of the heavens.

Our era is represented on the Aztec calendar by the date 4-Ollin, which means "movement and earthquake." The date 4-Quiahuitl, "end of the Rain Sun," is placed under the protection of Tláloc, god of rain. The third era ends in a rain of fire; Tláloc is not only the god of rain, but also the god of fire falling from the sky in the

Ages or Suns That Have Existed

It was related so, it was said
that 4 lives came before,
And that this is the Fifth Age.
Just as the old men knew it, in the Year 1-Rabbit
The Earth and the Heavens were founded.
And they knew it thus,
That when the Earth and Heavens were founded,
Four kinds of men had existed,
Four kinds of lives.

They also knew that each one of them
Had existed in a sun (in an age).
And they said that their god made the first men,
He forged them from ash.
This they attributed to Quetzalcóatl,
Whose sign is 7-Wind, he made them, he invented them.

The First Sun (Age) that was made,
Its sign was 4-Water and it was called the Sun of Water.
In it happened that water carried everything away.
The people became fish.

Then the Second Sun was made.
Its sign was 4-Jaguar, it was called the Sun of Jaguar.
In it happened that the heavens were weighed down,
The sun did not follow its course.
When the sun arrived at noon,
It then became night, and when it was dark

cont....

The jaguars ate people.
And giants lived in this sun.
The old men said that the giants greeted them thusly:
"Don't fall,"
because he who fell, fell forever.

Then the Third Sun was made.
Its sign was 4-Rain.
It was called the Sun of Rain (of Fire).
It happened that fire rained during it.
Those who lived in it were burned.
And during it sand also rained
And they said that in it
The little pebbles rained that we see
And that the *tezontle* stone boiled
And then the rocks turned red.

Its sign was 4-Wind, and the Fourth Sun was made.
It was called the Sun of Wind.
During it everything was carried away by the wind.
Everyone became monkeys.
They spread themselves through the mountains,
The ape men went away to live.

The Fifth Sun, 4-Movement its sign.
It is called the Sun of Movement,
Because it moves, it follows its course.
And as the old men say,
And in it will be earthquakes,
There will be famine and so we will perish.

In the Year 13-Reed, it is said that it came to exist,
The sun was born that exists now.
Then was when it illuminated, when it dawned,
The sun of movement that exists now.
4-Movement is its sign.
This is the Fifth Sun that was formed,
In it will be the movement of the Earth,
In it there will be famines.
This sun, its name 4-Movement,
This is our sun in which we live now,
And here is its trace
As the sun fell in the fire,
In the divine fire-box there in Teotihuacán.
Equally this was the sun
Of our prince in Tula,
That is to say, Quetzalcóatl.

English translation by M. C. Suárez Arriaga and S. Hodgson, from the Spanish translation by León Portilla in 1983, one of 11 Spanish translations of the poem. Transcribed in Náhuatl in 1558. (The dates in the poem correspond to the Aztec calendar.)

form of lightning and volcanic eruptions, equal to Tlequiahuitl, the rain of fire. Thus in four attempts, the world was born and destroyed from gigantic catastrophes, and this will happen once again.

In Mesoamerica the origin, existence, and development of the universe were the efforts of divine beings and came from divine energy in action. Therefore, religious understanding formed the basis of reality and Mesoamerican beliefs about geothermal manifestations were filled with fervent religious meaning with positive and negative dualities: both beneficent and harmful, constructive and destructive. The characteristics of three geothermal divinities illustrate this paradox.

Huehuetéotl was one of the principal gods, the oldest one, father of the other gods. He was the divinity who controlled fire, living in the "navel of the world," that is to say, at the center of the Earth. At the same time, he was the nearest

Top view of the basalt Coronation Stone of Moctezuma II, Stone of the Five Suns. The stone is remarkable for its colossal size (55.9 x 66 x 22.9 cm), technical refinement, and iconographic complexity. The Aztecs believed that four imperfect eras of the universe, each named after its destruction date, had preceded the present age. Located at the stone's four corners and read counterclockwise from the lower right, they are 4-Jaguar, 4-Wind, 4-Rain, and 4-Water. The name of the present era, 4-Movement, represented by the carved X at center, predicts a cataclysmic end by earthquakes.

Encased in a square cartouche below the 4-Movement sign, the year 11-Reed corresponds to the year 1503, while the 1-Crocodile carved above the sign corresponds to July 15, together the coronation day of Moctezuma II, the last Aztec ruler before the Spanish arrival.

From Tenochtitlán, Mexico, Aztec culture, c. 1503. Major Acquisitions Centennial Endowment, 1990.21.
Photograph by Robert Hashimoto. ©1996, The Art Institute of Chicago, all rights reserved

The underside of the Coronation Stone of Moctezuma II, Stone of the Five Suns, proclaiming the ruler's title to the Earth. The figure on the base represents 1-Rabbit, the mythical date marking the beginning of the present era in the Aztec calendar. The stone's dimensions are 55.9 x 66 x 22.9 cm.

From Tenochtitlán, Mexico, Aztec culture, c. 1503. Major Acquisitions Centennial Endowment, 1990.21. *Photograph by Robert Hashimoto.*

to humans, *Tloque Nahuaque*, "sir of the neighboring vicinity," of the immediate present, who lived directly with people.

To the Aztecs, Huehuetéotl was fire itself, which came from volcanoes, the heat of terrestrial origin and the oldest principal generator of life who resided in the center of the Earth. The representation of Huehuetéotl as an old man suggests the antiquity of the mountains and the volcanoes. The age of this god underlines the importance given by the Mesoamericans to geothermal heat.

Huehuetéotl, the old god of fire, one of the most ancient deities in Mesoamerica. During ceremonies, fire was kindled in the brazier on his head. The volcanic stone sculpture, 19 inches high, is from Teotihuacán, Mexico, 250 – 750 A.D.
Natural History Museum of Los Angeles County

"Mother of the gods, father of the gods, the old god,
The one that is in the navel of fire,
The one who is confined in turquoise."

—From Náhuatl elders, León Portilla, 1956

Huehuetéotl was also the god of time. The museum of the great temple in Mexico City has basalt sculptures of the god, characterized in a similar manner by the Aztecs and the Teotihuacanos. The two snails on the head are located above streams of water, clearly related to Tláloc, god of rain and volcanic eruptions.

Tláloc was at the same time malevolent and beneficent: he enriched the soil with water but brought floods and disasters from hail, thunder, lightning, and volcanic eruptions. He represented a rain of water or a rain of fire. Chalchiuhtlicue was his feminine part, the goddess of running waters.

Tlazoltéotl was the Aztec goddess of carnal love and birth but also the goddess of vapor baths and confessions of illicit relationships. She was the goddess of refuse, lewdness, and spiritual impurities; she also was named *Cihuacóatl* (“woman-serpent”), a form of the goddess of the Earth presiding over part of the underworld, the world of the dead, the sower, and childbirth.

Mesoamericans were obsessed by the flow of time and the oldest known writings of the Zapoteca and Maya (600 to 400 B.C.) are closely related to the calendar (Joyce, 1979). The Mesoamerican calendar has existed since 700 B.C., structured as a complex measuring system based on the idea of time as flow. The Aztecs called time *Cáhuitl*, “that which is leaving us” (León Portilla, 1961). Time began outside of human scope, at the beginning of the universe when vital activities of the gods created this world. The order in which the mythical actions occurred gave birth to the calendar, and every new being is named for the day it was created (López, 1990).

Cosmology is defined as a study of the philosophy of the universe as an ordered whole. Given the terrestrial nature of the oldest gods and the worldly origin of the sun and Earth, the symbols of the Aztec calendar representing days, months, years, and centuries reflect the influences of geothermal phenomena.

The calendar’s cycles were of distinct dimensions and significance. The major cycle of 52 years was attached to the Pleides constellation, at whose completion was celebrated the festival of New Fire. It was related to terrestrial heat and probably represented periodical eruptions or earthquakes. This cycle was divided into four groups of 13 years each. The ritual activities corresponded to the agricultural cycles and, given that the calendar was based on the solar year, the principal function of the religious cult was to regulate social and economic life.

The premier ring of 20 days shows the importance ancient Mesoamericans placed on concepts related to geothermal processes: rain (water, fire, or eruptions), movement (earthquake), and serpents (related to the Earth and to things beneath it). Most surprising is the concept related to water (*atl*), essential in all forms and whose patron saint was the god of fire, Xiuhtecutli (González, 1991). The Aztecs used the words *atl tlachinolli* to say “water in fire,” for which the geothermal relationship is evident.

The millennia of experience and wisdom accumulated by the Mesoamericans led them to a balanced existence between their societies and nature. Because of this, they revered water, corn, the volcano, the serpent, and the hummingbird. Thus Mesoamerican peoples studied and developed natural phenomena, including geothermal resources. Natural heat and its by-products, plus close observations of geothermal phenomena, influenced the shape of their cultural, ethical, religious, and social structures.

Selected References

Aguilera, C. 1979. Coyolxauhqui y la vía láctea. *CONACYT—Revista Ciencia y Desarrollo,* 24, 6-10.

Bernal, Díaz del Castillo. 1532. *Historia verdadera de la Conquista de la Nueva España.* Reprint 1961. Fernández Editores, Mexico, D.F.

Bernal, I. 1979. Una ciencia muy nueva, del Siglo XIX. Interview by M. García. *CONACTY—Revista Ciencia y Desarrollo,* 24, 22-26.

Broda, J. 1990. Calendarios y astronomía en Mesoamérica, su función social. *CIENCIAS—UNAM,* 18, 36-39.

Burgassi, P. D., R. Cataldi, J. L. Hernandez Galán, M. Moggi, R. Rubinovich, and J. Saldaña. 1992. Primera contribución al conocimiento historiográfico de la energía geotérmica en el area Mediterránea y en la America Latina. Contribution to *Simposio sobre la história de la geotermia,* Ciudad de Mexico. (This work has been published as a limited edition in the "grey literature"; it is available for consultation at selected places. Information on locations can be obtained from the first three authors of the publication).

Carmona M., M. Gómez, and S. Gómez. 1993. *El origen del hombre en América.* García and Valadés, eds. 2nd ed. Museo Nacional de Antropología e Historia, Mexico, D.F.

Cataldi, R., P. D. Burgassi, and C. Suárez. 1992a. La geotermia en el período Precolombino en las áreas Mediterránea y Mesoamericana. *Geotermia, Revista Mexicana de Geoenergía,* 8, no. 2, 145-175.

Cataldi, R., P. D. Burgassi, and C. Suárez. 1992b. La geotermia en el período Precolombino en las áreas Mediterránea y Mesoamericana. *Geotermia, Revista Mexicana de Geoenergía,* 8, no. 3, 251-303.

De la Garza, Mercedes. 1990. Cosmovisión de los Mayas antiguos. *CIENCIAS—UNAM,* 18, 33-35.

González, Yolotl. 1991. *Diccionario de mitología y religión de Mesoamérica.* Larousse. Mexico, D.F.

Gutierre Tibón and Martínez Negrete. 1967. *Mujeres y diosas de México, parviescultura Prehispánica en barro.* Instituto Nacional de Antropología e Historia, México, D.F.

Hernández Galán, J. L., J. Guiza, and C. Suárez. 1995. An overview of the historical aspects of geothermal energy in Mesoamerica. *Proceedings of the World Geothermal Congress,* Florence.

Hodgson, S. F. 1995. Heat over time, geothermal stories from Mexico. *Proceedings of the World Geothermal Congress,* Florence.

Joyce, M. 1979. Los orígenes de la escritura Meso-americana. *CONACYT—Revista Ciencia y Desarrollo,* 24, 35-52.

Krickeberg, W. 1956. *Altmexikanische Kulturen.* Springer Verlag, Berlin.

León Portilla, M. 1956. La filosofía Náhuatl, estudiada en sus fuentes. In *Monografía 10.* 7th ed. Instituto de Investigaciones Históricas, Serie Cultura Náhuatl, Universidad Nacional Autónoma de México, Mexico, D.F.

León Portilla, M. 1961. *Los antiguos Mexicanos a través de sus crónicas y cantares, colección popular no. 88.* Fondo de Cultura Económica, Mexico, D.F.

León Portilla, M. 1983. Origen del nuevo sol en Teotihuacán, informantes de Sahagún, códice matritense del real palacio. In *Antología de Teotihuacán a los Aztecas. Fuentes e Interpretaciones Historicas, CIENCIAS—UNAM,* 57-8.

León Portilla, Miguel. 1992. El agua: universo de significaciones y realidades en Mesoamérica. *CIENCIAS—UNAM,* 28, 7-14.

López, A. A. 1990. El tiempo en Mesoamérica. *CIENCIAS—UNAM,* 18, 28-34.

Lund, J. W. 1995. Historical impacts of geothermal resources on the people of North America. *Geo-Heat Center Quarterly Bulletin,* 16, no. 4, 7-14.

Museo Nacional de Antropología e Historia. 1996. *Personal communication.*

Paz, O. 1992. *El águila, el jaguar y la virgen.* CoNaCultA, Mexico, D.F.

Ruiz, F. 1987. *Los antiguos Michoacanos: visión de la prehistoria a la Conquista Hispana.* Morelia, Mich., Mexico.

Ruiz G., Juan Carlos. 1992. Los minerales en la mitología Mexica. *CIENCIAS—UNAM,* 28, 15-21.

Sahagún, B. de. 1575. *Historia general de las cosas de Nueva España.* Reprint 1956, Porrúa, ed. Mexico.

Sodi, D. M. 1992. *Las grandes culturas de Mesoamérica.* Panorama, Mexico, D.F.

Soustelle, J. 1940. *La pensèe cosmologique des anciens Mexicains.* Hermann, Paris.

Soustelle, J. 1955. *La vie quotodienne des azteques a la veille de la Conquete Espagnole.* Lib. Hachette, Paris.

Suárez, M. C. and R. Cataldi. 1993. La geotermia en Mesoamérica y regiones colindantes. *Geotermia, Revista Mexicana de Geoenergía,* 9, no. 1, 5-58.

Vaillant, G. C. 1941. *Aztecs of Mexico, Origin, rise and fall of the Aztec nation.* Doubleday & Company, Inc., Garden City, New York.

The Authors:

Mario César Suárez Arriaga
Mexican Geothermal Association
Michoacán University
E-mail: msuarez@zeus.ccu.umich.mx

Raffaele Cataldi
55, Via del Borghetto
56124 Pisa, Italy
Telephone/Fax: 39.050.59.81.07
E-mail: rafcat@tin.it

Susan F. Hodgson
Ca. Dept. of Conservation
Division of Oil, Gas,
and Geothermal Resources
801 K Street, MS 20-20
Sacramento, CA 95814-3530 USA
Telephone: 916.445.9686
Fax: 916.323.0424
E-mail: shodgson@consrv.ca.gov

Basalt sculpture from Mexico of a female deity reflected in an Aztec obsidian mirror with a wooden frame, 10 ¼" in diameter. ***Courtesy Department of Library Services, American Museum of Natural History (neg. 323348; photograph by L. Boltin, October 1953)***

32. An Overview of the Historical Aspects of Geothermal Influences in Mesoamerica

by
José Luis Hernández Galán
Jorge Guiza Lámbarri
Mario César Suárez Arriaga

Abstract: Mesoamerica is a zone with ubiquitous geothermal manifestations that have always influenced the lives of its inhabitants through the increased fertility of volcanic soils and the damage inflicted on lands, settlements, and crops by volcanic eruptions and earthquakes. Some materials of igneous origin, like obsidian, were of utmost importance for the region's peoples. However, the influence of geothermal phenomena went beyond material implications to affect even religious matters.

The Europeans coming to this area reacted to the geothermal manifestations with awe, which was soon followed by attempts to explain their characteristics with scientific theories current at the time or to exploit them. We summarize the interactions, both on utilitarian and nonmaterial planes, between mankind and geothermal phenomena in Mesoamerica.

Reprinted by permission of the International Geothermal Association, from the *Proceedings of the World Geothermal Congress,* 1995, pp. 423-426.

Introduction

Industrial exploitation of geothermal resources did not start in Mesoamerica until the second half of the 20th century, so we will refer to geothermal energy in its broadest sense, mentioning the utilization its inhabitants made of the Earth's heat or igneous materials, and also the way in which phenomena, such as earthquakes or volcanic eruptions, influenced the lives of those peoples, their societies, and their beliefs.

To establish the early history of geothermal influences in Mexico, Central America, and the Antilles is a very difficult task for various reasons, mainly because our sources are fragmentary and often distorted. Besides, it was not until the studies of Knorosov and Proskouriakoff that Maya inscriptions were deciphered properly. Before that, we had only second-hand transcriptions made by early missionaries who considered the beliefs of the aboriginal populations to be the work of demons. On the other hand, although the native inhabitants of

Mesoamerica left proof of their perception of geothermal phenomena in their place names and in their theogony, they were not much interested in utilitarian aspects, so they left very few records of the world that surrounded them.

The Peopling of Mesoamerica

In 1589, the Spanish Jesuit José de Acosta advanced the idea that American Indians had first come from Asia by land and, perhaps, by a short sea voyage. This is still the opinion commonly held, with the majority of experts dating the first human occupation to about 11,000 years ago, and some of them fixing it as far back as 25,000 years ago.

In any event, America was first colonized by modern humans that had already acquired many skills and presumably had developed systems of beliefs. Evidence from archaeological findings, anthropological measurements, analyses of mitochondrial DNA, and studies of comparative linguistics all agree that Mesoamerica was originally settled by the *Amerind* peoples on their migrations from Asia. After an archaic epoch in which the land was settled for the first time, the interval of human occupation is divided into three periods. The earliest one, termed *Preclassic* (or *Formative*), was from about 4,000 to 1,800 years ago. The next period, the *Classic,* extends to about 1,100 years ago, and the *Postclassic* period ends with the Spanish Conquest in the 16th century.

The Region

Mesoamerica is a land of very intense tectonic activity, so its first colonizers would have become acquainted with seismic, volcanic, and geothermal manifestations very soon. Chronologically, their first contact with these phenomena should have occurred in the Imperial and the Mexicali Valleys, where they could not have overlooked the conspicuous hot springs, mud volcanoes, steam jets, and mineral efflorescence. Similar manifestations also should have been evident to those arriving at the Neovolcanic Belt further south, with its 3,000 volcanic cones, abundant hydrothermal deposits, and numerous hot springs (Suárez Arriaga and Cataldi, 1993).

In Central America, the situation is comparable: Guatemala, with peaks such as Santa María and Fuego and lakes formed in volcanic craters, like Atitlán and Amatitlán; Honduras, with Conchagua and others; El Salvador, with its central valley so affected by seismic activity that it is known as the *Valley of the Hammock*; Nicaragua, with Momotombo, Negro, and Santiago volcanoes; and Costa Rica, with the steaming grounds of Rincón de la Vieja and the peaks of Poas and Irazú, to name only a few from a long catalogue of tectonic and thermal manifestations. We know that they were of importance to the early settlers in the region, because close to the manifestations it is common to find offerings dating from the very early Preclassic period.

From the ***Popol Vuh, The Mayan Book of the Dawn of Life***, written in Quiché by Maya elders at the time of the Spanish Conquest.

> "And this Seven Macaw has two sons: the first of these is Zipacna, and the second is the Earthquake.
>
> "And this Zipacna, this is the one to build up the great mountains: Fire Mouth, Hunahpu, Cave by the Water, Xcanul, Macamob, Huliznab, as the names of the mountains that were there at the dawn are spoken. They were brought forth by Zipacna in a single night.
>
> "And now this is the Earthquake. The mountains are moved by him; the mountains, small and great, are softened by him."

Note: *Fire Mouth* is thought to be Volcán de Fuego, 19 km southwest of Antigua; *Hunahpu* to be Volcán de Amatenango, 5 km north of Volcán de Fuego; *Cave by the Water* to be Volcán de Agua, 11 km south of Antigua; and *Xcanul* to be Volcán Santa María, 9 km south of Quezaltenango. *Xcanul* is also the generic term for *volcano* in the western part of the Quiché-speaking region.

John L. Stephens wrote in 1834 on a trip to Antigua, "The last eruption of the Volcán de Fuego took place about 12 years ago, when flames issued from the crater and ascended to a great height...but it can never burst forth again; its crater is no longer la Boca del Infierno, or the Mouth of the Infernal Regions, for, as a very respectable individual told me, it has been blessed by a priest."

Influence of Geothermal Activity on Mesoamerican Peoples

WEATHERED VOLCANIC SOILS, BECAUSE OF THEIR ENHANCED FERTILITY AND ABILITY TO RETAIN moisture, allow the growth of a dense and varied flora that supports abundant animal life. These elements make an area attractive to people, from nomadic hunter-gatherers to intensive farmers (Grayson and Sheets, 1979). The end result is that people concentrate around places with geothermal activity and, in so doing, expose themselves to potentially disastrous volcanic eruptions and earthquakes. The impacts of these events on the environment and on the societies established around them vary, depending on the magnitude of the event and its precise nature and characteristics. The impacts also depend on components related to the social organization of the settlers, their hazard perception, land use, and so on.

The pre-Columbian Mesoamerican societies were no exception. The influence of geothermal activity is exemplified by the volcanic eruptions of Xitle, Arenal, and Ilopango. The latter's eruption five centuries ago may have ended southeastern Maya civilization and allowed people from Northern Guatemala to seize lucrative trade routes. Some Maya lowland sites, like Barton Ramie in Belize, apparently received refugees from that event, whose sudden arrival may have produced not only agricultural adaptations, but greater political centralization and class differentiation as well (Sheets, 1979).

The Maya living 1,400 years ago in Cerén, in what is now El Salvador, also experienced the impacts of geothermal activity. A rapid series of explosive eruptions took place within two kilometers of their village when rising magma encountered ground water. Buildings were demolished and then buried by ash and pumice, preserving details of domestic Maya life: dirty dishes, pots full of food, and footprints in ash—all suggesting people fled the town (Fink, 1993). These cases agree with the studies made in Costa Rica by Sheets (1984) at Arenal, and Hurtado de Mendoza and Alvarado (1988) at Miravalles regarding the possible influences of volcanic eruptions on the socioeconomic, agricultural, and artisanal development of the pre-Columbian local inhabitants. Their research has shown how human populations are driven away during and immediately after major volcanic episodes and then return, reattracted to the area.

Besides anthropological evidence, we have direct indications of how pre-Columbian Mesoamerican inhabitants were aware of geothermal phenomena. For example, many volcanoes in the area still bear autochthonous names, such as Iztaccíhuatl (*iztac,* "white"; *huatl,* "woman"), Popocatépetl (*popoca,* "steaming"; *tepetl,* "hill"), Citlatépetl (*citlali,* "star"; *tepetl,* "hill"); and some of the extant codices, like the *Nutall*, the *Tellerianus Remensis*, and the *Cuauhtinchán*, bear these representations in hieroglyphics. Many towns have Indian names referring to some geothermal characteristic of the place: for example, Atotonilco (*a[tl],* "water"; *totoni[li],* "hot"; *co,* "place") signifies "place in the hot water" in Náhuatl. In Mexico, there are about 25 villages with that name. Also, there is a town named Puruándiro and a zone called Pathé; both names mean "place in the hot water" in the P'urhépecha and Otomí languages, respectively.

Codices hold expressive hieroglyphics for place names and for Mexican volcanoes. For Atotonilco ("place in the hot water") a drawing is used of a pot placed over fire. The volcano Popocatépetl is a simple bell-shaped curve with lava erupting from the top. In one drawing, the lava, shooting straight into the air, resembles leaves colored in and fastened asymmetrically to a corn-like stalk. A second drawing is more free-flowing. The erupting lava "leaves" are outlined—not colored in, unaffixed to a stem, and depicted as tear-like curves shot out against the sky. Eruptions of Popocatépetl were caused by the "departed spirits of wicked rulers whose fiery agonies, in their prison house, caused the fearful bellowings and convulsions in times of eruption," according to a legend recounted by William Prescott.

Two hieroglyphics for Citlatépetl, which means "star hill," depict a plan view of the volcano. On the first, the caldera with erupting lava is shaped as an eight-pointed star, colored in, and engulfed by a decorative circle representing the slopes of the volcano. The second hieroglyphic of the volcano has two parts. The plan view just described is drawn at the top of a vertical line seemingly suspended in air and swaying slightly. At its base, the line enters a small rectangular shape with tiny asymmetrical blocks at the surface depicting an earthquake. The hieroglyphic reads, "Citlatépetl was smoking when the Earth trembled."

Influence of Geothermal Activity on Mesoamerican Religion

The overpowering magnificence of geothermal phenomena and their connections with areas of bountiful resources, or else with death and devastation, made the ancient inhabitants of Mesoamerica quite naturally associate these manifestations with powerful gods who had to be propitiated. Consequently, archaeological caches frequently are found near places of geothermal activity. For example, Stone (1941b) describes a place called *Peroles Calientes* in the Black River Valley on the northern coast of Honduras, which "must have been a very important one to the natives, for the quantity of pottery to be found there is immense."

Food seems to have been the principal article offered to the hot water gods, because pot upon pot was placed upright, some within others, with the best pieces directly positioned under a hot waterfall. On the southern flank of Miravalles volcano in Costa Rica, offertory remains were discovered close to a lava outcrop that the ancient inhabitants revered. An additional example is that of the cache found under a stratum of volcanic sand on the slopes of Irazú volcano, also in Costa Rica. It amounted to more than 400 objects, including male and female figures, offering tables, and heads made from volcanic rock, plus some 100 wooden articles preserved from weather and insects by the layer of volcanic sand (Stone, 1941a).

These examples do not show whether these people believed there was a god residing at the place with the geothermal manifestations, or if they worshiped the phenomena themselves. Then again they may have perceived the geothermal manifestations as attributes of a god or representing the forces and elements the god controlled. This is what we have with Huehuetéotl, the oldest of the gods, who controlled fire from his dwellings at the "world's navel" that is, the Earth's center, or with Xiuhtecúhtli, who united the fires of Earth and Sun. Tláloc was the rain god, but he could bring volcanic eruptions also, so he could produce rains of water or fire. Tezcatlipoca, whose name means "Smoking Mirror (Obsidian Mirror)," was the omnipotent deity of fate, both beneficial and destructive. Tlazotéolt Ixcuina was the goddess of carnal love and natality, who ruled over steam baths, and Itzpapálotl, whose name means "Obsidian Butterfly," was the goddess associated with Earth and fertility. These deities were not perceived by pre-Columbian peoples as clear-cut individuals, but as complexes of closely related gods or goddesses and attributes.

Raw Materials of Volcanic Origin

Many of the artifacts manufactured by the early Mesoamericans were made from volcanic materials: silica, flint, basalt, and very especially, from obsidian. Obsidian, a volcanic glass that fractures in flakes with sharp edges, was particularly good for making knives, spear or arrow points, scrapers, needles, and similar utensils. It is interesting to note that the Náhuatl name for obsidian is *itztli,* which means knife. Obsidian blades have been found at practically all Mesoamerican sites. When obsidian was not available locally, it was traded, as the following examples show. In El Salvador at Chalchuapa, the remains were discovered of a large shop for manufacturing knives. However, the obsidian flakes found there are from Ixtepeque volcano in Guatemala, some 50 kilometers away. Also, the obsidian knives found in Honduras at Copán were imported because no natural deposits of obsidian are known in that area.

Volcanic rock was used widely, as is exemplified by the cache already mentioned found on the slopes of the volcano Irazú, the tombs with lids made out of laminar volcanic formations *(lajas)*, and the sepulchral stelae carved on columnar basalt on the flanks of Miravalles volcano in Costa Rica (Ryder, 1983).

Indirect References

Spanish missionaries and chroniclers have left transcriptions of Indian legends and traditions that they collected, and in many of these one can find references to geothermal phenomena. For example, a myth recounting the migration of peoples searching for Tamoanchán says they "were looking for mountains: some white mountains and some smoking mountains." Would it be too farfetched to believe that these people were consciously trying to find for themselves a country fertilized by volcanic deposits and well irrigated by water from melted snow? In another example, a number of writers from the beginning of the Spanish colonization, like Fray Bernardino de Sahagún (1534) and Diego Muñoz Camargo (1540), tell in very similar words of signs that forewarned King Moctezuma II of the coming of the Spaniards. One of the omens, the *Fifth Prodigy*, seems to correspond to a phreatic explosion in the lake on the outskirts of the Aztec capital Tenochtitlán, now Mexico City (Burgassi et al., 1992).

The Aztecs believed that our world had passed through different epochs, called "Suns," and that each previous epoch ended with a cataclysm. They knew the last epoch as the *Fifth Sun*, which presumably is the one in which we are now living, and they said that "in it, there will be earthquakes." A prior epoch, the *Third Sun*, was the "Sun of the Fire Rain. And it happened that during it, there rained fire. And those who were living at the time burned out. And at the time, it rained sand also. And they say that in that Sun, it rained the pebbles that we now see, that the *tezontle* stone boiled and that the boulders reddened." It is hard not to conclude that we have here the description of a devastating volcanic event, one narrated by a frightened survivor and later transmitted orally for generations.

The Post-Columbian Era

The Mediterranean basin has been tectonically active since the most remote times, so the first Europeans to arrive in Mesoamerica were already acquainted with geothermal manifestations, either first-hand or by reference. Most of them probably had never been exposed to phenomena of the magnitude and frequency of those they witnessed in the New World. Their reaction was awe, as can be seen in the books of Juan de Cárdenas (1591), Bernal Díaz del Castillo (1568), Francisco López de Gómara (1552), and Suárez de Peralta (1589). An interesting narration is the one by Pedro Castañeda Nájera (1560) relating the discovery by the Melchor Díaz expedition of geothermal manifestations adjacent to a lagoon now named la laguna Vulcano. The area is near geothermal power plants now at Cerro Prieto Geothermal Field, the largest geothermal installation in Mesoamerica. According to the chronicler, "as they were walking, they arrived into dunes of boiling ash, so that nobody could advance through them...the land trembled."

Sometimes, the Spanish writers tried to provide an explanation of an event's origin, one in many cases of real scientific value, as happened with the description of the geothermal gradient offered, probably for the first time in history, by Gonzalo Fernández de Oviedo (1526). In other instances, they did not content themselves with theorizing about geothermal phenomena, but sought to gain first-hand knowledge. A few examples were the attempts to sample the lava of the Masaya volcano by Fray Blas de Iñesta in 1551, or the different ascents of Popocatépetl while in eruption, the first one by soldiers of Diego de Ordaz, then the unsuccessful one by Antonio de Betanzos, and the one by Fray Bernardino de Sahagún, who, accompanied by four or five Spaniards, ascended to its very mouth "and there were very big flames, and it boiled as a cauldron, and this bad thing expelled a large amount of sulfur stone...and, from very far away, one sees the smoke leaving, as in a furnace, except that the amount was huge."

The influence of geothermal events on the lives of Mesoamerica's inhabitants continued after the Conquest. On repeated occasions, earthquakes and related catastrophes demolished colonial cities. That was the case with Antigua, or Santiago as it was originally known, the capital of Guatemala that was destroyed by earthquakes in 1773; the present-day Guatemala City was inaugurated capital of the country in 1776. Cartago, the former capital of Costa Rica, was so ruined by an earthquake in 1841 that only 100 houses and a church were left standing and the capital was reestablished in San José. After San Salvador, the capital of El Salvador, was destroyed by an earthquake in 1854, the capital was moved to Santa Tecla, which then took the name of Nueva San Salvador. Yet, illustrating the principle that people dismiss the risks from a potentially disastrous event if they are obtaining some benefit by doing so, the nation's capital was relocated again at San Salvador in 1859. The city again suffered heavy damage from earthquakes in 1873 and 1907. Special mention shall be made of Mount Pelée on the Caribbean island of Martinique. After two innocuous active events in 1792 and 1851, the volcano erupted violently on May 8, 1902, completely destroying the City of Saint-Pierre and killing 30,000 people.

Regarding the exploitation of minerals and heat associated with geothermal manifestations, we know of a few notable instances, the most famous one being the mining of sulfur deposits on Popocatépetl. Suárez de Peralta reported in 1589 that the Indians obtained this material from the volcano skirts, and it is legendary that soldiers under Diego de Ordaz took from its crater the sulfur needed by Cortés' army to manufacture gunpowder. These deposits were mined until February 19, 1919, when the company working them exploded a dynamite load that was too large. The explosion provoked an earthquake and a volcanic eruption, resulting in the death of 17 workers. In another curious case, the Nicaraguan authorities extended a permit in 1551 to one Juan Alvarez to excavate the volcano Masaya and extract the molten gold that, he thought, was collected in the depth of its crater.

Mining Sulphur from Popocatépetl

"The Spaniards, five in number, climbed to the very edge of the crater, which presented an irregular ellipse at its mouth, more than a league in circumference. Its depth might be from eight hundred to a thousand feet. A lurid flame burned gloomily at the bottom, sending up a sulphurous steam, which, cooling as it rose, was precipitated on the sides of the cavity. The party cast lots, and it fell on Montaño himself [their leader] to descend in a basket into this hideous abyss, into which he was lowered by his companions to the depth of four hundred feet! This was repeated several times, till the adventurous cavalier had collected a sufficient quantity of sulphur for the wants of the army. This doughty enterprise excited general admiration at the time. Cortés concludes his report of it [in the early 1500s] to the emperor with the judicious reflection that it would be less inconvenient, on the whole, to import their powder from Spain." From *History of the Conquest of Mexico*, by William Prescott, 1873

Also, pools of thermal waters were used for their therapeutic properties by the original settlers and Spanish conquerors, alike. For example, the Baños del Peñón ("Baths of the Boulders"), located in the boundaries of the present-day Mexico City airport, were renowned for healing many ailments.

Subsequent Testimonies

During the 19th century, and notwithstanding the problems in the Mesoamerican region, many foreigners came to visit for different reasons, and many of them left reports about the notable volcanic and seismic activity. For example, Henry Ward (1827), the first British ambassador to Mexico, confirmed that Popocatépetl is an active volcano and speculated on the effects of a sudden thawing of its snow cap due to a "violent eruption." He offered as endorsement to his ominous conclusions the report of a similar event witnessed by Alexander von Humboldt in Ecuador at Cotopaxi in 1802. In the course of a trip that Mr. Ward made in 1827 to inspect mines in the central part of Mexico, he crossed the Pathé River. Around it, he

Great Square of Antigua, Guatemala.

Antigua Destroyed
by John L. Stephens, 1841

"Yet this city [Antigua, Guatemala], surrounded by more natural beauty than any location I ever saw, has perhaps undergone more calamities than any city that was ever built.... At the corner of the street was the ruined church of Santo Domingo, a monument of the dreadful earthquakes which had prostrated the old capital, and driven the inhabitants from their home.

"On each side were the ruins of churches, convents, and private residences, large and costly, some lying in masses, some with fronts still standing, richly ornamented with stucco, cracked and yawning, roofless, without doors or windows, and trees growing inside above the walls. Many of the houses have been repaired, the city is repeopled, and presents a strange appearance of ruin and recovery. The inhabitants, like the dwellers over the buried Herculaneum, seemed to entertain no fears of renewed disaster. I rode up to the house of Don Miguel Manrique, which was occupied by his family at the time of the destruction of the city, and, after receiving a kind welcome, in company with Señor Vidaury walked to the plaza.

"The print above [by Frederick Catherwood] will give an idea, which I cannot, of the beauty of this scene. The great volcanoes of Agua and Fuego look down upon it; in the centre is a noble stone fountain, and the buildings which face it, especially the palace of the captain general, displaying on its front the armorial bearings granted by the Emperor Charles the Fifth to the loyal and noble city, and surmounted by the Apostle St. James on horseback, armed, and brandishing a sword; and the majestic but roofless and ruined cathedral, three hundred feet long, one hundred and twenty broad, nearly seventy high, and lighted by fifty windows, show at this day that La Antigua was once one of the finest cities of the New World, deserving the proud name which Alvarado gave it, the city of St. James of Gentlemen.

"This was the second capital of Guatemala, founded in 1542 on account of the destruction of the first by a water volcano. Its history is one of uninterrupted disasters.... 'Many severe shocks of earthquakes were felt at different periods; the one in 1565 seriously damaged many of the principal buildings; those of 1575, 76, and 77 were not less ruinous. On the 27th of December, 1581, the population was again alarmed by the volcano, which began to emit fire; and so great was the quantity of ashes thrown out and spread in the air, that the sun was entirely obscured, and artificial light was necessary in the city at midday.

"'The year 1585 and 6 were dreadful in the extreme. On January 16th of the former, earthquakes were felt, and they continued through that and the following year so frequently, that not an interval of eight days elapsed during the whole period without a shock more or less violent. Fire issued incessantly, for months together, from the mountain, and greatly increased the general consternation. The greatest damage of this series took place on the 23d of December 1586, when the major part of the city again became a heap of ruins, burying under them many of the unfortunate inhabitants; the earth shook with such violence that the tops of the high ridges were torn off, and deep chasms formed in various parts of the level ground.

"'On the 18th of February, 1651, about one o'clock, afternoon, a most extraordinary subterranean noise was heard, and immediately followed by three violent shocks, at very short intervals from each other, which threw down many buildings and damaged others; the tiles from the roofs of the houses were dispersed in all directions, like light straws by a gust of wind; the bells of the churches were rung by the vibrations; masses of rock were detached from the mountains; and even the wild beasts were so terrified, that, losing their natural instinct, they quitted their retreats, and sought shelter from the habitations of men.

"'The year 1717 was memorable; on the night of August 27th the mountain began to emit flames, attended by a continued subterranean rumbling noise. On the night of the 28th the eruption increased to great violence, and very much alarmed the inhabitants. The images of saints were carried in procession, public prayers were put up, day after day; but the terrifying eruption still continued, and was followed by frequent shocks, at intervals, for more than four months. At last, on the night of September 29th, the fate of Guatemala appeared to be decided, and inevitable destruction seemed to be at hand. Great was the ruin among the public edifices; many of the houses were thrown down, and nearly all that remained were dreadfully injured; but the greatest devastation was seen in the churches.

"'The year 1773 was the most melancholy epoch in the annals of this metropolis; it was then destroyed, and, as the capital, rose no more from its ruins.... About four o'clock, on the afternoon of July 29, a tremendous vibration was felt, and shortly after began the dreadful convulsion that decided the fate of the unfortunate city.... On the 7th September there was another, which threw down most of the buildings that were damaged on the 29th of July; and on the 13th December, one still more violent terminated the work of destruction.... A meeting was convoked for the purpose of collecting the sense of the inhabitants on the subject of the removal.... It was resolved to make a formal translation of the city of Guatemala to the Valley of Las Vacas. The king gave his assent to this resolution on the 21st of July, 1775.... On the 29th of July, 1777, a proclamation was issued in Old Guatemala, commanding the population to remove to the new city within one year, and totally abandon the remains of the old one.'

"Such is the account given by the historian of Guatemala concerning the destruction of this city; besides which, I saw on the spot Padre Antonio Croques, an octogenarian, and the oldest canonigo in Guatemala, who was living in the city during the earthquake which completed its destruction. He was still vigorous in frame and intellect, wrote his name with a free hand in my memorandum book, and had vivid recollections of the splendor of the city in his boyhood, when, as he said, carriages rolled through it as in the streets of Madrid. On the fatal day he was in the Church of San Francisco with two padres, one of whom, at the moment of the shock, took him by the hand and hurried him into the patio; the other was buried under the ruins of the church. He remembered that the tiles flew from the roofs of the houses in every direction; the clouds of dust were suffocating, and the people ran to the fountains to quench their thirst. The fountains were broken, and one man snatched off his hat to dip for water. The archbishop slept that night in his carriage in the plaza. He described to me the ruins of individual buildings, the dead who were dug from under them, and the confusion and terror of the inhabitants; and though his recollections were only those of a boy, he had material enough for hours of conversation.

"In company with the cura we visited the interior of the Cathedral. The gigantic walls were standing, but roofless; the interior was occupied as a burying-ground, and the graves were shaded by a forest of dahlias and trees seventy or eighty feet high, rising above the walls. The grand altar stood under a cupola supported by sixteen columns faced with tortoise shell, and adorned with bronze medallions of exquisite workmanship. On the cornice were once placed statues of the Virgin and the twelve apostles in ivory; but all these are gone; and more interesting than the recollections of its ancient splendor or its mournful ruins was the empty vault where once reposed the ashes of Alvarado the Conqueror."

From Incidents of Travel in Central America, Chiapas and Yucatan, Vol. I., *courtesy Dover Publications, Inc.*

noticed many " hot springs of mineral waters from which dense sulfurous vapors rose." Precisely in this area, the first geothermal power plant of Mesoamerica was installed in 1959.

Although there are many other references to the geothermal phenomena of Mesoamerica published in the 19th century, we will end by mentioning one especially well known, the study on "Les Geysers de Ixtlán" that Paul Waitz presented at the 10th International Geological Congress in 1906. This paper was among those consulted by the first Mexican geothermicists, like Jorge Isita Septién, Luis de Anda, and Héctor Alonso, whose work made possible modern geothermal development in their country.

Conclusions

The distinct manifestations of terrestrial heat—hydrothermalism, volcanic eruptions, earthquakes, and igneous mineral deposits—always have affected the lives of Mesoamerica's inhabitants in definitive ways. At a fundamental level, the existence of geothermal activity in an area can decide whether or not its occupants will live or die. But geothermal manifestations also have less categorical influences. They can affect utilitarian aspects, as when people profit from the increased fertility of volcanic soils, exploit igneous materials, and use hot springs to bathe or heal. On a nonmaterial plane, they can foster the advancement of scientific theories to explain geothermal phenomena, the proposal of philosophical

Plumbing Fiery Liquor for Gold in Masaya

John L. Stephens, while traveling in Nicaragua, visited the volcano Masaya in 1841 and relates these two stories in his book *Incidents of Travel in Central America, Chiapas and Yucatan, Vol. II*. The story of the descent down the crater surely occurred around 1551. Perhaps one of the two Spaniards involved was Alvarez, the man permitted by the Nicaraguan government to explore it.

Stephens wrote, "Among the recorded wonders of the discoveries in America, this mountain was one; and the Spaniards, who in those days never stopped half way in any matter that touched the imagination, called it El Infierno de Masaya, or the Hell of Masaya.

"The historian, in speaking of Nicaragua, says, 'There are burning mountains in this province, the chief of which is Masaya, where the natives at certain times offered up maids, throwing them into it, thinking by their lives to appease the fire, that it might not destroy the country, and they went to it very cheerful.'

"In another place he says, 'There is another Mouth like that of a Well about a Bowshot over [from Masaya's crater], the distance from which to the Fire is about a hundred and fifty Fathoms, always boiling up, and that mass of Fire often rises and gives a great Light, so that it can be seen at a considerable Distance. It moves from one Side to the other, and sometimes roars so loud that it is dreadful, yet never casts up anything but Smoke and Flame.

"'The Liquor never ceasing at the Bottom, nor its Boiling, imagining the same to be Gold, *F. Blas de Iñesta*, of the Order of *St. Dominick* [*sic*], and two other *Spaniards*, were let down into the first Mouth in two Baskets, with a Bucket made of one Piece of Iron, and a long Chain to draw up some of that fiery Matter, and know whether it was Metal. The Chain ran a hundred and fifty Fathoms, and as soon as it came to the Fire, the Bucket melted, with some Links of the Chain, in a very short Time, and therefore they could not know what was below. They lay there that Night without any Want of Fire or Candles, and came out again in their Baskets sufficiently frighted.'" *Courtesy Dover Publications, Inc.*

developments related to the Earth's heat in a general way, and the creation of complete plutonic theogonies.

In recent years, specialists of diverse disciplines—but mostly geologists and engineers working with geothermal exploitation, anthropologists, and historians of science and technology—have addressed this subject, and there is a growing bibliography. We hope that the study of the history of relationships between people and geothermal phenomena will continue for a long time.

The Ornamented Baths of Peñón

Around 1840, Frances Calderón de la Barca, Scottish wife of the Spanish ambassador to Mexico, visited the ancient, famous Baths of Peñón. She describes the baths in her book, *Life in Mexico.* Today the baths are gone, their former site part of the Mexico City airport.

"We drove out to the Peñón, a natural boiling fountain, where there are baths, which are considered a universal remedy, a pool of Bethesda, but an especial one for rheumatic complaints. The baths are a square of low stone buildings, with a church—each building containing five or six empty rooms, in one of which is a square bath. The idea seems to have been to form a sort of dwelling-house for different families, as each bath has a small kitchen attached to it.... People still flock there for various complaints. When one goes there to bathe, it is necessary to carry a mattress, to lie down on when you leave the bath, linen, a bottle of cold water, of which there is not a drop in the place, and which is particularly necessary for an invalid in case of faintness—in short everything that you may require. A poor family lives there to take charge of the baths, and there is a small tavern where they sell spirits and pulque; and occasionally a padre comes on Sunday to say mass in the old church.

"We were amused by meeting there with General ___ and his family, who had brought with them a whole coach-load of provisions, besides mattresses, sheets, etc. The road to the Peñón crosses the most dreary plain imaginable. Behind the baths are two volcanic hills; and the view of Mexico and of the great volcanoes from this is magnificent. It is the most solitary of buildings; not a tree to be seen in its environs; these volcanic rocks behind—Mexico fronting it—the great lakes near it—to the right Guadalupe—to the left San Ángel, San Agustín, and the mountains which bound the valley. The Indian family who live there are handsome...and the girl who attended me at the bath spoke an extraordinary jargon, half Spanish, half Indian.... The water is extremely warm, and my curiosity to try its temperature was very soon satisfied.

"These boiling springs are said to contain sulphate of lime, carbonic acid, and muriate of soda, and the Indians make salt in their neighbourhood, precisely as they did in the time of Moctezuma, with the difference, as Humboldt informs us, that then they used vessels of clay, and now they use copper caldrons. The solitary-looking baths are ornamented with odd-looking heads of cats or monkeys, which grin down upon you with a mixture of the sinister and facetious rather appalling."

Life in Mexico, *by Frances Calderón de la Barca, edited and translated by Woodrow Borah. Copyright by the Regents of the University of California, published by the University of California Press*

Selected References

Burgassi, P. D., R. Cataldi, J. L. Hernández-Galán, M. Moggi, R. Rubinovich, and J. J. Saldaña. 1992. *Primera contribución al conocimiento historiográfico de la energía geotérmica en el Area Mediterránea y en la América Latina.* Pisa. (Prepared for and presented at the Symposium on the History of Geothermal Energy, Mexico City, 1992. Limited publication. For further information, contact the first three authors at the addresses in this volume).

Calderón de la Barca, F. 1843. *Life in Mexico.* University of California Press, Berkeley, California.

Cárdenas, J. 1591. Primera parte de los problemas y secretos maravillosos de las Indias. In *Historia de la ciencia en México Siglo XVI.* Reprint 1983, E. Trebulse. Fondo de Cultura Económica, México, D.F.

Castañeda de Nájera, P. 1560. Relato de la expedición de Coronado. In *La última jornada de Melchor Díaz.* Reprint 1973, R L. Ives. *Revista CALAFIA*, yr. 2, no. 2, 18-21.

Díaz del Castillo, B. 1568. *Historia verdadera de la Conquista de la Nueva España.* Fernández Editores, S.A., México, D.F.

Fernández de Oviedo, G. 1526. *Sumario de la natural historia de las Indias.* Fondo de Cultura Económica, México, D.F.

Fink, J. 1993. Down under the volcano. *Nature,* 366, 108.

Grayson, D. K. and P. D. Sheets. 1979. Volcanic disasters and the archeological record. In *Volcanic activity and human ecology.* Academic Press, Inc.

Hurtado de Mendoza, L. and G. Alvarado. 1988. Datos arqueológicos y vulcanológicos de la región del Volcán Miravalles, Costa Rica. *Vínculos,* 14, 77-89.

Longyear, J. M. III. 1952. *Copán ceramics.* Carnegie Institution of Washington, Washington, D.C.

López de Gómara, F. 1552. *Hispania victrix, Primera y segunda parte de la historia general de las Indias.* Ediciones Orbis, S.A., Barcelona.

Muñoz Carmago, Diego. 1540. *Historia de Tlaxcala.* Reprint 1963. In *Visión de los vencidos,* M. León Portillo and A. M. Garibay, eds. Fondo de Cultura Económica, México, D.F.

Prescott, William H. 1873. *History of the conquest of Mexico, vol 2.* David McKay, Philadelphia, Pennsylvania.

Ryder, P. 1983. Guayabo de Bagaces. *Vínculos,* 14, nos. 1-2, 121, 125.

Sahagún, B. 1534. Historia general de las cosas de Nueva España. Reprint 1963. In *Visión de los vencidos,* M. León Portillo and A. M. Garibay, eds. Fondo de Cultura Económica, México, D.F.

Sharer, R. J. 1979. *Sucesos terminales en el sureste de Guatemala.*

Sheets, P. D. 1979. Volcanic eruption effects. In *Volcanic activity and human ecology.* Academic Press, Inc.

Sheets, P. D. 1984. Proyecto prehistórico Arenal during 1984: Summary and conclusions. *Vínculos,* 10, nos. 1-2, 207-223.

Stephens, J. L. 1841. *Incidents of travel in Central America, Chiapas and Yucatan, vols. 1-2.* Dover Publications, New York.

Stone, D. 1941a. *Precolumbian man finds Central America: The archeological bridge.* Peabody Museum Press, Boston.

Stone, D. 1941b. Archeology of the north coast of Honduras. *Peabody Museum Transactions,* 9, no. 1, 25-29.

Suárez Arriaga, M. C. and R. Cataldi. 1993. *Geotermia,* 9, no. 1, 9-58.

Suárez de Peralta, J. 1589. *Tratado del descubrimiento de las Indias.* Consejo Nacional para la Cultura y las Artes, México, D.F.

Tedlock, D. 1985. *Popul Vuh, The definitive edition of the Mayan book of the dawn of life and the glories of gods and kings.* Simon & Schuster, Inc., New York.

The Authors:

José Luis Hernández Galán
FIDE
León Tolstoi 22-5
México 11590, D.F.

Jorge Guiza Lámbarri
Wimbledon 8, casa 7
Lomas de Cuernavaca
Temisco 62589, Mor.
México

Mario César Suárez Arriaga
The Mexican Geothermal Association
Michoacán University
E-mail: msuarez@zeus.ccu.umich.mx

Geothermal stories are included in the oldest existing manuscript describing P'urhépecha life in Michoacán, Mexico, written between 1539 and 1541, about 20 years after the Spanish Conquest. This illustration from the frontispiece depicts a Franciscan friar (probably the author) presenting the manuscript to the Spanish Viceroy Mendoza, who had commissioned it. Important P'urhépecha leaders in Spanish dress stand behind the friar. *The Library of Congress*

33. Heat over Time, Geothermal Stories from Mexico

by
Susan F. Hodgson

Abstract: Mexican geothermal stories, ancient and modern, written and oral, record the history of geothermal Mexico. The stories in this chapter span hundreds of years. Some are from a 1541 Spanish manuscript about the just-conquered P'urhépecha, an indigenous people in central Mexico. The rest are oral transcriptions made in 1993 and 1994 at Mexican geothermal sites.

The stories offer historical commentary and drama shaped by magic realism. They include P'urhépecha political and religious ceremonies, Maya beliefs, Spanish conquistadors, the Mexican War of Independence, and country life after the Conquest.

Reprinted by permission of the International Geothermal Association, from the *Proceedings of the World Geothermal Congress,* 1995, pp. 411-416.

The Beginning

There are many geothermal stories in Mexico—remembered, forgotten, oral, and written. Here are a few. Some have characters and plots; others do not. All show how geothermal settings influence lives, the story of heat over time.

The P'urhépecha stories are first, as accounts from this indigenous Mexican people started my search. The P'urhépecha still live in the heart of Mexico, in the modern State of Michoacán. Today, their lands are bordered on the east by the State of Mexico and Mexico City, once the home of the Aztecs, an ancient enemy. Much of P'urhépecha culture was destroyed with the Spanish Conquest, which came to Michoacán in 1521.

Soon afterwards, from 1539 to 1541, a Franciscan friar, probably Jerónimo de Alcalá, interviewed the P'urhépecha and compiled the only known survey from the time of their government, customs, and traditions (see illustration on the left). He called it the *Relación de las Ceremonias y Ritos y Población y Gobierno de los Indios, de la Provincia de Michoacán,* or *The Story of the Ceremonies and Rites and People and Government of the Indians, of the Province of Michoacán.*

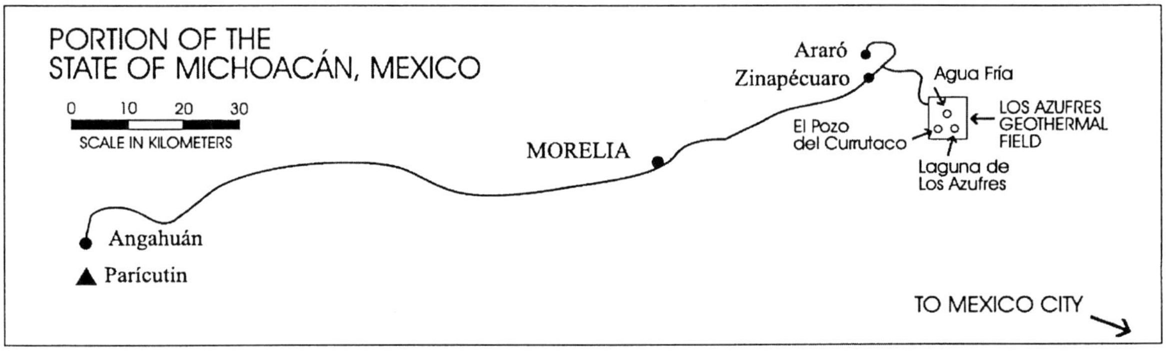

Geothermal sites from the text in the State of Michoacán, Mexico.

In the *Relación,* the hot springs and volcanoes of Michoacán are woven into religious ceremonies and stories, as are many sites and cities existing today, including the volcanic hills of Zinapécuaro and the hot springs of Araró. These towns are mentioned in the following story about the P'urhépecha feast of *Sicuíndiro*, which means "the skinning" and honors the goddess of rain, who made rain clouds in Zinapécuaro and kept them as vapor in the sacred hot springs of Araró. The portion that follows retains its original cadence from the oral tradition when read aloud.

Sicuíndiro (The Skinning)

Narrative of the P'urhépecha

Relación, *translated by S. Hodgson*

(Probably) compiled by Friar Jerónimo de Alcalá

Michoacán, Mexico, 1539-1541

"Five days before this feast,
the priests arrived from the towns mentioned, with their gods, and they came to the
feast,
and dancers called *cesquárecha* entered the houses of the priests
and two other priests called *hauripitzípecha*,
and they fasted until the day of the feast,
and the day before the feast,
the priests marked the chests of two slaves or criminals that
they had to sacrifice the day of the feast,

and they sacrificed the slaves mentioned,
and in taking the hearts, they performed their ceremonies with them,
and while they were still warm,
they carried them to the hot springs of the town of Araró from the town of Zinapécuaro,
and they threw them in a small hot spring,
and they covered them with wooden tablets,
and they threw blood in all the other hot springs that are in the town mentioned,
that were dedicated to other gods that were there,
and those hot springs gave off vapor,
and they said that from there the clouds left to rain,
and this goddess called Cuerauáperi was in charge of them,
and that she sent them from the east, where she was,
and for this reason they threw that blood in the springs."

Araró, which means "the perforation site," received its name from the feast when great warriors received lip rings, nose rings, and earrings. Blood let during the insertion was thrown in the hot springs (Corona Nuñez, 1977). Two great P'urhépecha warriors standing behind the Franciscan friar in the drawing at the front of the chapter have rings in their lower lips, most likely inserted at this ceremony.

I visited Zinapécuaro and Araró, searching for geothermal stories.

At Zinapécuaro

In Zinapécuaro, on the outskirts of town, a sign advertises hot springs at a resort of therapeutic and recreational renown, *La Reina Atzimba*—the Queen Atzimba.

Stopping nearby, I asked a family about geothermal stories. They directed me to María Guadalupe Avalos, a life-long resident of the town. Señorita Avalos spoke about Queen Atzimba, calling her Princess Atzimba, and this is what she said.

María Guadalupe Avalos, narrator. *Photos by S. Hodgson*

Princess Atzimba

Brief narrative of María Guadalupe Avalos

Zinapécuaro, Michoacán, Mexico, May 1994

> "Princess Atzimba was the sister of a P'urhépecha king named Zuangua. His palace is on a hill called *Las Clavellinas* that is by the resort of *La Reina Atzimba*, on the edge of town."

This was the hot spring resort I had passed on the way. Although archaeological ruins found here predate the Conquest, the main temple of Cuerauáperi was elsewhere. The *Relación* states:

> "Cuerauáperi had her priests in the town of Araró and other towns,
> and her main idol in a temple
> that is in the town of Zinapécuaro, on top of a hill,
> where today it is demolished."

Lopez Lara writes that Cuerauáperi's temple was built and torn down on the hill today at the center of modern Zinapécuaro, the site where the Spaniards built the city's main church in 1570. He thinks the church was built on the ruins of her temple, following the Spanish dictum, "temple over temple."

I thought of other Mexican cities with their main churches on flat, central plazas, and Zinapécuaro has a lot of flat terrain. Yet it is true that the main church stands on a singular rise about two blocks higher than the terrain around it. Why weren't the flatter areas used, unless the main temple for Cuerauáperi had stood on that hill?

The approach to the main church in Zinapécuaro.

To approach the church, you walk for about two blocks up a straight narrow street ending in two tiers of steps. The walkway rises up to the church in a carefully graded way, paved with stones perhaps especially well placed. Her old temple entranceway?

On to Araró

Leaving Zinapécuaro, I drove about six kilometers north to Araró, to the field of hot springs where Cuerauáperi kept the clouds. In town I met Carlos Ferrer, a young man about 15 years old, who led the way through modern Araró out to the ancient field of hot springs described in the *Relación.* As we drew near, we passed thriving plots of corn, and Carlos explained that thermal vapors helped the corn to grow, that sometimes hot springs were used for cooking the ears of corn.

Finally we came to the ancient field of hot springs and the two modern resorts at the edge, Balneario Los Hervideros and Balneario Hingo. We parked and entered the large flat expanse a few square kilometers in size, covered with brown meadow grass and about 17 hot springs, by Carlos' estimate. He said that once a geyser, called *El Géiser*, had shot straight up and an apparition had appeared here, the "Señor de Araró." He pointed to an especially large hot spring resembling a collapsed cave, named *La Cueva del Diablo*, or "The Devil's Cave."

Hot spring in the foreground, salt collection mound at photo center, highway construction in the background, at Araró.

In the field I saw several small sites for collecting salt. At each, a wooden funnel about 1½ meters wide and high was erected over a few covered troughs. First the funnels were filled with soil from the field. Then water was poured into them, leaching out the salts and dripping into the troughs. Here the water evaporated into a salty residue. Carlos said that salt collection is an old practice, that his grandfather had done it.

Much of the field looked natural and untouched, perhaps not too different from the days of Cerauáperi. Then I noticed highway construction crews working at the far edge, building the new superhighway across the State of Michoacán to connect Mexico City with Guadalajara. The superhighway crosses Araró's field of hot springs, and special efforts were made to save hot springs in its path.

A hot spring preserved under the superhighway at Araró.

We went to take a closer look. The superhighway rises like a bridge across the field. Between the columns that support its spans are small passageways three to four meters wide. Some passageways have holes in their concrete floors about a meter wide. Here preserved hot springs bubble and steam away, still storing the clouds.

From Araró to Los Azufres

The P'urhépecha believed in omens and dreams, which warned them of the Conquest at least four years before the Spaniards arrived. I believe the pine forests, volcanoes, and thermal areas of Los Azufres Geothermal Field in Michoacán are included in a dream recounted in the *Relación.* Drawings and text from the *Relación* support this, but not conclusively.

In the dream, a P'urhépecha woman is flown on the back of an eagle over steep mountain slopes covered by pine trees. The topography and vegetation of the slopes, illustrated in Plate

XLII from the *Relación*, are identical—albeit in rough-sketch format—to the topography and vegetation at Los Azufres Geothermal Field today.

On the journey, they fly over a mountain with a hot spring whose description matches that of a famous hot spring, *Laguna de Los Azufres* ("Lagoon of Sulfurs"), found today on the southern edge of Los Azufres Geothermal Field. The *Relación* describes the hot spring as surrounded by sulfur-bearing rock. The lagoon is indeed surrounded by such rock, which was mined for its sulfur in the 1800s. Here is the dream.

"Of the Omens and Dreams This People Had before the Spanish Came to This Province"

Relación, *Chapter XIX, translated by S. Hodgson*
Narrative of the P'urhépecha
Michoacán, Mexico, 1539-1541

Portion of Plate XLII at the beginning of Chapter XIX, the *Relación.* The comet at the top is one of two that warned the P'urhépecha of the Conquest. The mountain covered with pine trees on the right depicts the topography and vegetation of the Los Azufres area. The figures are of gods in the dream, whose divine natures are shown by exaggerated features and teeth.

"The goddess Cuerauáperi, mother of all earthly gods,
came and took that woman from her own house,
guided her awhile toward the road from Mexico,
and turned her toward the road to Araró,
And that woman went by the road,
and on the road she met an eagle that was white,
and it had a large cape in front,
and the eagle began to whistle, and to arch its feathers,
and with some large eyes that they said to be the god
Curicaueri, and the eagle greeted her, and told her that
she was welcomed, and she also greeted him,
and told him: 'Sir, you come just in time.'
The eagle told her, 'Climb up here, on top of my wings,
and don't be afraid of falling.'
And just as the woman climbed up,
the eagle rose up with her, and began to whistle,
and carried her to a mountain, where there is a hot
spring,

and sulfur in the rock,
and carried her by that mountain flying with her,
and now dawn was breaking,
when he carried her to the foot of a very high mountain,
that is near there, called Xanoato-hucatzio,
and carried her high,
and that woman saw
that all the gods of the province were seated."

Solemnly, the gods warn the woman that invaders are coming, that "everything will be gone, other men will be throughout the land." They say she must warn everyone, including King Zuangua, that the fates of the P'urhépecha people and the gods themselves depend on him.

Though warnings were issued, King Zuangua ignored them and the Conquest came. It began in 1519 to the southeast, where many indigenous peoples were attacked, including the Aztecs. The Aztecs appealed twice for help from the P'urhépecha and were refused both times. First they petitioned aging King Zuangua, who would die of smallpox in 1520. They next asked his heir, but to no avail. By 1521, the Aztecs were conquered and the Spaniards had moved into P'urhépecha lands, where the onslaught continued (Bravo Ugarte, 1963).

Now at Los Azufres

TODAY, 473 YEARS LATER, JUST ON THE northern edge of Los Azufres Geothermal Field, Doña Celia Escalante Arreola runs a small hot spring resort on lands where her family has lived at least as far back as her great-grandparents. On her front porch, she told me a very short story of an Aztec queen, *Xóchitl*, which means "flower." The Los Azufres area was a boundary between P'urhépecha and Aztec lands, and her account reflects the cultural interface. This is what she said.

Celia Escalante Arreola, narrator.

Queen Xóchitl

Brief narrative of Celia Escalante Arreola
Michoacán, Mexico, May 1994

"When I was a child, Queen Xóchitl would bathe here in the thermal waters. She did this because the waters were beautiful and kept you young."

As she spoke, Doña Celia nodded to the southeast, indicating the location of the hot spring—the same direction as the nearby famous Lagoon of Sulfurs, although she didn't mention it by name. She said that the Virgin of Guadalupe has appeared there as well.

The author with narrators Celina Silva and Arturo Gonzalez Salazar, in Agua Fría.

The next two stories feature Los Azufres Geothermal Field and the Lagoon of Sulfurs at its southern border. The stories are by Arturo González Salazar, a geologist who has worked extensively in Mexican geothermal fields. Comments by Celina Silva and others are included.

The History of Los Azufres

Narrative of Arturo González Salazar
Morelia, Michoacán, Mexico, November 1993, May 1994

Additional narrative by Celina Silva
Agua Fría, Michoacán, Mexico, May 1994

"In 1968, I arrived in the high pine forests of eastern Michoacán," said Arturo González Salazar. "I explored the geology of a place now called Los Azufres Geothermal Field in the Parque Nacional de Los Azufres [Los Azufres National Park]. The Comisión Federal de Electricidad [CFE] sponsored my work and I lived by the famed Lagoon of Sulfurs, today on the field's southern edge. This hot spring area, known through centuries for steamy vapors and curative waters, is just south of the CFE's main field camp and a volcano called Las Humaredas, so named by people mistaking its vapors for smoke.

Rock patterns at Las Humaredas.

"When I came to Los Azufres, almost no one was there. One day I climbed the slopes of Las Humaredas, and near the top in a flat grassy area, I found rock patterns laid out on the ground and some rocks piled into a triangle.

"Small funerary rock piles and triangular rock mounds were by the Lagoon of Sulfurs as well, and I asked about them. The country people said the rocks were from the past, part of P'urhépecha religious ceremonies when a young girl entered puberty.

"From the time of the P'urhépecha, there are stories of women seeking fertility in the waters of Los Azufres, and of people bathing to alleviate rheumatism and arthritis. When I came, I saw many using the waters for cures. Poor rural people arrived on burros, bringing along their own food and blankets.

"For the P'urhépecha, the lagoon was a sanctuary and a center of religious life. Priests held ceremonies in the early morning when vapors from the hot waters were intense. We can imagine them colorfully dressed, standing in the hot coiling vapors, worshiping the sun.

The Lagoon of Sulfurs, with people bathing in the waters.

"The lagoon served other purposes. P'urhépecha society was very hierarchical and the country people told of large, periodical P'urhépecha migrations to the lagoon, where leaders would undertake the formal rites matched to their political status, mixing religion and power. Ceremonies were held at nearby Cerro Las Humaredas as well.

"All this ended with the Spanish. After Cortés began the Conquest of Mexico in 1519, the Spaniards moved quickly to alter native religious expression. At first this did not stop people from worshiping secretly in small groups, continuing to speak with their gods, but then it did. One day at Los Azufres, I found an ancient Spanish coin from that time.

"The end for Mexico of almost 300 years of Spanish rule began on September 15, 1810, in the State of Guanajuato with Miguel Hidalgo's famous cry, 'El Grito de Dolores.' His proclamation is considered the Declaration of Independence for Mexico. The revolution spread south into Michoacán, which quickly became a center of the conspiracy and a home for revolutionary leaders.

"Because the revolutionaries had no gunpowder, they made their own. They mined sulfur, the main ingredient, at Los Azufres from the rock surrounding the Lagoon of Sulfurs, perhaps increasing the size of the hot spring in the process. Walking around the lagoon today, you pass several old mine shafts filled with water. The revolutionaries drained the water through a channel cut into solid rock.

"Once the sulfur-rich rock was mined, it was brought to a nearby gunpowder factory built where the CFE has placed its main field camp at Los Azufres. The factory itself

Fragment of the gunpowder factory wall at Los Azufres.

was over a stream, the Arroyo de Agua Fría, a source of water in case of attack. At the base of a fragment of wall still standing, a beautiful stone archway marks where the creek flowed through it, passing into the building.

"The rocks with sulfur were pulverized at the factory by large grinding wheels. Today one of the wheels is displayed on a rise overlooking the field camp. The gunpowder factory was among the first in Mexico and the source of the first gunpowder used in the revolution. When the war was over and no more gunpowder was needed, the factory was abandoned."

Grinding wheel used to pulverize rock.

Celina Silva told Arturo González Salazar and me that the sulfur miners used large pots, called *casos*, to process the sulfur into large pieces called *marquetas.* Some miners were injured or killed by fumes of hydrogen sulfide. The factory had a special area for storing explosives.

Celina Silva grew up at Los Azufres, and both she and Doña Celia said it was once part of the Hacienda Agua Fría and the home of Melchor Ocampo, a prominent Mexican political figure in the mid-1800s. Celina said that the father of the revolutionary priest, Miguel Hidalgo, had lived at Los Azufres. I asked the CFE if old land records could confirm this, but unfortunately nothing in the files could. I was disappointed, for this could be the way revolutionaries had learned of sulfur at the Laguna de Los Azufres.

Arturo González Salazar continued, "When I came to Los Azufres, just a few sections of factory wall remained. A man living nearby told me about the ghosts, saying that in the vapors of Los Azufres he had seen the ghosts dressed in costumes from the days of the Conquest and from the time of the gunpowder factory. He and others never went out at night for fear of meeting them.

"I was sent to Los Azufres because Luis de Anda, in the newly formed CFE, had placed the area on a list he made in the 1940s of geothermal sites to be developed for electrical generation. Thus, I saw the famous thermal features and forested hillsides almost untouched by mankind.

"Today Los Azufres is a national park and beautiful, but at the same time some things have passed away and changes have come. Some are from geothermal development, of course, though Los Azufres Field is a good geothermal project and great care has been taken.

"Even so, for a long time I didn't go back; I didn't want to return. I feared the ghosts in the vapors, the sight of Los Azufres as it used to be."

The next story, also by Arturo González Salazar, is a geothermal legend about a hot spring a few kilometers away from the Lagoon of Sulfurs, called *El Pozo del Currutaco*.

El Pozo del Currutaco

> "A *currutaco* is an old-fashioned pejorative in Spanish meaning a man with exaggerated elegance, a dandy. A hot spring in Los Azufres Geothermal Field is called *El Pozo del Currutaco*, and this is how it got its name.
>
> "At the turn of the century, a *currutaco* came to a small community not so far from the hot spring where he fell in love with a woman who had another suitor.
>
> "The *currutaco* was very elegant and perhaps even handsome. The woman loved him for this, and forsook the first suitor who was humble and poor.

El Pozo del Currutaco, "The Dandy's Hot Spring."

"One day, an outing was arranged into the countryside at the hot spring. The party included the *currutaco*, the woman, and her humble suitor.

"At a propitious moment, the humble suitor pushed the *currutaco* into the hot spring. The water was so hot, the *currutaco* couldn't get out and he died.

"Thus through crime, the humble suitor changed his fate and came to marry the woman he loved. Ever since the hot spring has been called *El Pozo del Currutaco*.

"Today the scalding waters are still famous. People come from miles around. Kneeling by the hot spring, wary of splattering bubbles, they reach inside to dig out clay for beauty masks, poetic justice for the handsome *currutaco*."

Chiapas

This story by Arturo González Salazar is from the State of Chiapas. Though modern, it may be the oldest of all, an evocative paradox of nature, mankind, and geothermal resources, and a volcano named Chichonal.

My Days at Chichonal

Arturo González Salazar, narrator.

"About 20 years ago in the Mexican State of Chiapas, a governor phoned the Comisión Federal de Electricidad [CFE]. He said he had flown over a mountain on fire, smoke billowing from its sides. The governor asked the CFE to come to Chiapas and explore the mountain, a volcano called Chichonal, for he feared something serious could happen.

"The CFE sent a geochemist, Rafael Molina Berbeyer (now deceased), a geologist (myself), and a technician, Blas González Mendez. After a long journey, we finally arrived in the town of Pichucalco in Chiapas. From there, we drove with great difficulty to a little town called Volcanes, a site very isolated in the heart of tropical jungle. It was a dangerous trip and we crossed several wide rivers by jeep.

"When we arrived in this small farming commune (whose members were Mayan, of a tribe called the Zoque), the town authorities let us stay in the school. We spread out our belongings and that evening talked to people who returned from the fields. We explained why we had come and asked them to take us to the volcano Chichonal, now not so far away. We could see the 'smoking fumaroles' from the town.

"To our great surprise, nobody would go. This was very unusual for in general the people we met were helpful. Finally, we convinced a small group to guide us up the slopes of the volcano. The leader was a man named Cansino, who agreed on condition that we would walk ahead of him.

"The next day, we penetrated an untouched jungle of rich, tropical vegetation without roads or trails. We went first and our hesitant guides stayed a distance behind, nervous and fearful. From time to time, passing carefully through dense foliage and stands of huge leaves, Cansino and I spoke. Gradually I realized that he feared visiting the volcano because he had heard its loud noises. He had watched birds flying near fumaroles fall dead from the sky and feared he, too, would die.

"I found it no mystery what had killed the innocent birds—the hydrogen sulfide gases breathed in from the vapors.

"Hours later we came to the fumaroles but stayed briefly, for the hydrogen sulfide fumes were intense. Then we returned to Volcanes. Here, Cansino convinced others that the situation was not perilous, that they should help establish our camp at the top of the volcano. With great organization they did so, bringing us food and water each day.

"With time more people began to help, while we explored one of the most remarkable geothermal zones I have ever known. The temperatures were very high, there were immense areas of fumaroles, and more enchanting, perhaps because it was unvisited by any human being, the zone was a verdant and natural sanctuary for animals I had known in books but had never seen.

"On the days I stayed there I saw many deer, tapirs, peccaries, snakes of many kinds, and birds of all colors and sizes. Many times I felt like Christopher Columbus in the New World, finding a paradise full of thriving plants and animals. It was something extraordinary.

"In the nighttime, I meditated. I knew breaking into this sanctuary would have mortal consequences, because teaching the people to enter here without fear would increase man's predation. I have always regretted this, but it was our job to explore the volcano.

"Nonetheless, I listened with hope to Cansino, their leader, who said that hunters would not be permitted to kill in this place, although the taboo was gone. For seeing us move calmly through the vapors, our guides had lost their fear.

"After 45 days we were done. Working in heat over 40° C and in constant rains and cyclones from the Caribbean Sea, we had sampled vapor, measured the temperature of land, and mapped the area. The hydrogen sulfide was so concentrated that we often had headaches, but clearly great geothermal promise existed.

"In my report, I wrote that we had felt tremors from five to six earthquakes each day, that perhaps the volcano of Chichonal was awakening. Thus with several kilos less of weight, without our boots, and with dirty clothing, our mission was ended and we returned home.

"Years later, in 1982, the volcano Chichonal erupted, spewing volcanic ash that changed the climate of the world and destroyed nearby areas, including the sanctuary with the flora and fauna I had so admired. With time, they have told me that almost all my friends and guides of Chichonal died in the catastrophe, which in the end brought destruction, death, and sorrow to the Zoque tribe.

"As a final note, perhaps it is worthwhile here to remember the geologist Salvador Soto Pineda, who died in one of the eruptions while trying to save a community. It was when the volcano was in its full eruptive cycle and some people living in a little town nearby did not wish to evacuate their homes. Salvador Soto Pineda went to convince them to leave, for the risk of death was clear. He and others arrived by helicopter in a humanitarian effort. Once in the town, all were surprised by an eruption and buried forever under tons of volcanic ash. For this I consider Salvador Soto Pineda a hero of volcanology."

Geothermal Stories

Thus geothermal forces destroy lives, and geothermal sites enhance them. Just where devastation reigns from earthquakes, eruptions, lava, hot water, fumes, and fire, there the gods and mortals commune, ailments are alleviated, beauty perfected, crop growth quickened, food cooked, sulfur and salt mined, resorts visited, and energy made. Geothermal stories record it all.

But our affinity to geothermal forces goes deeper. A quote from Benito Bravo helps illustrate the connection. Benito lives in the P'urhépecha town of Angahuán, which is near Parícutin, the volcano that rose in 1943 in the middle of a cornfield in Southwestern Michoacán.

Angahuán remained untouched by lava flows covering nearby areas. I asked Benito what people in Angahuán think about Parícutin, which has so influenced their lives and he said, "It is important because we saw it when it was born and when it died."

Birth and death. Like us, geothermal features begin and end, moving through cycles of their own. We draw towards them, lured by change, beauty, and an unusual cast of the familiar—water, rocks, and heat. We search them for answers to mysteries in our own lives, like birth and death. We have done this through time, and geothermal stories are the archives of our quest.

Benito Bravo, narrator.

ACKNOWLEDGMENTS

MANY PEOPLE OFFERED A LOT OF HELP, FOR WHICH I AM MOST GRATEFUL. ESPECIALLY, I WANT TO thank the Comisión Federal de Electricidad for its support, including Gerardo Hiriart Le Bert, Luis Carlos Gutiérrez Negrín, Raul Sánchez Velasco, and Javier Gaona.

The assistance of Arturo González Salazar was fundamental to this task. In his career with the CFE, he was among the first to visit geothermal areas. He not only looked at geology but talked with people and remembered what they said. His personal narratives, assistance with interviews, and interest in what we saw made a big difference.

Thanks also to María Guadalupe Avalos, Carlos Ferrer, Celia Escalante Arreola, Celina Silva, and Benito Bravo, who shared their geothermal stories.

Selected References

Alcalá, Fray Jerónimo de (most probably). 1541. *Relación de las ceremonias y ritos y población y gobierno de los Indios, de la provincia de Michoacán.* Reprint 1977. Reproduction of the Ms. c. IV. 5. de El Escorial, Spain. Balsal Editores, S.A., Morelia, Michoacán.

Bravo Ugarte, José. 1963. *Historia sucinta de Michoacán, Tomo II, Provincia Mayor e Intendencia.* Ed. Jus, S.A. Mexico.

Corona Nuñez, José. 1977. Preliminary study. In *Relación de las ceremonias y ritos y población y gobierno de los Indios, de la provincia de Michoacán.* Balsal Editores, S.A., Morelia, Michoacán.

Lopez Lara, Ramón. 1984. *Zinapécuaro, Tres épocas de una parroquia.* Tercera Edition. Editado por Fimax Publicistas, Morelia, Michoacán.

Refugio Cabrera V., Ma. del and Benjamín Pérez González. 1991. *El estado P'urhépecha y sus fronteras en el siglo XVI.* Carrasquilla Editores, S.A. de C.V., Morelia, Michoacán.

The Author:

Susan F. Hodgson
Department of Conservation
Division of Oil, Gas, and Geothermal Resources
801 K Street, MS 20-20
Sacramento, CA 95814-3530 USA
Telephone: 916.445.9686
Fax: 916.323.0424
E-mail: shodgson@consrv.ca.gov

Osorno volcano and the Petrohué River in the Chilean lake country. *Drawing by Jim Spriggs*

34. Andean Cultures and Geothermal Phenomena in Historical Chronicles

by
Gonzalo Rico Calderón

Abstract: Active volcanoes, salt deposits originating from hot brines, hydrothermal minerals, and other geothermal by-products contributed significantly to the formation and development of indigenous cultures in the Andean Cordillera of South America prior to the Spanish Conquest in the 16th century. Based on archaeological findings and written descriptions dating from colonization, this chapter explains the use of natural heat and its by-products by the Andean people. It also describes habits and myths that formed in this region because of volcanic eruptions, earthquakes, and high-temperature manifestations. These phenomena, associated with imposing volcanic mountains, exerted an influence on the Andean peoples, both in the formation of religious sentiments and the development of a unique conception of the world.

The Andean Cultures and the Physical Environment

"The Inca called their sprawling realm Tahuantinsuyu, or 'Land of the Four Quarters,' and on the eve of Columbus' Caribbean landfall it probably surpassed Ming China and the Ottoman Empire as the largest nation on Earth [see map above]. Stretching down the mountainous Andean backbone of South America for more than 5500 km, it was the biggest native state to arise in the Western Hemisphere, and also the largest empire of antiquity ever to develop south of the equator. By dint of armed conquest the masters of Tahuantinsuyu governed the most rugged mountain chain on the face of the Earth, second only to the Himalayas in height and harshness. To the west their sovereignty reached over the dry Atacama Desert; to the east it included the flanks of the Amazon rainforest. Inca legions—like their Roman counterparts—marched far beyond the frontiers of civilization to dominate barbarian tribes and heterogeneous societies. At its height, the imperial capital of Cuzco exercised rule over Northern Chile, upland Argentina, Bolivia, Peru, Ecuador, and Southern Colombia. No contemporary Andean nation compares in magnitude or prosperity, and the great wealth of Tahuantinsuyu fostered its downfall" (reprinted with permission from *The Incas and Their Ancestors* by Michael E. Moseley, 1992, published by Thames and Hudson Inc., copyright 1992 Thames and Hudson Ltd., London).

Andean cultures that grew under Inca rule between the 2nd and the 16th centuries A.D. developed over a very wide territory, covering the Andean highlands of South America from Southern Colombia to the Maule River in Chile at an elevation of 3500-4000 m, including the western and eastern slopes of the Andean Cordillera and the adjoining lowlands. These are referred to as the Pacific and Amazonian Yungas, respectively. The Andean Cordillera was thus an imposing cultural barrier between the eastern and western lowland regions. The desert-like region of Atacama in Northern Chile was also under Inca rule.

The vast territory of the Inca Empire corresponds, as a whole, to a sector of the Pacific Ring of Fire. Of the many active volcanoes found here, the most important are Tolima, Puracé, Cotopaxi, Chimborazo, Sangay, El Misti, Ubinas, San Pedro, Ollaü, San José, Cerro Azul, and Llaima in the northern segment, and Villarrica, Riñihue, Nilahue, Puyehue, Osorno, and Calbuco in the far south. The volcanoes resulted from the collision between the Nazca and the South America plates, a collision that also triggered countless earthquakes—many disastrous—in recent geological times and in the present.

Prehistoric Times

The first human settlements in South America probably occurred in the Middle Paleolithic, between 50,000 and 70,000 years ago. The most ancient sites identified so far are those of Viscachani (Bolivia), Taltal (Chile), and Patagonia (Argentina), dating back 30,000 years. Handmade objects found at these sites include scrapers, arrowheads, axes, and many other stone tools. Radiocarbon dating of organic materials at the sites places human populations at about 37,000 years ago.

Among the Paleolithic pre-ceramic deposits, Viscachanensis, Ayampitensis, and Quila Quilense have been analyzed in detail. The stone tools found there are about 30,000 years old and show that people of that time were already using silica, quartzite, black flint, obsidian, and other hard

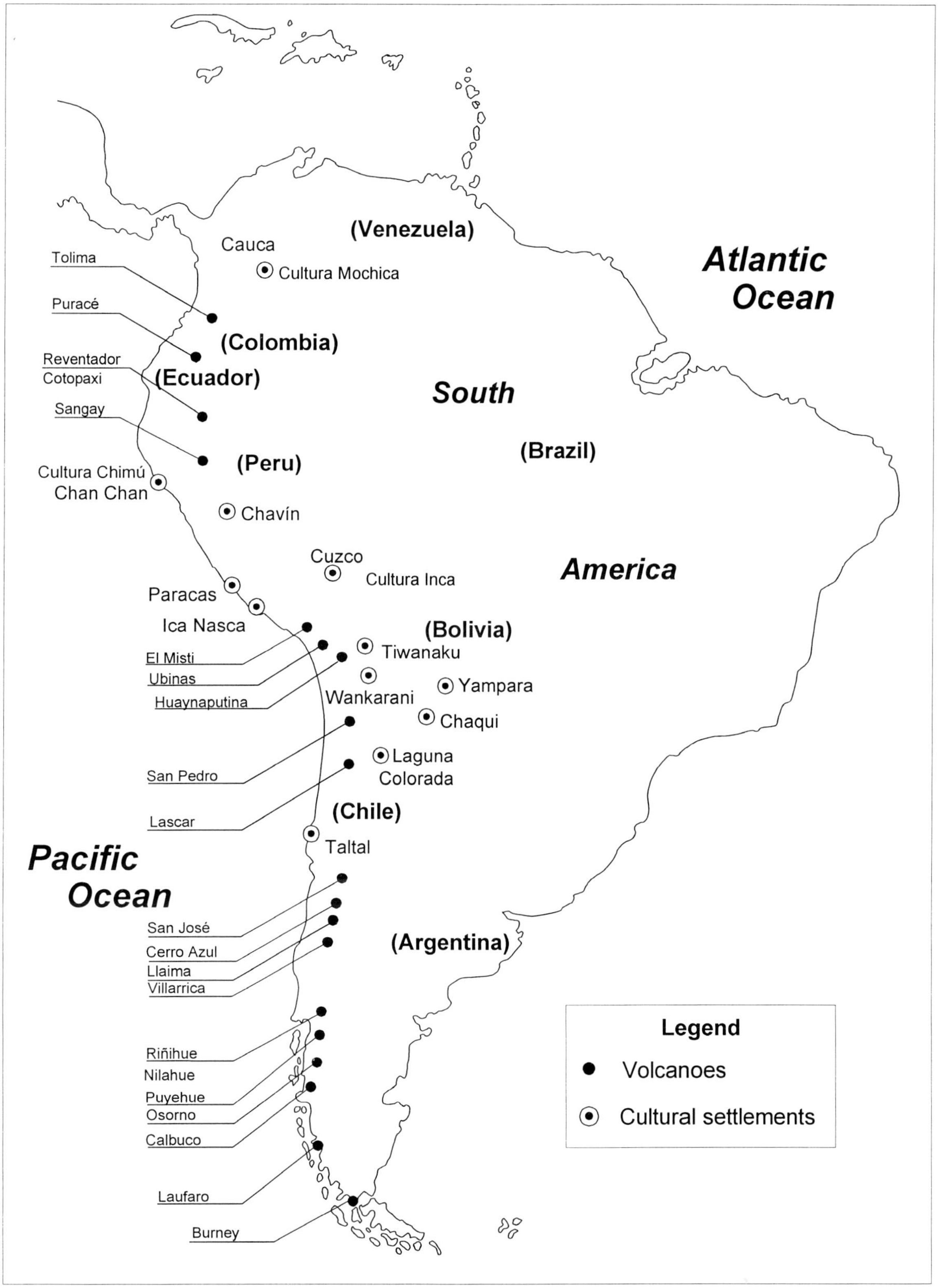

Main volcanic complexes and principal prehistoric and Inca cultural settlements in South America.

rocks to manufacture arrow points, choppers, spear points, knife blades, and one-sided polished splinters. These objects resemble those found in North American sites (Sandia, Trenton, Tule Springs, and Nevada) and suggest that the same manufacturing techniques were used (Ibarra Grasso, 1956 and 1958).

The Andean Cultures, Where and When They Developed
(Ibarra Grasso and Querejazu Lewis, 1986; Escalante, 1994)

- Viscachani (pre-ceramic), Bolivia; 28,000-8000 B.C.
- Mojocoyense, Bolivia: 8000
- Ayampitense, Argentina: 4000-3000
- Laguna Colorada (Lipez pre-ceramic), Bolivia: 3000
- Quila Quilense, Bolivia: 2000(?)
- Valdivia, Ecuador: 1000
- Kotosh, Peru: 2000-1000
- Chavín, Peru, 1000
- Tiwanaku (rustic), Bolivia: 1000-300
- Chiripa, Bolivia: 500-300
- Wankarani, Bolivia: 500-300
- Tiwanaku (classic urban and expansive imperial), Bolivia: 100 B.C.-1000 A.D.
- Mochica, Peru: 200
- Nasca, Peru: 300
- Wari, Bolivia; 500
- Los Maitas, Chile: 500
- Chimú, Peru: 1000
- Kolla, Lupaka, Pakaxa, Mollo, Karanga, Charca, Chicha, Huruqui, Chaqui, Yura, Collas, and Incas, various Andean countries: 1200-1500

Thus a number of products of volcanic origin, together with other materials directly or indirectly related to geothermal resources, played an important role in the development of early Andean cultures of South America. Moreover, igneous rocks such as diorite, andesite, and basalt were used to make many items and to construct many great pre-Hispanic monuments, such as those at Tiwanaku and Cuzco, which date back to 374 and 724, respectively (Ponce Sanjinés, 1970; Avila Salinas, 1975).

Geothermal Phenomena, Hot Baths, and the Use of Geothermal By-Products

Indirect proof of the relationship established between some ancient Andean peoples and the surface manifestations of geothermal resources is suggested by the location of several sites in areas with important natural manifestations. This is the case, for instance, of Viscachani, Laguna Colorada (Lipez cultures), Wankarani, and Chaqui (Ibarra Grasso and Querejazu Lewis, 1986; Ponce Sanjinés, 1970). It is easy, therefore, to argue that native peoples of the Andean region had already started to use natural heat and a number of geothermal by-products.

At several locations—Cuzco and Cajamarca in Peru; Huscarán and Puente del Inca, near Mendoza in Argentina; and Tarapaya in Bolivia—important Inca findings and construction works have been studied carefully. These sites offer direct proof of the care, knowledge, appreciation, and value the Incas granted the hot springs because of their healing properties. Evidence also suggests that the hot springs influenced Inca social customs and religious practices.

Not only the thermal springs but volcanic eruptions, hydrothermal deposits, and earthquakes must have had an important effect in attracting or discouraging the settlement of Inca villages, at least in the cold highlands where these types of manifestations frequently occur. Many travelers to the Andes chronicled the abundant geothermal events. These self-appointed historians included Catholic missionaries and monks, fortune seekers, and others who followed the Spanish conquistadors. López de Gómara, for example, was personal secretary to Hernando Cortés and one of the more prolific chroniclers. He described the 1533 eruption of the Pichincha volcano in Peru: "For many days ash expelled from the volcano of Quito fell over [the Inca villages nearby] and was thrown a distance of more than eighty leagues [a league is 5572.7 meters].... It throws so much fire and makes so much noise when it boils that it can be seen at a distance of more than one hundred leagues; and, as [the Incas] say, [the noise of the explosion] frightened people more than one caused by shocking thunder and lightning..." (1553).

De Gómara also described other geothermal phenomena, such as the 1541 earthquake in Quije, close to the Cayambe volcano in present-day Ecuador. This is where Gonzalo Pizarro, half-brother of Francisco Pizarro, conqueror of Peru, had arrived during his ill-fated quest for El Dorado, the "land of cinnamon and gold": "On the road to Quijos, north of Quito...the Earth trembled terribly and more than sixty houses crumbled in ruins...the Earth was fractured in many places.... There was so much thunder and lightning that [Pizzarro's soldiers] were frightened..." (1553).

Some of the writers were indigenous Incas, such as Garcilaso de la Vega, son of a Spanish soldier and an Inca princess, who wrote about the earthquakes and volcanic eruptions that occurred in Chile in 1599 and in Peru in 1600. Thick ash, he explained, blanketed the area surrounding Arequipa for about 20 days, measuring about one meter thick in some spots, more than two meters in other places, and everywhere "at least one fourth of a yardstick." In addition to heavy ash fall, residents suffered severe damage to crops, cattle, and orchards. Vineyards, wheat fields, and fruit trees—barren of branches and fruit— were buried in locations thousands of kilometers from Arequipa, and all livestock—cows, sheep, goats, and pigs—perished, either buried by ash or starved due to lack of pasture. In addition, many houses collapsed under the ash load, although some were saved because the owners diligently shoveled ash from the roof.

The eruption was accompanied by "great lightning and thunder," audible at distances of 150 to 200 kilometers around Arequipa. The heavy ash fall brought darkness to the land: "the sun in those days had darkened so much, that even at noon candles had to be lighted..." (1609).

In describing the territories conquered by the Inca Capac Yupanqui, de la Vega wrote about the hot springs of Tapacari, a province with many springs of water so hot "that nobody can keep his hand in it even for the time it takes to say *Ave Maria*." The odoriferous sulfur steam was visible from great distances. In addition to the hot springs, the area also contained "other springs with very cold and savory water," from which the Cochabamba River originated.

Cieza de León, a self-taught seeker of fortune, extensively traveled the Andean Cordillera and recorded his observations in *Crónica del Perú* (*A Chronicle of Peru*), written in the mid-16th century. He described several geothermal phenomena, particularly volcanic eruptions and the presence and use of hot brines in many Andean localities. In the province of Quimbaya, for example, a volcanic peak in the Sierra Nevada, the great Andean mountain range, threw "a great quantity of smoke" into the air. This mountain is the source of many rivers, including the Tacurumbi and the Cegue. Between these rivers lay brine sources, which formed fanciful formations as they flowed into the rivers.

Indeed, de León seemed fascinated by the indigenous peoples' methods for salt extraction and their uses of hot brines. From the catchment basin of the Santa Marta River, populated by Coconucos Indians, he witnessed "many *bocas de fuego* ['fire mouths' or erupting volcanoes] on the tops of the Sierra...." Hot water flowing down the slopes was a source of salt for the Indians, as were the rivers in the interior Popayán province. The local people extracted the salt and used it for cooking. Salt was so plentiful in the lakes and rivers of Peru, "that they can meet with the demand of all kingdoms of Spain, Italy, France, and other major parts of the world..." (1550).

De León also embarked on a series of study tours. On one trip from Cartegena, with graduate Juan de Vadillo, they were running out of water and supplies when they reached a small lake with black, salty water. The travelers, "in the need that teaches great things to men, took some water from the lake and poured it into hot pots, and this gave flavor to [their] meals...." Later, they learned that the Indians took the waters from this lake in large pots and boiled it to produce a black salt "not so good tasting" but one useful for cooking meals.

In Cori, located on the outskirts of Villa de Acerma, the Indians found brine ponds next to a rushing river. They boiled the water to produce salt cakes. These were of "very pure salt, excellent and as special as that obtained from the salt mines in Spain...." In the nearby province of Caramanta, a local hot spring flowing inside a fresh water river was also a source of salt, and springs in the towns of Mungia and Canupata yielded hot brines. In fact, noted de León, production was prolific: "they obtain so much salt, that we saw houses almost full of it..." (1550). Near Juanja, a "whitish rock outcrop," probably hydrothermal encrustations, marked the mouth of a hot brine source.

De León's travels took him ever deeper into the Peruvian countryside. On a journey between Villa de Pasto and Quito, he came upon the ruins of an Inca fortress called *Lumichaca* or *Rumichaca*, which means "stone bridge" in Quechua, the Inca language. Nearby he found a hot spring with scalding water: "if one puts his hand into [the water] in no way can he afford to keep it in for a long time due to the high heat with which the water flows out..." (1550).

Closer to Quito, de León visited Mulahalo, a former recruitment center for Inca warriors, which, in former times, was a bustling town with many inns and supply depots. A volcano was nearby, and the Indians told stories of destructive eruptions with stones and ash leveling towns in the surrounding area. Legend has it that, "before the eruption, one saw infernal visions and heard frightful voices...." This was confirmed by Don Pedro de Albarado, leader of a Spanish army in Peru, who witnessed an eruption with his soldiers.

Like their counterparts in other geothermally active areas of the world, the Incas used hot springs for thermal bathing. The chroniclers described the myriad palaces and temples erected near thermal sites: "After the Incas conquered this province of Cajamarca...they constructed a number of palaces and a temple to the Sun.... In these palaces there were many things to see, especially some baths, very good ones, where the lords and the dignitaries were used to taking baths while staying there...." While it is not known whether these baths were supplied with natural hot water, it is likely that the nearby thermal springs fed the baths. Inca buildings and a hot water pool in a stone enclosure were near another town, Jauja. The Inca lords and their wives used the water for daily bathing (de León, 1550).

De León found a similar arrangement in Bilcas, the central part of the Inca Empire located between Guamaga and Cuzco. A stone temple to the Sun dominated the town's main square, which was bisected by a lined " huge ditch channeled with much delicacy." These were "secret baths" for the Inca lords and their ladies. While de León did not specify whether or not the ditch served as a canal for natural hot water, the proximity of thermal springs suggests that it was likely used for collecting and channeling thermal waters. The baths may also have represented a centralized balneological establishment for Inca aristocracy.

Overall, de León concluded that thermal bathing for the Incas was ubiquitous: "Likewise, in many localities of the Inca Empire there were countless sources of hot water that the natives used, and still use at present, to take baths.... I myself have seen many of them everywhere I walked..." (1550).

Garcilaso de la Vega also wrote about Inca thermal bathing. When the Spanish conquered King Atahuallpa, in the province of Cajamarca in December 1531, they discovered a sumptuous royal bath occupied by "five thousand women who, despite being sad and helpless, enjoyed themselves with the Christians..." (1609). The baths were probably supplied with hot water from nearby thermal springs. In addition, they found tents, clothes and linens, and beautiful silver and golden artifacts that they plundered.

According to Francisco de Jerez, personal secretary to Francisco Pizarro, the royal palace of Cajamarca also included hot pools used for bathing: "four rooms surrounded the internal patio of this palace where a decorated pool could be seen, which was fed by two conduits: one conveying cold water and the other supplying natural hot water from a mountain source. During the day, the royal family stayed in one of the four rooms.... The walls were varnished with a brilliant red lacquer and the framework of the ceiling was painted the same color. Close to the royal residence there was another pool, also supplied with hot and cold water, surrounded by stone steps. It was used by the people for bathing..." (Baudin, 1943).

At least three writers referred to the Incas' therapeutic use of thermal waters. Vasquez de Espinosa, in writing about the valley of Cochabamba, noted that three villages, Santiago del

Thermal Baths, Rich and Splendid

"All the royal palaces had gardens and orchards for the Inca's recreation. They were planted with all sorts of gay and beautiful trees, beds of flowers, and fine and sweet-smelling herbs found in Peru. They also made gold and silver models of many trees and lesser plants: they were done in natural size and style with their leaves, blossoms, and fruits, some beginning to sprout, others half-grown, and others in full bloom....

"There were also large and small animals carved or hollowed out of gold and silver: rabbits, mice, lizards, snakes, butterflies, foxes, and wild cats (they had no domestic cats). There were birds of all kinds, some perched on trees as if they were singing, while others were flying and sucking honey from the flowers. There were, too, deer and stags, lions, tigers, and all the other animals and birds that bred in the country, each being set in its natural surroundings to give greater similitude.

"In many palaces, or even in all of them, they had baths, consisting of great tubs of gold and silver, for washing in: water was brought to them through pipes of silver and gold. And wherever there were natural hot springs, they also made very rich and splendid baths. Their other wonders included piles and heaps of firewood, done in gold and silver, as if they were stocks for use in the palace."

Royal Commentaries of the Incas and General History of Peru, by Garcilaso de la Vega, trans. Harold Livermore.

Paso, Sipi Sipi, and Tiquipaya, had "very good and healthy thermal waters." Ill people bathed in them "to recover their health..." (Guzmán, 1973).

Another chronicler, Vicente Cañete y Dominguez, described the mineral waters of the Tarapaya lagoon located atop a small hill about 1.7 km from Potosí. With a circumference of about 400 meters, the lagoon had a small creek flowing from it: "The lake is very deep; its waters are supplied by healthy hot springs that serve for bathing. At the center of the lagoon a boiling source can be seen, whereas on both sides of the creek there are other sources of hot water. These have a lower temperature, but they still are warm enough for bathing.... It is said that this lagoon swallows down all those who venture to take a bath in it.... For sure, these hot waters are mineral and healthy; however, it is not known why they have such healing properties and how they have kept their heat for so many centuries..." (Guzmán, 1973).

Garcilaso de la Vega described high-temperature manifestations and hydrothermal by-products, including a hot water lagoon in the Lipez region, a sector of the Andean highland covering Southern Bolivia. At Tumaquifa, about 2 km from the mines of San Cristobal de Achocolla, a small lagoon on a low hill was used for dysentery treatments:

> "In the center of the lagoon the water boils frenetically, and water jets with steam can be seen that are sometimes small and other times high with a great and dreadful noise.... I was truly frightened just to see the tumultuous boiling of the water, but there are some people who dare even to reach its border. The water is so turbid that it looks like mud. [The lagoon] has a small outflow through which the water flows, encrusting the rocks with much red salt; then it streams into a *guayco*

['gully'] where it rushes away very fast. The red encrustations are a kind of very strong salt that, because of its metal content, is much stronger and more beneficial than common salt; it has also been tested as a good cure for dysentery. It may be that [this salt] contains a mixture of red alum that makes it more efficacious.... Near this lagoon, a vein of 'Judaic stone' [probably malachite] outcrops and plenty of copper minerals are around..." (1609).

In addition to curative properties, some thermal waters had a ritualistic significance. For example, the fortress of Pukaramarka, about 60 km from Cuzco near the banks of the Urubamba River, had a special bathing place reserved for use by *ñust'as* ("virgins"). This bath was fed by natural thermal water and bathing here probably had some ritualistic significance for the young ladies before their marriages (Lara, 1976).

Some events related to geothermal manifestations were regarded as ominous portents and thus received the chroniclers' attention. Montecinos, secretary to the Viceroy of Peru, wrote a monograph entitled *Viejas Memorias Históricas del Perú* (*Old Historical Memories of Peru*) that described the pre-Inca kings of Peru called the Amautas. Referring to the great-grandchild of the famous King Tupaj Yupanqui, Montecinos stated that: "At the times of the 61st [pre-Inca] king...many bad omens, comets and earth tremors had occurred.... According to legend, there is a celebrated cave in the area where it is said with all certainty that earthquakes were never felt, nor were pestilence or tremors." Nonetheless, by the reign of the 63rd king, nobody could remain in Cuzco because of the strong earthquakes that destroyed many buildings, both in the city and surrounding areas. The quakes caused flooding as rivers were rerouted, destroying some cities that had withstood the earthquakes. As a final coup, the area

Llicancahur volcano and Laguna Verde on the Southern Bolivian plateau. Here and in the neighboring lands of Southern Bolivia, the Lipez culture developed some 5,000 years ago.
G. Rico Calderón

was then devastated by a plague that killed "countless people." These events caught people unawares, since they subscribed to the superstitious belief of the kings that Cuzco "was the only place where no natural catastrophe and pestilence could occur" (Bingham, 1972).

The Incas used hydrothermal minerals for trade and currency. An excavation at Cuzco, according to Bingham (1972), yielded 29 obsidian pebbles, slightly larger than regular marbles, that were apparently used for counting. Since oral tradition does not account for an eruption near this location, the volcanic pieces were imported from another place. Obsidian, along with chalcedony and rock crystal, was also a currency in Peru and Chile during Inca times (Latcham, 1909). These geothermal resources—such as obsidian, silica, and rock crystal—were traded during the time of the Classic and Expansive Tiwanaku culture (100 B.C.-1000 A.D.), and there was, as Baudin explained, "widespread commerce of the ceramics made in Bolivia, which were largely exported to present-day Ecuador. Moreover, large quantities of decorated pieces were exported to the [Pacific] coastal areas..." (1943).

Religious Aspects and Geothermal Phenomena

As in other areas of the world, the first religious feelings in the Andean region probably developed after observing exceptional natural events. The tremendous force of volcanic eruptions and earthquakes contributed to the formation of popular belief in the existence of supernatural forces ruling these phenomena. Early Andean populations probably believed in and revered superhuman entities capable of triggering or stopping, at will, volcanic eruptions and earthquakes.

These initial religious stirrings were gradually refined over centuries until they became the basis of a complex pantheon that included not only the entities governing geothermal phenomena but many others as well. During Inca times, these beliefs became an integrated perception of the whole world: a true model of the universe in which each natural phenomenon had its own divinity and was explained by the logic of a supernatural sphere.

A major god of the Inca pantheon was Pachacamac, or Pachak'amaj, the creator of the world (Lara, 1976), or, as Garcilaso de la Vega more poetically stated, the god "who gives soul to the universe" (1609). In earlier times, Pachak'amaj was also regarded as *wak'a,* "lord of the earthquake," as a way to explain the frequent earthquakes along the Pacific coast. Pre-Inca populations built a grand temple dedicated to him on the Peruvian coast, in a town which now bears the god's name. Lara (1976) recounts an old legend relating the god to the Earth's movement: "The earthquake was under his governance, and to produce an earthquake it was enough for him to only move his head. When he becomes irritated, the whole world moves...however, he does not move his face at all. Should he only turn his body half around, this would suffice to cause the end of the world. The myth says that one of the sons of the *wak'a*, named Llayawankupa, was proud to proclaim: 'I am the son of Pachak'amaj, the god who shakes the Earth....'"

Pachak'amaj produced earthquakes when he was displeased or irritated by the behavior of people (Guzmán, 1973). To placate the god and ensure his protection, the Incas sacrificed animals—up to 10,000 llamas in one ceremony at Cuzco—and people. According to Juan Polo de Ondegardo, by royal decree many young women were sacrificed "under the necessary condition that they had to be virgins.... Moreover, extraordinary sacrifices were made by the Incas during that main feast called Tyri...if there had been an earthquake, or a noticeable sun or moon eclipse, or a volcanic eruption..." (Guzmán, 1973). Victims were entombed in mountain caves or hurled from cliffs or high peaks.

Another religious precept important to Inca theology was *huaca*. *Huaca* was, according to Baudin (1943), "everything that was above the ordinary: a high mountain top, a strange beast, a rare flower...." A *huaca* was sacred, "a divine manifestation in everything, a miracle in every phenomenon." De la Vega mentioned that the Sierra Nevada range was *huaca*, as were the very highest mountains in it, and therefore "most admirable to all those who look at it with attention..." (1609). Antonio de La Calancha, an Augustinian monk born in Bolivia and educated in Lima, wrote that, "Besides others, the idols worshiped by these Indians were the permanent

The Payachata volcanoes, near the border of Bolivia and Chile. Volcanoes and permanent snow lines, here above 5000 m, were two natural features worshiped by the Incas. *R. Iriarte*

snow lines of the mountains, and especially those having the highest altitude.... In particular...they used to make many [human] sacrifices and worship the Illimani Cullcachata [the volcanic massif of the Cordillera Real, near La Paz, Bolivia].... In these mountains the devil gave the Inca his answers, and the oracles issued from those mountains were continuous..." (Guzmán, 1973).

According to Murra (1975), the snow-capped volcanoes of Cotopaxi and Chimborazo played an important role in the myths of the creation of all volcanic complexes called *Colorados*: "The tempest is a god who lives in the high mountains of the Andes and roars when he becomes drunk.... He has a sword that he draws to produce lightning.... The Indians of Quito worshiped the high mountains of their land, considering them deities, animated with the same feelings as a human being...."

Finally, worthy of mention is an event from the bloody battle of Tiocajas between the Incas and the Spaniards: "Huaina Capac, a son of the Inca king...had succeeded in reconquering Cajamarca.... The battle then continued all day long, and the Spaniards would have retreated if a providential eruption of Cotopaxi had not scared the Indians, who withdrew..." (Baudin, 1943). This behavior suggests that the Incas considered the volcanic eruption as a warning sign addressed to them by a supernatural entity that deserved great respect and veneration.

Concluding Remarks

Until the mid-16th century, written documents tracing the development of Andean cultures during Inca and pre-Inca times were unavailable. For these times, the only sources of information on lifestyles, traditions, and relationships that Andean natives established with their physical environment are the results of relatively recent archaeological work.

The first written accounts of these relationships, particularly the rapport between Andean populations and geothermal phenomena, are chronicles by missionary friars, geographers, and others who followed the Spanish armies on their military expeditions in the northern and western sectors of South America. These chronicles, however, were only partially based on direct observations of actual situations; in many instances, they were distorted and embellished with descriptions of prior events gleaned from the oral histories of local people. Furthermore, one must consider that the European chroniclers were not accustomed to natural phenomena of such imposing magnitude, particularly the violent force of some geothermal manifestations: catastrophic earthquakes, huge volcanic eruptions, and geyser-like jets of water and steam.

There is little doubt that all these phenomena, associated with the harsh environmental conditions of the cold highlands and thick-forested lowlands, have deeply influenced the development of the Andean cultures since prehistoric times. Two forms of geothermal influence can be identified, one pragmatic and the other spiritual. The pragmatic influence includes using thermal waters for bathing, processing hot brines for salt, manufacturing volcanic products into tools, using volcanic materials for buildings, and extracting hydrothermal compounds for practical applications. The spiritual influence is reflected in the meanings attributed by the Andean populations to manifestations of the Earth's energy, which were thought to represent supernatural entities. Such reactions to geothermal phenomena were gradually internalized, taking root in people's minds and spirits. They contributed significantly to the development of a unique, complex model of the Andean universe.

Cited and Selected References

Ahlfeld, F. E. 1969. *Geografía física de Bolivia.* Los Amigos del Libro, Cochabamba.

Arguedas, A. 1975. *Historia general de Bolivia.* Gisbert & Cia, La Paz.

Arguedas, J. M. 1970. *Mitos, leyendas y cuentos peruanos.* Casa de La Cultura del Perú. Lima.

Avila Salinas, W. 1975. *Análisis espectrográfico semicuantitativo de algunas obsidianas de Bolivia, Argentina, Perú y Chile.* Instituto Nacional de Arqueología, La Paz.

Betanzos (de), J. 1992. Suma y narración de los Yngas. In *Culturas aborígenes de América, vol. 2.* Fondo Rotatorio, Cochabamba.

Baudin, L. 1943. *El imperio socialista de los Incas.* José Antonio Arze. Santiago.

Bingham, H. 1972. *Machu Picchu—La ciudad pérdida de los Incas.* Ediciones Rodas, Madrid.

Cieza de León, P. 1550. *La crónica del Perú.* Reprint 1973. Editorial Zig Zag S.A., Santiago.

Emmerich, F. 1986. *Leyendas chilenas; Programa Andrés Bello.* Santiago.

Escalante, M. J. 1994. *Guía arqueológica de Bolivia.* CIMA, La Paz.

Fellmann Velarde, J. 1976. *Historia de la cultura boliviana.* Editorial Los Amigos del Libro, Cochabamba.

García Rossel, R. 1908. Departamento de Piura. In *Revista histórica.* Sociedad Geográfica, Lima.

Garcilaso de la Vega. 1609. *Crónica de la conquista.* Reprint 1973. Buenos Aires.

Garcilaso de la Vega. 1609?. *Comentarios reales de los Incas I y II.* Reprint 1991. Fondo de Cultura Económica, México.

Girard, R. 1976. *Historia de las civilizaciones antiguas de América, vols. 1-3.* Colegio Universitario de Ediciones Istmo, Madrid.

Gonzáles, R. 1926. *Aguas minerales—Medicinales de Hurmiri.* Edición Municipal, La Paz.

González de la Rosa, M. 1908. Ensayo de cronología Incáica. In *Revista histórica.* Lima.

Guisbert, T. 1994. *Iconografía y mitos indígenas en el arte.* Editorial Guisbert, La Paz.

Guzmán, A. 1973. *Antología colonial de Bolivia.* Editorial Los Amigos del Libro, Cochabamba.

Hohagen, M. L. 1908. Las divinidades de Pachacamac, Museo del Trocadero. In *Revista histórica.* Lima.

Humboldt, A. 1814. *Voyage aux régions équinoxiales du nouveau continent, part 1.* Librairie Greque-Latine-Allemande, Paris.

Ibarra Grasso, D. E. 1956. Decubrimiento de puntas paleolíticas en el altiplano de Bolivia. *Revista de Cultura,* no. 2, 186-221.

Ibarra Grasso, D. E. 1958. El paleolítico inferior en América. *Revista de Cultura,* no. 3, 103-141.

Ibarra Grasso, D. E. and R. Querejazu Lewis. 1986. *30.000 años de prehistória en Bolivia.* Los Amigos del Libro, Cochabamba.

Kinder, H. and H. Werner. 1988. *Atlas histórico mundial, part 1.* Ediciones ISTMO, Madrid.

Lara, J. 1973. *Mitos, leyendas y cuentos de los Quechuas.* Los Amigos del Libro, Cochabamba.

Lara, J. 1976. *La cultura de los Incas.* Los Amigos del Libro, Cochabamba.

Latcham, R. E. 1909. *El comercio precolombino en Chile y en otros países de América.* Impr. Cervantes, Santiago de Chile.

Lopez de Gómara, F. 1553. Hispania victrix. In *Primera y segunda parte de la historia general de las Indias.* Amigos del Círculo de Bibliofilo, Zaragoza, Madrid.

Lucas Jaimes, J. 1975. *La villa imperial de Potosí.* Ediciones Camarlinghi (re-edición de 1903), La Paz.

Macera, P. 1970. *Historia del Perú.* Wirakipu Edit., Lima.

Marquez Ostria, J., S. Kussman, and P. K. Hormann. 1975. *Estudio petrográfico del Cerro Querimita y su comparación con artefactos líticos.* Centro de Investigaciones Arqueológicas, La Paz.

Martinez, A. and N. Vela. *Historia de la villa imperial de Potosí.* Reprint 1975. Biblioteca Sesquicentenario de la República, La Paz.

Mille, M. and C. Ponce Sanjinés. 1968. *Las andesitas de Tiwanaku.* Academia Nacional de Ciencias de Bolivia, La Paz.

Murra, J. V. 1975. *Formaciones económicas y políticas del mundo Andino.* Instituto de Estudios Peruanos, Lima.

Ponce Sanjinés, C. 1970. *Las culturas Wankarani y Chiripa y su relación con Tiwanaku.* Los Amigos del Libro Edit., Cochabamba.

Querejazu Lewis, R. 1989. *Bolivia prehispánica.* Editorial la Juventud, La Paz.

Terán Erquicia, V. 1966. *Chiwanwayus Achankaras: Flores de leyendas Queswas.* Renovación Edit., La Paz.

Uhle, M. 1929. Influencias del país de los Incas. In *Revista histórica.* Sociedad Geográfica, Lima.

Villamil de Rada, E. 1939. *La lengua de Adán y el hombre del Tiaguanaco.* Ministerio de Educación, Bellas Artes y Asuntos Indígenas, La Paz.

Wiener, C. 1993. *Perou et Bolivie.* Instituto Francés de Estudios Andinos, Lima.

Willey, G. R. 1956. *Prehistoric settlement patterns in the New World.* Werner-Gren Foundation, New York.

The Author:

Gonzalo Rico Calderón
General Manager
SYNERGIA S.R.L.
Avenida Libertador Bolivar 1227
P.O. Box 2632
Cochabamba, Bolivia
Telephone: 42.84.901
Fax: 11.71.09
E-mail: ghrc@ghrico.bo

Volcanic rock was used to fashion *moai*, Easter Island statues (whose natures were conjured by the words of Chilean poet Pablo Neruda in his poem, *The Rose Apart, Easter Island,* 1973) and *ahu*, the platforms on which the statues were raised. *Craig McNamara, 1971*